International Youth Scientific Conference on Heat and Mass Transfer in the Thermal Control System of Technical and Technological Energy Equipment (HMTTSC 2017)

MATEC Web of Conferences Volume 110 (2017)

Tomsk, Russia
26 - 28 April 2017

Editors:

Geniy V. Kuznetsov
Alena O. Zhdanova

ISBN: 978-1-5108-4275-5

Printed from e-media with permission by:

Curran Associates, Inc.
57 Morehouse Lane
Red Hook, NY 12571

Some format issues inherent in the e-media version may also appear in this print version.

Printed by Curran Associates, Inc. (2017)

For additional information, please contact EDP Sciences – Web of Conferences at the address below.

EDP Sciences – Web of Conferences
17, Avenue du Hoggar
Parc d'Activité de Courtabœuf
BP 112
F-91944 Les Ulis Cedex A
France

Phone: +33 (0) 1 69 18 75 75
Fax: +33 (0) 1 69 28 84 91

contact-edps@webofconferences.org

Additional copies of this publication are available from:

Curran Associates, Inc.
57 Morehouse Lane
Red Hook, NY 12571 USA
Phone: 845-758-0400
Fax: 845-758-2633
Email: curran@proceedings.com
Web: www.proceedings.com

TABLE OF CONTENTS

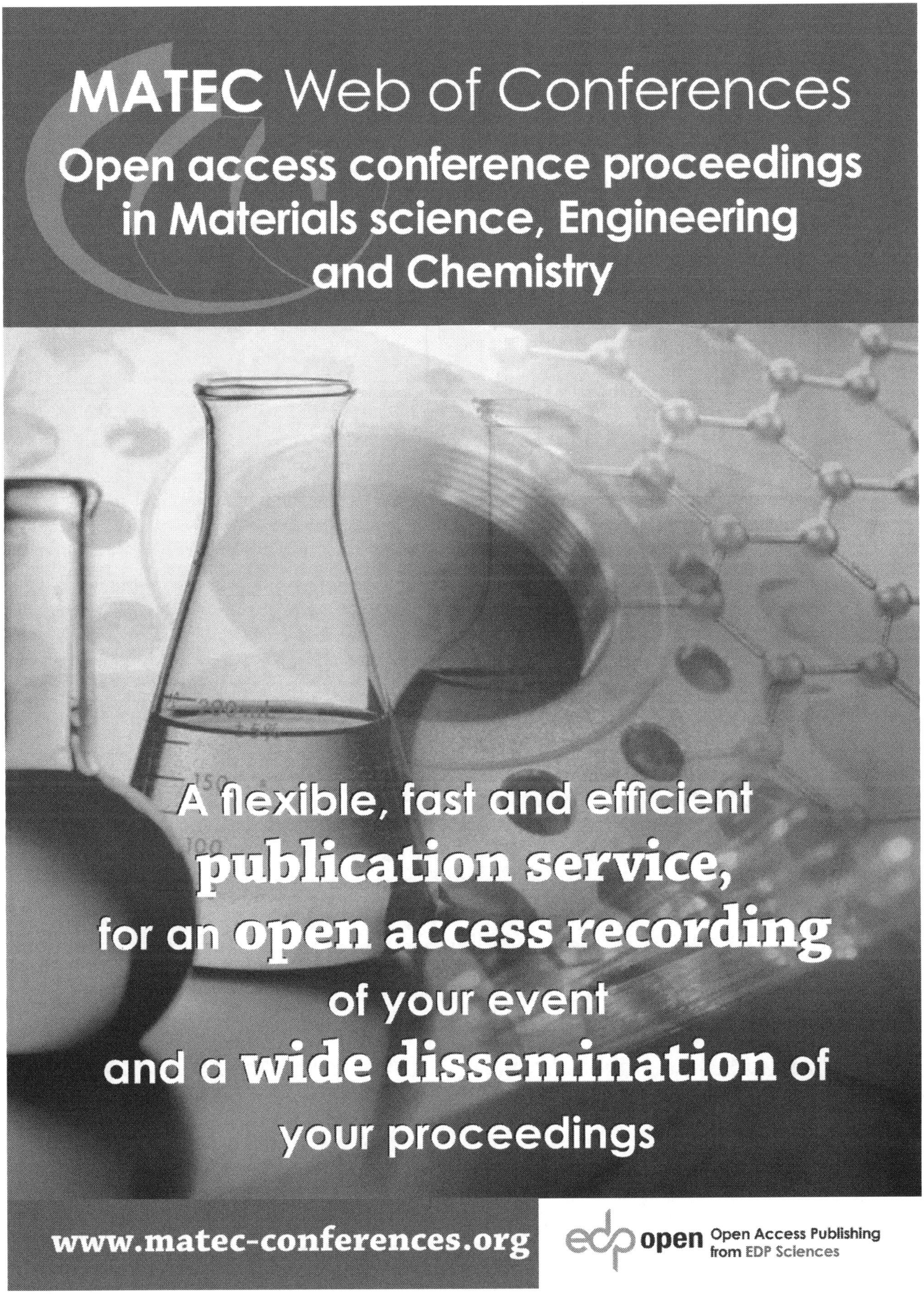
MATEC Web of Conferences
Open access conference proceedings in Materials science, Engineering and Chemistry
A flexible, fast and efficient publication service, for an open access recording of your event and a wide dissemination of your proceedings
www.matec-conferences.org
edp open Open Access Publishing from EDP Sciences

edp sciences

MATEC Web of Conferences

Statement of Peer review

In submitting conference proceedings to MATEC Web of Conferences, I certify to the Publisher that I adhere to the **Policy on Publishing Integrity** of the journal in order to safeguard good scientific practice in publishing.

1. All articles have been subjected to peer review administered by the proceedings editors.
2. Reviews have been conducted by expert referees, who have been requested to provide unbiased and constructive comments aimed, whenever possible, at improving the work.
3. Proceedings editors have taken all reasonable steps to ensure the quality of the materials they publish and their decision to accept or reject a paper for publication has been based only on the merits of the work and the relevance to the journal.

Title, date and place of the conference

International Youth Scientific Conference "Heat and mass transfer in the thermal control system of technical and technological energy equipment"-2017, 26-28 April 2017, Tomsk, Russia

Proceedings editor(s):

Geniy V. Kuznetsov

Alena O. Zhdanova

Date and editor's signature

01/06/2017

17, avenue du Hoggar - PA de Courtabœuf – BP 112 - 91944 Les Ulis Cedex A (France)
Tél. : 33 (0)1 69 18 75 75 - Fax : 33(0)1 69 07 45 17 – www.edpsciences.org

MATEC Web of Conferences **110**, 01001 (2017) DOI: 10.1051/matecconf/201711001001
HMTTSC-2017

Effect of surfactant on the evolution of bubble motion

Aleksandra Antonnikova[1,*], *Anna* Usanina[1]

[1]National Research Tomsk State University, 634050 Tomsk, Russia

Abstract. Results of an experimental study of the process of ascent of a single bubble of air in the presence of a surfactant are presented. An empirical dependence for drag coefficient of the bubble that emerges in a liquid containing a surfactant in the region of small Reynolds numbers has been obtained. Theoretical studies of the effect of surfactants on the velocity of motion of a bubble in a nonstationary regime have been carried out. A theoretical analysis of the effect of nonstationary forces on the dynamics of bubble rising in the nonstationary regime has been conducted.

1 Introduction

In two-phase flows containing deformable particles of a dispersed phase (liquid-drop, bubble flows) the properties of the interface, in particular the coefficient of surface tension, play an important role. One of the physical factors that affect the boundary conditions is the presence of a surfactant that can change the criteria for the motion of a particle of the dispersed phase significantly. On the free boundary of a bubble or drop moving in a liquid with a surfactant, tangential capillary Marangoni forces arise contributing to an increase in the particle drag coefficient. Problems arising in the study of the laws of motion and the deformation of bubbles (or drops) in a liquid containing a surfactant are formulated in [1-8]. These problems are associated with a variety of types of interfacial interactions and processes (deformation, break down, collision of bubbles, capillary effects, coagulation, heat exchange between phases, adsorption, adhesion). The understanding and adequate description of the dynamics of the behavior of two-phase systems taking into account the physical and chemical properties of the dispersed and dispersive phases is extremely difficult and requires new experimental data (empirical dependencies and criterial relationships) on elementary processes occurring at the phase boundary during their interaction. In particular, at the present time there is a limited amount of experimental data that allows estimating the effect of nonstationary effects on the parameters of particle motion in a flow in the presence of surfactants in a dispersion medium. Nonstationary effects (the force associated with the attached mass, the "hereditary" Basset force) transforms the differential equation of the particle motion into an integro-differential one that significantly complicates the solution of the problem of the nonstationary motion of a particle. An experimental study of the particle motion in a nonstationary regime is also difficult because of the small characteristic times and distances of the process.

* Corresponding author: leva@niipmm.tsu.ru

DOI: 10.1051/matecconf/201711001001

The purpose of this paper is to investigate the laws of the ascent of a single air bubble in a liquid containing a surfactant.

2 Experimental setup and study technique

To investigate the process of the ascent of a single air bubble in a liquid containing a surfactant, an experimental method and setup were used that met the following requirements: providing the precise measurement of all the main parameters (linear dimensions of the process and the physical properties of the fluid: density, coefficient of dynamic viscosity); control over the change in the characteristics of the ascent of the bubble due to the use of multi-angle video.

The experimental setup consists of a vertically mounted transparent tank with plane-parallel walls filled with a liquid solution, a device for generating bubbles, and a system for visualizing the ascent of a single bubble. Glycerin, water-glycerol solution and distilled water were used as a liquid; liquid soap was used as a surfactant. Mass fraction of surfactant in the liquid varied in the range z=(0.01 - 0.13)%.

Air bubbles were obtained by means of a syringe. A hollow aluminum tube with a diameter of 3 mm was fixed to the outlet hole of the syringe. At the end of the tube a replaceable socket with a diameter (0.4-7.0) mm located in lower part of the tank is assembled. The separation of the bubble from the outlet hole did not occur due to the change in pressure (on exposure to the air flow) but in a natural way, when the resultant of all forces exceeded zero.

The visualization system included the light sources (two 18W fluorescent lamps) installed on the rear panel along the entire height of the tank, a digital video camera "Panasonic HDC-SD60" and a high-speed video camera "Citius C100". The first camera was used to determine the size of the bubble; the shooting field was 5×5 cm with a twofold increase. The second one fixed the ascent velocity of the bubble. The video was taken with a spatial resolution of 384×790 pixels with camera speed of at least 300 frames per second and exposure time (0.5-2.0) ms. To control the distance traveled by the bubble a scale rule was used with a division value of 1 mm that was calibrated taking into account the optical distortion due to the walls of the tank.

3 Experimental results

Experimental studies were carried out in the following range of parameters: bubble diameter D=(0.82 ÷ 24) mm; the ascent velocity of the bubble u=($3.9\cdot10^{-3}$ ÷ 0.22) m/s; liquid density ρ_f=(1000÷1260) kg/m^3; coefficient of dynamic viscosity of liquid μ_f=(10^{-3}÷2.44) Pa·s; Reynolds number Re=($3.5\cdot10^{-3}$÷211); the content of the surfactant z=(0.01÷0.13)%.

The experimental data on the stationary ascent velocity of a single spherical bubble in the presence and without surfactant are compared. It is found that the velocity of a bubble emerging in a liquid with a surfactant is less than the velocity of a bubble rising in a clean liquid.

In the range of Reynolds number Re=$3.5\cdot10^{-3}$÷0.6 an empirical dependence for the drag coefficient of a spherical bubble rising in the presence of a surfactant is obtained: $C_D = 24.06/\mathrm{Re}$. The coefficient of determination of the obtained dependence is equal to R^2=0.99. Calculations of the drag coefficient for nonspherical bubbles rising in the presence of a surfactant were also made. The equivalent spherical diameter was used as the diameter (Fig. 1, 2).

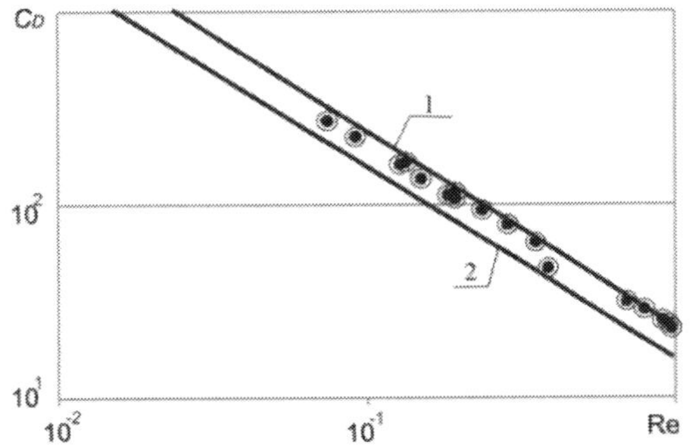

Fig. 1. Dependence of the drag coefficient of a nonspherical bubble on Reynolds number: • - with surfactants; 1 - Stokes curve (for a solid sphere); 2 - the Hadamard-Rybchinsky curve (for the liquid sphere).

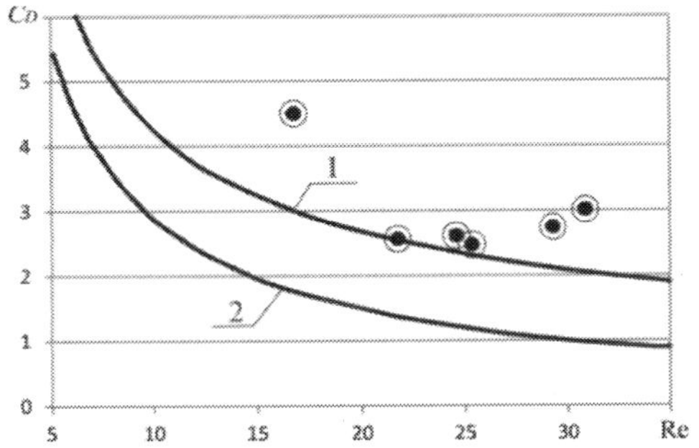

Fig. 2. Dependence of the drag coefficient of a nonspherical bubble on Reynolds number: • - with surfactants; 1 - Klyachko curve (for a solid sphere); 2 - curve C_D=48/Re (for the liquid sphere).

Theoretical calculations of the bubble velocity taking into account the nonstationary forces (the forces associated with the attached mass and the "hereditary" Basset force) for a non-stationary ascent regime. A comparative analysis of the theoretical and experimental dependences has shown that at calculating the dynamics of the ascent of a single spherical bubble it is necessary to take into account nonstationary forces.

The contribution of the viscous drag force F_{st}, the force associated with the attached mass F_{ad}, and the "hereditary" Basset force F_B to the total drag force F_Σ of a single spherical bubble rising in the presence of surfactant is estimated. The Basset force is calculated from the analytical formula of L.D. Landau [8]. Calculations of the forces acting on the bubble were carried out for the following experimental conditions: D=2.33 mm, ρ_l=1218 kg/m^3, μ_l=0.099 Pa·s, Re=1.15. The forces acting on the bubble were calculated by the formulas:

$$F_A = \frac{\pi D^3}{6}\rho_l g\,,\ F_{ad} = \frac{\pi}{12}\rho_l D^3 a\,, F_{st} = -C_D\frac{\pi D^2}{4}\frac{\rho_l (at)^2}{2}\,,$$

$$F_B = 3aD^2\sqrt{\pi\mu_l\rho_l t}\,,\ F_\Sigma = F_B + F_{st} + F_{ad}\,,$$

where F_A is the Archimedes force; C_D=24.06/Re, a=0.18 m/s^2 is the bubble acceleration, calculated from the experimental data.

Calculations showed that the main contribution to the total drag force on the nonstationary region of the bubble trajectory is made by the viscous drag force and the "hereditary" Basset force (Fig. 3). At $t < t_1 \approx 0.1\,t_*$ the value of the Basset force is insignificantly different from the drag force ($F_B \approx F_{st}$), and at $t_1 < t < t_*$ Basset force becomes less than the viscous drag force and at $t = t_*$ Basset force is equal to $F_B \approx 0.3F_{st}$. The force associated with the attached mass is ~ 0.5% from the sum of all the forces acting on the bubble. It is constant on the non-stationary part of the bubble motion. A similar estimate of the forces for a bubble rising in a pure liquid showed a qualitative agreement of the results for a bubble rising in the presence and without surfactant.

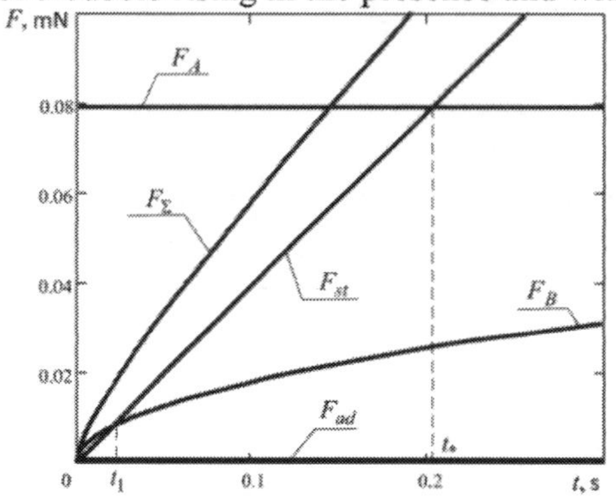

Fig. 3. Estimation of the forces acting on the bubble at uniformly accelerated motion.

4 Conclusion

The results of the experimental-theoretical study of the dynamics of the ascent of a single bubble of air in the presence of surfactant allow to make the following conclusions.

• The presence of surfactant leads to an increase in the drag coefficient of 33% at Reynolds number Re<1 and approximately of 50÷60% at Re>1 and, accordingly, to a decrease in the ascent velocity of the bubble.

• An empirical dependence for the drag coefficient of a single spherical bubble rising in a liquid in the presence of a surfactant for Reynolds numbers Re <1 is obtained.

• The theoretical analysis of the problem of the bubble rising in liquid containing surfactants showed that the "hereditary" Basset force and the viscous drag force make the main contribution to the total drag force on the non-stationary region of the bubble rise.

The study was supported by the Russian Science Foundation (No 15-19-10014).

References

1. V.G. Levich, *Physico-chemical hydrodynamics* (Prentice-Hall, New Jersey, 1959)
2. A.N. Frumkin, V.G. Levin, Russian Journal of Physical Chemistry A, **21**, 1183 (1947)
3. R. Clift, J.R. Grace, M.E. Weber, *Bubbles, Drops and particles* (Academic Press, London, 1978)
4. R. Griffith, Chem. Eng. Sci. **17**, 1057, (1962)
5. R. Davis, A. Acrivos, Chem. Eng. Sci. **21**, 681 (1966)
6. J. Harper, Appl. Sci. Res. **38**, 343 (1982)
7. Y. Wang, D. Papageorgiou, C. Maldarelli, J. Fluid Mech. **390**, 251 (1999)
8. L.D. Landau, E.M. Lifshits, *Course of Theoretical physics. Fluid Mechanics* (Pergamon Press, Oxford, 1987)

Effect of external conditions and steam parameters on performance of combined-cycle power plants with back-pressure steam turbine

Aleksandra Antonova[1,*], *Artem* Uvarov[1]

[1]National Research Tomsk Polytechnic University, 634050 Tomsk, Russia

Abstract. The cogeneration combined-cycle gas turbine plant with back-pressure steam turbine is considered. A simple circuit of the plant increases its reliability, and the use of optimal temperatures of the heating and return water allows obtaining the highest thermal efficiency of the plant taking into account the ratio of its electric and heat power. The technical and economic efficiency of the combined-cycle plant with back-pressure steam turbine is studied to estimate the capital investment. The parameters and conditions to maximize the net present value growth relative to the base version are determined.

1 Introduction

Combined-cycle technologies have occupied a leading position in heat power engineering due to high heat efficiency and environmental friendliness. Generating equipment in domestic thermal power plants and in the world energy is mainly implemented through combined-cycle power plants (CCPP) [1]. In the most efficient binary-cycle CCPP with bottoming cycle and is called a combined cycle gas turbine (CCGT) plant. The exhaust heat of gas turbine (GT) is recovered in a heat recovery steam generator (HRSG) for generation of superheated steam, which then flows to the steam bottoming cycle. The efficiency of these CCGTs based on high-temperature GT reaches 60% and higher [2,3]. Higher levels of heat efficiency are obtained in cogeneration combined-cycle plants supplied with different cycles and circuits. One of the versions of this plant is CCGT with a back-pressure steam turbine (CCGT-BP) [4].

2 CCGT with a back-pressure steam turbine

In CCGT-BP, the exhaust GT gas heat is recovered in HRSG in two sections of the heating cycle: upon supply of the superheated steam (hot water, steam generation, superheating) and during heating the return water in the HRSG back-end surfaces. The heat generated by the exhaust steam is transferred to the steam or to the water supplied directly to the network.

* Corresponding author: anton@tpu.ru

CCGT-BP is very simple in its design since no several steam pressure levels and reheating are required. Fig. 1 shows a CCGT-BP schematic diagram to supply heat to heat and hot water consumers and for household needs.

The efficiency of HRSG and that of the whole plant can be increased through deep heat recovery due to reduced exhaust gas temperature during transmission of heat to the return heating water in the gas-water network heater (GWNH).

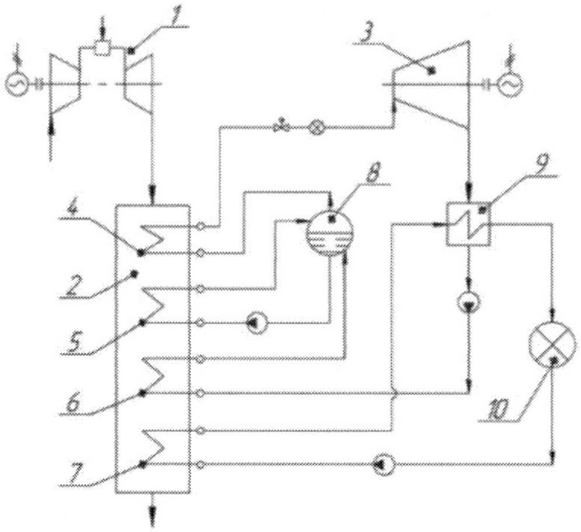

Fig. 1. Schematic diagram of CCGT with a back-pressure steam turbine: GT (1); HRSG (2); steam turbine (3); steam superheater (4); evaporator (5); economizer (6); GWNH (7); drum (8); network heater (NH) (9); heat consumer (10).

The temperature chart of the heat supply system is crucial for the performance of the CCGT-BP steam section, since the temperature of the heating water affects the pressure of the turbine exhaust steam; its value is close to the atmospheric pressure or exceeds it.

3 Thermal efficiency of CCGT-BP

The thermal efficiency of CCGT-BP can be estimated using the so-called total efficiency of CCGT, which shows the portion of the fuel heat utilized in the plant.

Utilized heat is the sum of the CCGT electric power N_E, and heat power transferred to heat consumers Q_{HC}. The total efficiency of CCGT

$$\eta_{CCGT} = \frac{N_E + Q_{HC}}{Q_{CC}}, \tag{1}$$

where Q_{CC} is the fuel heat input to the working fluid in the GT combustion chamber.

In turn, the CCGT electric power is the sum of two summands

$$N_E = N_E^{GT} + N_E^{ST}, \tag{2}$$

where $N_E{}^{GT}$ and $N_E{}^{ST}$ are the electric power of gas and steam turbine.

Transform the formula of the CCGT total efficiency

$$\eta_{CCGT} = \frac{N_E^{GT}}{Q_{CC}} + \frac{N_E^{ST} + Q_{HC}}{Q_{CC}} = \eta_{GT} + \frac{Q_{CC} \cdot (1-\eta_{GT}) \cdot \eta_{HRSG}}{Q_{CC}} = \eta_{GT} + (1-\eta_{GT}) \cdot \eta_{HRSG} \tag{3}$$

Thus, the CCGT total efficiency is

$$\eta_{CCGT} = \eta_{GT} + \left(1 - \eta_{GT}\right) \cdot \eta_{HRSG} \tag{4}$$

This formula lacks the efficiency of the steam turbine or any other factor that indicates its heat efficiency. Hence, the effectiveness of the combined-cycle plant with a back-pressure turbine does not depend on the steam cycle (SC) efficiency. This can be attributed to the fact that SC works without cycle losses: the exhaust steam condensation heat is fully utilized to cover the consumers' heat demands.

In addition to the total efficiency, the most important characteristic of the CCGT-BP heat efficiency is the power ratio N_E^{ST}/Q_{HC}.

The aim is to study the effect of a number of factors on CCGT-BP heat efficiency. These factors are as follows:

- temperature of the heating water t_{hw},
- temperature of the return water t_{rw},
- initial steam pressure p_0,
- initial steam temperature t_0.

In accordance with the recommendations [5], a number of variables are taken:

- evaporator temperature difference 5 °C,
- subcooling to the saturation temperature of 5 °C in network heater,
- GWNH temperature difference of 5 °C,
- initial steam temperature t_0.

The design conditions for CCGT-BP imply its operation at external temperature of –40 °C (temperature specified for the heating system design in Tomsk). The key GT characteristics are calculated with regard to its energy parameters; air heating to 3.4 °C through air recirculation at the compressor outlet is considered. The Siemens SGT5-4000F gas turbine was used as the CCGT-BP basis. The GT performance parameters at external temperature of –40 °C are as follows:

- GT load is 100%,
- electrical efficiency is 35.37 %,
- electric power is 247.6 MW,
- exhaust gas temperature is 594.8 °C,
- exhaust gas flow is 644.9 kg/s.

The values of the performance parameters for the steam section for the base case are summarized in Table 1.

Table 1. Main parameters of the steam section of CCGT-BP and its efficiency for the base case.

No	Parameter, dimension	Value
1	Initial steam pressure, MPa	5.0
2	Initial steam temperature, °C	540
3	Electric power, MW	68.51
4	Heat power, MW	302.9
5	Heating water temperature, °C	110
6	HRSG efficiency	0.828
7	CCGT total efficiency	0.885
8	Total capital investments, mln. USD	156.3
9	Net present value (NPV), mln. USD	547.9

The model was made in "MathCad" software and tested in the special "Thermoflex" complex.

In this study, the total capital investment cannot be taken as a criterion for the solution choice since the electric power of the steam turbine varies considerably. Therefore, to choose the best option, a net present value or net value gain relative to the base case should be used [6].

4 Results of the design efficiency of CCGT-BP

The effect of the external temperature on the main annual technical and economic indicators for CCGT-BP indicates the dependences shown in Figures 2,3.

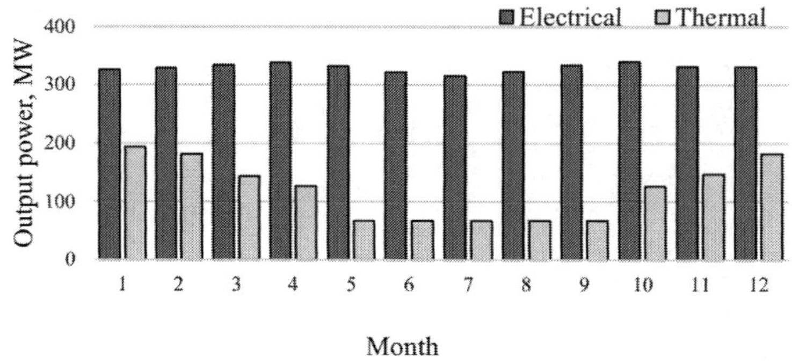

Fig. 2. Change in electrical and thermal capacity of CCGT-BP throughout the year.

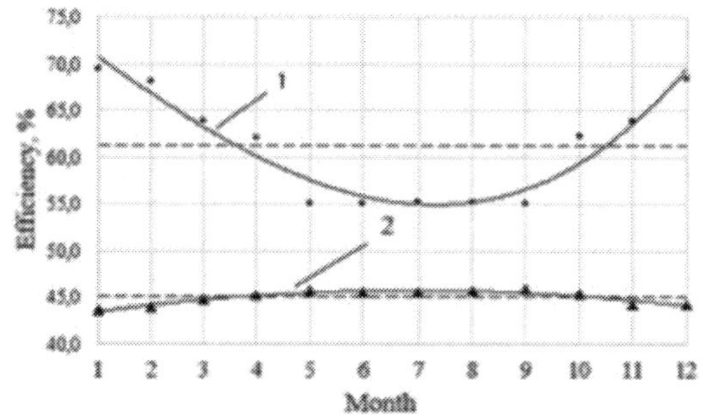

Fig. 3. Change in the total efficiency (1) and electrical efficiency of CCGT-BP (2) throughout the year. Dashed lines show average annual values of the efficiency of electrical output (45%) and total efficiency of CCGT-BP (61.2%)

Thermal load during the year varies significantly under change in external temperature thus changing in the total efficiency of CCGT-BP.

4.1 Effect of the heat network chart

Dependences of the main parameters of CCGT-BP on temperature of the heating water are presented in Fig. 4. The return water temperature is 60 °C.

Reduced temperature of the heating water results in increased efficiency of the investment in the project as a result of increased electric power of the steam turbine unit and increased gain in the net present value ΔNPV. The values of the CCGT-BP total efficiency and that of HRSG do not depend on temperature of the heating water since useful power of HRSG does not change due to constant temperature of gases at its inlet and outlet. The results of the calculations carried out for different values of the return water temperature are shown in Fig. 5. The heating water temperature is 110 °C.

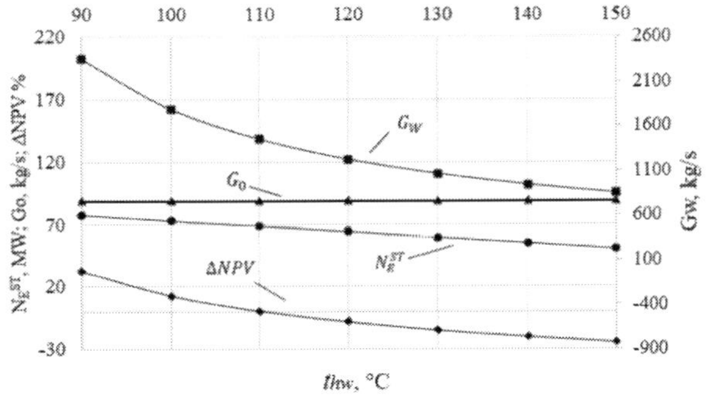

Fig. 4. Variation in CCGT-BP parameters against the temperature of the heating water.

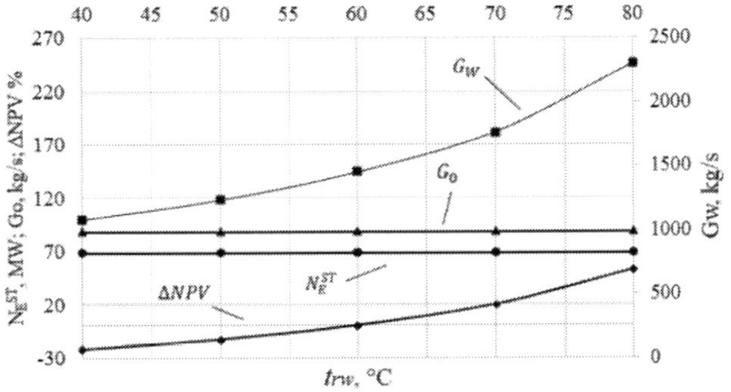

Fig. 5. Variation in CCGT-BP parameters against the temperature of the return water.

When the return water temperature grows from 40 °C to 80 °C, the HRSG efficiency drops by 7 %, which reduces the amount of the delivered heat by 8.5 % and reduces the total value of the CCGT-BP efficiency by 4.3 %. In addition, as temperature grows, the net present value ΔNPV changes by 74.8 %.

4.2 Effect of initial parameters of steam on CCGT-BP efficiency

The results of calculations for the CCGT-BP steam section performed at different heating water temperature are shown in Fig. 6. The return water temperature is 60 °C; the initial parameters of steam: 5.0 MPa and 540 °C. This effect is not significant.

The growth of the initial temperature in that range (by 7.4 %) leads to a decrease in the live steam flow. At increased temperature, the electric power slightly increases regardless of the initial pressure level, and the electric power gain is 1.6–1.9 MW.

At increased initial steam pressure, the parameter of the heat transferred to each surface of HRSG, except for GWNH, decreases. As a result, the steam flow to the turbine reduces and the electric power decreases. At increased pressure, the total capital investment remains virtually unchanged: at pressure increased in the above range, it reduces by 0.73 %–0.75 %. The value of ΔNPV relative to the base version is negligibly negative in all the cases, and it does not virtually change.

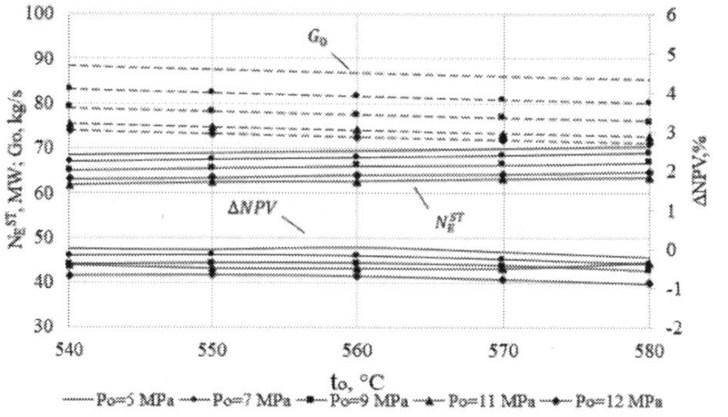

Fig. 6. Variation in CCGT-BP parameters against initial parameters of steam.

Taking into account the impact of ecological benefits of the installation can be using the software complex [7].

5 Conclusions

• The efficient performance of the steam section does not affect the CCGT-BP heat efficiency since bottoming cycle runs without cycle losses. In addition, simple circuit of the plant due to single-loop bottoming cycle and lack of steam reheating increases the CCGT-BP reliability.
• External temperature significantly affects the CCGT-BP operation, in particular, the amount of the produced electric energy and the parameters of the steam generated by waste heat recovery boiler.
• The temperature chart of the heat network is a crucial factor for the CCGT-BP parameters since the temperature of the heating network water determines the back pressure.
• Initial steam parameters do not virtually affect the main CCGT-BP parameters. If no additional electrical power is needed, the initial vapor pressure can be decreased to 5 MPa. The study complements the results obtained in [8].

References

1. G. Olkhovsky, V. Trushechkin, Thermal power plants, **1**, 2 (2013)
2. A.M. Antonova, A.V. Vorobiev, A.S. Orlov, MATEC Web Conf. **23**, 01069 (2015)
3. A. Uvarov, A. Antonova, A. Vorobiev, MATEC Web Conf. **37**, 01062 (2015)
4. A. Gushchin, V. Torzhkov, Turbiny i dizeli, **7**, 16 (2007) [In Russian]
5. A. Trukhniy, *Parogazovyye ustanovki elektrostantsiy* (MEI, Moscow, 2013) [In Russian]
6. A. Devyanin, *Optimizatsiya parametrov teplovyh skhem trekhkonturnyh parogazovyh ustanovok* (Abstract of diss., Moscow, 2009) [In Russian]
7. A.V.Vorobiev, E.M. Dutova, A.S. Latyshev, MATEC Web Conf. **72**, 01125 (2016)
8. A. Uvarov, A. Antonova, MATEC Web Conf. **91**, 01013 (2016)

Optimization of Construction of the rocket-assisted projectile

Vladimir Arkhipov[1], *Ksenia* Perfilieva[1,*]

[1]National Research Tomsk State University, 634050 Tomsk, Russia

Abstract. New scheme of the rocket motor of rocket-assisted projectile providing the increase in distance of flight due to controlled and optimal delay time of ignition of the solid-propellant charge of the SRM and increase in reliability of initiation of the SRM by means of the autonomous system of ignition excluding the influence of high pressure gases of the propellant charge in the gun barrel has been considered. Results of the analysis of effectiveness of using of the ignition delay device on motion characteristics of the rocket-assisted projectile has been presented.

1 Introduction

Rocket-assisted projectile (RAP) is the artillery shell that has appearance of normal (active) shell in the body of which the jet motors is built [1].

The special thing about external ballistics RAP is that at the time of a departure from in the gun barrel the shell has the maximum (muzzle) speed [2]. Thus, the shell takes maximum resistance from environment, proportional to the square of its speed. Engaging of the rocket-assisted projectile of solid-propellant rocket motors (SRM) directly at the time of shell departure from barrel does not allow using the thrust of SRM for increase in flying range of RAP effectively. To increase the efficiency of RAP it is worth to realize start of SRM on flight trajectory of rocket-assisted projectile with using various ways and delay devices time of ignition (inhibitor) [3-4].

2 The scheme of the rocket-assisted projectile

To increase the RAP flying range the new scheme of autonomous ignition system is presented [5]. The scheme of RAP is presented in Fig. 1. The rocket motor rocket-assisted projectile operates as follows. When driving a shell in a gun barrel under the influence of high pressure of propellant gases of a propellant charge the protruding edge of a flange adaptor 10 is cut off, and the rod 14 with the console 18 fixed on it moves towards the pierced membrane 4 and squeezes a spring 20. At the same time the conic section 11 of a rod 14 interfaced to a the conic cavity 9 in the basis 8, interfering with breakthrough of powder gases in a self-contained cavity 5. After shell exits from the gun barrel pressure of

* Corresponding author: k.g.perfiljeva@yandex.ru

powder gases upon the cut-off cover of a nozzle 13 sharply decreases, at the same time the rod 14 with the attached console 18 fixed on it under the influence of a spring 20 moves back and influences strikers 16 in the pyrocartridge 15. The products of combustion of the pyrocartridge, through perforations 17 in the console 18 received to a self-contained cavity 5 and ignite face surfaces of inhibitor charges 21. Thermo insulation fillers 22 interfere with penetration of powder gases into prenozzle volume from a self-contained cavity 5 before the complete combuston of the inhibitor charges 21. After a burnout of charges of a inhibitor 21 during the given time t_{ign} products of combustion through pierced membrane 4 come from a self-contained cavity 5 to prenozzle volume and set fire to an igniter 23. The products of combustion of the ignitor 23 ignite the charge of solid propellant 2. At achievement in a combustion chamber of the predetermined pressure the cut of turning 7 the cut-off covers of a nozzle 13 is cut off, providing a departure from the diffuser of a nozzle of 6 all elements of system of ignition (membrane 4, sleeving 19, basis 8, etc.). Thus, the launch of the SRM on flight path through tightly controlled time t_{ign} determined by the burning rate and the height of the charges of the inhibitor 21.

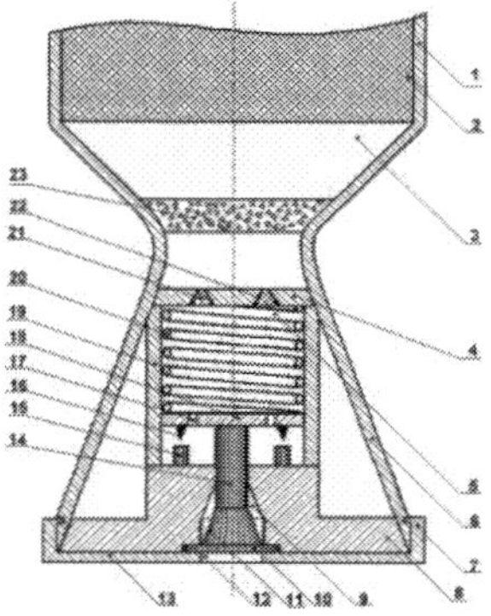

Fig. 1. The scheme of the rocket motor rocket-assisted projectile.

Thus, the scheme of the rocket motor ARS provides the increase in flying range due to providing a controlled and optimum delay time of ignition of the solid-propellant charge of the SRM and increase in reliability of initiation of the SRM by the expense of the independent ignition system excluding influence of high pressure gases of a propellant powder in the gun barrel.

3 Determination of the optimal value of ignition delay time of the solid-propellant charge

Let's consider the motion of ARS with a caliber of 76 mm (ZIS–3) [6] with SRM tubular-channel charge armoring at the face end surfaces of size 40/8–150 mm.

The main characteristics of the solid-propellant rocket motor ARS received by results of intra ballistic calculation of SRM operated by a technique [7] given in table 1.

Table 1. The main characteristics of the solid-propellant rocket motor ARS.

Parameter	d_{cr}, mm	d_a, mm	p_k, MPa	P, N	G, kg/s	J, m/s	t_*, s
Powder H	9.2	27.8	8.0	778	0.35	2214	0.82

Motion of an inert body in the air environment determinate by the system of equations of external ballistics [7]:

$$\frac{dU}{dt} = -\frac{F}{m} - g\sin\theta, \quad \frac{d\theta}{dt} = -g\frac{\cos\theta}{U},$$
$$\frac{dy}{dt} = U\sin\theta, \quad \frac{dx}{dt} = U\cos\theta,$$

where U is shell speed; F is force of a head resistance; m is mass of a shell; g is free fall acceleration; θ is a corner between a vector of speed and the line of the local horizon; y is vertical coordinate; x is horizontal coordinate.

The system of equations of motion of ARS in the active site includes the modified equation for the velocity of the given thrust of the rocket motor and the additional equation for mass change ARS with the combustion SRP:

$$\frac{dU}{dt} = \frac{P}{m} - \frac{F}{m} - g\sin\theta;$$

$$\frac{dm}{dt} = -\rho_T u_T S_T,$$

where P is force of draft of SRM; ρ_T is density of solid propellant; u_T is burning rate of solid propellant, S_T is surface area of combustion of solid solid propellant.

Force of a head resistance was calculated by the formula:

$$F = C_x \rho_b S_m \frac{U^2}{2},$$

where C_x is coefficient of a head resistance; ρ_b is density of air; S_m is area midship section of the shell.

The coefficient of head resistance was calculated by approximating equations depending on Mach number [7].

Conduction series of external ballistics calculations of flying range of an inert shell (without SRM) weighing 6.6 kg and with SRM (the mass of fuel is equal 0.29 kg or 4.3% of the mass of a shell) with tign ignition delay time variation, at a shot from the tool with the initial (muzzle) velocity of $U_0 = 680$ m/s, a corner $\theta_0 = 45^o$.

The results of the calculations, the range of inert shell ZIS-3 (no SRM) is $x_\kappa = 13.7$ km, which well coincides with the results of full-scale tests ($x_\kappa = 13.3$ km) [6].

The results of parametric calculations of the motion of the rocket-assisted projectile with SRM for different delay times of ignition t_{ign}, are given in table 2 (x_κ is range capability, t_κ is time of flight).

Table 2. The results of parametric calculations of the motion of the RAP with SRM.

t_{ign}, s	0	5	10	15	18	19	20	25
x_κ, km	15.12	15.37	15.66	16.35	16.50	16.53	16.52	16.17
t_κ, s	64.06	64.37	64.55	64.69	64.21	63.94	63.63	61.69

The calculated dependence of the rate of motion of the RAP from time is shown in Fig. 2, and the trajectory motion of the RAP shown in Fig. 3: 1 is inert shell (without charge SRP); 2 is charge with SRM at t_{ign}=0 s; 3 is charge with SRM at t_{ign}=19 s.

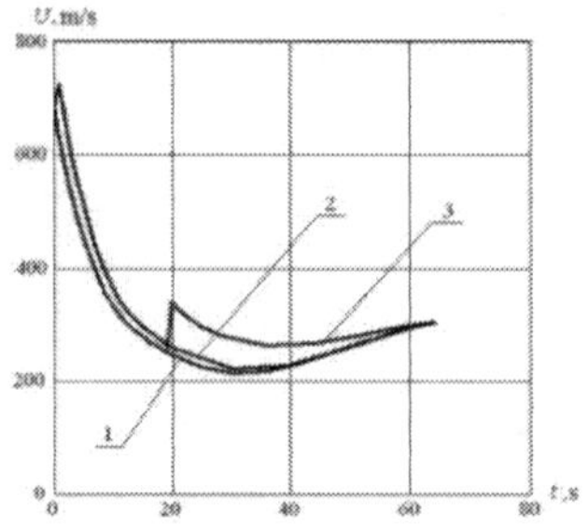

Fig. 2. Dependence of the rate shell from time.

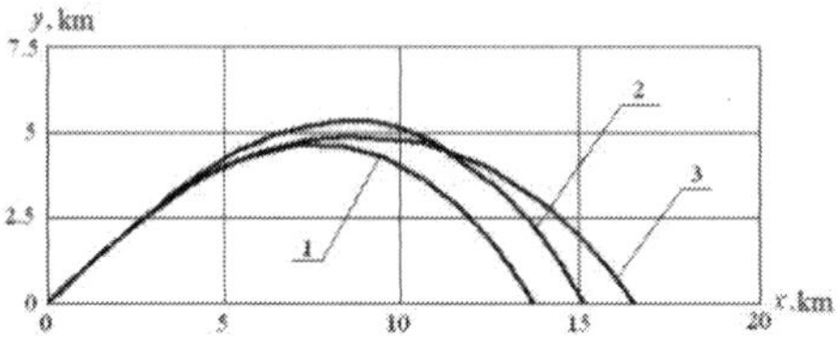

Fig. 3. Trajectory motion of the rocket-assisted projectile.

4 Assessment of the height of the inhibitor charges

The height of the charges of the inhibitor provides the optimal value of delay time of ignition of solid propellant charge in which the range of the ARS maximum [8].

To determine the height of the charges of the inhibitor, consider the equation of conservation of mass in a closed cavity between the pierced membrane and cut-off cover of a nozzle:

$$\frac{d}{dt}\left(\rho_g V\right) = \rho S u, \tag{1}$$

where t is time; ρ_g is density of the gaseous products of combustion inhibiter; ρ is density inhibite cherge; V is volume close cavity; S is the total surface combustion charge of the inhibitor; u is rate combustion inhibitor.

In accordance with the state equation Mendeleev – Clapeyron

$$\rho_g = \frac{p}{RT}, \tag{2}$$

where p is the pressure in the closed cavity; R is the gas constant of the combustion products of the inhibitor; T - temperature of the combustion products of the inhibitor.

Substituting (2) into (1), we get

$$\frac{dp}{dt} = \frac{\rho SRT}{V} u. \tag{3}$$

The law of burning rate inhibitor will take the form [8]:

$$u = u_0 \left(p/p_0\right)^{\nu}, \tag{4}$$

where u_0 is the burning rate inhibitor at atmospheric pressure; p_0 is atmospheric pressure; ν is the exponent in the dependence of the combustion rate inhibitor from the pressure.

Substituting (4) into equation (3) and carrying out the integration, we get:

$$p = p_0 \left[1 + \frac{(1-\nu) u_0 t \rho S R T}{p_0 V} \right]^{\frac{1}{1-\nu}}. \tag{5}$$

The height of the charges of the inhibitor is determined by the equation:

$$h = \int_0^{t_{ign}} u(dt).$$

Taking into account (4), (5), we obtain the desired relation:

$$h = \left\{ \frac{\rho S R T}{p_0 V} \left[u_0 (1-\nu) t_{ign} \right]^{\frac{1}{\nu}} \right\}^{\frac{\nu}{1-\nu}}. \tag{6}$$

Calculate the height h of the charge of the inhibitor (medium diameter 5mm), which provides the necessary time delay ignition of the charge of SRM. The main characteristics of the inhibitor composition of the TCS-20 are given in table 3 [9].

Table 3. The main characteristics of the inhibitor composition of the TCS-20.

ρ, kg/m^3	RT, kJ/kg	u_0, mm/s	ν
1600	280	0.6	0.06

The results of the calculations by equation (6) the height h of the inhibitor depending on the required magnitude of time delay of ignition t_{ign}, for different values of the free volume of the closed cavity are shown in Fig. 4 For time t_{ign}=19s, the height of the charge of the inhibitor is equal to h=14 mm in the free volume V=10 sm^3.

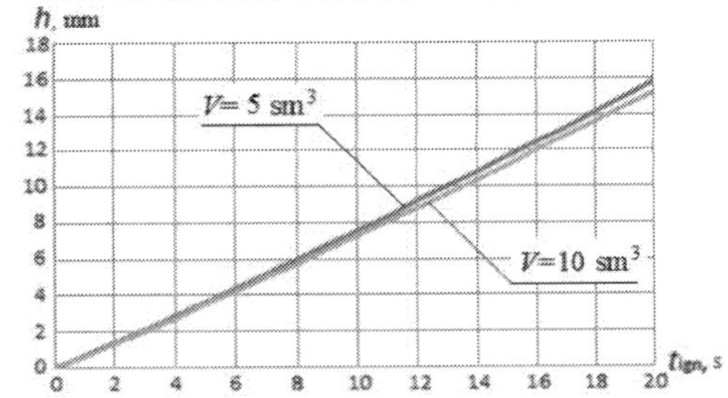

Fig.4. The dependence of the height h of the inhibitor depending on the required inhibitor of time delay of ignition t_{ign}, for different values of the free volume of the closed cavity.

The results of the calculations by equation (5) of value of the maximal pressure p in the closed cavity during the combustion of the inhibitor charges for different values of the free volume are shown in Fig. 5. The pressure value essentially depends on the free volume and V=10 sm^3 is p=12.7MPa.

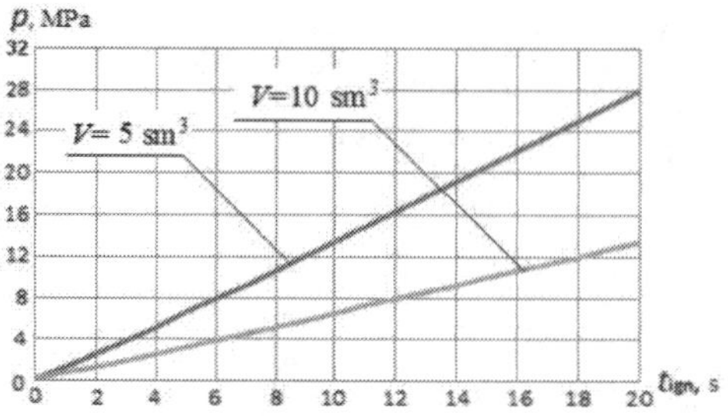

Fig. 5. Dependence for the values of the maximum pressure p in the closed cavity at combustion of the inhibitor charges for different values of the free volume.

5 Conclusion

- The modified scheme of the rocket motor rocket-assisted projectile is presented.
- It is defined that using the tubular and channel charge of ballistite (gunpowder H) propellant in SRM rocket-assisted projectile, increase in flying range of a shell by 10% (is provided at t_{ign}=0 s) and for 20% (at t_{ign}=19 s) in comparison with an inert shell.
- Inhibitor charge height (h=14 mm) at the free volume of V=10 sm^3 for an optimum delay time of ignition is calculated.
- The pressure in a self-contained cavity during combustion of charges of a decelerator is determined (p=12.7MPa) with a free volume of V=10 sm^3.

The work it is executed at financial support of the Ministry of education and science of the Russian Federation in the framework state task, project No. 9.9063.2017/BP.

References

1. M.E. Serebryakov, *Vnutrennyaya ballistika* (Barongis, Moscow, 1949) [In Russian]
2. B.A. Raizberg, B.T. Erokhin, K.P. Samsonov, *Osnovy teorii rabochikh processov v raketnykh sistemakh na tverdom toplive* (Mechanical Engineering, Moscow, 1972) [In Russian]
3. Patent US No 7360355 B1, IPC F02K 9/36. Long range artillery shell / M.H. Guillaume, J.L. Gadiot – Date of patent: 22.04.2008
4. Patent US No 9207054 B2, IPC F42C 11/06, F42B 15/10, F42B 15/36, F42C 11/02. Solid state ignition safety device / V. Jeffrey, A. James, V. Robert, M. James – Date of patent: 08.12.2015
5. The application on patent RF No 2017113324, IPC F02K 9/08. Method the increase in distance of flight due rocket-assisted projectile / V.A. Arkhipov, S.S. Bondarchuk, A.I. Konovalenko, K.G. Perfilyeva. – Filing date: 17.04.2017 [In Russian]
6. A. Ivanov, *Artillery of the USSR in World War II* (OLMA of Media groups, Moscow, 2003) [In Russian]
7. V.P. Stepanov, *Dynamics of flight of rocket projectiles. Manual* (Publishing house of the Tomsk university, Tomsk, 1977) [In Russian]
8. A.P. Hardt, *Pyrotechnics* (Post Falls, Idaho, 2001)
9. D.B. Demyanenko, A.S. Dudyrev, I.G. Strakhov, M.N. Tsynbal, The letter of the St. Petersburg state technological university, **42** (2012) [In Russian]

MATEC Web of Conferences **110**, 01004 (2017) DOI: 10.1051/matecconf/201711001004
HMTTSC-2017

Analytical evaluation of energy efficiency potential enhancement of heat-generating units by flow variation in the boiler circuit

Pavel A. Artamonov[1,*], *Nikolai* I. Kurilenko[1], and *Gennady* Ya. Mamontov[2]

[1]Tyumen State Industrial University, 625000 Tyumen, Russia
[2]National Research Tomsk Polytechnic University, 634050 Tomsk, Russia

Abstract. Three-dimensional models of the most common boiler schemes are developed and the hydraulic characteristic of each scheme is built. The calculation results of the main technical and economic indicators of the schemes are done. They are energy, metal consumption, capital costs and operating costs. According to the results of the analysis, there is a technical solution to reduce the energy consumption of the boiler circuit pumping equipment. The complex evaluation of the options in the context of the boiler manufacturer and operation is made, the investment attractiveness is calculated.

1 Introduction

Nowadays the great attention is paid to the energy saving and energy efficiency enhancement. All over the world the laws, strategies and programs are passed to stimulate the energy conservation. Due to the necessity of the energy resources efficient use, new technologies are developed, the energy saving equipment is implemented and this results in energy saving [1]. One of the initial stages of heat-generating units is a selection of the basic operation circuit. Optimization of thermal and hydraulic conditions inside the heat-generating units is one of the ways to enhance the energy efficiency [2]. The terms of a heat-generating unit and a boiler station are equal in this work.

2 Problem statement an solution methods

Let us consider two most widespread types of boiler circuits of the heat-generating units that use an independent heating scheme. These types will be examined in terms of the boiler heat capacity that is 9 Mega Watts (three KCB-3,0 boilers produced by "MAXTHERM") with a HH-62-16/3-167- TKTM32 plate heat exchanger produced by "Ridan" (heating system circuit) and a HH-42-16/1-33-TKTM25 plate heat exchanger produced by "Ridan" (hot-water supply system circuit) that are 6 and 3 Mega Watts

* Corresponding author: P.Artamonovv@yandex.ru

respectively. The temperature in the boiler circuit delivery pipeline is 105 °C, in the return one – 80 °C.

The first type is a boiler circuit with pumps that are immediately near each boiler at the return pipeline (fig. 1a). In this case, each pump operates together with a boiler and turns on under its start-up conditions. The boilers are operated in a cascade mode and the amount of the supplied heat is controlled by triple valves (condition 5).

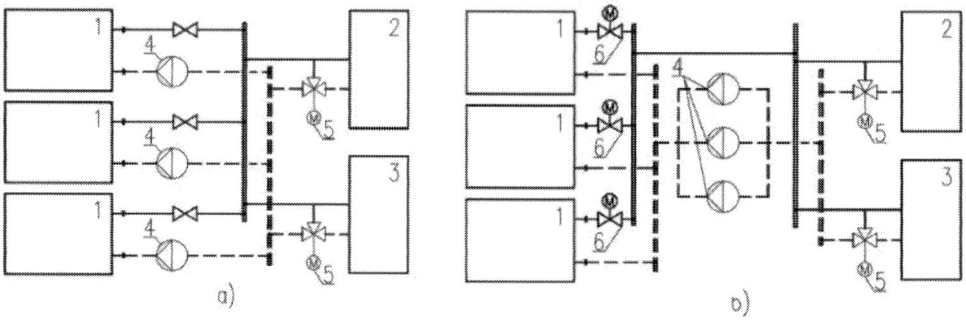

Fig. 1. The basic boiler circuit: a – type 1, b – type 2. 1 – boiler having a capacity of 3 Mega Watts; 2 – heat exchanger of the heating system; 3 – heat exchanger of the hot-water supply system; 4 – boiler circulation pump; 5 – triple valve; 6 – motor-operated valve.

In the second type, the boiler capacity and the amount of the supplied heat are controlled in a similar way as in the first one, but the boiler pumps are installed at the return pipeline (fig. 1b). To avoid the coolant circulation via an idle boiler, an electrical motor operated valve is provided (condition 7).

To carry out an overall comparison, the authors have developed two 3D boiler circuit models of the heat-generating units (fig. 2, 3) that correspond to the schemes shown in figure 1.

Based on the developed models the authors have analyzed each scheme in three basic operation modes:

- mode No 1 – one boiler is in use;
- mode No 2 – two boilers are in use;
- mode No 3 – three boilers are in use.

Each mode corresponds to a certain heat rate in the boiler circuit. If the predetermined temperature difference at the return and delivery pipeline is 25°C, the mean heat capacity of a heat transfer agent equals 4,19 kJ/kg and a density is 963,85 kg/m^3, the heat rate in the boiler circuit in operation modes No 1, 2, 3 will be equal to 103, 206, 309 m^3/h respectively.

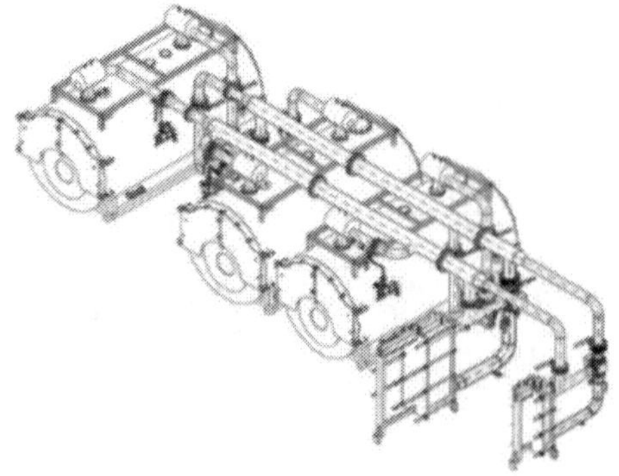

Fig. 2. 3D model developed according to the basic scheme of the first type.

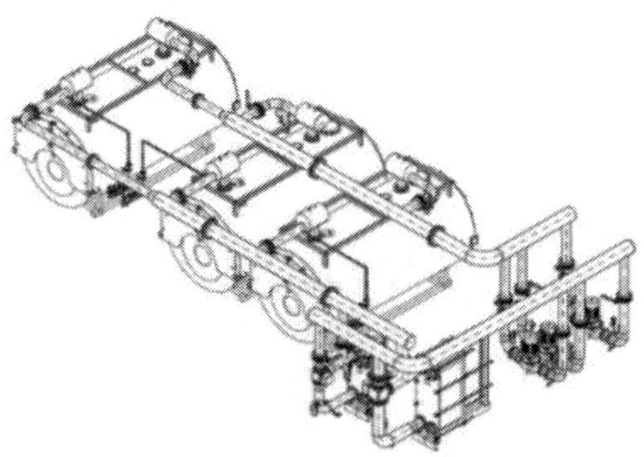

Fig. 3. 3D model developed according to the basic scheme of the first type.

The friction pressure loss was calculated by Darcy – Weisbach's equation, the restriction losses were determined by Weisbach's equation [3]. The local loss coefficients are specified according to [4]. Equipment losses (gate valves, nonreturn and triple valves, air collectors, etc.) were determined on the basis of the flow factor mentioned in the test certificate. The pumping machinery selection was carried out based on the hydraulic design findings (see Table 1).

Table 1. Hydraulic design findings.

	Pressure losses, w.c.m.			Pump
	Mode No 1	Mode No 2	Mode No 3	
Type 1	8.77	9.34	9.94	WILO BL 125/185-5,5/4
Type 2	9.14	9.7	10.77	WILO IL 125/210-5,5/4

The main purpose of the energy efficient operation of the heat-generating units is a fine control of the generated heat amount [5]. The authors have calculated the consumed heat rate throughout the year at the average monthly temperature of Tyumen outside air by the procedure set in design and construction specifications 41-104-2000 "Design of independent heat supply sources". Based on the findings the boiler operation in various year periods is determined. From April to October the boiler operates in mode No 1 to provide the necessary heat amount to consumers. From October to March – in mode No 2. The boiler operates in mode No 3 to provide the boiler peak load. The power control of the boiler thermal rating is done by the flame modulation in the burner, and the supplied heat is controlled by the triple valve. Based on the annual heat consumption value, the annual energy consumption was calculated. For the first type ("a" fig. 1) it is 64215,43 kWh, for the second one ("b" fig. 1) - 68791,9 kWh. The higher energy consumption of the second type is connected to the distinctive features of pipelining (see Table 1). Thus, the boiler circuit scheme of the second type is characterized by high energy consumption and a high maintenance cost as well.

As it was aforementioned, the boiler thermal rating is controlled by the flame modulation. In other words, when the heat-generating unit operates in the first mode and the power is controlled by the flame modulation from 40 to 100%, the boiler efficiency can vary from 1200 to 3000 kilowatt. Therewith, the heat rate is fixed regardless of the burner capacity. It means that if we neglect the pump operating point fluctuations when the triple valve position changes, we can state that the boiler pump energy consumption is fixed in each operation mode.

The future oriented way to reduce the pump unit energy consumption and thus, the maintenance cost reduction is a coolant rate control in the boiler circuit by the rotation frequency change of the pump rotor when the number of pump motor axis rotations is changed [6, 7, 8].

The authors suggested the boiler circuit operation scheme with the pump variable-frequency control (type 3). This scheme is different from the second type by the flow rate meter (condition 9), temperature detectors (condition 7) and a frequency converter (fig. 4).

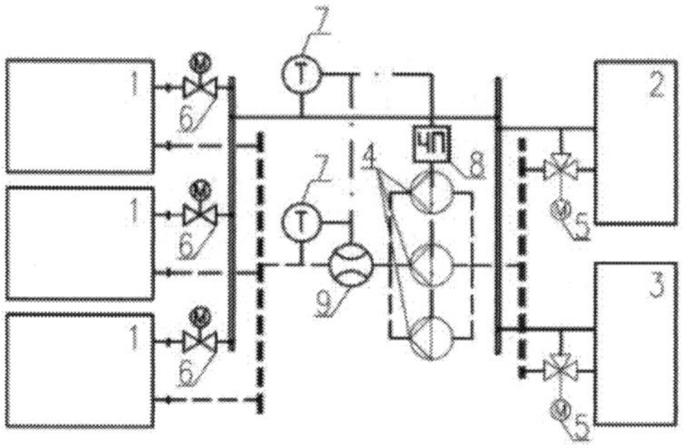

Fig 4. Boiler circuit basic scheme with the pump variable-frequency control. 1, 2, 3, 4, 5, 6 – similar to figure 1; 7 – temperature detector; 8 – frequency converter; 9 – flow rate meter.

The 3 D model is similar to the second type apart from the flow rate meter (see fig. 3). Based on the hydraulic design findings, the IL 100/220-5,5/4 pump produced by "Wilo" was selected. The more express pump was selected in terms of the pressure resistance in the boiler circuit when the flow rate meter is implemented.

The pump motor rotation frequency change is carried out in the following way: the heat-generating unit controller calculates the required boiler heat rate value in virtue of the coolant parameters of the line-operated circuit and a temperature dependence diagram of the delivery pipeline as it leaves the heat-generating unit on the outside air temperature. Based on the flow rate meter (condition 9) and temperature detector (condition 7) findings, the controller calculates the real value of the boiler heat rate. The differential of these values must be close or equal to zero. If the differential is higher than zero, i.e. the value of the supplied heat energy is higher than the target one, the controller sends out a signal to reduce the pump motor rotation frequency. If the differential is lower than zero, the frequency converter receives the signal to increase the frequency. To avoid the flow rate ramp change, the control logic should include the proportional integral derivative controller [9, 10]. Concurrently, it is necessary to control the boiler thermal rating and its cascading. The target value of the temperature differential is constant at the detectors (condition 7). Otherwise, the pumps will balance the low temperature differential by the flow rate increase that leads to the excess demand energy.

3 Results and discussion

The developed algorithm of the boiler circuit control of the heat-generating units allowed scaling the coolant flow rate and the consumer heat load change at a specified value of the temperature differential that is 25 °C at the delivery and return pipeline. The energy consumption was calculated on the ground of the average monthly heat consumption values (fig. 5). The annual energy consumption for the third type is 21264 kWh. The obtained results show that the boiler circuit operation allows reducing the annual energy consumption by a factor of 3,23 (by 47527,9 kWh) over the second type and by a factor of 3,02 (by 42951,43 kWh) over the first type.

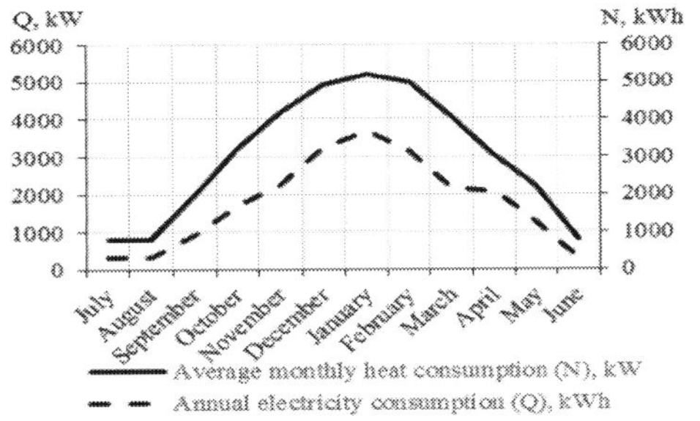

Fig. 5. Consumed heat and electric energy.

To evaluate all studied types of the boiler circuit operation of the heat-generating units the investment prospects will be analyzed. To calculate the capital costs the authors considered only the component and material cost of the boiler circuit manifold. Other capital cost items (wiring cost, transport charges, procuring and storage charges, etc.) are equal for all types and that allows neglecting them. The capital investments and capital energy costs consumed by pumps are presented (when one of the three types of the boiler circuit is implemented) in Table 2. The electricity tariff is 2,5 rub/ kWh.

Table 2. Comparison of the critical parameters.

Parameters	Type 1	Type 2	Type 3
Capital costs, RUB, in thousands	781.78	972.19	1000.94
Energy costs, consumed by boiler pumps, RUB, in thousands /year	160.54	171.98	53.16
Specific quantity of metal per structure, kg/ Mega Watts	210.47	256.1	259.82
Pay-back period, year	2	2.3	2.13

The calculation results show that the pay-back period for the heat-generating unit operation of 1, 2 and 3 type is 2, 2,3 and 2,13 years respectively. After a year from the date of operation even if the capital costs are the highest (see Table 2) the third type will exceed the consolidated profits obtained when the second type is used and in three years it becomes the most profitable of all studied types of the boiler circuit operation.

4 Conclusion

Analysis of the suggested method to change the flow rate showed the high potential to reduce the energy consumption of the heat-generating units and is the most investment attractive.

The models studied in this work provide insight into the overall perspectives of the flow rate proportional change in heat-generating units. Despite the high potential of energy saving, the use of the suggested method should be based on the detail analysis as the reduction in heat-generating unit capacity leads to the reduction of the total value of the saved physical resources and to the pay-back period change.

The work was supported by the Russian President's grant (Scientific School project - 7538.2016.8).

References

1. B.M. Nizameev, Bulletin of St PbSIT(TU), **34**, 97 (2016) [In Russian]

2. S.B.Nekhoda, Plumbing system, heating, air-conditioning, **2**, 110 (2011) [In Russian]
3. A.D. Altshul, *Hydraulics and aerodynamics* (Stroyizdat, Moscow, 1965) [In Russian]
4. I.G. Staroverov, *Sanitary-engineering installations* (Stroyizdat, Moscow, 1990) [In Russian]
5. A.M. Suleimanov, Ufa state oil technical university, 381 (2016) [In Russian]
6. A.V. Kozhukhova, Bulletin of Voronezh state technical university , **11-3**, 95-97 (2016) [In Russian]
7. J. Bujak, Energy, **34**, 1260–1270 (2009)
8. F.A.Amirov, Equipment and technologies for oil and gas complex, **4**, 34-35 (2012) [In Russian]
9. J.Quevedo, T.Escobet, Proceedings of the IFAC Workshop, Eds. Terrassa, **5**, 618 (2000)
10. K.J. Strom, T. Hagglund, *Advanced PID control* (ISA - The Instrumentation, Systems, and Automation Society, 2006)
11. P.N. Kamenev, E.I. Tertichnik, *Ventilation system: study guide* (ASV, Moscow, 2008) [In Russian]

DOI: 10.1051/matecconf/201711001005

Influence of thermal and mass transfer, the properties of steam-oxygen oxidizer and structure of coke on the composition of the generating synthesis-gas at the gasification of fuel

Roman N. Aryukov[1], *Alexander* N. Subbotin[1,*]

[1]National Research Tomsk Polytechnic University, 634050 Tomsk, Russia

Abstract. A mathematical model of the solid fuel conversion process which is used to analyze the synthesis-gas that is the result of coke gasification is proposed. During the conversation of coke in an environment of high-temperature water vapor, synthesis-gas consisting of hydrogen, oxide and carbon dioxide is obtained. That is shown in the article. Hydrogen in the produced gas at any oxidizer temperatures is always over 50% of total volume. The ratio of oxide and carbon dioxide depends on the temperature of the oxidant. The lower the oxidant temperature, the greater the proportion of carbon monoxide. It is established that it is economically more advantageous to use water vapor with the addition of oxygen as an oxidizer in the gasification of fuel. With the use of such an oxidizer, synthesis-gas consisting of 100% of the combustion gas is obtained. The temperature of the steam-oxygen oxidant after the fuel has ignited can be significantly reduced practically before the beginning of the process of vapor condensation. The synthesis-gas produced in this gasification process will contain in volume proportions about 25% hydrogen and 75% carbon monoxide.

1 Introduction

Solid fuel in the total amount of fuel consumed in the energy sector takes a significant share. In particular, in Europe and the US, coal accounts for about 60% of the energy sector. In the energy sector of Russia, the use of coal is approximately 18%. According to the development strategy of the Russian coal industry, it is planned to significantly increase the export of coal for the coke-chemical industry. Such an increase in coal production and planned exports will lead to an even greater increase in accumulated waste, which, as a rule, consists of slimes and fines. Burning of these wastes is ineffective, since their ash content is high, up to 50% and higher.

*Corresponding author: subbot@tpu.ru

The universal method of processing such wastes is their gasification. In modern gas generators, as a rule, steam-air or steam-oxygen blowing is used. This technology has both advantages and disadvantages. In particular, the disadvantages include the content of ballast admixtures in the synthesis gas, the part of the carbon burns in the gas generator. In connection with this the creation of a mathematical model of a gasifier for the conversion of solid fuels in an environment of various oxidants is an urgent task. Such a mathematical model can serve as a tool for estimating the composition of gasification products and for finding the optimal regimes for the gasification process.

2 Mathematical formulation of the problem

We consider a cylindrical capsule filled with coke, through which a heated steam heated up to a high temperature or a steam-oxygen oxidizer is pumped. In accordance with the work [1] and proceeding from experimental studies conducted at the department of the Kuzbass State Technical University [2], the following chemical reactions were considered:

$$\begin{gathered} C + H_2O = CO + H_2 - 130.5\ \text{kJ/mol}, \\ C + CO_2 = 2CO - 175.5\ \text{kJ/mol}, \\ C + O_2 = CO_2 + 395\ \text{kJ/mol}, \\ 2C + O_2 = 2CO + 219\ \text{kJ/mol}, \\ 2H_2 + O_2 = 2\,H_2O + 219\ \text{kJ/mol}, \\ 2CO + O_2 = 2CO_2 + 571\ \text{kJ/mol}, \\ CO + H_2O = CO_2 + H_2 + 40.4\ \text{kJ/mol}. \end{gathered} \tag{1}$$

Research objective: The determination of the synthesis-gas in the conversion of coke, the analysis of the composition of the gas formed, depending on the conditions of heat and mass transfer, the structure of the coke sample under study and the properties of the oxidant.

The simulation of the non-stationary mode of heat and mass transfer during coke conversion was carried out on the basis of the mathematical model [3]. It was assumed that there was no free moisture in the pores during the conversion, then the volume fraction of moisture (φ_2) is zero, and since the coke was considered, the volume fraction of solid fuel (φ_1) is also zero. Therefore, according to [3] and the reactions (1) we have the following system of equations:

- conservation of coke mass

$$\rho_3 \cdot \frac{\partial \varphi_3}{\partial \tau} = -R_C\,, \quad R_C = R_{C1} + R_{C2} + R_{C3} + R_{C4}\,; \tag{2}$$

- continuity of the gas phase

$$\frac{\partial\, \rho_5 \varphi_5}{\partial \tau} + v \cdot \frac{\partial\, \rho_5 \varphi_5}{\partial z} = R_C\,; \tag{3}$$

- continuity α – of the gas phase component

$$\frac{\partial(\rho_5\varphi_5 c_\alpha)}{\partial \tau} + v\frac{\partial(\rho_5\varphi_5 c_\alpha)}{\partial z} = \frac{\partial}{\partial z}(\rho_5\varphi_5 D_\alpha \frac{\partial c_\alpha}{\partial z}) + R_{5\alpha} + R_{S\alpha},\ \alpha = \overline{1{,}4}\ , \tag{4}$$

here parameter with an index $\alpha = 1$ corresponds to the oxygen (O_2), $\alpha = 2$ – CO, $\alpha = 3$ – CO_2, $\alpha = 4$ – H_2, $\alpha = 5$ – H_2O;

- conservation of energy of the gas phase

$$\rho_5\,\varphi_5\,c_{P5}\left(\frac{\partial T}{\partial \tau}+v\cdot\frac{\partial T}{\partial z}\right)=\frac{\partial}{\partial z}\left((\lambda_5+\lambda_R)\,\varphi_5\,\frac{\partial T}{\partial z}\right)-A_s(T-T_S)$$
$$+q^1_{CO}\,R^1_{CO}+q^2_{CO}\,R^2_{CO}+q_{H_2}R_{H_2}-\frac{2}{r_o}\alpha_1\,\varphi_5\,(T-T_S)\,; \tag{5}$$

- conservation of energy of the condensed phase (coke)

$$\sum_{i=3}^{4}\rho_i\,\varphi_i\,c_{pi}\frac{\partial T_S}{\partial \tau}=\frac{\partial}{\partial z}(\lambda_S\frac{\partial T_S}{\partial z})+q_{C1}\,R_{C1}+q_{C2}\,R_{C2}-q_{C3}\,R_{C3}-q_{C4}\,R_{C4}+$$
$$A_S\,(T-T_S)+\frac{2\cdot r_1}{r_0^2}\,\alpha_2\,(T_S-T_E)+q_R\,; \tag{6}$$

- the gas motion in the pores (in the form of a quadratic filtration law) and the equation of state

$$v=\frac{k}{\mu+k\cdot\beta\cdot\rho_5\,|V|}\cdot\frac{\partial p}{\partial z},\quad p=\frac{\rho_5\cdot R\cdot T}{M_5}\,. \tag{7}$$

The system of equations (2) - (7) was solved under the boundary conditions:

$$\tau=0:\quad \varphi_i=\varphi_{iH},\ i=\overline{3,4};\ \ c_\alpha=c_{\alpha H},\ \alpha=\overline{1,4},\ \ T=T_S=T_H,\ \ p=p_H\,; \tag{8}$$

$$\mathrm{z}=0:\quad T=T_G,\ \ (pv)_w=const,\ \ \lambda_S\cdot\frac{\partial T_S}{\partial z}=\alpha_1\cdot(T_S-T_G),\ \ c_\alpha=c_{\alpha G},$$
$$\mathrm{z}=\mathrm{h}:\quad \frac{\partial c_\alpha}{\partial z}=\frac{\partial T}{\partial z}=0,\quad p=p_H,\ \ \lambda_S\frac{\partial T_S}{\partial z}=0. \tag{9}$$

In the equations (2) - (7) the following notation is introduced: $c_{p5}=\sum_{\alpha=1}^{5}c_{p\alpha}\,c_\alpha$, $\lambda_S=\sum_{i=3}^{4}\lambda_i\,\varphi_i$, $\lambda_R=\dfrac{16\,\sigma\,T^3}{s}$, $M_5=1/\sum_{\alpha=1}^{7}\dfrac{c_\alpha}{M_{4\alpha}}$, $R_{C1}=s\,\rho_5\,\varphi_5\,c_7\,k_{S1}\,\varphi_3\,\exp\left(\dfrac{-E_{C1}}{R\cdot T_S}\right)$, $R^2_{CO}=k^2_{CO}\left(c_1\,M/M_{51}\right)^{0.25}\left(c_2\,M/M_{52}\right)\cdot T^{-2.25}\exp\left(\dfrac{-E^2_{CO}}{R\cdot T}\right)$, $R_{C2}=s\,\rho_5\,\varphi_5\,c_3\,k_{S2}\,\varphi_3\,\exp\left(\dfrac{-E_{C2}}{R\cdot T_S}\right)$, $R_{C3}=s\,\rho_5\,\varphi_5\,c_1\,k_{S3}\,\varphi_3\,\exp\left(\dfrac{-E_{C3}}{R\cdot T_S}\right)$, $R^1_{CO}=k^1_{CO}c_2\,c_7\,\exp\left(\dfrac{-E^1_{CO}}{R\cdot T}\right)$, $R_{H_2}=k_{H_2}c_1\,c_4\cdot\exp\left(\dfrac{-E_{H_2}}{R\cdot T}\right)$, $R_{52}=\gamma_5 R_{C1}+\gamma_6 R_{C2}+\gamma_7 R_{C4}$, $R_{54}=\gamma_{13}R_{C1}+\gamma_{14}R^1_{CO}-\gamma_{15}R_{H_2}$, $R_{57}=\gamma_{16}R_{H_2}-\gamma_{17}R_{C1}-\gamma_{18}R^1_{CO}$, $R_{51}=-(\gamma_1 R_{C2}+\gamma_2 R_{C3}+\gamma_3 R_{H_2}+\gamma_4 R^2_{CO})$, $R_{53}=\gamma_8 R_{C3}-\gamma_{10}R_{C2}+\gamma_{11}R^1_{CO}+\gamma_{12}R^2_{CO}$, $R_{C4}=s\,\rho_5\varphi_5\,c_1\,k_{S4}\,\varphi_3\,\exp\left(\dfrac{-E_{C4}}{R\cdot T_S}\right)$. All other notations coincide with the notation introduced in [3, 4 and 6].

3 Analysis of numerical results

The comparison of experimental and numerical studies [5] showed that good agreement on the percentage content of gasification products after sample heating is achieved. The temperature in the sample layer, found numerically, also agrees with the results of laboratory studies. Thus, this mathematical model adequately simulates the process under investigation and can be used for analysis of the gas composition when studying the processes of conversion of organic fuel. The values of thermophysical and kinetic properties at which numerical calculations were performed are given in [3, 4].

With the help of the mathematical model proposed above, studies of the yield of synthesis gas depending on the temperature of the supplied vapor (T_G), on the initial porosity and the specific surface of pores of the gasified solid fuel (coke) were conducted. The results of all numerical calculations for different values of porosity, specific surface of pores and three values of the temperature of the blown oxidant (water vapor) are given in tables 1, 2, 3.

In Tables φ_{5H} – initial porosity of coke, S_* – specific surface of pores. The set of these two parameters determine the structure of coke. Analyzing the results presented in Tables, we come to the conclusion that as the porosity decreases at any specific surface area of the pores $S_*$the fraction of CO in the synthesis gas decreases, and the fraction of CO_2 and H_2 increases.

Table 1. Composition of synthesis gas at a water vapor temperature T_G = 873 K.

φ_{5H}	Volumetric composition of gas synthesis,%	S_*=5·10^2, 1/m	S_*=5·10^3, 1/m	S_*=5·10^4, 1/m
0.9	CO	41.5	46.9	51.2
	CO_2	6.0	1.7	0.3
	H_2	52.5	51.4	48.5
0.5	CO	39.3	45.2	47.1
	CO_2	6.4	2.4	1.0
	H_2	54.3	52.4	51.9
0.1	CO	35.4	42.1	44.6
	CO_2	10.0	3.8	1.6
	H_2	54.6	54.1	53.7

Analyzing the results given in the tables, it can be concluded that when the fuel is converted, the yield of the gas composition depends significantly less on the porosity than on the specific surface of the pores.

Table 2. Composition of synthesis gas at a water vapor temperature T_G = 1173 K.

φ_{5H}	Volumetric composition of gas synthesis,%	S_*=5·10^2, 1/m	S_*=5·10^3, 1/m	S_*=5·10^4, 1/m
0.9	CO	7.4	37.0	47.8
	CO_2	27.2	6.9	0.6
	H_2	65.4	56.1	51.6
0.5	CO	3.2	28.5	44.9
	CO_2	29.3	11.8	2.3
	H_2	67.5	59.7	52.8
0.1	CO	1.4	184	37.6
	CO_2	29.2	16.3	6.2
	H_2	69.4	65.3	56.2

It also should be noted that for the same porosity, the composition of the gas is strongly dependent on the specific surface area of the pores (pore size). If the pores are very small (S_*=5·10^4, 1/m), the synthesis gas by more than 90% will consist of carbon monoxide and hydrogen. For samples with big pores (S_*=5·10^2, 1/m) the resulting synthesis gas will have about 30% of carbon dioxide and about 70% of hydrogen and a negligible amount of CO (Tables 2 and 3).

It should be noted that during gasification of coke in a water vapor environment, the proportion of combustion gas in the gasification products is the higher, the higher the oxidant temperature. In particular, at an oxidizer temperature of 873 K the carbon dioxide fraction is the lowest (less than 10%) for any porosity and specific surface area of the pores. Numerical calculations have shown that the method of gasification using high-temperature water vapor is not economically profitable. Because even at the temperature of the supplied steam of 1473 K, the volume fraction of the combustion gas at the outlet from the gasifier is only about 3% and 97% is water vapor.

Table 3. Composition of synthesis gas at a water vapor temperature T_G = 1473 K.

φ_{5H}	Volumetric composition of gas synthesis,%	S_*=5·10^2, 1/m	S_*=5·10^3, 1/m	S_*=5·10^4, 1/m
0.9	CO	1.3	21.5	43.4
	CO_2	30.5	15.6	1.8
	H_2	68.2	62.9	54.8
0.5	CO	0.5	14.9	41.8
	CO_2	30.7	18.7	3.8
	H_2	68.8	66.4	54.4
0.1	CO	0.3	12.2	33.7
	CO_2	30.3	19.6	8.9
	H_2	69.4	68.2	57.4

At an oxidizer temperature of 873 K, water vapor is 99%, and combustion gases are only 1%. Therefore, the resulting combustion gases are not enough even for heating to the desired temperature of the oxidant. In connection with this, further studies on gasification were carried out in an oxidizer environment consisting of water vapor and oxygen.

Table 4 presents the results of calculating the coke gasification in a steam-oxygen oxidant environment with the same initial porosity as for pure water vapor, the mass concentration of oxygen was 0.4 and the mass concentration of water vapor was 0.6.

Table 4. The composition of the synthesis gas produced at the initial oxidant temperature T_G = 873 K.

φ_{5H}	Volumetric composition of gas synthesis,%	S_*=5·10^2, 1/m	S_*=5·10^3, 1/m	S_*=5·10^4, 1/m
0.9	CO	72	75.3	76
	H_2	25	24.5	24
	CO_2 / O_2	0.4 / 2.6	0.05 / 0.15	0 / 0
0.5	CO	731	75.8	76.2
	H_2	26.8	24.2	23.8
	CO_2 / O_2	0.1 / 0	0 / 0	0 / 0
0.1	CO	73.5	75.1	76.6
	H_2	26.5	24.9	23.4
	CO_2 / O_2	0 / 0	0 / 0	0 / 0

Analyzing the results given in Table 4, we come to the conclusion that irrespective of the porosity and specific surface of the pores in the steam-oxygen gasification, we obtain a combustion gas consisting in volume approximately by (23-27) % of hydrogen and by (77-73) % of carbon monoxide A small amount of oxygen and carbon dioxide remains unreacted for samples with large porosity and (or) a small specific surface of the pores. It should be noted that under these gasification conditions, water vapor reacts completely, and there is no need to separate the vapor from the gasification products. From the reactor comes carbon monoxide diluted in volume by about a quarter by hydrogen. In some cases (with a large porosity of gasified fuel), a small amount of carbon dioxide and not consumed oxygen appears.

The results on the effect of the proportion of oxygen in the oxidant on the composition of synthesis gas are presented in Table 5. Analyzing the data in the table, we come to the conclusion that when the mass concentration of oxygen C_1 varies from 0.1 to 0.3 and the mass concentration of water vapor C_5 varies from 0.9 to 0.7, the volumetric composition of synthesis gas varies rather weakly, the maximum temperature of the gasification process increases quite strongly, especially the temperature of the coke. Therefore, by changing the oxidizer composition, it is possible to control the gasification temperature and consequently the conversion rate of the coke. The composition of the resulting synthesis gas will vary slightly in this case.

Table 5. The gas composition and temperature of the gasification products at different ratios of oxygen and water vapor in the oxidizer ($\varphi_{5H} = 0.1$, $S_* = 5\cdot 10^4$, 1/m).

The composition of the oxidizer C_1 / C_5		0.1 / 0.9	0.15 / 0.85	0.2 / 0.8	0.25 / 0.75	0.3 / 0.7
Volumetric composition of gas synthesis,%	CO	61.7	62.5	65.9	67.8	70.5
	H_2	38.3	37.5	34.1	32.2	29.5
Maximum temperature, K	gas	1216	1356	1586	1723	1907
	coke	1250	1527	2038	2368	2753

In addition to the results presented above, studies of the influence of the initial temperature of the oxidizer on the gasification process were carried out. Figure 1 shows the distribution of temperature in the coke along the depth of the fuel for three values of the initial temperature of the oxidizer. Distribution 1 was obtained at an initial oxidant temperature of 873 K.

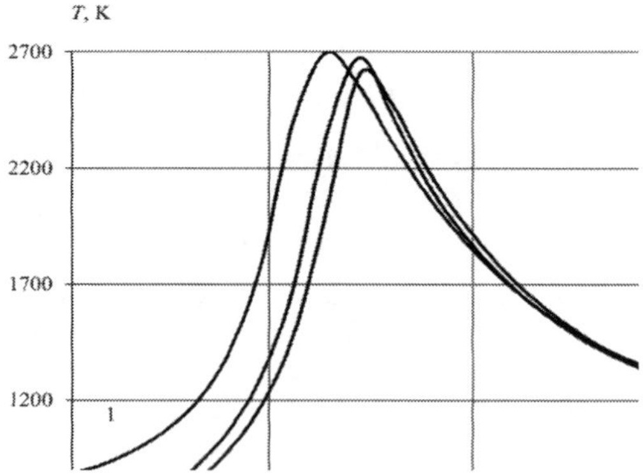

Fig. 1. Temperature profiles in coke at different initial oxidizer temperatures.

If the initial temperature of the oxidizer is further reduced, the temperature profile will move to the right (Fig. 1, curve 2). The maximum temperature will decrease slightly, and the profile will shift deep into the gasified fuel. If the oxidizer temperature drops to 373 K, the maximum temperature will slightly decrease and shift to the right (Fig. 1, curve 3). In all three cases, the composition of the synthesis gas produced is practically the same.

Consequently, after the coke has ignited and the gasification process has begun, the oxidizer temperature can be reduced to a minimum.

4 Conclusions

Based on the numerical calculations performed, the following conclusions can be drawn: the formulated mathematical model describes the process of conversion of solid fuel with sufficient accuracy for practice and it can be used to predict the composition of synthesis-gas in the gasification of solid fuels. During steam gasification the change in the temperature of the oxidizer, the speed of its supply and the porosity of the fuel affect the composition of the combustion gas formed. The temperature of the oxidizer exerts the greatest influence on the yield of synthesis gas during steam gasification.

In the case of steam-oxygen gasification, the influence of the fuel structure (porosity and specific surface of the pores) is small. The determining parameter of the process is the proportion of oxygen in the oxidant. By changing it, different regimes can be obtained, both in speed and in gasification temperature.

References

1. V.V. Pomerantsev, K.M. Arefyev and others, *Bases of the practical theory of burning: the Manual for high schools* (Energoatomizdat, Moscow, 1986) [in Russian]
2. S.A. Shevyrev, A.R. Bogomolov, Power system, **12**, 33 (2013)
3. A.N. Subbotin, News of Tomsk polytechnic university, **4**, 23 (2008)
4. R.N. Kulesh, A.S. Mazanik, A.N. Subbotin, News of Tomsk polytechnic university, **4**, 25 (2014)
5. A.N. Subbotin, S.A. Shevyrev, News of Tomsk Polytechnic University, **4**, 107(2012)
6. R.N. Aryukov, A.N. Subbotin, MATEC Web Conf. **72,** (2016)

Evolution process of fine fraction of condensed combustion products of aluminized condensed systems: experimental study

Valery A. Babuk[1,*], *Nikita* L. Budnyi[1]

[1]Baltic state technical university "VOENMEH", 190005 Saint Petersburg, Russia

Abstract. The experimental technique and the results of experimental study of the fine fraction of condensed combustion products (CCP) of aluminized condensed systems during their evolution are presented. An analysis was conducted and the interpretation of experimental data is given. The physical picture of formation and evolution of the fine fraction of CCP is proposed.

1 Introduction

Aluminized condensed systems (CS) are widely used in different power and propulsion systems nowadays. The formation of condensed combustion products (CCP) near the burning surface is an important feature of the burning process of these systems [1]. Condensed combustion products take part in a set of physical and chemical transformations during motion in a combustion products flow (called "evolution process" [2-3]) and, as a result, final CCP are formed. Condensed combustion products of aluminized CS consist of two fractions: the coarse – agglomerates, and the fine – smoke oxide particles (SOP) [1]. Agglomerates are products of a coalescence of condensed substances in a surface layer of a burning CS: they have relatively big size (from tens to hundreds and thousands of micrometers) and consist of aluminum and its oxide. Smoke oxide particles are fine Al_2O_3 particles (from tens of nanometers to tens of micrometers). In this work the results of the experimental investigation of CCP fine fraction evolution gathered in BSTU "VOENMEH" are presented and discussed.

2 Experimental study

2.1 Experimental technique

During experimental study, the CS specimens were burnt in the constant volume bomb (CVB) (fig. 1A) with the following collection and analysis of the CCP [2].

* Corresponding author: babuk@peterlink.ru

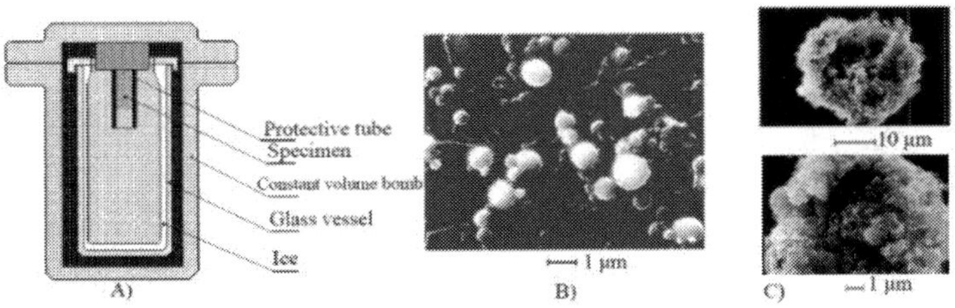

Fig. 1. A) Experimental set-up; B) Individual SOP; C) Coagulated SOP.

The constant volume bomb was pressurized with the inert gas. The glass vessel with a layer of ice on its inner surface was placed inside the CVB. The time of staying of the CCP in a flow of the combustion products was set through placing the specimen in the protective tubes of various lengths. This technique provides a possibility to analyze CCP at various distances from the burning surface. The constant volume bomb was sealed for at least 2 hours after combustion to provide a deposition of small particles. The fractions were separated using sieves. The individual SOP are shown on fig. 1B. The characteristics of dispersity were obtained during analysis of the whole totality of collected SOP using micro-analyzers "COULTER 4" and Horiba LA-950. The collected SOP undergo an ultrasonic processing with an addition of sodium acid pyrophosphate for destruction of undesirable coagulants (рис.1C). The special technique was developed for determination of the optimal ultrasonic processing duration, which provides a destruction of the coagulants and prevents formation of new ones [1]. The chemical analysis and the qualitative analysis using microscope visualization also were conducted. The "ice tubes" (made of water and hydrogen peroxide) were using for the reduction of coagulation due to decomposition of protective tubes material (plexiglass and polytetrafluoroethylene) in the case of using extended tubes. The "ice tubes" were placed at the inner surface of protective tubes. The chemical analysis allowed to make a conclusion about almost complete absence of any substances besides Al_2O_3 in SOP.

The following SOP characteristics were studied: SOP mass share in CCP, presented by the fraction of initial metal in the propellant used to form SOP Z_m^{SOP}, and parameters of dispersity – the mass distribution density function $f_m(d)$ and the mass-mean diameter d_{43}.

2.2 Experimental results

Table 1 presents the results of experimental study of SOP evolution for the ammonium perchlorate-based formulation N1-N5 [1], CSP-X4 [4], and for the ammonium dinitramide-based formulation CSP-X1 [4]. Figure 2 shows some examples of $f_m(d)$ for various formulations at various distances from a CS surface.

The dispersion analysis of the collected particles allowed to make important conclusions:

- As a rule, SOP grow during evolution – d_{43} increases during moving away from the surface. This phenomenon was observed for various formulations and pressure levels.
- SOP growth intensity can be various and depends on CS formulation, combustion conditions, agglomerates characteristics.
- Mass function of SOP size distribution density can be similar to unimodal or bimodal distribution, but there is a tendency to merging of modes or to disappearing of mode during evolution.
- SOP can reach large diameter (up to several microns).
- Fraction of initial metal in the propellant used to form SOP Z_m^{SOP} increases during moving away from surface.

Table 1. Experimental results on SOP dispersity during evolution.

Formulation	Pressure, MPa	Distance from CS surface, mm	d_{43}, μm
N1	1	*	0.35±0.07
		70	0.535±0.09
	6	*	0.40±0.08
		70	0.42±0.09
N4	6	*	0.55±0.05
		30	0.62±0.05
		70	0.67±0.06
N5	6	*	0.80±0.16
		70	1.35±0.20
CSP-X1	4	*	0.52±0.05
		30	0.61±0.06
		70	0.68±0.07
CSP-X4	4	*	0.54±0.05
		30	0.63±0.06
		70	0.70±0.07

* - combustion without protective tube

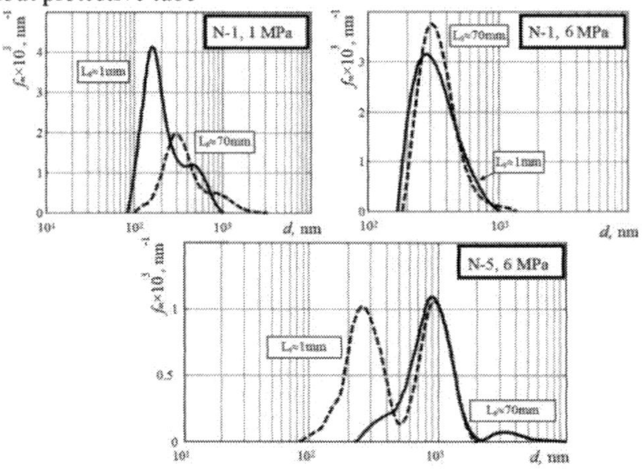

Fig. 2. Smoke oxide particles $f_m(d)$ for the ammonium perchlorate-based CS on various distances from surface (L_s – distance from CS surface).

3 Interpretation of results: physical picture of SOP formation and evolution

The common scheme of SOP formation and evolution is offered based on the analysis of experimental data and references on burning of aluminum particles in active environments (fig.3).

DOI: 10.1051/matecconf/201711001006

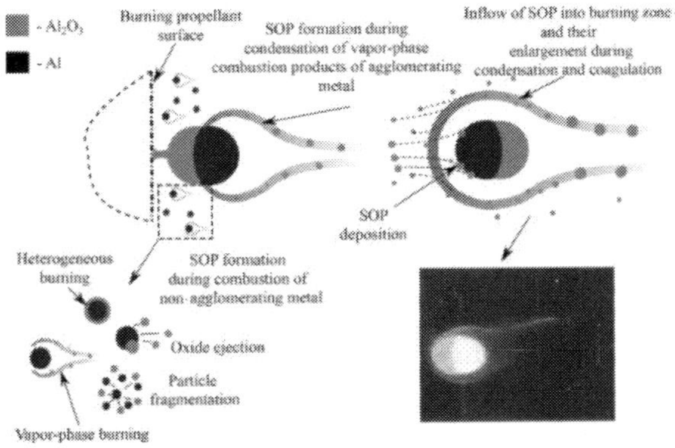

Fig. 3. Scheme of SOP formation and evolution.

The smoke oxide particles formation at the burning surface of CS is a result of combustion of non-agglomerating particle and burning of agglomerating particles. According to the common hypothesis, bimodal SOP distribution is a result of competition of oxide formation during vapor-phase burning and oxide accumulation on surface of particles. Word "mode" has to be understood conditionally - every "mode" in fact can be "polymodal". Apparently, the finest oxide particles (first mode) are formed during condensation of vapor-phase combustion products of non-agglomerating metal particles. The second mode, apparently, are formed during oxide accumulation on surfaces of non-agglomerating metal particles as a result of heterogeneous burning and, probably, diffusion and surface condensation of combustion products from zone of vapor-phase burning. The structure and characteristic particle sizes of the second mode are determined by dynamics of the accumulated oxide dispergation (or its absence). After oxide melting and starting of metal evaporation, there are some possible "scenarios": oxide ejection from particle surface (one or several droplets), "explosive" fragmentation of particle, transformation of oxide film into oxide "cap" without fragmentation or ejection. The final part of process is a complete combustion of residual metal core and formation of oxide particle as a result. The vapor-phase burning of agglomerating particles and the heterogeneous condensation of combustion products also can make a "contribution" to the second mode. The carbon particles (soot), which are formed during CS combustion, can act as nuclei in the heterogeneous condensation process [5]. Probably, the condensation behavior (homogeneous or heterogeneous) is determined by a concentration and size of nuclei in gas phase ("extraneous" particles, ions, impurities etc.) and also by size of burning particles. In other words, the condensation mechanism is determined by probability of presence of "extraneous" condensation centers in zone of vapor-phase burning.

The gas phase and the CCP are transformed during combustion products evolution [2, 3]. The essential phenomenon of the combustion products evolution is an agglomerate metal burning. Smoke oxide particles evolution is connected with this phenomenon – process of metal vapor-phase burning is a source of aluminum sub-oxides which condense into Al_2O_3. In [6] SOP growth during evolution explained in the following way: during burning of agglomerate metal sub-oxides are being produced; SOP, which are in the velocity equilibrium with gas, can flow into agglomerates burning zones due to their velocity lag, and then SOP can act as nuclei (condensation centers). Sub-oxides in agglomerate burning zone condense into Al_2O_3 on existed SOP surfaces that resulting in a SOP growth. Particles coagulation is also possible. Growth rate during evolution depends on many factors: gas velocity, SOP concentration and size, agglomerates concentration and size, their velocity lag relative to gas, and combustion environment properties, which have an influence on CS and agglomerates combustion dynamics.

The smoke oxide particles deposition on agglomerates can cause additional transformation of SOP concentration and dispersity. Deposition intensity increases with increasing of an agglomerate velocity lag and their concentration.

In a combustion chamber SOP growth can be probably more intensive due to effects of a turbulent flow leading to coagulation.

4 Conclusions

The experimental results on SOP evolution in combustion products flow are presented. It is assumed that the observed SOP growth is caused by inflow of SOP into the burning zones of agglomerates where particles grow during condensation of aluminum sub-oxides on their surfaces.

Authors are grateful to M. S. Malakhov and P. A. Naslednikov, who made a significant work in the development of experimental technique and experimentation in different years.

References

1. V.A. Babuk, V.A. Vasilyev, M.S. Malakhov, J. Propul. Power. **15**, 783 (1999)
2. V.A. Babuk, V.A. Vasilyev, P.A. Naslednikov, *Combustion of Energetic Materials* (Begell House, New York, 2002)
3. V.A. Babuk, V.A. Vasilyev, J. Propul. Power. **18**, 814 (2002)
4. V.A. Babuk, V.A. Vassiliev, V.V. Sviridov, Combust. Sci. Technol. **163**, 261 (2001)
5. V.A. Babuk, N.L. Bydnyi, Khimicheskaya fizika i mezoskopiya, **17**, 39 (2015) [In Russian]
6. V.A. Babuk, N.L. Bydnyi, Khimicheskaya fizika i mezoskopiya, **19**, 5 (2017) [In Russian]

MATEC Web of Conferences **110**, 01007 (2017) DOI: 10.1051/matecconf/201711001007
HMTTSC-2017

Dynamics and Heat and Mass Transfer of Liquid-Droplet Cloud in the Emergency Discharge of Aviation Fuel into the Atmosphere

Evgeny Maslov[1, 2,*], *Amgalan* Badmaev[2], *and Irina* Zharova[1]

[1]National Research Tomsk State University, 634050 Tomsk, Russia
[2]National Research Tomsk Polytechnic University, 634050 Tomsk, Russia

Abstract. A mathematical model for the propagation of liquid-droplet cloud in the atmosphere during the emergency discharge of aviation fuel is presented. The results of a numerical analysis of the dynamics and heat and mass transfer of a liquid-droplet cloud in the emergency discharge of aviation fuel into the atmosphere, depending on the season-climatic and meteorological conditions of specific regions of Russia are given.

1 Introduction

To reduce the possible risks in the event of emergency situations during flight it is the practice of dumping fuel into the atmosphere from filling tanks before an urgent landing (Fig. 1) [1].

Fig. 1. Emergency discharge of fuel from the aircraft Northwest Airlines Boeing 747.

Since the mass of jet fuel to be discharged is tens of tons, it is necessary to provide emergency discharge conditions in which the fuel droplets deposition on the earth's surface will be minimized. Currently, the conditions for the dumping of aviation fuel in emergency situations are regulated by the rules of the International Civil Aviation Organization (ICAO). The minimum height of fuel discharge is 1850 m for flights over Western Europe territory in accordance with ICAO rules. These conditions ensure that the bulk of the fuel evaporates in the atmosphere, and only ~ 8 % of the mass of the fuel reaches the ground surface. It should be noted that when flying over the territories of other regions whose climatic conditions are significantly different from those in Western Europe, these regimes

* Corresponding author: maslov_eugene@mail.ru

of discharging do not always guarantee that there is no significant amount of fuel on the earth's surface.

The aim of the presented paper is to analyze the dynamics and heat and mass transfer of a liquid-droplet cloud in the emergency discharge of aviation fuel into the atmosphere, depending on the season-climatic and meteorological conditions of specific regions of Russia.

2 Mathematical modeling of dynamics and heat and mass transfer of liquid-droplet cloud in the emergency discharge of aviation fuel into the atmosphere

Numerical analysis was carried out under the following assumptions. At the initial time, the volume of fuel injected into the atmosphere was considered as a drop of spherical shape (a liquid-droplet cloud) with a diameter equal to the characteristic width of the impact jet (Fig. 1). The initial velocity of the liquid-droplet spherical cloud was equal to the speed of the aircraft. It was believed that the liquid-droplet cloud propagates in the atmosphere under the influence of gravity, the force of aerodynamic drag and wind force. The processes of repeated crushing of drops, evaporation and heat exchange with the ambient air were taken into account. Within the framework of the assumptions made, the mathematically considered process is described by a system of equations [2–4]. The motion of a drop in a Cartesian coordinate system was described by the equations of motion in the velocity field of the steady gas flow with allowance for the forces of viscous drag and gravity:

$$\begin{aligned}
&\frac{dV_{px}}{dt}=\frac{3\rho_f}{4\rho_p d_p(t)}C_D\left|U_f-V_p\right|\left(U_{fx}-V_{px}\right),\\
&\frac{dX_p}{dt}=V_{px},\\
&\frac{dV_{py}}{dt}=\frac{3\rho_f}{4\rho_p d_p(t)}C_D\left|U_f-V_p\right|\left(U_{fy}-V_{py}\right)-g,\\
&\frac{dY_p}{dt}=V_{py},
\end{aligned} \tag{1}$$

where t is time; d_p is the droplet diameter; ρ_f, ρ_p are the density of air and droplet substances, respectively; U_{px}, U_{py}, V_{px}, V_{py} are projections of the velocity of air movement and the velocity of the droplet movement on the axis of the Cartesian coordinate system; X_p, Y_p are the current coordinate of the drop movement along the x and y axes, respectively; C_D is coefficient of aerodynamic drag, $g = 9.81$ m/s^2 is acceleration of gravity.

The thermal state of a drop was described by the thermal conductivity equation in a spherical coordinate system [3–5]:

$$\rho_p(T_p(r,t))\cdot c_p(T_p(r,t))\cdot\frac{\partial T_p(r,t)}{\partial t}=\frac{1}{r^2}\cdot\frac{\partial}{\partial r}\left(\lambda(T_p(r,t))\cdot r^2\cdot\frac{\partial T_p(r,t)}{\partial r}\right), \tag{2}$$

where r is the spherical coordinate; r_p is the radius of a drop; T_p is drop temperature; c_p and λ are the specific isobaric heat capacity and the coefficient of thermal conductivity of the droplet substance, respectively.

Taking into account the processes of evaporation and convective heat exchange with air, equation (2) was supplemented by the following initial and boundary conditions [3, 6]:

$$t=0,\ 0\le r\le r_p:\ T_p(r)=T_{p0}, \tag{3}$$

$$t>0,\ r=0:\ \frac{\partial T_p(t)}{\partial r}=0, \tag{4}$$

$$t>0,\ r=r_p:\ \lambda\left(T_p(t)\right)\cdot\frac{\partial T_p(t)}{\partial r}=\alpha_f\cdot\left(T_f-T_p(t)\right)-q_{vap}\cdot w_{vap}(t), \tag{5}$$

where T_{p0} is the initial drop temperature; T_f is ambient temperature (air); q_{vap} is the heat of evaporation of drop substance; $w_{vap}(t)$ is the rate of drop evaporation; α_f is the coefficient of heat exchange with air. The coefficient of heat transfer α_f in (5) is determined by the formula for flow past a spherical particle:

$$\alpha_f(t)=\mathrm{Nu}\cdot\frac{\lambda_f}{d_p(t)}, \tag{6}$$

where λ_f is the thermal conductivity coefficiente of the air; Nu is the Nusselt number.

The Nusselt number in (6) is determined by the Rantz-Marshall relationship [5] under the conditions considered:

$$\mathrm{Nu}=(2+0.6\cdot\mathrm{Re}_p^{1/2}\cdot\mathrm{Pr}_f^{1/3}), \tag{7}$$

where Re_p is Reynolds number; Pr_f is Prandtl number.

The Reynolds and Prandtl numbers for the flow around a spherical drop by air are determined by the relations

$$\mathrm{Re}_p=\frac{\rho_f\cdot\left|U_f-V_p\right|\cdot d_p}{\mu_f},\quad \mathrm{Pr}_f=\frac{\mu_f c_{pf}}{\lambda_f}, \tag{8}$$

where μ_f, c_{pf}, and λ_f are the dynamic viscosity coefficient, the specific isobaric heat capacity and air thermal conductivity coefficient.

The change of a drop diameter due to evaporation was determined by the formula [2]:

$$\frac{d_p(t)}{dt}=\frac{2\cdot G_p(t)}{\rho_p(t)\cdot S_p(t)}, \tag{9}$$

where $d_p(t)$ is a drop diameter; $G_p(t)=w_{vap}(t)\cdot dt$ is the mass of the evaporated substance in a time dt; $S_p(t)$ is the surface area of a spherical drop.

The crushing of drops under the action of inertia forces (Rayleigh-Taylor instability) was taken into account by calculating the value of the Bond's number

$$\mathrm{Bo}=\frac{\Delta\rho\cdot|\omega|\cdot d_p^2(t)}{\sigma_p}, \tag{10}$$

where $\Delta\rho=\rho_p-\rho_f$; $\Delta\rho$ is the surface tension coefficient of a drop; $|\omega|$ is the modulus of acceleration vector of a drop.

The crushing of drops under the action of aerodynamic drag (Kelvin-Helmholtz instability) was taken into account by calculating the value of the Weber's number

$$\mathrm{We} = \frac{\rho_f \cdot \left|U_f - V_p\right|^2 \cdot d_p^2(t)}{\sigma_p} \tag{11}$$

It was assumed that when the critical value of the Bond number is $\mathrm{Bo}_* = 22.5$ or the critical Weber number $\mathrm{We}_* = 17$, the drop is split into two spherical drops of equal mass with an initial velocity equal to the velocity of drop initial [2].

When the mathematical model (1)–(5) was parametrized, it was taken into account that in the conditions of emergency fuel discharge from the filling tanks, the liquid-droplet cloud initially falls into the "trace" – into dragged by co-current aircraft wing flow. Thus, at the initial moment of time, the liquid-droplet cloud has a velocity equal to the airflow velocity in the "trace". The air velocity profile in the "trace" is described by the equation [7] in case of an axisymmetric co-current:

$$\frac{u_f}{U_{f0}} = \left(1 - \eta^{3/2}\right), \tag{12}$$

where u_f is the airflow velocity in the "trace"; U_{f0} is the speed of the aircraft; $\eta = y_s/\delta$, y_s is the distance from the "trace" axis, δ is the half-width of the "trace".

The half-width of the "trace" δ was estimated according to [7]:

$$\delta \approx \left(\frac{F_x}{\rho_f \cdot U_{f0}^{\;2}}\right)^{1/3}, \tag{13}$$

where F_x is the lift force of the wing, U_{f0} is the aircraft speed at the moment of fuel discharge.

The mathematical model (1)–(5) is implemented as a software package.

3 Analysis of the results of numerical modelling

A numerical study was made of the effect of the most significant factors limiting the crushing of droplets during the propagation of a liquid-drop cloud in the atmosphere on the basis of the mathematical model (1)–(6). Based on the results of the analysis of the preliminary calculations, the size of the drops formed during crushing during gravity deposition was estimated, depending on the initial size of the liquid-drop cloud. The initial size of the liquid-drop cloud was varied in the range d_p=(0.01÷0.5) m. The speed of the aircraft varied in the range U_{f0}=(600 ÷ 900) km/h at the time of emergency fuel discharge.

It is found that drops form of diameter $d_p \sim 10^{-3}$ m during the time of ~ 1 s in the whole range of speed of the aircraft, regardless of the initial size of the liquid-drop cloud. At U_{f0} = 600 km/h, the minimum drop size was $d_p \approx 1.25 \cdot 10^{-3}$ m, at U_{f0} = 900 km/h – $d_p \approx 0.98 \cdot 10^{-3}$ m. The aerodynamic crushing is completed when the drop is vertically moved to a distance of < 10 m, depending on the initial size of the liquid-droplet cloud. The drop size decrease is determined by the evaporation process in the process of further propagation in the atmosphere. The minimum height $h_{\min}$ of the emergency fuel discharge, which ensures complete evaporation of fuel droplets in the atmosphere were determined, depending on the season-climatic conditions. Analysis is performed for the regions of Russia located in different climatic zones: Kaliningrad and Novosibirsk Regions, Yakutia (Sakha Republic). The atmospheric parameters averaged over time are taken from [8]. The results of the minimum height calculating $h_{\min}$ are showed in Fig. 2–4.

DOI: 10.1051/matecconf/201711001007

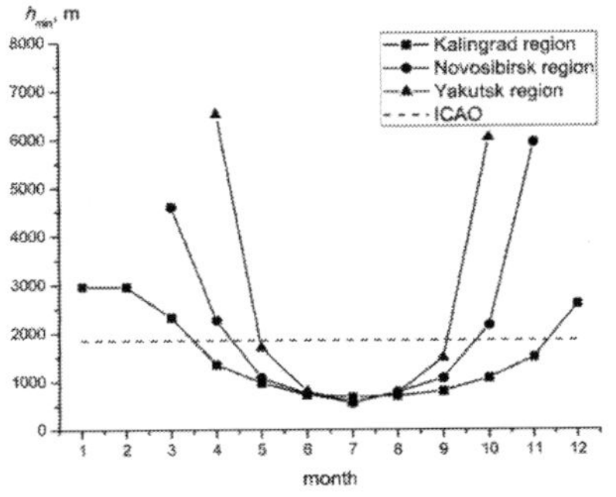

Fig. 2. Minimum heights of emergency fuel discharge for different regions of Russia, U_{f0}=600 km/h.

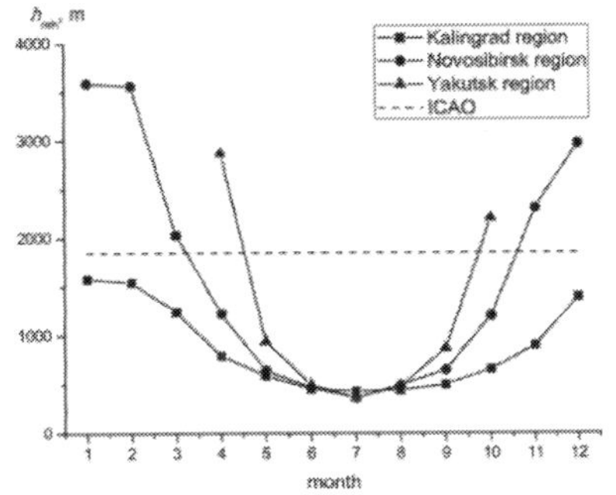

Fig. 3. Minimum heights of emergency fuel discharge for different regions of Russia, U_{f0}=900 km/h.

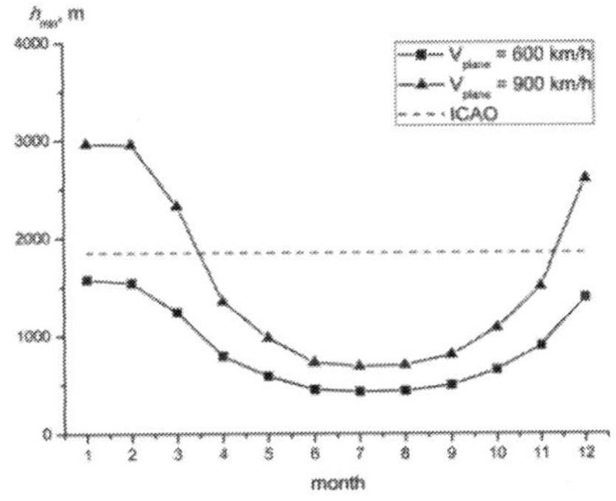

Fig. 4. The minimum heights of fuel discharge above the territory of the Kaliningrad region depending on the speed of the aircraft.

From the data ones shown in Fig. 2 it follows that when speed of aircraft is 600 km/h, the minimum fuel discharge heights, ensuring no precipitation of aviation fuel to the earth's surface, do not meet the requirements of the ICAO in the cold months of the year. This period is from January to April and from October to December for the territory of the Novosibirsk region, from January to May and from September to December for Yakutia, from January to March and December for the Kaliningrad region. There is a risk of precipitation of unevaporated fuel to the surface of the earth in these periods.

Emergency discharge of aviation fuel with an increase in aircraft speed to 900 km/h allows to reduce ecologically dangerous periods of the year (Figure 3) for the regions of Yakutia and the Novosibirsk region. The minimum heights of the emergency discharge above the territory of the Kaliningrad region ensure the evaporation of fuel droplets in the atmosphere throughout the year at this speed. The obtained results are in satisfactory agreement with ICAO rules developed for flights over the territory of Western Europe (Fig. 4). This is explained by the compliance of the seasonal and climatic conditions of the Kaliningrad region with the conditions of the European regions. Seasonal and climatic conditions of Yakutia and the Novosibirsk region are significantly different from European by continuance of cold periods and by air temperature ranges during these periods.

Based on the mathematical model (1)–(5), the values of the mass fraction of aviation fuel, falling on the earth's surface in ecologically dangerous periods depending on the speed of the aircraft at the time of discharge were determined. In particular, it was obtained that ~ 0.5 mass of fuel is deposited on the earth's surface during emergency discharge from an airplane over the territory of Yakutia in the coldest period of the year (from December to February).

4 Conclusion

Analysis of the results of mathematical modeling of dynamics and heat and mass transfer of a liquid-droplet cloud in the emergency discharge of aviation fuel, depending on the season-climatic and meteorological conditions of specific regions of Russia, showed that the speed of the aircraft, at which an emergency fuel discharge occurs, is one of the most significant factors affecting the processes of crushing and evaporation of drops in the atmosphere. It seems advisable to develop regional rules, similar to ICAO rules, regulating the organization of fuel dumping in accidents, taking into account the seasonal and climatic conditions of specific regions. The development of regional rules can be carried out on the basis of mathematical modeling using the proposed model.

This study was funded by grant from the Russian Science Foundation (Project No. 15-19-10014).

References

1. Electronic resource: the website of news agency Newsru, http://www.newsru.com/arch/world/27sep2005/usa80.html
2. V.A. Arkhipov, E.A. Kozlov, I.K. Zharova, A.S. Tkachtnko, Atmospheric and Oceanic Optics, **26** (2013)
3. Ye.A. Maslov, I.K. Zharova, N.S. Fedotova, Selected Reports of All-Russian Scientific and Technical Conference London, **5** (2014)
4. V.A. Arkhipov, E.A. Kozlov, S.S. Titov, A.S. Tkachenko, A.S. Usanina, I.K. Zharova, Arabian Journal of Geosciences, **9**,114 (2016)
5. M.A. Pakhomov, V.I. Terekhov. Flow, Turbulence and Combustion, **98** (2017).
6. A. Semenov, G. Kuznetsov, D. Zaitsev. Evaporation of an intensively heated sessile droplet into the open atmosphere, MATEC Web Conf., **84** (2016)
7. R.I. Nigmatulin. *Dynamics of Multiphase Media* (Hemisphere Publishing Corporation, New York, 1991)
8. *New Aeroclimatic Reference Book of the Free Atmosphere over the USSR* (Gidrometeoizdat, Moscow, 1990) [In Russian]

Energy Resources Consumption Minimization in Housing Construction

Alexey Balastov[1,*], *Nikolay* Kachalov[1], *Elena* Senkiv[1] *and Olga* Kachalova[2]

[1]National Research Tomsk Polytechnic University, 634050 Tomsk, Russia
[2]The Russian Presidential Academy of National Economy and Public Administration, 634050 Tomsk, Russia

Abstract. The article deals with the energy savings analysis during operation of buildings, provides the heat balance of residential premises, considers options for energy-efficient solutions for hot water supply systems in buildings. As technical facilities that allow the use of secondary heat sources and solar energy, there are also considered the systems with heat recovery of "gray" wastewater, heat pumps, solar collectors and photoelectric converters.

1 Introduction

Studies conducted in Western Europe show that there is a decrease in economic costs for heating and hot water supply up to 35% of total energy consumption.

In Siberia, the result of heated facilities energy saving can be the creation of energy-saving life support systems for both residential buildings and thermal buildings modernization. It is necessary to find a solution in the construction of such residential premises, where the renewable energy sources are also widely used. In this direction scientists from Russia, Germany, Austria and other countries successfully work.

The idea of saving energy during the buildings operation is popular and is being realized in the whole world [1-11]. There have already been built the first experimental buildings with zero energy consumption (Null-Energie) [1, 3] and with their positive balance (Plus-Energie) [4, 8].However, the economically feasible construction of such buildings requires additional research.

2 Energy-efficient buildings

At the end of the last century, in the Western world, the first steps were taken in order to save heat in the operation of premises and residential buildings, which we today call energy efficient. There should be considered a multifactor aspect while the construction of energy-efficient buildings, the choice of a technical solution, especially using alternate solar, wind, and geothermal energy sources. In the design and construction of energy-efficient buildings with extremely low heat losses in Western Europe, they turn to a systematic approach, but

* Corresponding author: lexiusb@gmail.com

the climatic and social characteristics and economic conditions of the construction regions are not always taken into account.

The level of heat losses is reduced, the characteristics of an energy efficient building are improved as new engineering equipment is connected, the cost-saving ratio with the obtained energy savings remains optimal during the operation of an energy efficient building. For existing technical and economic operating conditions, this can be called an open energy system with an optimum level of heat consumption, and by connecting new energy efficient modules with the possibility of further lowering this level [12].

3 Buildings heat balance

As a rule, there aren't shared the irrevocable and irreversible heat losses when considering the thermal losses of buildings. It should be borne in mind that some losses in the building are subject to certain regulations, and this, in turn, allows a new approach to the possible optimization of the resistance to heat transfer of the buildings enclosing structures.The heat loss with exhaust air from buildings is returnable. 100% return heat of exhaust air ensures high air tightness of premises and the highly efficient heat exchangers application. Through the building enclosures, the transmission loss of heat remains irrecoverable.

4 Characteristics of the heat consumption minimum level

As for the heat load redistribution between individual apartments and the amount of the building heat losses there are affected the following impacts:

a) climatic conditions, with regard to which there is formed the heat balance during the building operation process,

b) socio-urban conditions that affect the heat losses level,

c) housing stock structures and the power of thermal emissions in the building, as a result of the construction process that determine their compactness

d) tenants accidental exposure on the parameters of the apartments microclimate.

Building enclosure structures resistance to the heat transfer can be taken via high-efficiency heat exchangers in order to utilize heat loss during the ventilation and based on the equality of the total energy value of internal heat sources with the average transmission heat loss in the building during the heating season.

The building enclosures heat transfer resistance comparative analysis of the regulatory requirements values in countries of Western Europe and in Siberia suggests that the resistance to heat transfer is somewhat greater for virtually all types of enclosing structures, except for the floor slabs above the basements.

The window structures heat transfer resistance value exceeds the same value for all countries except for Siberia and Finland. The heat transfer resistance should, in our opinion, be close to the optimal value for specific economic and climatic conditions.

The heat transfer resistance of the enclosing structures for the medium and high storeys buildings is below the optimum value, so it can be adopted as a normative one. The heat economy for medium and high storeys buildings ventilation and heating should be transferred to the heat utilization area of the ventilation emissions from the heated buildings and engineering equipment improvement. Ventilation systems automation at the same time, gives the opportunity to save energy by controlling system modes and thermodynamic parameters depending on the time of day and climatic characteristics

5 Options for energy-efficient solutions of buildings hot water supply systems

In modern buildings in Western Siberia, the heat cost for heating buildings is lower than the cost for domestic hot water.

There is needed more energy saving in the development of energy efficient hot water systems, including individual hot water meters\counters installation.

In addition to the secondary heat sources and solar energy use, we will consider other options for energy-efficient solutions for hot water systems:

- "gray" wastewater heat utilizing systems;
- heat pumps
- solar collector
- photoelectric converters.

Toilets and kitchens individual drainage system should be allocated for the "gray" wastewater heat utilization, that undoubtedly makes it possible to reduce the current and capital costs for the wastewater treatment system [13].The waste heat recovery system allows to provide the first stage of tap water heating.

Designed after the waste water heat recovery to utilize the low-potential gray" wastewater heat the Heat Pumps are the second-stage utilizer. The heat pump provides the possibility of environmentally friendly and economical heating, using the warmth of the environment.

Mounted mainly in the south on the buildings roof solar collector is designed to convert the heat energy of solar radiation into heat, that is further used in the second stage of tap water heating.

Photoelectric converters are designed to convert solar energy into electrical energy, while ensuring the compression heat pump operation.

The provided analysis of the above considered options has shown the absolute economic efficiency of the waste water heat energy utilization application. The year-round use points to the advantage of this system to a solar collectorThe solar collector and waste heat exchanger combination increases not only the energy savings, but also the capital costs. However, only the third option provides full thermal energy costs coverage for hot water supply.

6 Conclusion

1. The energy-efficient technologies used in the buildings design and construction, as well as the technical means and methods used to reduce the energy operating costs in the residential buildings operation in Western Europe and Siberia, have common regularities despite the different climatic conditions.
2. Heat exchangers application helps to utilize heat losses during air exchange. Based on the equality for the heating season of the transmission heat losses mean value to the total value of the solar energy and the energy of the heat internal sources in the building, it is possible to take building enclosing structures heat transfer resistance.
3. Practically for all types of building enclosures, the heat transfer resistance value in Siberia is greater than or equal to the performance statutory indicator.The only exceptions are the floor slabs above the basements
4. For specific economic and climatic conditions, the heat transfer resistance should be close to the optimum value.
5. The economy of ventilation heat and high and medium-level storeys buildings heating should be transferred to the field of buildings engineering technical equipment improvement.

References

1. A. Costa, Internationale Passivhaustagung, **485** (2010)
2. Z. Szalay, Internationale Passivhaustagung, **491** (2010)
3. A.E. Elokhov, Technologies for Designing and Building Energy-Efficient Buildings Passive House, **11-12**,115 (2012) [in Russian]
4. H. Ronacher, Internationale Passivhaustagung, **525** (2010)
5. P. Regner, Internationale Passivhaustagung, **203** (2010)
6. O.I. Panitkov, Technologies for the design and construction of energy efficient buildings Passive House, **6** (7), 23 (2011) [In Russian]
7. V.A. Leonova, Technologies for the Design and Construction of Energy-Efficient Buildings Passive House, **6** (7), 26 (2011) [In Russian]
8. A.E. Elokhov, Technologies of designing and construction of energy efficient buildings Passive House, **3** (4), 57 (2013) [In Russian]
9. S. Glushkov, N. Kachalov, E. Senkiv, V. Glushkova, MATEC Web Conf. **92** (2017)
10. O. Seppanen, Energy saving, **7**, 42 (2010) [In Russian]
11. B. Anderon, *Solar Energy (Building design fundamentals)* (Stroiizdat, Moscow, 1982) [In Russian]
12. V.G. Gagarin, ABOK, **1,** 10 (2009) [In Russian]
13. I.M. Kalinin, Ecological systems, **6,** 12 (2003) [In Russian]

Modeling Combined Heat Transfer in the Operating Area of the High Power Heating Test Facility

Riza Baldzhiev[1,*], *Pavel* Prosuntsov[1]

[1]Bauman Moscow State Technical University, 105005 Moscow, Russia

Abstract. The operating temperatures for the current aerospace heat shield materials exceed 2000 K [1], which makes their experimental testing a complex and expensive issue requiring specialized testing facilities capable to obtain the temperature level in question. In most cases, the thermal tests of the materials samples and thermal shield elements are conducted on the radiative heating test facilities. Generally, these facilities use tungsten halogen lamps (THL) [2, 3]. However, maximal operating temperatures for such type facilities do not exceed 1500 K due to the tungsten-halogen cycle disruption. The main challenge is to comply with the bulbs thermal conditions, which must remain below 1270 K [1]. This paper outlines the layout of the working area of the THL radiative heating facility, where the test specimen temperature is increased to 2100 K due to the compressed air flow over the bulbs, which for a considerable part solves the problem of the thermal behavior simulation.

1 Modeling combined heat transfer

The facility design layout required solving the complex problem of simulating the thermal processes in the operating area, which must take into account for the three interrelated heat transfer processes: heat conductivity in the test specimen, heat insulation, heaters, bulbs and glass partitions; convection in the flowing gas medium and radiation in the semi-transparent bodies and the gas medium. Apart from this, the spectral and temperature functions of the thermal physical and optical characteristics were taken into account.

Since the longitudinal dimension (250-400 mm) of the heating sources exceeds their lateral dimension (11 mm) by an order of magnitude, the heat transfer problem can be solved two-dimensionally, which will mean a significant decrease in the required processing capacity. For the radiative heat transfer simulation we used the discrete-ordinates method with 16 angular dimensions, which enabled accurately accounting for the radiative heat transfer in the quartz glasses [4] and THL casings. It was assumed that the gas flow is turbulent in nature and is described by K-eps model. The modelling was conducted in the finite-element package ANSYS Fluent [5] (license 339001). The characteristic dimension of the elements was about 1 mm, and the total amount of elements

*Corresponding author: lexiusb@gmail.com

– about 30000 pcs. The time increment was 0.01 s, and the total process duration was 600 s, the duration of computations varied from 20 to 30 hours on advanced computers (Intel Core i7-4790 processor).

The operating area of the high-power heating test facility (Fig. 1) comprises the following components: heat-insulating blocks (1, 7), radiative screen from the polished heat-resistant steel (2), nine HTLs, each 2.0 kW in capacity, a system of glass partitions (4, 5), the test specimen (6).

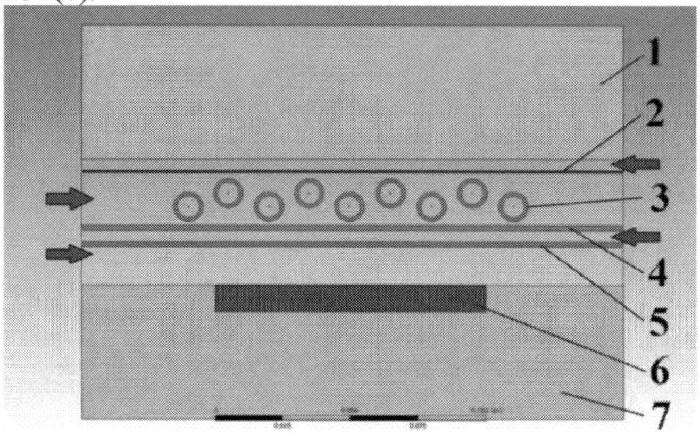

Fig. 1. Geometric model of the test facility working area: 1, 7) upper and lower heat insulation blocks; 2) radiative screen: 3) THL; 4, 5) upper and lower glass partitions; 6) test specimen.

The working area was divided into several zones, so that the flow and the direction of the gas fluxes can be varied (the fluxes directions are indicated with arrows in Fig. 1). The choice of the fluxes directions and the symmetrical positioning of the heaters (relative to the test specimen) helps to achieve the sufficient uniformity of the temperature field over the specimen surface. The effective heat transfer away from the bulbs was a result of the THL checkerboard positioning at a relatively small distance to the partitions and screen result.

The modelling results (Fig. 2, 3) demonstrate that the specimen temperature reaches 2100 K, with 4% non-uniformity of the temperature distribution. The temperature of the THL bulbs as well as the temperature of other components do not exceed the permitted values (1200 K for the THL bulbs and glass partitions, 1300 K for the heat-resistant steel, 1500 K for the heat insulation blocks). Maximum values of the gas flux velocity are found in the clearances between the THL bulbs and the glass partitions and (or) the radiative screen, constituting 20 m/s, the mass-average velocity is around 7 m/s. The total gas flow on the facility elements cooling is about 0.05 kg/s.

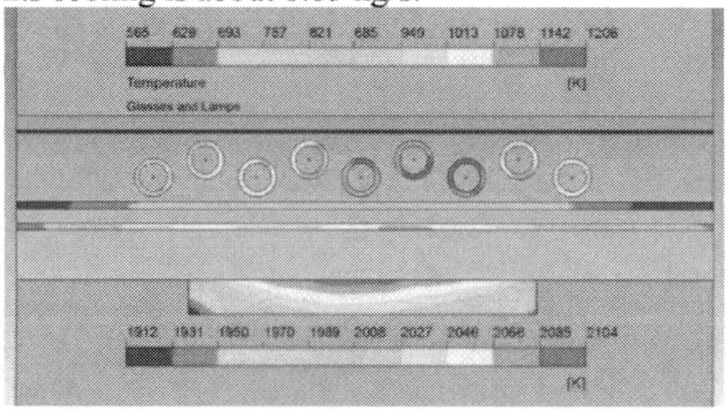

Fig. 2. Temperature fields in the bulbs, glass partitions and the specimen, K.

DOI: 10.1051/matecconf/201711001009

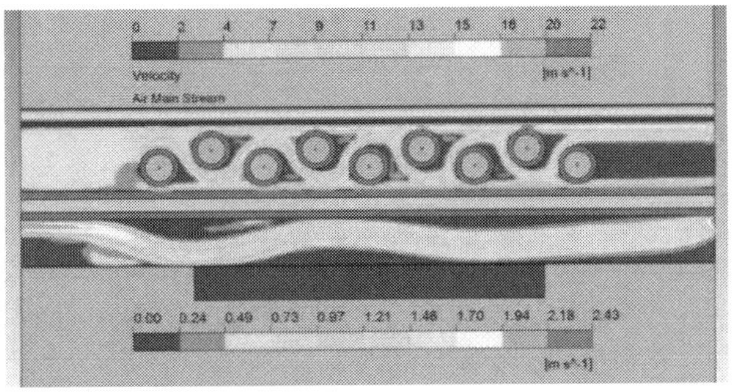

Fig. 3. The gas velocity in the model, m/s.

One of the advantages of the proposed layout is the possibility to simulate the high-power cooling of the specimen by supplying gas into the zone above it. Fig. 4 shows the temperature change in specimen surface centre as a function of time. The maximum heating rate was 140 K/s, and the maximum cooling rate was 110 K/s. It was considered that the heating was conducted in 300 seconds, after which the THLs were switched off and the gas was supplied into the zone above the test specimens at 20 m/s.

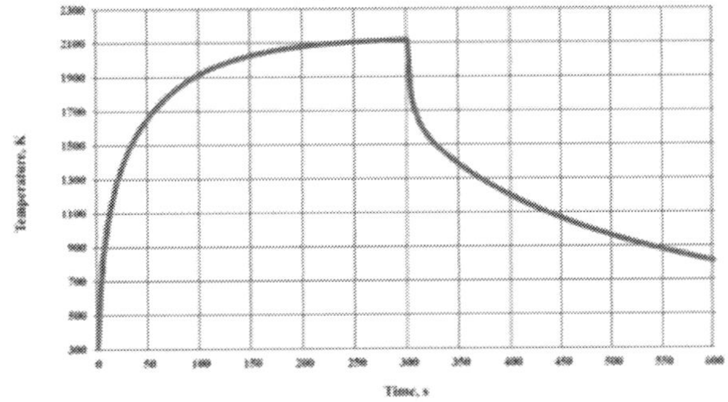

Fig. 4. Temperature on the specimen surface as a function of time.

2 Modification of the operating area

The temperatures distribution over the specimen surface can be more uniform if the amount of THL is increased up to 13 pcs (Fig. 5). The heaters are then divided into four groups: three groups of 3 lamps and one group of 4 lamps. The output of the leftmost group is specified for this type THL constituting 2.0 kW, the output of the next two groups is 50% and 60% of the nominal, while the rightmost is 75% of the nominal output. The other parameters were remained the same. Thus, the non-uniformity of the surface temperature distribution can be reduced to 2.5%. The modelling results are presented at Fig.6.

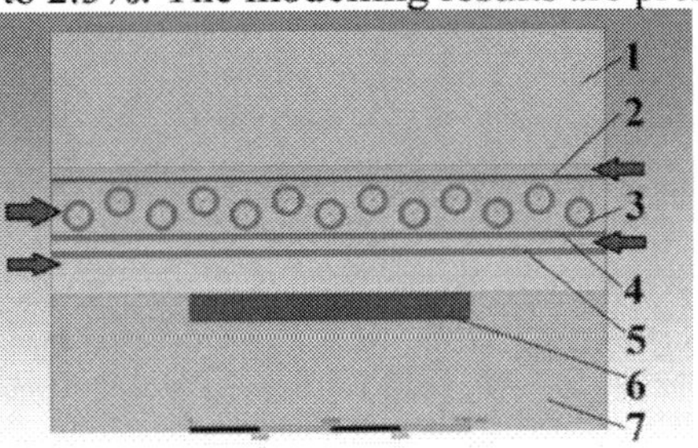

Fig. 5. Geometric model of the test facility working area with added THL: 1, 7) upper and lower heat insulation blocks; 2) radiative screen: 3) THL; 4, 5) upper and lower glass partitions; 6) test specimen.

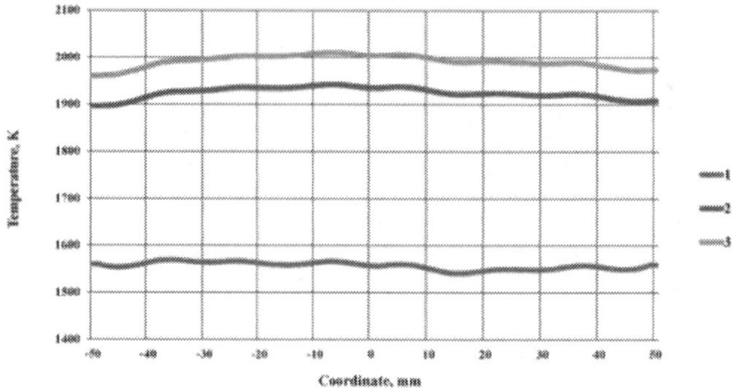

Fig. 6. Surface temperature distribution at various points in time, K: 1) 60 s; 2) 180 s; 3) 300 s.

3 Pneumatic system

To enable the reliable operation of described test facility, the gas should be supplied at a steady rate with the specified velocity, with the hot gas effectively removed from the working area. This challenge required designing a pneumatic system (Fig. 7) which comprises a gas supply duct decreasing the non-uniformity of the velocity fields at the input to no more than 5%, and a hot gas disposal duct decreasing the mass-average temperature at the facility exit to 365 K. The temperature behavior of the system elements was simulated and the requirements to the pneumatic system materials were specified.

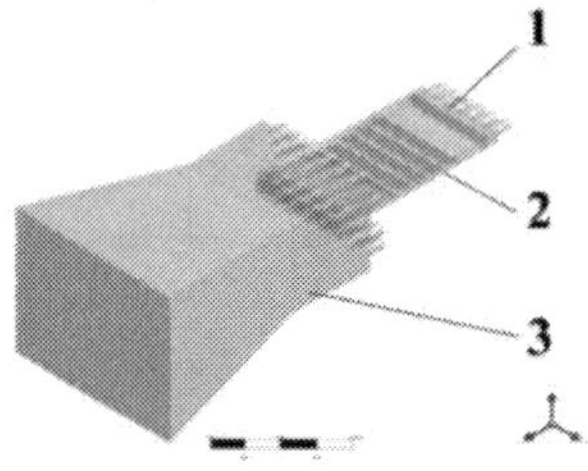

Fig. 7. Pneumatic system: 1) gas supply duct; 2) working area; 3) hot gas disposal tract.

4 Conclusion

Thus, it was verified theoretically that the power of the THL heating facilities could be increased by means of the dynamic air flow. The operable model of a perspective facility was proposed for testing heat shield materials with operating temperatures up to 2100 K.

References

1. V.N. Yeliseev, V.A. Tovstonog, Vestnik MGTU im. Baumana. Ser. Mashinostroenie, **1**, 57 (2011) [In Russian]
2. Yu.V. Polezhaev, S.V. Reznik, A.N. Baranov, *Materials and coatings under extreme condition* (Bauman MSTU Publ., Moscow, 2002)
3. A.N. Baranov, A.G. Belozerov, Yu.S. Il'in, V.F. Kut'inov, *Static strength test of supersonic aircraft* (Mashinostroenie, Moscow, 1974) [In Russian]
4. P.V. Prosuntsov, I.A. Mayorova, A.V. Zuev, Thermal Processes in Engineering, **6**, (2) 317 (2014)
5. ANSYS Fluent [Electronic resource] / Ansys. – 2017 – Access mode: http://www.ansys.com/Products/Fluids/ANSYS-Fluent.

Mathematical simulation of the forest fire thermal effect on root system fragments of coniferous tree

Nikolay Baranovskiy[1,*], *Dmitry* Zakovryashin[1*]

[1]National Research Tomsk Polytechnic University, 634050 Tomsk, Russia

Abstract. We present the numerical simulation results of heat transfer in a layered structure of the root branch when exposed to a high temperature from a forest fire in this paper. The problem is solved by the method of finite differences in a cylindrical coordinate system using a one-dimensional formulation. Typical exposure times and temperatures in the forest fire front are considered. The temperature distributions are presented for the system "root fragment-soil-forest fuel-air". We revealed the character of the temperature distribution and the depth of the soil layer heating to the values which are dangerous for the coniferous tree root system.

1 Introduction

Analyzing the events of recent decades, we can say that forest fires have turned from a natural regulatory factor into catastrophic phenomena leading to environmental, economic and social damage [1, 2]. High temperatures of forest fires have a negative impact on forest stands, leading to their damage and even death [3, 4]. Besides predicting forest fire danger [5], it is extremely important to develop technologies that predict the damage caused by forest fires to stands.

At present, there are a lot of papers devoted to the analysis of statistics of forest fires in the context of their environmental impact [6]. There are a number of articles devoted to the mathematical modeling of heat transfer in soil [7] during a forest fire, however, only the depth of warming up the soil layer to temperatures lethal for microorganisms living in the soil considered in this work. At the same time, many articles have been published to date, describing the results of full-scale studies of the effect of forest fires on the soil, which make it possible to estimate the temperature of warming up the soil layer to the marks dangerous for the roots of trees [8, 9]. Despite this, creation of computer systems for geo-ecological monitoring of forest areas exposed to forest fires, based on physically meaningful mathematical models of heat transfer, the should become the main direction of development.

The purpose of this work is a numerical study of heat transfer in the structure of the root fragment of a coniferous tree exposed to high temperature from a forest fire.

* Corresponding author: firedanger@tpu.ru

2 Mathematical model

Fig. 1. presents photo of bare root of a coniferous tree. The roots of these tree species mainly located at a depth of 0,2 m. The pronounced root is absent, the core root is not deep, the greater part of the root system grows in the near-surface layer of the soil and extends not inland, but in the sides, occupying an area of up to 47 m^2. These circumstances make the roots of coniferous trees such as spruce, pine and cedar very sensitive to the effects of grassland forest fires.

To study the processes of heat transfer, we consider a separate fragment of the root, lying at a depth of 0.2 m, located horizontally with respect to the soil surface.

Fig. 1. The root of the coniferous tree.

The boundary-value problem of conductive heat transfer in the conjugate formulation is considered. The geometry of the solution area shown in Fig. 2.

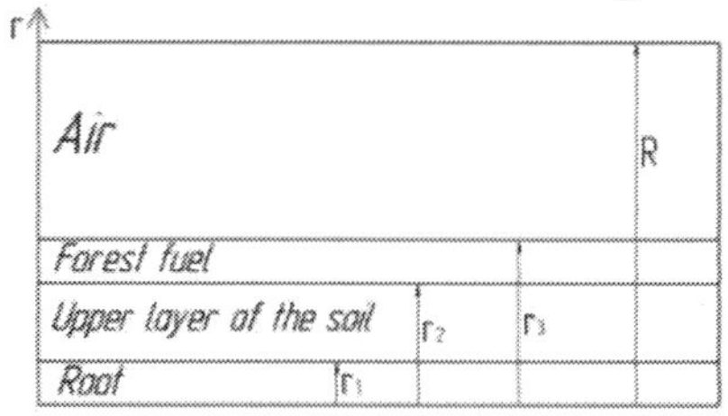

Fig. 2. The solution area.

The solution region is represented as a layered structure: core, a layer rich in organic matter, a layer of forest combustible material on the surface of the soil (subject to combustion), and a layer of ambient air. Boundary conditions of the 4th kind are set on the media interfaces. A high temperature corresponding to the a grass-roots fire is given in the layer of forest combustible material.

The following assumptions accepted: the thermophysical properties of the organic layer of the soil, the layer of forest fuel and air do not depend on temperature, heat transfer in the air is due to conduction.

According to the proposed physical model, the process under investigation is described by nonstationary heat conduction equations for each element of the solution area:

- Heat transfer equation for the root of a tree.

$$\rho_1 c_1 \frac{\partial T_1}{\partial t} = \frac{\lambda_1}{r}\frac{\partial}{\partial r}\left(r\frac{\partial T_1}{\partial r}\right) \quad (1)$$

- Heat transfer equation for the organic layer of the soil.

$$\rho_2 c_2 \frac{\partial T_2}{\partial t} = \frac{\lambda_2}{r}\frac{\partial}{\partial r}\left(r\frac{\partial T_2}{\partial r}\right) \quad (2)$$

- Heat transfer equation for the forest fuel layer.

$$\rho_3 c_3 \frac{\partial T_3}{\partial t} = \frac{\lambda_3}{r} \frac{\partial}{\partial r}\left(r \frac{\partial T_3}{\partial r} \right) \tag{3}$$

- Heat transfer equation for the layer of ambient air.

$$\rho_4 c_4 \frac{\partial T_4}{\partial t} = \frac{\lambda_4}{r} \frac{\partial}{\partial r}\left(r \frac{\partial T_4}{\partial r} \right) \tag{4}$$

The initial conditions for equations (1) - (4) defined as:

$$T_i\big|_{t=0} = T_{i0} \tag{5}$$

The boundary conditions for equations (1) - (4) defined as:
- On the outer boundaries of the solution area.

$$r = R,\ -\lambda_4 \frac{\partial T_4}{\partial r} = \alpha_4 \left(T_e - T_4 \right) \tag{6}$$

$$r = 0,\ -\lambda_1 \frac{\partial T_1}{\partial r} = 0 \tag{7}$$

- On the interfaces of the layers.

$$r = r_3,\ -\lambda_3 \frac{\partial T_3}{\partial r} = -\lambda_4 \frac{\partial T_4}{\partial r},\ T_3 = T_4 \tag{8}$$

$$r = r_2,\ -\lambda_2 \frac{\partial T_2}{\partial r} = -\lambda_3 \frac{\partial T_3}{\partial r},\ T_2 = T_3 \tag{9}$$

$$r = r_1,\ -\lambda_1 \frac{\partial T_1}{\partial r} = -\lambda_2 \frac{\partial T_2}{\partial r},\ T_1 = T_2 \tag{10}$$

where $T_i, c_i, \rho_i, \lambda_i, r_i$ - temperature, heat capacity, density, heat conductivity, coordinate along the axis r (1 - root, 2 - soil layer, 3 – forest fuel layer, 4 - air layer); α_4 - heat transfer coefficient; t - time coordinate.

In the numerical simulation, the following initial data were used: $\rho_1 = 520\ kg/m^3$, $\rho_2 = 1070\ kg/m^3$, $\rho_3 = 500\ kg/m^3$, $\rho_4 = 0{,}03\ kg/m^3$, $c_1 = 2700\ J/kg \cdot K$, $c_2 = 976\ J/kg \cdot K$, $c_3 = 1400\ J/kg \cdot K$, $c_4 = 1200\ J/kg \cdot K$, $\lambda_1 = 0{,}11\ W/m \cdot K$, $\lambda_2 = 0{,}5\ W/m \cdot K$, $\lambda_3 = 0{,}102\ W/m \cdot K$, $\lambda_4 = 0{,}1\ W/m \cdot K$, $\alpha_4 = 80\ W/m^2 \cdot K$.

3 Results and discussion

Currently, forest fires divided into several types, each of them characterized by a certain temperature in the area of the combustion. In this paper, a high temperature in the forest fuel layer equal to 1073 K assigned to simulate the process of heating from the focus of forest fire, which according to the classification of forest fires corresponds to the temperature of the surface forest fire. This type of forest fire chosen for modeling due to the prevalence.

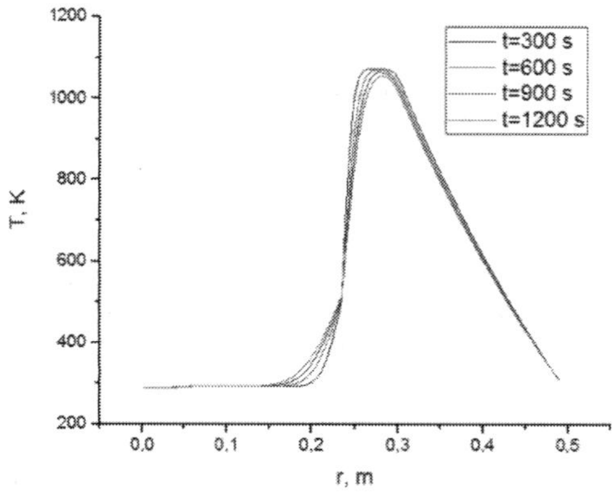

Fig. 2. Distribution of temperatures in the "root-soil-forest fuel-air" system at different time of exposure to the fire.

This task solved by the method of finite differences [10,11]. The double-sweep method used to solve finite-difference analogues. Fig. 2. shows the temperature distribution in the "root - soil - forest fuel – air" system, Fig. 3. presents distribution in "root-soil" system with an exposure time of 300 s, 600 s, 900 s, and 1200 s. Fig. 3. shows that lethal for the root system a value of temperature 330 K observed at a depth of the soil layer equal 0,08 m at a maximum exposure time 1200 s, while the root of the tree lies at a depth of 0,2 m. According to this, we can say that the forest fire does not affect the roots located deeper than 0,08 m and is safety to the stand in terms of thermal damage to the roots.

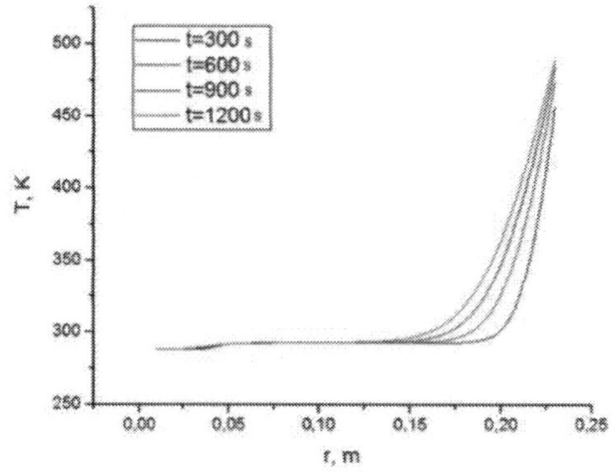

Fig. 3. Distribution of temperatures in the "root-soil" system at different time of exposure to the fire.

4 Conclusions

A simplified one-dimensional mathematical model presented to assess the effect of forest fire on a root fragment. A layered structure consisting of a separate root fragment of a coniferous tree, a soil layer, a forest fuel layer and an air layer is considered. The temperature distributions in the "root-soil-forest fuel-air" system obtained for different exposure times of the surface forest fire: 300 s, 600 s, 900 s, 1200 s. A high temperature 1073 K established in the forest fuel layer to simulate the heating process. It is found that the lethal value of temperature for the tree root observed at a maximum depth of 0.08 m. The obtained data are in qualitative agreement with field observations. In the future, the results of this work can be used to create systems for geo-ecological monitoring of forest areas with the aim of predicting mortality and fragmentation of forest stands after forest fires.

The work was supported by the Russian President's grant (Scientific School project - 7538.2016.8).

References

1. G.M. Byram, *Combustion of forest fuels, Forest fire control and use* (McGraw-Hill, N.Y., 1959)
2. D.V. Korobkina, N.V. Baranovsky, Cloud of Science, **1** (2014)
3. W.F. Laurance, P. Delamonica, S.G. Laurance, H.V. asconcelos, T.E. Lovejoy, Nature, **404** (2000)
4. R.C.G. Mesquita, P. Delamonica, W.F. Laurance, Biological Conservation, **91** (1999)
5. E.P. Yankovich, N.V. Baranovskiy, K.S. Yankovich, IFOST (2014)
6. J.Barlow, C.A.Peres, Biological sciences, **359** (2004)
7. N.V. Baranovsky, A.E. Nee, Engineering Physics and Thermophysics, **87** (2014)
8. E.N.Krasnoshchekova, I.V.Kosov, G.A.Ivanova, Conifers of boreal zone, **3-4** (2008) [In Russian]
9. E.N. Valendik, I.V. Kosov, Forestry, **5** (2008) [In Russian]
10. A.A. Samarskii, P.N. Vabishchevich, *Computational Heat Transfer. Mathematical Modelling.* (Wiley, Chichester, 1995)
11. A.A.Samarskii, P.N.Vabishchevich *Computational Heat Transfer. The Finite Difference Method.* Vol. 2 (Wiley, Chichester, 1995)

DOI: 10.1051/matecconf/201711001011

NUMERICAL SIMULATION OF HEAT AND MASS TRANSFER IN A LIQUID FILM MOVING OVER A HEATED HORIZONTAL SURFACE UNDER THE ACTION OF A GAS FLOW

Maria V. Bartashevich[1,*], *Maxim* G. Vlasenko[1], *Andrey* A. Pil'nik[1], and *Andrey* A. Chernov[1]

[1]Kutateladze Institute of Thermophysics, 630000 Novosibirsk, Russia

Abstract. Heat and mass transfer during desorption on a horizontal film of lithium bromide water solution flowing on a heated wall under the action of shear stress is numerically investigated in this paper. The shear stress on the film surface is set by the motion of surrounding saturated water vapor. It is shown that at low values of heat flux the film temperature and vapor concentration in the solution downstream decreases due to desorption. However, with an increase in heat flux, general film heating and desorption slowing down are observed.

1 Problem statement

In the paper, combined heat and mass transfer in the process of desorption on a liquid film moving on a heated horizontal substrate under the action of shear stress is numerically investigated. The liquid film is a water solution of lithium bromide, the shear stress on the film surface is set by the motion of surrounding saturated water vapor. Desorption takes place in various energy devices, for example, in the absorption heat pump, consisting of generator (desorber), condenser, evaporator, and absorber [1,2]. It is assumed that the shear stress applied to the interface will allow intensification of the process of heat and mass transfer in comparison with the fixed-vapor regime. The investigation of thin liquid film flowing down a vertical wall in the roll-wave regime in presence of heat and mass transfer through the free surface was presented in [3,4]. The absorption and desorption processes under the conditions, corresponding to the operating modes of thermal transformers are described in [5]. The analytical solution to the problem of conjugate heat and mass transfer in a laminar falling liquid film with a linear velocity profile is presented in [6]. The results of the numerical simulation of absorption and their comparison with the experimental data are presented in [7].

We consider the two-dimensional stationary flow of a laminar liquid film of thickness h over a horizontal plate. The surface of liquid is in contact with moving vapor. The shear

* Corresponding author: bartashevichmv@gmail.com

stress on the film surface is τ. Desorption is considered in the framework of ordinary assumptions [1]. The thermal-physical properties of solution are assumed constant.

Let us introduce the Cartesian coordinate system with Ox axis in the direction of the flow and Oy axis directed along the normal to the plate. The velocity profile in the film takes form

$$u = y\tau / \mu .$$

The process of heat and mass transfer at film desorption is described by the equations of heat conductivity and diffusion:

$$u\partial T/\partial x = a\partial^2 T/\partial y^2 , \quad (1)$$

$$u\partial C/\partial x = D\partial^2 C/\partial y^2 , \quad (2)$$

a is thermal diffusivity, m^2/s; D is diffusion coefficient, m^2/s. The boundary conditions at the inlet at $x = 0$ are:

$$T = T_i, C = C_i .$$

Similarly to [1], the equilibrium of solution-vapor system occurs on the interface at $y = h$:

$$C_d = f(T_d), \ -\lambda\partial T / \partial y = -\chi\partial C / \partial y .$$

where $\chi = L\rho D/(1-C_i)$, L is latent heat of phase transformation, J/kg; ρ is solution density, kg/m^3; λ is heat conductivity, W/(mK). This equilibrium condition joins equilibrium temperature T_d with concentration C_d. We set on the wall at $y = 0$:

$$-\lambda\partial T/\partial y\big|_{y=0} = q_w, \partial C/\partial y\big|_{y=0} = 0,$$

where q_w is heat flux on the wall. Let us turn to dimensionless variables $\xi = x/l$, $l = \tau h^3/(\mu a)$, $\eta = y/h$, $\theta = (T-T_i)/(T_s-T_i)$ and $T_s < T_i$, $\gamma = (C-C_i)/(C_s-C_i)$ and $C_s < C_i$ under the conditions of desorption. Here, $C_s = f(T_i)$ is equilibrium concentration, corresponding to initial temperature of solution T_i; T_s is equilibrium temperature corresponding to initial concentration C_i ($C_i = f(T_s)$), similarly to [1]. The system of equations (1) – (2), reduced to the dimensionless form, is rewritten in the following way:

$$\eta\partial\theta/\partial\xi = \partial^2\theta/\partial\eta^2 , \quad (3)$$

$$\eta\partial\gamma/\partial\xi = \mathrm{Le}\,\partial^2\gamma/\partial\eta^2 , \quad (4)$$

here $\mathrm{Le} = D/a$. The boundary conditions at the inlet for $\xi = 0$:

$$\theta = 0, \ \gamma = 0 .$$

Similarly to [1], on the interface at $\eta = 1$ and linear character of function $f(T) = k_1 - k_2 T$, we obtain:, similarly to [1]:

$$\theta = \theta_d, \gamma = \gamma_d,$$

$$\theta_d + \gamma_d = 1, \ (\mathrm{KuLe}(C_s - C_i)/(1 - C_i))\partial\gamma/\partial\eta = \partial\theta/\partial\eta. \tag{5}$$

Here $\mathrm{Ku} = L/(C_p(T_s - T_i))$ is Kutateladze number; C_p is specific heat capacity of solution, J/(kg/K). On the wall at $\eta = 0$, we obtain:

$$\partial\theta/\partial\eta\big|_{\eta=0} = -\bar{q}_w, \partial\gamma/\partial\eta\big|_{\eta=0} = 0,$$

where $\bar{q}_w$ is dimensionless heat flux.

2 Numerical calculations

Numerical calculations were carried out by the finite difference method. The results of calculations of desorption on a film moving under the action of a gas flow over a horizontal wall are given in Fig.1 – Fig.4. The calculations were performed at $(C_s - C_i)/(1 - C_i)$ = -0.4, Ku = -25, Le = 0.01 and $\bar{q}_w$ from 0.01 to 0.3.

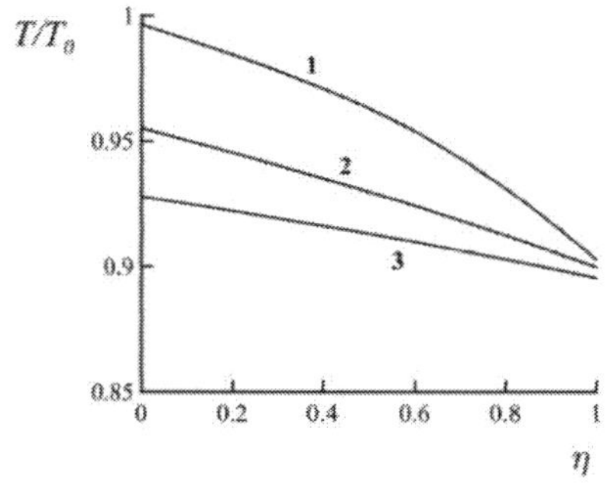

Fig. 1. Profiles of dimensionless temperature in different cross-sections of the film, $\bar{q}_w$ = 0.05. 1 – ξ = 0.1; 2 – ξ = 0.5; 3 – ξ = 1.0.

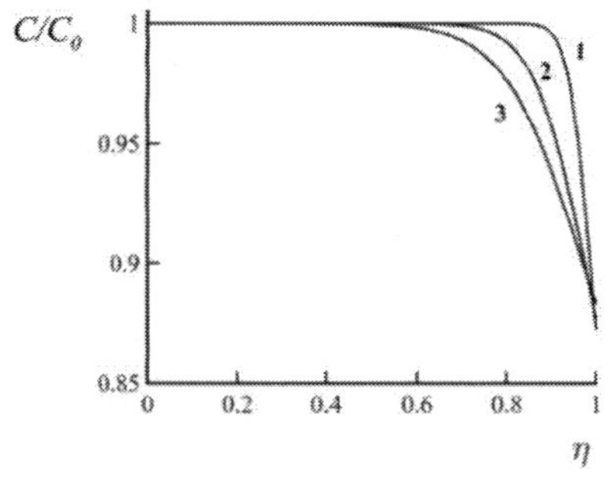

Fig. 2. Profiles of dimensionless concentration in different cross-sections of the film, $\bar{q}_w$ = 0.05. 1 – ξ = 0.1; 2 – ξ = 0.5; 3 – ξ = 1.0.

Figure 1 shows the calculated distributions of the dimensionless temperature at different distances from the inlet, and in Fig. 2 there are corresponding concentration distributions. Figure 3 shows the calculated distributions of the dimensionless temperature along the film surface for different heat fluxes on the wall. Corresponding concentration distributions vs. longitudinal coordinate are shown in Fig. 4.

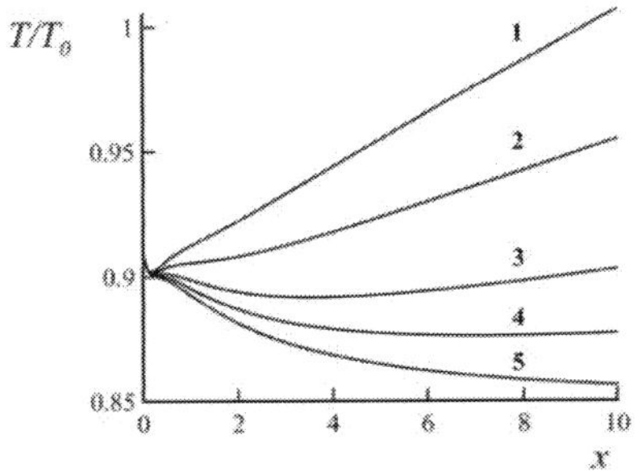

Fig. 3. Profiles of dimensionless temperature of the film surface at different heat fluxes:
1 – $\overline{q}_w = 0.3$; 2 – $\overline{q}_w = 0.2$; 3 – $\overline{q}_w = 0.1$; 4 – $\overline{q}_w = 0.05$; 5 – $\overline{q}_w = 0.01$.

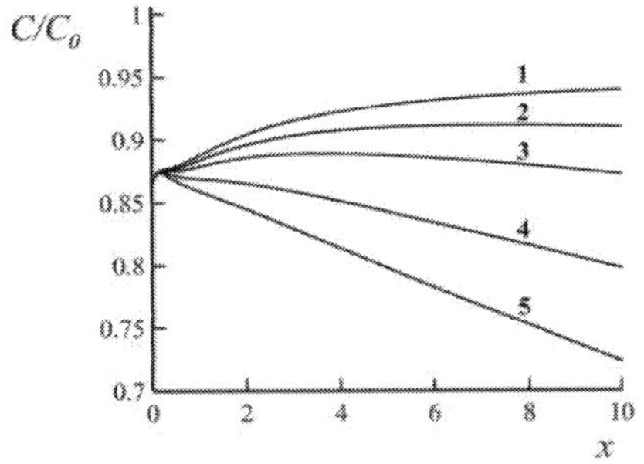

Fig. 4. Profiles of dimensionless concentration on the film surface at different heat fluxes:
1 – $\overline{q}_w = 0.3$; 2 – $\overline{q}_w = 0.2$; 3 – $\overline{q}_w = 0.1$; 4 – $\overline{q}_w = 0.05$; 5 – $\overline{q}_w = 0.01$.

3 Conclusions

The problem of combined heat and mass transfer at film desorption on a liquid film moving over the heated horizontal substrate under the action of shear stress for the given constant heat flux is numerically investigated. It is shown that at low values of $\overline{q}_w$ the film temperature and vapor concentration in the solution downstream decreases due to desorption. However, with an increase in $\overline{q}_w$, general film heating and desorption slowing down are observed.

This work was carried out at the Kutateladze Institute of Thermophysics SB RAS and financially supported by the Russian Science Foundation (project number 15-19-10025).

References

1. V.E.Nakoryakov, N.I.Grigorieva, *Non-Isothermal Absorption in Thermal Transformers* (Nauka, Novosibirsk, 2010)
2. V.E. Nakoryakov, N.I. Grigoryeva, M.V. Bartashevich, Int. J. Heat and Mass Transfer, **54,** 21 (2011)
3. V.E. Nakoryakov, V.V. Ostapenko, M.V.Bartashevich, Int. J. Heat and Mass Transfer, **55,** 23 (2012)
4. V.E. Nakoryakov, V.V. Ostapenko, M.V.Bartashevich, Int. J. Heat and Mass Transfer, **89** (2015)
5. M. Mittermaier, F. Ziegler, Int. J. Refrig, **59,** (1) (2015)
6. M. Mortazavi, S. Moghaddam, Int. J. Refrig, **66** (2016)
7. E. Garcia-Rivera, J. Castro, J. Farnos, A. Oliva, Int. J. Thermal Sciences, **109** (2016)

MATEC Web of Conferences **110**, 01012 (2017) DOI: 10.1051/matecconf/201711001012
HMTTSC-2017

Experimental Setup For Study Of Drop Deformation In Air Flow

Sergey Basalaev[1,*], *Anna* Usanina[1,2], and *Yana* Dubkova[1]

[1]National Research Tomsk State University, 634050 Tomsk, Russia
[2]Federal Research Center for Cool and Cool Chemistry of Siberian Branch of the Russian Academy of Sciences, 650000 Kemerovo, Russia

Abstract. Experimental study for study of deformation of drops in air flow is considered. Experimental setup includes a module for obtaining the drops, an air flow system and measuring system. Module for formation of drops is in the form of vertically arranged dropper with capillary with the possibility of formation of fixed drops. Air flow supply system comprises an air pump coupled conduit through a regulating valve with a cylindrical pipe, installed coaxially with dropper. The measuring system includes the video camera located with possibility of visualization of drop and the Pitot gage for measurement of flow rate of air located in the output section of branch pipe. This experimental setup allows to provide reliable and informative results of the investigation of deformation of drops in the air flow.

1 Introduction

Processes of loss of stability of the shape of drops in the blowing gas flow leading to their deformation and crushing play important role in hydrogasdynamics of two-phase flows [1]. These processes are of practical importance in meteorology (the formation of a spectrum of drop sizes of atmospheric precipitation [2]), in engine building (the dispersion of fuel droplets in internal combustion engines and liquid rocket engines [3, 4]), in ecology problems (evolution of a cloud of drops of toxic components liquid rocket fuels, formed during depressurization in the atmosphere of fuel tanks of carrier rockets [5]) and in a number of other branches of engineering and technology. In this paper, we consider an experimental setup for studying the deformation of a stationary droplet in an air flow.

2 Experimental facility and methods of measurement

In order in investigate the deformation of drops in air flow, the experimental setup has been developed, including a module for producing of drops, a system for supplying the air blowing into the drop, and measuring system. Module for formation of drops is in the form of vertically arranged dropper with capillary with the possibility of formation of fixed

* Corresponding author: tarm@niipmm.tsu.ru

drops. Air flow supply system comprises an air pump coupled conduit through a regulating valve with a cylindrical pipe installed coaxially with dropper. The measuring system includes the video camera located with possibility of visualization of drop and the Pitot gage for measurement of flow rate of air located in the output section of branch pipe.

The basic criterion for determining the deformation of a drop in the gas stream is the Weber number [1]:

$$\mathrm{We} = \frac{\rho_g \left|\vec{u}_l - \vec{u}_g\right|^2 D_0}{\sigma}, \tag{1}$$

where ρ_g is the gas density; $\vec{u}_l$ is the drop velocity vector; $\vec{u}_g$ is the velocity vector of the gas flow blowing in drop; D_0 is the diameter of initial spherical drop; σ is the coefficient of liquid tension.

Weber' number characterizes the relation of forces of dynamic pressure of gas to force of surface tension. With increase in *We* the deformation of drop increases, and at achievement of some critical value of Weber number the crushing of drop occurs due to development of instability of Kelvin-Helmholtz [1].

In an experimental study of patterns of deformation and crushing of drops in flow along with registration of the characteristic sizes of the deformed drop it is necessary to determine Weber number (1) by the values measured in experiment ρ_g, $\vec{u}_l$, $\vec{u}_g$, D_0, σ.

To improve the accuracy of measuring the degree of deformation of the drop and the parameters included in the Weber number, by providing strictly controlled blow-off conditions for the initial droplet, an installation scheme is proposed for studying droplet deformation in a flow [6]. The scheme of setup for research of deformation of drops in flow is given in fig. 1.

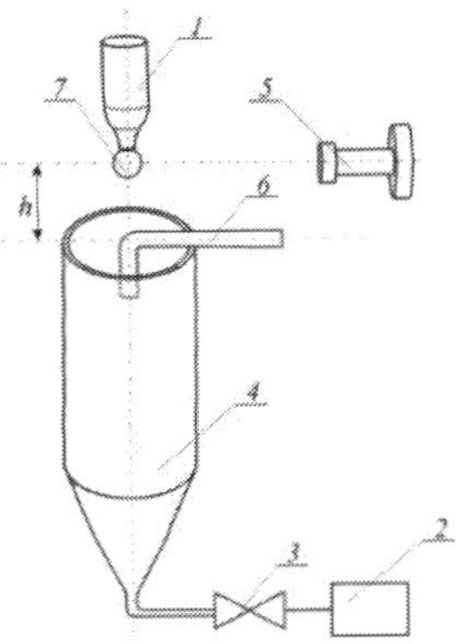

Fig. 1. The scheme of setup for studying the deformation of drops in flow.

The experimental setup includes a vertically located dropper with a capillary *1*, system of giving directed vertically up blowing in the hanging drop of flow of air and measuring system. The system of giving of flow of air contains the air pump *2* connected by the pipeline through the regulating valve *3* to the cylindrical branch pipe *4* established coaxially with dropper *1*. The measuring system includes the video camera *5* located with possibility of visualization of drop and the Pitot gage *6* for measurement of flow rate of air located in the output section of branch pipe *4*. To register the shape of the drop *7* "Citius C 100" video camera was used. Video registration was carried out with the space permission 384×790 pixels with rate of 300 frames per second and time of exposure (0.5÷2.0) of ms.

Experimental setup works as follows. By means of the air pump *2* and valve *3* the set rate of flow of u_g, air that is measured by the Pitot gage *6* is established. By means of dropper with capillary *1* the motionless drop of working fluid forms (water and glycerin solutions, silicone oil, castor oil, etc.).

After forming of initial drop *7* it is influenced by uniform flow of air from branch pipe *4* under the influence of which the drop is deformed. Dimensions of the deformed drop *7* are recorded by a video camera *5*. Then by means of the regulating valve *3* higher u_g levels are established and similar measurements are taken.

3 Measurement results

Let's consider results of measurement of deformation of drop of glycerin in air flow at the room temperature. Physical characteristics of air and glycerin, necessary for calculations, at temperature 20°C are provided in table 1.

Table 1. Physical characteristics of air and glycerin.

Parameter	Air	Glycerin
ρ, kg/m^3	1.205	1260
μ, Pa·c	$1.808\cdot10^{-5}$	–
σ, H/m	–	$63\cdot10^{-3}$

Distance between cut of dropper and output section of branch pipe of $h \le 10$ mm.

Diameter of capillary is determined by equation:

$$d_\kappa \le \frac{1}{6}\sqrt{\frac{0.8\cdot\sigma \mathrm{Bo}^3_{\kappa p}}{g\cdot\rho_l}}, \tag{2}$$

where d_k is the diameter of capillary of dropper; $\mathrm{Bo}_{\kappa p} = 4.5$ critical value of number of the Bond; g is the acceleration of gravity; ρ_l is the liquid density; h is the distance between cut of dropper and output section of branch pipe [7].

Thus, diameter of capillary has to be no more $d_\kappa \le 3.21$ mm. For experiments, the value $d_\kappa = 2.5$ mm was chosen.

Diameter of initial drop is calculated on equation:

$$D_0 = \sqrt[3]{\frac{0.8\cdot 6\sigma d_k}{\rho_l\cdot g}}. \tag{3}$$

Video frames of the initial and deformed droplets are shown in fig. 2 for different values of the blowing rate and Weber numbers calculated by equation (1). Degree of deformation is calculated by processing the results of video shooting using the equation:

$$\varepsilon = \frac{D_m}{D_0}, \tag{4}$$

where D_m, D_0 diameter of transverse section of the deformed and initial drop.

Drop shape	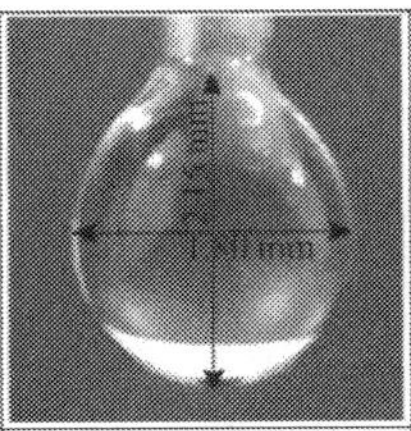	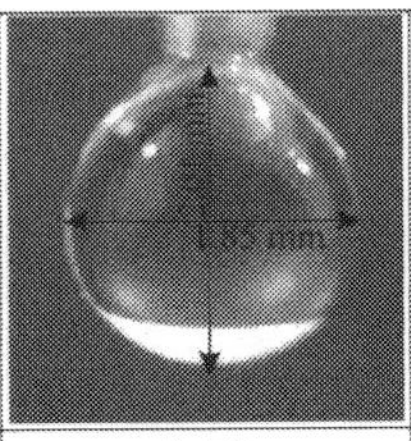	
ε	1	1.03	1.05
u_g,m/c	0	0.7	3.6
We	0	0.018	0.49

Fig. 2. Video frames of the initial and deformed drops.

4 Conclusion

The considered scheme of the experimental setup for the investigation of drop deformation in flow provides an increase in the accuracy of measuring the degree of deformation of the drop and the parameters included in the Weber number due to the providing the strictly controlled blow-off conditions for the initial drop with a significant simplification of the experimental equipment of the apparatus and the alignment of the measuring apparatus.

References

1. R.I. Nigmatulin, *Dynamics of multiphase media*. Part 1 (Science, Moscow, 1987) [In Russian]
2. A.T. Matveev, *Fundamentals of General Meteorology* (Gidrometeoizdat, Leningrad, 1965) [In Russian]
3. A.P. Vasiliev, V.M. Kudryavtsev, V.A. Kuznetsov, *Fundamentals of the theory and calculation of liquid rocket engines* (Higher School, Moscow, 1983)
4. V.Arkhipov, A.Antonnikova, S.Basalayev, I.Zharova, S.Orlov, EPJ Web of Conferences, 119591 (2015)
5. V.A. Arkhipov, E.A. Kozlov, S.S. Titov, A.S. Tkachenko, A.S. Usanina, I.K. Zharova, Arabian Journal of Geosciences, **9**, 114 (2016)
6. Application for a patent of the Russian Federation No. 2016149916. *Experimental setup for investigation of the deformation of drops in stream of air.* V.A Arkhipov,. A.S. Usanina, S.A. Basalaev, K.G. Perfilieva.
7. R.Clift, J.R.Grase, M.E.Weber, *Bubbles, drops and particles* (Acad. Press, N.Y., 1978)

Modification of aluminum alloy surface properties by wave-long laser texturing

Kseniya Batishcheva[1,*], *Dmitriy* Feoktistov[1], *Yuliya* Shanenkova[1], and *Ekaterina* Kirichenko[2]

[1]National Research Tomsk Polytechnic University, 634050 Tomsk, Russia
[2]Institute of Thermophysics Siberian Branch Russian Academy of Sciences, 630090 Novosibirsk, Russia

Abstract. Change of the static contact angle (SCA) of 10 μL distilled water droplet on laser textured aluminium alloy (AMG-6) substrates was studied. The texture was deposited by a laser system based on a fiber laser. An increase in the power of laser radiation is found to lead to a decrease in the SCA measured on the first day after texturing. Change of dispersion and polar surface energy components of textured substrates is determined. Under the influence of the environment, the surface properties of AMG-6 change with time. SCA increased and reached stable state on the twentieth day after texturing.

1 Introduction

Control of wetting attracts the researchers' attention [1-4] because of the possibility of a wide range of applications: the creation of self-cleaning, antimicrobial, anti-corrosion, waterproof materials, reducing frictional forces, surface resistance. Research in this area is of interest to enterprises engaged in the development of equipment for transportation, separation of liquids, sounding devices, antennas, and radars in the space industry, in the design of satellites.

Currently the ways to create surfaces with controlled properties are under search. When choosing a method, decisive factors are the cost-effectiveness of the installation, ease of use, the possibility of obtaining samples with different texture, roughness. The following methods of texturing are known:

1 Chemical deposition from gas phase and sublimation [5]

It is a method of obtaining nanotubes or nanorods arrays by means of their deposition from the gas phase (chemical vapor deposition, CVD). This method can produce ordered carbon nanotubes arrays of the same diameter, a texture matrix that simulating the surface of the lotus leaf. To make such surface a superhydrophobic, an additional chemical modification of the surface is necessary. In addition to carbon nanotubes, the method makes it possible to obtain surfaces on the basis of soft silicone films without the need for further hydrophobization. A related method based on the deposition of matter from the gas phase is

* Corresponding author: bka1801@mail.ru

sublimation, but unlike CVD, completely unordered rough surfaces forming a porous texture are formed after sublimation.

2 Electrochemical methods [6]

The method is used to deposit clusters of metals or metal oxides. The roughness of the resulting surface depends on the deposition time and the applied voltage. In [6], a superhydrophobic surface was obtained by electrochemical deposition of gold clusters on a multilayer polyelectrolyte matrix.

3 Methods based on the use of sol-gel materials [7]

To obtain porous textures with a high roughness, sol-gel materials based on metal alkoxides are used [7]. Controlling hydrolysis and condensation during the formation of the sol makes it possible to control the process of obtaining foams, and, consequently, the roughness of the obtained surfaces.

4 Template methods [8]

Membranes with an ordered system of one-dimensional pores of various diameters are used. Such an array is anodic aluminum oxide [8]. As precursors, which crystallize in the form of nanofibers inside such pores, polymers are used, for example, polyvinyl acetate or polyacrylonitrile. The main disadvantage of this method is the limitation of the maximum pore diameter.

5 Chemical modification of the surface [9]

Usually textured surfaces without further processing are not hydrophobic since their material is hydrophilic. Based on Wenzel and Cassie models, additional texturing of the surface will strengthen their hydrophilic properties. Depending on the material from which the surface consists, there are several options for hydrophobization. The deposition of an additional hydrophobic layer can occur by chemical modification of the surface with the necessary molecules through the gas phase in solution or by physical adsorption. Depending on the material, hydrophobic derivatives of alkanedioles, silanes, fatty acids, and others are used to hydrophobize the surface.

6 Reactive ion etching [10]

It is based on the use of reactive plasma etching material unprotected chemically. Part of the surface which does not need to be etched is chemically protected with coatings. Plasma is formed in a vacuum under the influence of an electromagnetic field. The type and amount of used gas depends on the material to be etched. For example, plasma based on sulfur hexafluoride is usually used for silicon surfaces.

7 Double texturing [11]

It is based on the production of superhydrophobic surfaces combining two textures – in micro- and nanoscale. The surfaces allow to increase the contact angle of wetting, to reduce the contact angle hysteresis, and also to increase the stability of the Cassie state. Microtexture determines the functional properties of the surface, for example, the length of sliding. Nanotexture is used to improve the wetting of surfaces.

8 Lithographic methods [12]

They are based on the use of templates. In other words, the topology of the surface is predetermined by the template form embedded in the control program: when etched by an electron beam; mask printed in advance for photolithography; a solid template which form is copied to another material by printing.

9 Laser irradiation [13]

It is based on a change in the surface morphology by the action of pulsed laser radiation. The deposition of laser radiation on the surface makes it possible to control wetting and spreading. By changing the parameters of the laser (power, diameter of the laser beam, and ablation rate of the substance from the surface), it is possible to create several samples with controlled hydrophobic and hydrophilic properties from one material.

Based on the results of analysis of known methods for creating surfaces with controlled properties, the most promising is laser radiation. It has several advantages in comparison with others: non-contact, low-cost, simple and cheap in operation, with the possibility to obtain a wide range of textures (limited only by the parameters of the laser installation) and to process various materials.

The purpose of this work is to experimentally determine the dependence of the static contact angle on time on the treated aluminum surfaces by the laser radiation.

2 Materials and methods

Scheme of an experimental setup is presented in Fig. 1.

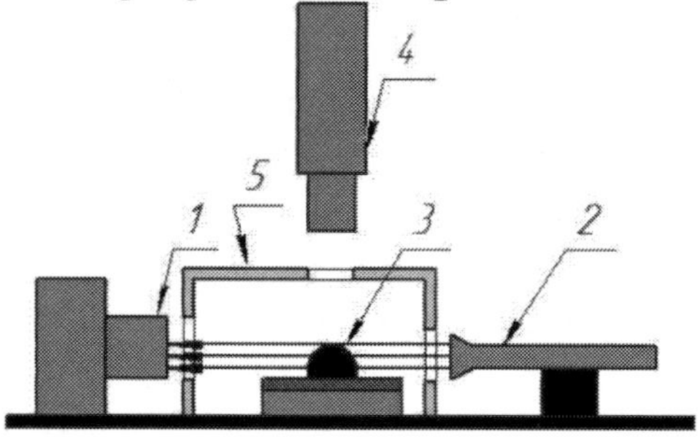

Fig. 1. Scheme of an experimental setup: 1 – photographic camera; 2 – source of plane parallel light; 3 – test cell, 4 – video camera; 5 – transparent box.

The main elements of setup: photographic camera Nikon D7100 with lens "Sigma AF 105mm f/2.8 EX DG OS HSM MacroNikon F" 1; source of plane parallel light 2; test cell 3 consisting of goniometer to level the surface relative to the horizontal in two planes and the substrate; high-speed video camera Fastvideo-500M with lens "Nikon 85mm f/1.4D AF Nikkor" with a shooting speed up to 500 frames per second and a resolution of 1280 x 1024 pixels 4; transparent box 5. The setup includes an optical shadow system. A detailed description of the operational principles is given in [14-15].

The experiment was conducted using six substrates made of aluminum alloy AMG-6. Their surfaces were polished. After polishing five substrates were treated with laser radiation, one remained unchanged. The texture is deposited with "Mini Marker 2 M 20" laser system. Table 1 shows the main technical characteristics of the laser system used for aluminum surfaces treatment.

Table 1. Technical characteristics of "Mini Marker 2 M 20" laser system used for surface texturing.

Frequency of the laser beam, Hz	99
Travel speed of the laser beam, 10^{-3} m/sec	800
Power of the laser radiation, W	2, 6, 10, 14, 18

The average arithmetic deviation of the surface profile of the substrates (Ra) was determined on "HOMMEL TESTER T1000" profilometer. Photos of the microstructure were obtained on "Hitachi-3000M" microscope. All samples were examined by X-ray diffractometry using a Shimadzu XRD 7000S diffractometer. The polar and dispersion components of the surface energy of the substrates were determined.

A drop of distilled water with a volume of 10 μl was placed on the surface by a high-precision dispenser. The SCA was measured for twenty days with an interval of five days.

3 Results and discussion

DOI: 10.1051/matecconf/201711001013

The SCAs values of laser textured substrates were obtained (Fig. 2).

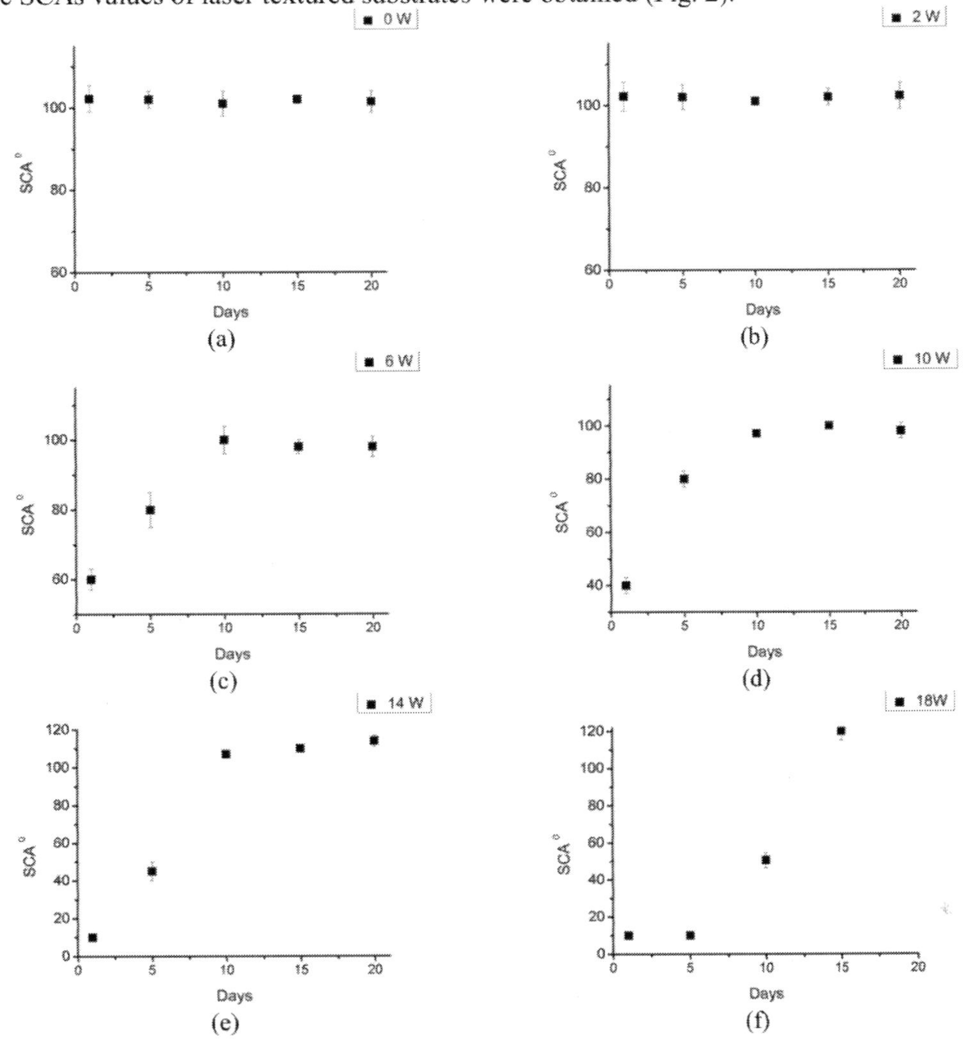

Fig. 2. Dependencies of SCAs on time after (a) – polishing surface; (b)-(f) – texturing by laser with a power of 2, 6, 10, 14 18 W, respectively.

The SCAs values of laser textured substrates are obtained. Texturing by laser with a power of up to 2 W does not lead to a change in the SCA. Its value is equal to the value of the polished surface (102°). The surface energy remains constant 34 ± 4 mN/m. The effect of laser radiation with a power of up to 2 W does not affect the crystal lattice structure of the surface layer of the metal. The SCA is stable over time.

The SCA of laser textured substrates with a power of 6, 10, 14, 18 W increased and reached a stable state after twenty days. An increase in the power of laser radiation is found to lead to a change in the structure of the crystal lattice. Twenty days later the SCA of laser treated substrates with a power of 6, 10 W reached the value of a polished surface (102°). The laser treatment with a power of 14, 18 W made it possible to obtain a porous texture formed by a network of crystals of various sizes and shapes. The depth of the pores does not exceed 2±0.19 μm. SCAs were 114° and 120°, respectively. Over time the sessile drop under the action of gravity spread out filling the pores. SCA decreased to 70°.

4 Conclusions

It has been found that the value of the SCAs of laser textured substrates with power of 6, 10, 14, 18 W varies with time. A stable condition occurs on the twentieth day after laser texturing. The increase in the SCA is due to a change in the crystal lattice and a change in the ratio of the polar and dispersion components of the surface energy of the substrates. Laser texturing with power of 14, 18 W is found to improve the hydrophobic properties of aluminum alloy AMG-6 surface.

The reported study was supported by the grant of President of Russian Federation for the government support of young Russian scientists (MK-6810.2016.8).

References

1. D. Zaitsev, D. Kirichenko, O. Kabov, Technical Physics Letters, **41**, 6 (2015)
2. S. Misyura, Appl. Therm. Eng. **113**, 472-480 (2017)
3. A. Sivkov, A. Ivashutenko, Y. Shanenkova, I. Shanenkov, Adv. Powder Technol. **27**, 4 (2016)
4. R. S. Volkov, G. V. Kuznetsov, J. C. Legros, P. A. Strizhak, Int J Heat Mass Transf. **95**, 184 (2016)
5. K. Lau, J. Bico, K. Teo, M. Chhowalla, G. Amaratunga, W. Milne, G. McKinley, K. Gleason, NanoLetters, **3**, 12 (2003)
6. X. Zhang, F. Shi, X. Yu, H. Liu, Y. Fu, Z. Wang, L. Jiang, X. Li, J. Am. Chem. Soc. **126**, 10 (2004)
7. N. Shirtcliffe, G. McHale, M. Newton, C. Perry, Langmuir, **19**, 14 (2003)
8. L. Feng, S. Li, H. Li, J. Zhai, Y. Song, L. Jiang, D. Zhu, Angew. Chem. Int. Ed. **41**, 7 (2002)
9. Z. Guo, F. Zhou, J. Hao, W. Liu, J. Am. Chem. Soc. **127**, 45 (2005)
10. C. Pakpum, N. Pussadee, Surf. Coat. Technol. **306**, 194–199 (2016)
11. N. Patankar, Langmuir, **20**, 19 (2004)
12. H. Zhao, K. Park, K. Law, Langmuir, **28**, 42 (2012)
13. F. Shi, X. Chen, L. Wang, J. Niu, J. Yu, Z. Wang, X. Zhang, Chem. Mater. **17**, 24 (2005)
14. G. Kuznetsov, D. Feoktistov, E. Orlova, J Eng Phys Thermophy, **89**, 2 (2016)
15. G. Kuznetsov, D. Feoktistov, E. Orlova, K. Batishcheva, Colloid J. **78**, 3 (2016)

DOI: 10.1051/matecconf/201711001014

Research destruction ice under dynamic loading. Part 1. Modeling explosive ice cover into account the temperature

Gennady N. Bogomolov[1,*], *Victor P.* Glazyrin[1], and *Maxim Yu.* Orlov[1]

[1]EDP Sciences, Tomsk State University, 634050 Tomsk, Russia

Abstract. In the research, the behavior of ice under shock and explosive loads is analyzed. Full-scale experiments were carried out. It is established that the results of 2013 practically coincide with the results of 2017, which is explained by the temperature of the formation of river ice. Two research objects are considered, including freshwater ice and river ice cover. The Taylor test was simulated numerically. The results of the Taylor test are presented. Ice is described by an elastoplastic model of continuum mechanics. The process of explosive loading of ice by emulsion explosives is numerically simulated. The destruction of the ice cover under detonation products is analyzed in detail.

1 Introduction

At the present time, the study of the behavior of ice under explosive and shock loads is actual and complex scientific and technical problem. This is due to the need for the development of northern territories, the elimination of ice jams ion the Siberian rivers, the extraction of natural resources in the Far North, the creation of new protections against micrometeorites, etc. The main difficulty is that the ice is a poorly studied natural material under dynamic loads. Concepts of ice destruction are only being developed [1, 2].

In the present research, complex theoretical and experimental studies of the behavior of ice under explosive loading have been carried out. Theoretical studies consist of the numerical modeling of the underwater explosion of the ice cover. Experimental studies consist of full-scale tests on the explosion of the ice cover by emulsion explosives.

2 Investigation of the dynamic loading of ice

Research behavior of ice under dynamic loads is held in the Tomsk State University. The subject of the study is the state of ice under shock and explosive loading. Mathematical model of ice under dynamic loads is developing. Software package for solving modern dynamic multicontact problems of solid mechanics have been developed [3]. Modern dynamic multicontact problems are problems of deep penetration, perforation and explosion

* Corresponding author: orloff_m@mail.ru

DOI: 10.1051/matecconf/201711001014

of homogeneous and multilayer structures. Lagrangian numerical method was modified by algorithm of splitting nodes and algorithm destruction of triangular elements. The latest modification was improved algorithm for calculating the contact surfaces and free surfaces [4]. Mobile laboratory "Explosive destruction of natural materials" was organized for experimental research. The experimental results are summarized in a relational database. For the development of industries, some applied problems have been solved [5].

2.1 Experimental studies of the explosive loading of ice

The analytical review of the authors confirms a small amount of experimental data on the explosive loading of ice. Some experimental data do not agree with each other, and some are already a bibliographic rarity. Organized five years ago, the mobile laboratory was designed to solve this problem. The mobile laboratory is developing as an alternative to the US research program ScIceExe. At the present moment, the mobile laboratory can perform an snap analysis of the behavior of natural materials after the explosion. In the last year, profiles of explosive craters in limestone have been identified, depending on the type of explosive [6]. Expeditions were conducted in various regions of the Siberian Federal District and blast up more than 1000 kg of TNT.

The current paper presents the results of full-scale experiments on the explosion of river ice by emulsion explosives. The explosion is produced in the water under the ice. Full-scale experiments were carried out jointly with KuzbasSpetsVzryv company and the support of the MES of Russia. The object of study was river ice average thickness. The ice cover was snow covered, the snow thickness was about 15-30 cm. The air and water temperature was about 4 °C. The explosive was chosen emulsion explosive Emulast AS-FP-90 (Russia).

Experimental tests of explosion ice under water were performed at the Mobile laboratory "Explosive destruction of natural materials" of the Tomsk State University. Scheme of the experiment remained constant since 2013. Explosive charge weight of 4 kg was placed beneath the ice. At the time of the blast up, it was perpendicular to the ice cover. The initiation point was at top of sample charge. The subject of the study was the diameter and edge of the ice line, the morphology of the destruction of ice including the cloud of ice fragments after the explosion.

Figure 1 shows the lane in the ice after the explosion of 4 kg of emulsion explosives. Photo was received five minutes after the blast up by S. Kulkov from a special boat. Before the blast, the diameter of the hole was 16 cm.

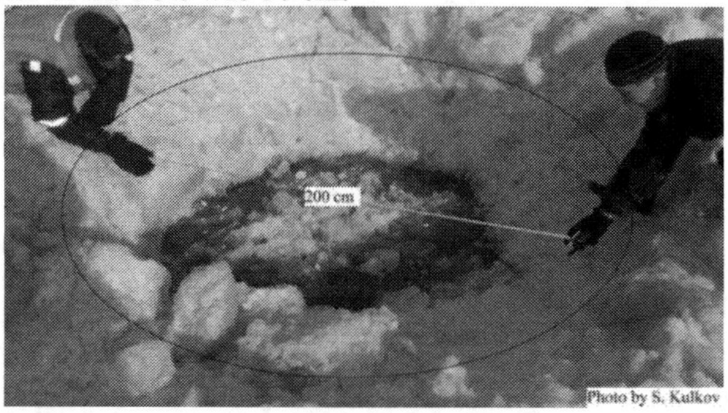

Fig. 1. The lane in a snow covered ice cover. Full-scale test 2017.

On the photo: Officer of MES and Maxim Orlov.

Thus, the results of the full-test experiments of 2017 practically coincided with the results of 2013. In both cases the average diameter of the lane in the ice was about 200 cm. It should be noted that an ice cork weighing was knocked out under the action of detonation products. Damage on the front surface of the ice cover was not detected. The back side of the ice cover has not been investigated. Most of the fragments were of medium size (15-25 cm).This fact was mentioned in [7]. In 2013, the thickness of snow on ice was less by 5-10 cm than in 2017.

2.2 Numerical studies of the explosive loading of ice

The basic system of equations is based on the fundamental laws of conservation of mass, momentum and energy. A complex model of continuum mechanics used to describe the material behavior under dynamic load. Ice modeled a porous, compressible medium, taking into account the strength properties, shock-wave phenomena, as well as joint formation of spall and shear damage and temperature [8].

The system of equations for two-dimensional axisymmetric case is solved by the modified method of Johnson G.R. It is known that this method based on lagrangian approach, but it allows solve the problem of penetrating and perforating of heterogeneous structure, including modern safety structure. For this incorporated the algorithm destruction of triangulation elements and algorithm splitting nodes, the algorithm for constructing the free surface, etc.

Before the numerical research were carried out test calculations. Taylor's test was chosen as first test calculation. Ice cylinder height was 20 mm and a diameter was 6.88 mm. The initial velocity was varied from 50 to150 m/s.

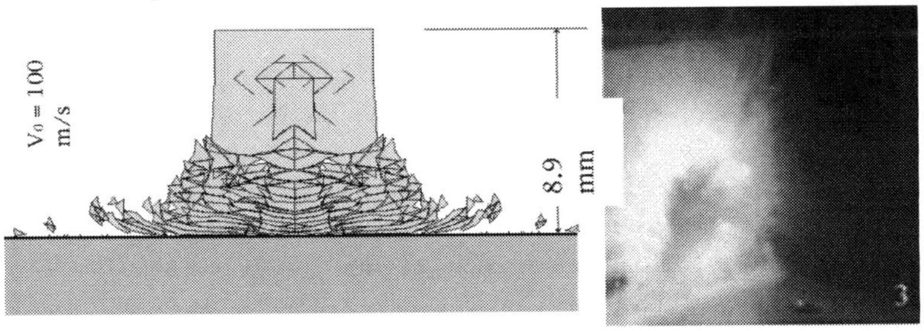

Fig. 2. Impact ice cylinder on rigid wall.

Figure 2a shows the calculated configuration "Ice cylinder – rigid wall" at an initial speed of 100 m/s. It is seen that the nose of the cylinder is almost completely destroyed. Figure 2b illustrates the experimental results of this test [1]. In both cases, the ice cylinders were destroyed. Also, the calculation results are compared with known analytical solutions of Rankin – Hugonion and other experimental data. It is possible to determine mechanical properties of ice, such as shear modulus, yield strength, and others.

In this paper, the problem of the blast up of the ice cover by emulsion explosives is solved. The thickness of the ice cover was 70 cm. The mass of the emulsion explosive was 4 kg ($\approx$ 3.3 TNT equivalent). The length of the explosive cartridge is 70 cm. The length of the ice cover was 800 cm. In calculations, the location of the charge was perpendicular to the ice cover. The depth of the water under the ice was more than 500 cm. The number of calculated elements of the field was about 10000 . The equilateral triangle was used as the design element. Calculations are carried out in a two-dimensional axisymmetric case using the computer code 'Impact'.

Figure 3 illustrates the configurations of ice at the initial and final time. Water and detonation products are absent in the figure. It was revealed that the first foci of destruction are formed by 100 μs. Ones are oriented perpendicular to the 'Ice – Water' contact boundary.

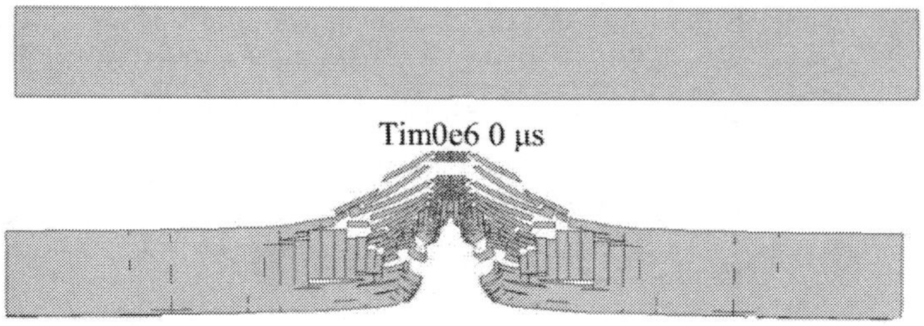

Fig. 3. Configuration of the ice cover after blast up 4 kg of emulsion explosives.

Under detonation products, foci propagate to the face contact surface. At 500 microseconds, the destruction covers a 200-centimeter zone of ice cover. After 2 milliseconds, a gap is formed between water and ice. This is a consequence of the action of detonation products. The ice seems to go up. After 4 milliseconds at a distance of 150 cm from the lateral surface, perpendicular cracks are formed.

It is established that after 6 milliseconds the picture of ice destruction does not change. The field occupied by the detonation products had a triangular shape. In calculations, the diameter of the lane in the ice was approximately 210 cm. The velocity of the leading fragments of ice was 120 m/s.

3 Conclusions

1. The performed experimental studies allowed to draw conclusions. Full scale tests 2013 and 2017 gave the same results. The explosion of the emulsion explosive formed a 200-cm-diameter lane in the ice cover. The shape of the lane is close to the circumference. Edge of the lane in the ice was not smooth. The height of the expansion of the cloud of ice fragments was about 10 meters. There were no annular and radial cracks on the surface of the ice. On the experimental site, fragments larger than 50 cm were observed.
2. Test calculations were carried out, including the Taylor test, the shock of the ice cylinder on the aluminum plate, and the explosion of the emulsion explosive in the water under ice. The article gives the results of the Taylor test only. Comparison with the analytical solution and other experiments were given in [10] Test problem on the explosion of emulsion explosives in water under ice is given in [4]. The difference between calculation and experiment is satisfactory.
3. The destruction of the ice cover was simulated under the action of detonation products. It is found that the first foci of fracture is formed in the contact zone of ice-water and explosives. Foci of destruction developed in the main crack. Most of the ice cover moves in the axial direction under the action of detonation products. The estimated diameter of the lane in the ice was 210 cm. The discrepancy with the experiment was 7%. The ice cork is almost completely destroyed.

The work was supported by RFBR (project 16-38-00515).

References

1. K.S. Carney, Int. J. Solids and Structures, **43**, 7820 (2006)
2. J. Pernas-Sanchez, J.A. Artero-Guerrero, D. Varas, J. Lypez-Puente, Int. J Impact Engineering, **96**, 1 (2016)
3. V.P. Glazyrin, Yu. N. Orlov, Ju. N. Orlova, Izvestija Vuzov. Fizika, **52**, (7) 2 (2009)
4. Yu.N. Orlova, Study of ice dynamic loading, Ph.D. Thesis Tomsk, Russia (2014)

5. M.Yu. Orlov, Yu.N. Orlova, V.F. Tolkachev, Journal of Physics: Conference Series, **653,** 012038 (2015)
6. M.Yu. Orlov, Yu.N. Orlova, G.N. Bogomolov, The behavior limestone under explosive load Journal of Physics: Conference Series, **774,** 012052 (2016)
7. M.Yu. Orlov, Yu.N. Orlova, G.N. Bogomolov, Proceedings of the Third International Scientific Conference, 192 (2016)
8. G.N. Bogomolov, V.P. Glazyrin, M.Yu. Orlov, MATEC Web Conf. **72**, 01013 (2016)
9. A. Combescure, Y. Cyuzel-Marmot, J. Fabis, Int. J. Solids and Structures, **48**, 2779 (2011)
10. Yu.N. Orlov, Study of high-speed deformation and destruction of projectiles, Ph.D. Thesis Tomsk, Russia (2007)

Comparative Analysis of Optimal Temperature Distributions in the Responsible Sections of Load Bearing Structures

Mikhail Yu. Livshits[1,*], *Boris B.* Borodulin[1]

[1]Samara State Technical University, 443000 Samara, Russia

Abstract. This work presents a comparative analysis of the temperature distribution for different thermal conductivity coefficients. The analysis of the general problem shows that temperature distribution in the critical sections of the load bearing structures significantly affects of the diagnostic instrumentation's data accuracy placed on them. Temperature distribution control optimization allows to reduce influence of temperature factors on the accuracy of diagnostic instrumentation. All this requires a thorough analysis.

1 Problem statement

The temperature distribution in the critical sections of the load bearing structures significantly affects the accuracy of diagnostic instrumentation data placed on them, especially the opto-mechanical equipment. In [1] for a load bearing structure in the form of a rectangular prism is justified of the optimal accuracy of maintaining the required temperature T_{tar} and speed of operation nonsmooth boundary value problems. Taking into account the specific operation of the optical equipment placed on the supporting structure can be modified to spatially one-dimensional nonsmooth boundary value problems.

2 Numerical calculations

Semi-infinite optimization on the line in a given section $l_y^* \in l_y$, $l_y^* = const$; $l_z^* \in l_z$, $l_z^* = const$; in relative (l_x, l_y, l_z) Cartesian coordinates.

The solution of the problem is determination of the optimum law of vector control change $U^{(\beta\beta)}(\varphi, l_x, l_y, l_z) = [q_x^{(\alpha\alpha)}(\varphi, l_y, l_z), q_y^{(\alpha\alpha)}(\varphi, l_x, l_z), q_z^{(\alpha\alpha)}(\varphi, l_x, l_y)]$, $\beta = 1,2$, providing $J^{opt}(q_{x(opt)}^{(\alpha\alpha)}, q_{y(opt)}^{(\alpha\alpha)}, q_{z(opt)}^{(\alpha\alpha)}) = \min\limits_{q_x^{(\alpha\alpha)}, q_y^{(\alpha\alpha)}, q_z^{(\alpha\alpha)}} J$, $J = \left\| \Theta(l_x, l_y^*, l_z^*, \varphi_k) \right\|_{L_\infty[l_x \in L]}$. at the specified value

* Corresponding author: entcom@samgtu.ru

$\varphi = \varphi_k$ - maximum accuracy task (β=1) or $\min\limits_{q_x^{(\alpha)}, q_y^{(\alpha)}, q_z^{(\alpha)}} \varphi_k$ at the specified value $J = J_{des}$ - maximum performance task (β=2) in terms of restrictions on the allowable temperature $\Theta_d = const$.

$$\Theta(\rho,\varphi) = T(\rho,\varphi) - T_{зад} \le \Theta_d, \varphi \in (0,\infty), \rho \in \overline{\Omega}_\rho \tag{1}$$

and resource control:

$$\begin{aligned} &\left|q_x^{(\alpha\alpha)}(\varphi, l_y, l_z)\right| \le 1 \quad ;\alpha = 3,4; l_x = 0; l_x = 1; \\ &\left|q_y^{(\alpha\alpha)}(\varphi, l_x, l_z)\right| \le 1 ;\alpha = 5,6; l_y = 0; l_y = l_{y_{\max}}; \\ &\left|q_z^{(\alpha\alpha)}(\varphi, l_x, l_y)\right| \le 1 ;\alpha = 1,2; l_z = 0; l_z = l_{z_{\max}}; \end{aligned} \tag{2}$$

when exposed to a control object of the internal and external disturbances. $\varphi_k = \dfrac{\lambda \tau_k}{Cx^2_{\max}}$ - the relative end time of the process, λ, C - thermal conductivity and specific heat of the material of construction. In this example (Fig. 1) the temperature of the inner surface of the outer shell of the supporting structure has a constant time value T_{out_i} . The temperature of the outer shell affects the temperature of the faces of the plate by radiant heat transfer. From opposite sides of the supporting structure has a rectangular area, S1 and S2, which, through radiant heat transfer affects the temperature of the environment. On the lateral surfaces of construction installed heat-emitting devices, occasionally included in the work and controlled heat sources 1,2,3,4,5,6 attached to the surface areas S1 and S2 on opposite faces $\alpha = 3$ and $\alpha = 4$ of the supporting construction.

A valid control $\left|U^{(\beta\beta)}(\varphi, l_x, l_y, l_z)\right| \le 1$ in this example can be represented in a multiplicative form: $q_x^{(\alpha\alpha)}(\varphi, l_y, l_z) = U_x^{(\alpha\alpha)}(\varphi) V_x^{(\alpha\alpha)}(l_y, l_z)$, $\alpha = 3,4$; $\alpha = 1,2,5,6$, $q_x^{(\alpha\alpha)}(\varphi, l_x, l_y) = q^{(\alpha\alpha)}(\varphi, l_x, l_y) \equiv 0$, fixed in advance the spatial distribution $V_x^{(\alpha\alpha)}(l_y, l_z)$ of the controllable heat sources on two opposite faces $\alpha = 3,4$ with a specified coordinate location $V_x^{(\alpha\alpha)}$, $\alpha = 3,4$. Under these conditions, the optimal control of autonomous $\overline{U}^{(\beta\alpha)}(\varphi)$ groups of heat sources each α-th face:

$$\overline{U}^{\beta\alpha}(\varphi) = \left[U_x^{(\beta 3)}(\varphi), U_x^{(\beta 4)}(\varphi), U_y^{(\beta 5)}(\varphi), U_y^{(\beta 6)}(\varphi), U_z^{(\beta 1)}(\varphi), U_z^{(\beta 2)}(\varphi)\right] \tag{3}$$

$$U_{opt}^{(1\alpha)}(\varphi) = \arg \min_{\overline{U}^{(1\alpha)}(\varphi)} J \; ; \; U_{opt}^{(2\alpha)}(\varphi) = \arg \min_{\overline{U}^{(2\alpha)}(\varphi)} \varphi_k \; ; \tag{4}$$

in the problem of maximum precision (β=1) and task performance (β=2) using the method of moments [2,3] it is possible to parameterize and to reduce to the problem of finding a finite number of parameters $\Delta_\delta^{(i)}$, δ=1,2, *i; i*=1,2,..*J;* where $\Delta_\delta^{(i)}$ the duration δ–th interval of constancy of the total number i, where each vector $\overline{U}^{(\beta\alpha)}(\varphi)$ component alternately takes its maximum value at the level of the constraints (2).

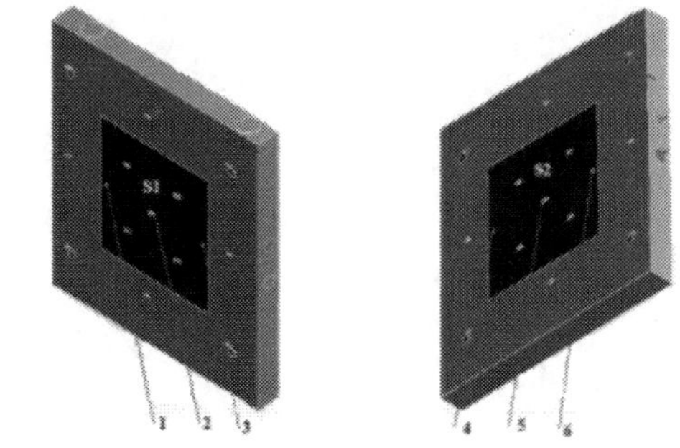

Fig. 1. The location of the heat sources on the supporting structure (example).

The duration $\Delta_{\delta}^{(i)}$ and the desired number of intervals i, is determined by the alternance method [3] from the system of transcendental equations [6]:

$$\Theta(\rho_k,\varphi_k)\Big|_{\rho_k\in L_k} = \pm\varepsilon^{(i)}, grad\Theta(\rho_k,\varphi_k)\Big|_{\rho_k\in L_k} = 0 \qquad (5)$$

where $\Theta(\rho_k,\varphi_k)\Big|_{\rho_k\in L_k}$ - numerical solution of boundary value problems in the final time φ_k points $\rho_k(l_x^{(k)},l_y^*,l_z^*)\in L_k(l_x^{(k)},l_y^*,l_z^*,\varphi_k)\subset L(l_x,l_y^*,l_z^*)$ of the considered line section $L(l_x,l_y^*,l_z^*)$, which constitute a countable set of limit points, which is the ratio (5).

The solution of the transcendental system (5) is carried out by a specially developed application package [5] based on the numerical solution of the boundary value problem in the ANSYS package. The processing of the results of the solution and its analysis were carried out using the developed application [5]. The results of calculations in the software package ANSYS are shown, which demonstrate the deviation $\Theta(L,\varphi_k)$ of the resulting temperature distribution $T(L,\varphi_k)$ from the given T_{tar} for different values of the thermal conductivity coefficient λ. Figure 2 shows the deviation $\Theta(L,\varphi_k)$ for the optimal control accuracy β = 1 for 1 and 2 interval control in the responsible line $L(l_x,l_y^*,l_z^*)$ $l_y^* = 0{,}5$, $l_z^* = 0{,}5$ of the section of the load-bearing structure The maximum temperature deviation on this line $L(l_x,l_y^*,l_z^*)$ does not exceed 4°C for a single-interval and 2°C for a two-interval temperature when exposed to the heaviest heat load option.

 DOI: 10.1051/matecconf/201711001015

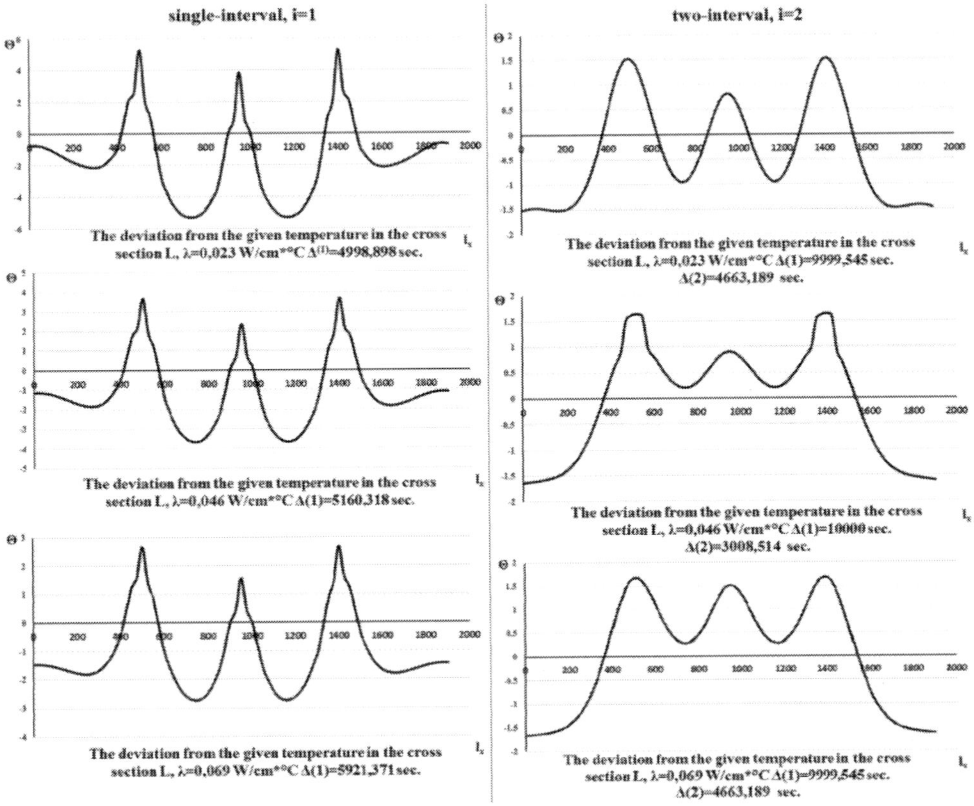

Fig. 2. Deviation from specified temperature on the section line $L(l_x, l_y^*, l_z^*)$, $l_y^* = 0{,}5$, $l_z^* = 0{,}5$ with the optimal control accuracy.

The dependence of the optimization results on the properties of the construction material's thermal conductivity coefficient λ is shown in Figures 3 and 4.

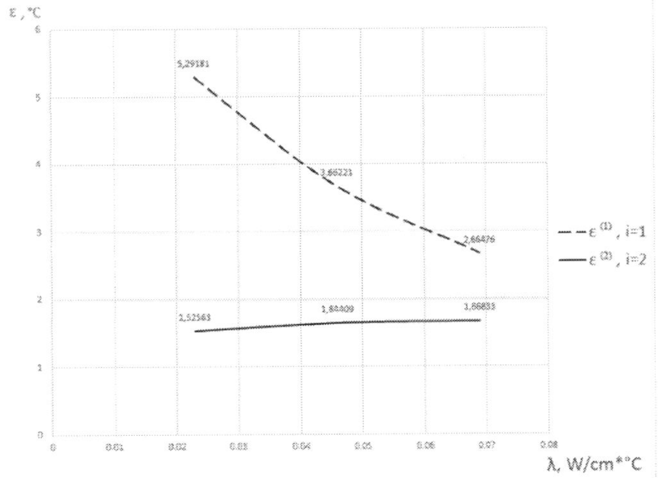

Fig. 3. The dependence of the accuracy of control on the value of thermal conductivity for single- and two-interval (i = 1, i = 2) controls.

 DOI: 10.1051/matecconf/201711001015

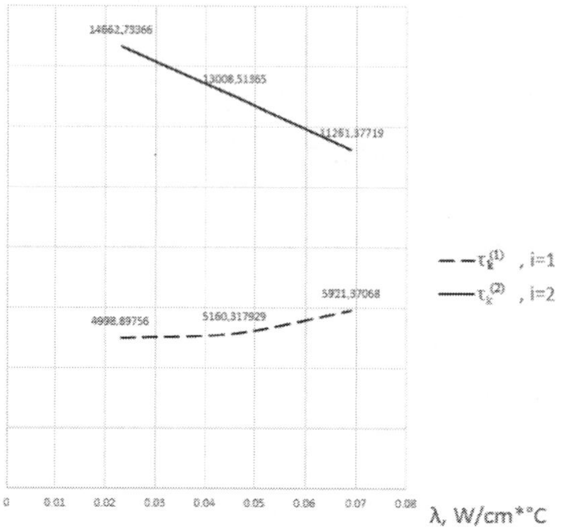

Fig. 4. The dependence of control time on the value of the coefficient of heat conductivity for single- and two-interval (i = 1, i = 2) controls.

3 Conclusions

The analysis of the available parameters has shown that the two-interval (i = 2) control in both problems has a significantly lower sensitivity to the coefficient of thermal conductivity λ with significantly higher accuracy of maintaining the set temperature. All this leads us to conclude that designer has greater freedom in choosing the material of the structure.

References

1. M.Y. Livshits, M.Y. Derevjanov, S.A. Kopytin, XIV Minsk International Forum on Heat and Mass Transfer: abstracts and reports, **1,** 719 (2012)
2. A.G. Butkovskii, *Theory of optimal control of distributed parameter systems* (Nauka, Moscow, 1965) [In Russian]
3. E'.Ya. Rapoport, *Optimal'noe upravlenie sistemami s raspredelenny'mi parametrami.* (Vy'sshaya shkola, Moscow, 2009) [In Russian]
4. B.B. Borodulin, M.Y. Livshits, EPJ Web of Conferences, **110** (2016)
5. A.P. Efimov, Proc. VIII int. Conf. Samara scientific centre of RAS, 195 (2006)
6. B.B. Borodulin., M.Y. Livshits, S.E. Korshikov, MATEC Web Conf. **92**, 26-28 (2016)

MATEC Web of Conferences **110**, 01016 (2017) DOI: 10.1051/matecconf/201711001016
HMTTSC-2017

Research on permeability of carbon nanotubes

Michael A. Bubenchikov[1,2,*], *Alexander* V. Malozemov[1], and *Alexander* A. Sherstobitov[1]

[1]National Research Tomsk State University, 634050 Tomsk, Russia
[2]LLC "Gazprom Transgaz Tomsk", 634029 Tomsk, Russia

Abstract. The present paper deals with a geometric object composed of energy sources interacting with test atoms and molecules in a gaseous state. The sources are stationary and they are located in a hexagonal manner on the surface of a circular cylinder. The permeability of the resulting model of a carbon nanotube with respect to helium atoms and hydrogen, oxygen and methane molecules is investigated. Further evaluation of impacts caused by thermal vibrations of the carbon skeleton atoms follows. It is shown, that a carbon nanotube can be used as a support element of a filter for gas mixtures separation.

1 Introduction

It is known, that natural gas from wells must be prepared for transportation before it is supplied to the end user. The need for such works is caused by natural gas containing impurities along with target components, which can create difficulties in transportation or use. One of the promising methods to separate natural gas components is the use of carbonic filters, ideally consisting of identically oriented carbon nanotubes [1], the permeability study of which is conveniently carried out using models of dynamic systems. Molecular dynamics methods are widely used in modern research for the purpose of simulating fluid motion in nanochannels [2, 3] and when calculating the diffusion coefficient of nanoparticles [4-7], as well as their interactions with each other [8]. These same methods are used in studying diamond-like phases [9-13] and discrete breathers [14], in analyzing severe plastic deformation of media containing crystal grains [15], as well as in investigating permeability of ultra-thin films in the framework of Newtonian dynamics [16-22] and on the basis of quantum-mechanical models [23, 24]. In this paper we research permeability of an open carboxylic nanotube for molecules and atoms of various substances in a gaseous state. We consider the influence of a tube's length on the motion of molecules in the gas phase, as well as the impact of thermal oscillations of carbon skeleton atoms on permeability of open nano-tubes. In respect of these graphene structures, the internal dynamics of hydrogen, oxygen, helium, and methane molecules is analyzed.

2 Numerical model

Studies of the interaction process between a moving molecule and atoms of a carbon nanotube are based on the use of the mathematical model and numerical methods of solution.

[*]Corresponding author: michael121@mail.ru

Interaction between individual molecules is determined by the classical Lennard-Jones potential. Interaction with the structure, in turn, is described by the law of action independence as the sum of each structure atom's effect on the test molecule under consideration. According to this law, if a material point is influenced by several forces, it experiences acceleration equal to the geometric sum of the accelerations these forces would cause acting separately.

In projections on the coordinate axis the motion equations of the molecule interacting with the tube can be written as follows:

$$M\frac{dU}{dt}=X',\ M\frac{dV}{dt}=Y',\ M\frac{dW}{dt}=Z',\ \text{here}\ X'=\sum_{j=1}^{N_p}X'_j\ ,\ Y'=\sum_{j=1}^{N_p}Y'_j\ ,\ Z'=\sum_{j=1}^{N_p}Z'_j \tag{1}$$

M is the mass of the molecule, X', Y', Z' are the projections of the Van der Waals forces resultant from the carbon structure atoms which are defined as simple sums of force contributions from the nodes of the crystalline lattice, N_p is the number of the nanotube's atoms. Next, we introduce into consideration the values of X, Y, Z: $MX = X'$, $MY = Y'$, $MZ = Z'$. Then, in the new variables, equations (1) can be rewritten as:

$$\frac{dU}{dt}=X,\ \frac{dV}{dt}=Y,\ \frac{dW}{dt}=Z,\ \text{here}\ X=\sum_{j=1}^{N_p}a_j\frac{x-x_j^0}{\rho_j}, Y=\sum_{j=1}^{N_p}a_j\frac{y-y_j^0}{\rho_j}, Z=\sum_{j=1}^{N_p}a_j\frac{z-z_j^0}{\rho_j} \tag{2}$$

Here the zero-index indicates the coordinates of the carbon skeleton nodes, a_j is the value of acceleration which is acquired by the test molecule under the influence of the j-th atom of the crystalline structure. This value is expressed through the parameters of the Lennard-Jones potential as follows:

$$a_j = 24\frac{\varepsilon}{M\rho_j}\left(\frac{\sigma}{\rho_j}\right)^6\left[2\left(\frac{\sigma}{\rho_j}\right)^6-1\right]. \tag{3}$$

In formula (3) ρ_j is the distance between the considered test molecule and the j-th atom of the carbon structure; σ and ε are the interaction parameters of substance pairs.

If equations (2) are supplemented by kinematic relations determining the velocity of the point:

$$\frac{dx}{dt}=U,\ \frac{dy}{dt}=V,\ \frac{dz}{dt}=W, \tag{4}$$

We obtain a system of six ordinary differential equations of the first order with respect to six unknowns: x, y, z, U, V, W. To solve this system we will use the Runge-Kutta method of the fourth order accuracy. In case the system under study consists of heterogeneous molecules (atoms), the parameters ε and σ follow the Lorentz-Berthelot mixing rules.

3 Calculation results

Initially, we considered a tube with the diameter of 0.92 nm and the length of 1.14 nm. If the test molecule is directed strictly along the axis of the tube, it moves straightforward inside the nanostructure, however, it comes out at a certain angle to the axis. Moreover, this angle depends on the length of the tube, as well as its radius. When the input of the molecule is not central, it is included in a complex oscillatory motion, in which, however, one can identify the frequency determined by the initial velocity of the molecule and the value of the initial displacement from the position of dynamic equilibrium, i.e. from the axis of the tube (refer with: Fig.1).

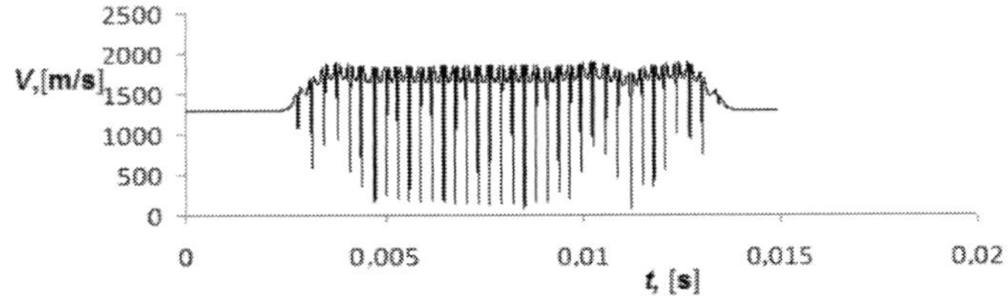

Fig.1 Changes in rate values of helium atom at non-central input.

As can be seen from Fig.1, the output velocity is equal to the input velocity as well as the average velocity of helium thermal motion at room temperature. The fact of such velocity equality is a consequence of the law of total energy conservation for a moving particle, which consists of its kinetic energy and potential energy of interaction with the carbon nanostructure.

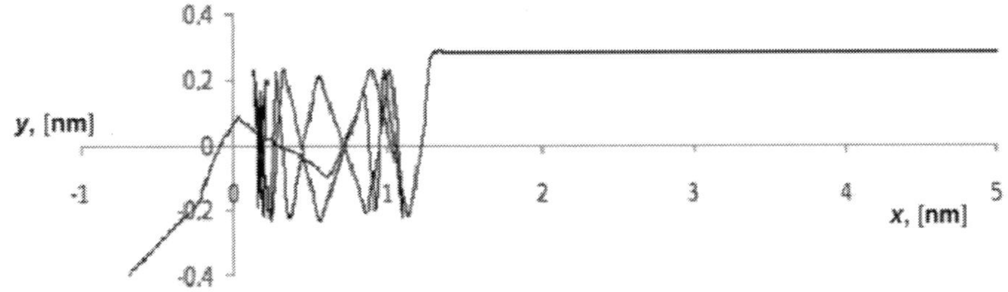

Fig.2 View of trajectory in axial section plane of tube.

The movement of a helium atom, as shown in Fig.2, starts at the position x=5 nm, y=0.22 nm, z=0 from the right to the left in the direction of the tube and then through the tube into the surrounding space.

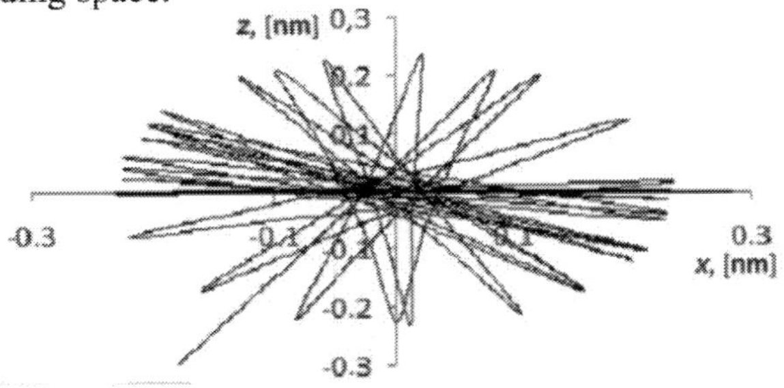

Fig.3 Radial movements of helium atom in short nanotube.

In addition, in the cross section perpendicular to the tube axis, precession oscillations of the helium particle are observed (refer with: Fig.3). Next, tubes of different diameters are considered. We carried out calculations in order to determine the filtering effect. We calculated the minimum allowed diameters for carboxylic tubes, which still permit penetration of particles through the structure. For example, for the case of methane, the minimum allowed diameter is equal to 0.69 nm. Tubes of smaller diameters are impermeable to molecules of this gas. Minimum allowed permeability diameters for all examined molecules are given in Table 1.

Table 1. Permeability diameters.

Interacting particles	He	H_2	O_2	CH_4
Permeability diameter	0.54 [nm]	0.59 [nm]	0.66 [nm]	0.69 [nm]

Thus, the use of tubes of different diameters allows using the filtering effect and selecting target components. For example, a tube with the diameter of 0.68 nm is permeable to helium atoms and not permeable to methane molecules.

Next, we examined nanotubes of different lengths. Calculations showed that the length of a grapheme structure does not influence the minimum permeability diameter.

4 Summary

A permeability study for an open carboxylic nanotube with respect to molecules and atoms of different substances in a gaseous state has been conducted. Considering these graphene structures, we have investigated internal dynamics of hydrogen, oxygen, helium, and methane molecules. The obtained results show the ability of carbon nanotubes of a given diameter to filter only target components. For each moving particle under consideration, the minimum allowable diameter of the nanotube has been defined and the sizes of permeability windows for tubes have been found. As shown by calculations, tube lengths have no influence on permeability. Thermal fluctuations, as expected, reduce the size of a permeability window.

This paper was written within the frame of Competitiveness Improvement Program of Tomsk State University and with financial support of the Russian Science Foundation, grant under Agreement N 16-19-00089.

References

1. N.B. Kondrikov and S.A. Tsarev, RU Patent, 2474466 (2013)
2. V.Ya. Rudyak, S.L. Krasnolutskii, D.A. Ivanov, J. Dok. Phys. **57,** 33 (2012)
3. V.Ya. Rudyak, A.A. Belkin, V.V. Egorov, D.A. Ivanov, Int. J. Multiphys, **5,** 145 (2011)
4. V.Ya. Rudyak, S.L. Krasnolutskii, D.A. Ivanov, J. Microfluid. Nanofluid, **11**, 501 (2011)
5. V.Ya. Rudyak, S.N. Dubtsov, A.M. Baklanov, Tech. Phys. Lett. **34**, 519 (2008)
6. V.Ya. Rudyak, S.L. Krasnolutskii, E.N. Ivashchenko, J. Eng. Phys. Thermophys. **81,** 520 (2008)
7. V.Ya. Rudyak, J. VDI-Bericht, **1803,** 187 (2003)
8. V.Ya. Rudyak, A.A. Belkin, Tech. Phys. Lett. **29**, 560 (2003)
9. K.A. Krylova, Y.A. Baimova, S.V. Dmitriev, R.R. Mulyukov, Phys. Sol. St. **58,** 394 (2016)
10. A.V. Savin, E.A. Korznikova, S.V. Dmitriev, Phys. Sol. St. **57** 2348 (2015)
11. K.A. Bukreeva, R.I. Babicheva, S.V. Dmitriev, K. Zhou, R.R. Mulyukov, A.I. Potekaev, Russ. Phys. J. **57**, 69 (2014)
12. J.A. Baimova, B. Liu, S.V. Dmitriev, N. Srikanth, K. Zhou, Phys. Chem. Chem. Phys. **16,** 19505 (2014)
13. E.A. Korznikova, S.V. Dmitriev, J. Phys. D: Appl. Phys. **47**, 345307 (2014)
14. E.A. Korznikova, S.Y. Fomin, S.V. Dmitriev, Mat. Sc. Forum. **845,** 211 (2016)
15. J.A. Baimova, S.V. Dmitriev, Mat. Sc. Forum. **838-839**, 361 (2016)
16. A.I. Potekaev, M.A. Bubenchikov, Russ. Phys. J. **54**, 211 – 220 (2011)
17. M.A. Bubenchikov, Russ. Phys. J. **54**, 102 (2011)

18. M.A. Bubenchikov, A.I. Potekaev, A.M. Bubenchikov, Russ. Phys. J. **56**, 785 (2013)
19. M.A. Bubenchikov, A.I. Potekaev, A.M. Bubenchikov, Russ. Phys. J. **56,** 341 (2013)
20. A.I. Potekaev, A.M. Bubenchikov, M.A. Bubenchikov, Russ. Phys. J. **55**, 1434 (2013)
21. M.A. Bubenchikov, A.I. Potekaev, A.M. Bubenchikov, O.V. Usenko, A.V. Ukolov, IOP Conf. S.: Mater. Sc. Eng. **87,** 012111 (2015) [In Russian]
22. M. A. Bubenchikov, A. M. Bubenchikov, O. V. Usenko, V. A. Poteryaeva, S. Jambaa, EPJ Web Conf. **110,** 01014 (2016)
23. A.M. Bubenchikov, M.A. Bubenchikov, A.I. Potekaev, E.E. Libin, Yu.P. Khudobina, Russ. Phys. J. **57**, 1126 (2014)
24. A.M. Bubenchikov, M.A. Bubenchikov, A.I. Potekaev, E.E. Libin, Yu.P. Khudobina, The potential field of carbon bodies as a basis for sorption properties of barrier gas systems, Rus. Phys. J. **58**, 882 (2015)

DOI: 10.1051/matecconf/201711001017

Effect of the shape of wet wood particles on ignition characteristics

Samen Syrodoy [1,*], *Ksenia* Bugaeva[1]

[1]National Research Tomsk Polytechnic University, 634050 Tomsk, Russia

Abstract The results of experimental studies of various geometric shape wood biomass single particles ignition processes (on the example of a cylinder and a cube) are presented. According to the results of the experiments, it has been stated that the shape of the fuel particle can influence the characteristics and conditions of ignition, but it is insignificant. It has been demonstrated, that in most cases cylindrical particles ignite faster than cubic particles. However, in this respect, two ignition modes are set: high temperature (the cube fires up faster) and low temperature (the cylinder ignites faster).

1 Introduction

In recent years, wood biomass has become increasingly attractive as a fuel for thermal power plants [1-3]. This trend persists not only in the United States, but also in Eastern Europe [4]. For example, the most modern thermal power plant Avedore (Denmark) [5, 6] completely proceded to the burning of wood pellets in the second half of 2016. The latter is explained by the number of undoubted advantages of wood fuel in comparison with traditional types of solid (coal), liquid (fuel oil) and gaseous (natural gas) types of energy carriers [7,8]. First of all, wood is the only renewable fuel [7] in many regions of the North and (especially) South America, Asia and Africa. It should also be noted that when burning wood pellets much less (5 times) oxides of nitrogen and sulfur are formed [9] (compared to coal). It is also important that woodworking industry waste can be used as a raw material for the production of fuel based on wood [10]. The storage of the latter causes great problems due to their high fire hazard. For this reason, the most expedient is the use of such wood as a solid fuel for small-sized boiler units (and, in the long term, for large thermal power plants). However, the theory allowing to conduct predictive modeling of the conditions and characteristics of even dry small, medium and large wood particles ignition in the high-temperature gases flow hasn't been developed yet. The latter is due to the fact that the specific properties of wood fuel (the type of wood, external and internal moisture, the anisotropy of wood structure, etc.) have a significant effect on the conditions of ignition. One of the most important parameters is the shape of the particle. As is known from the theory of heat conduction [11], the shape of the body can have a significant effect on the heat transfer characteristics. Accordingly, it can reasonably be assumed that the characteristics of wood particles ignition, for example cube- or cylinder-shaped, will differ

*Corresponding author: syrodoy@tpu.ru

substantially. The currently known studies [12] of the effect of the geometry of fuel particles on the dynamics of their ignition have been carried out for coal-based fuel compositions. At the same time, the questions of the difference in the characteristics of spherical, cylindrical and cubic shape wood particles ignition have not been given sufficient attention. However, we can say that the optimization of the shapes of fuel briquettes made of wood-chip waste is an urgent task for the energy industry. For this reason, the aim of the work is an experimental study of the processes of thermal preparation and of various geometric shapes moist wood particles ignition.

2 Experiment

Experimental studies were carried out at a stand similar in its methods of recording and processing experimental data [13]. A particle of wood was fixed on a ceramic holder and brought by a remote-controlled coordinate device into a hollow ceramic cylinder whose inner surface was heated to the temperature of 1270 K. The registration of the processes of heat and mass transfer and ignition of the fuel particle was carried out by a high-speed Photron FASTCAM CA4 5 video camera (video speed up to 20,000 fps). The air temperature inside the cylinder (T_e) in experiments varied over a fairly wide range (from 870K to 1270K). The period of time from the start of the heat exposure (the moment of entering the cylinder cavity) to the ignition (the appearance of a flame) was considered to be the ignition delay time (t_i). The particle sizes varied in the range of $1 < \delta < 3 \cdot 10$-3 m. The systematic error in determining the main measured parameters (Te, t_i) was no more than 5%. The confidence interval for determining tign at a confidence probability of 0.95 did not exceed 17%.

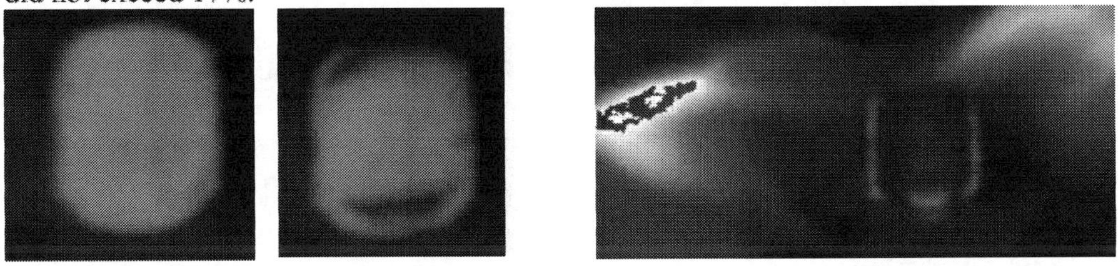

t=0 s. t=1.75 s. t=2.35 s.

Fig.1. Frames of a typical videogram of a moist wood cylindrical particle ignition processes.

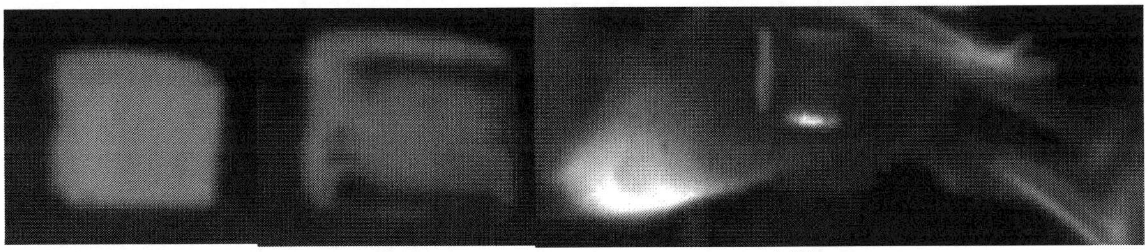

t=0 s. t=1.85 s. t=2.75 s.

Fig.2. Frames of a typical videogram of a moist wood cubic particle ignition processes.

3 The results

Fig. 1 shows the frames of a typical videogram of a moist wood cylindrical particle ignition processes. (diameter $\delta = 3 \cdot 10^{-3}$ m). The temperature of the external environment is Te = 1273K. The total time from the start of heating to the moment of ignition is $t_i \approx 2.35$ s. As experiments show, the process of thermal preparation can be conditionally divided into a number of successive interrelated stages (inert heating, water evaporation, thermal decomposition of the organic part of the fuel, ignition of volatiles, ignition of carbon). At the initial moment (frame a), a particle of wood enters a high-temperature environment. Inert heating initiates the water evaporation and thermal decomposition (frame b). A detailed (at a great zooming) analysis of frame b shows the characteristic darkening of the particle surface. In t = 2.35 s after the start of heating, the ignition of the volatiles occurs (frame c). It should be noted that a torch of burning volatiles is formed, whose axis direction corresponds to the direction of the wood fibers.

Fig. 2 shows typical frames of the videogram of the processes of moist wood cubic particles ignition. Having analyzed the frames, we can say that the main stages of thermal preparation are similar to Figure 1. It should be noted that the thermal decomposition process is localized mainly at the corners of the cube. The latter is explained by the large heat flows in these places. Also, the analysis of the frames of the videogram shows that when the volatiles are ignited, a torch is formed, whose axis direction coincides with the direction of the wood fibers. Accordingly, it can be said that the characteristics of the ignition of a particle essentially depend on the anisotropy of the thermophysical and filtration properties of the wood.

Figure 3 shows the dependences of the ignition delay times for moist wood particles of different shapes (cylinder and cube). The dependence analysis shows that the particle shape can have an effect on ignition characteristics. However, the degree of such influence is insignificant. It should be noted that the existence of two modes of ignition is possible: high-temperature and low-temperature with the conventional border of $T_e \approx 973K$. Under $T_e < 973K$, cubic particles ignite faster than cylindrical ones, at higher temperatures of the external environment, cylindrical fuel particles ignite earlier than cubic ones. The latter is most likely due to the fact that when the cylinder is heated, the diffusing volatiles form a uniformly distributed vapor-gas mixture, which is ignited at critical temperatures. Under conditions of low-temperature heating, heterogeneous ignition occurs as a rule. For this reason, the cubic particles ignite faster because of the concentration of heat flows in the zones close to the corner points of the cube.

4 Conclusion

According to the results of experimental studies, it has been stated that the shape of a particle of wood fuel can influence the ignition characteristics, but this effect is insignificant. It is shown that when igniting a volatile, the direction of its axis coincides

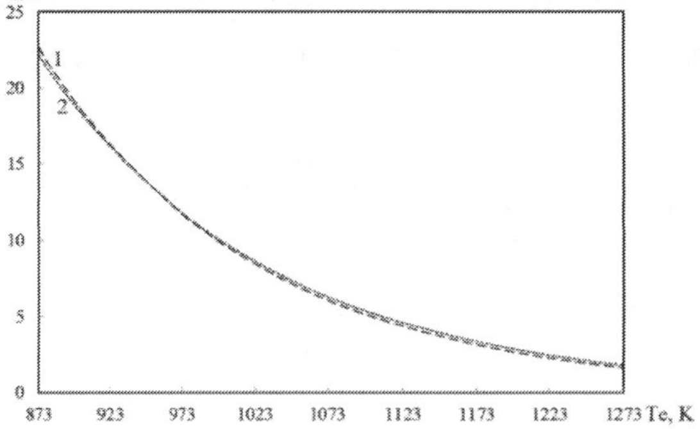

Fig. 3. Dependence of ignition delay on the temperature of the external environment for wood particles of various shapes: 1 - cylinder; 2 - cube.

DOI: 10.1051/matecconf/201711001017

with the direction of the fibers of the wood. Accordingly, it can be assumed that the characteristics of ignition substantially depend on the thermal and filtration anisotropy of the wood. According to the results of the experiments, two ignition modes are distinguished: high-temperature (Te> 973K) and low-temperature (Te <973K). It has been established that at Te> 973K, cylindrical particles ignite faster than cubic particles, at higher temperatures the situation is diametrically opposite (the cylinders are ignited earlier than the cubes).

The work was supported by the scientific schools grant NSH-7538-2016.8.

References

1. U. Mantau, Biomass and bioenergy, **79** (2015)
2. U. Mantau, U. Saal, K. Prins, F. Steierer, M. Lindner, H. Verkerk, *Real potential for changes in growth and use of EU forests. EUwood* (University of Hamburg Centre of Wood Science, Hamburg, 2010)
3. H. Weimar, P. Doring, U. Mantau, *Einsatz von Holz in Biomasse-Grobfeuerungsanlagen 2011,* Final report (University of Hamburg Centre of Wood Science, Hamburg, 2012)
4. P. Lauri, P. Havlik, G. Kindermann, N. Forsell, H. Bottcher, M. Obersteiner, Energy Policy, **66** (2014)
5. Fuel and Energy Abstracts. **37,** (96) 05114 (1996)
6. Y. Voytenko, P. Peck, Biomass and bioenergy, **38** (2012)
7. L.J.R. Nunes, J.C.O. Matias, J.P.S. Catalao, Renewable Energy, **85** (2016)
8. C. McGlade, J. Speirs, S. Sorrell, *A review of regional and global estimates of unconventional gas resources* (A Report to the Energy Security Unit of the Joint Research Centre of the European Commission, London, 2011)
9. T. Klason, X.S. Bai, Fuel, **86** (2007)
10. C. Antwi-Boasiako, B.B. Acheampong, Biomass and Bioenergy, **85** (2016)
11. H. Carslaw, J. Jaeger, *Conduction of Heat in Solids* (Oxford University Press, Oxford, 1959)
12. S.V. Syrodoi, G.V. Kuznetsov, V.V. Salomatov, Solid Fuel Chemistry, **49** (2015)
13. V.V. Salomatov, G.V. Kuznetsov, S.V. Syrodoy, N.Y. Gutareva, Appl. Therm. Eng. **106** (2016)

Features of mathematical modelling of the microclimate of the non-productive building

Maria Prorokova[1,*], *Vyacheslav* Bukhmirov[1]

[1]Ivanovo State Power University named after V.I. Lenin, 153003 Ivanovo, Russia

Abstract. The mathematical model of heat and mass transfer in a non-production building is presented in the article. Features of modeling of objects of the specified type are analyzed. A mathematical model of the microclimate in the ANSYS Fluent was developed. Verification of the model was carried out by comparing the values of the microclimate parameters obtained by calculation, with experimental data.

1 Introduction

Calculation of the main parameters of the microclimate in non-production buildings is done to determine the degree of comfort of the microclimate, so the calculation must be carried out both during construction and when the building structure is changed, including for energy saving during its operation (replacement of windows, thermal insulation a. o.). All calculations of microclimate parameters can be conditionally divided into two groups. The first group includes, so-called, engineering calculations, which are based on integral or balance models. Balance models are also called models with lumped parameters or zero-dimensional mathematical models. Balance models analyze the room as a whole, in which the temperature and humidity of the air are the same at all points in the room. All engineering calculations are based on the solution of the equations of thermal and material balances for the object, and the boundary conditions are set, as a rule, using the equations of stationary heat transfer through non-permeable walls.

Due to increase in power of computing tools and improving their mathematical apparatus, including numerical methods for solving differential heat transfer, mass, and momentum transfer equations, it became possible to predict the microclimate in rooms using multidimensional non-stationary mathematical models. This group of calculation methods is traditionally called mathematical modeling of microclimate [1].

For the numerical solution of the system of transport equations for the main parameters of a substance (heat, mass, and momentum), so-called, software-computational complexes (SCC) are developed, which can be universal or adapted for modeling physical and chemical processes in a specific device, for example, in a nuclear reactor. In this work the simulation of the microclimate in a non-production room was performed in the SCC ANSYS Fluent, which refers to the universal SCC, which allows modeling the features of physical and chemical processes in various objects. When using ANSYS Fluent is very

* Corresponding author: prorokova_mv@list.ru

DOI: 10.1051/matecconf/201711001018

important is the stage of mathematical formulation of the problem of calculating a particular process. At the stage of the mathematical description of the object, the assumptions that simplify the problem write, define the conditions for single-valuedness, and write the differential transfer equations.

2 Features of mathematical modeling of microclimate for non-production objects

In non-production premises, the microclimate is formed by convective, conductive and radiative heat transfer, and also by molecular and convective diffusion of air pollutants. Consequently, the mathematical model for the formation of a microclimate in a buildings room must include the energy transfer equations due to radiative and convective heat transfer, the momentum transfer equation, the continuity equation, and the mass transfer equation for each component of the mixture, taking into account the assumptions: air in the room is a four-component mixture, that consists of nitrogen, oxygen, carbon dioxide and water vapor; the components of the gas mixture and the mixture itself obey the laws of an ideal gas; in the room there is a thermo gravitational movement of air; gas in the room is a non-diathermic substance; diffusion transport of air through the solid elements of the enclosing structures (walls, linen of the window block) is absent; the airflow inlet is through the window, the outlet is through the ventilation openings and doors; sources of heat due to internal friction are small; scattering and refraction of radiant energy are not taken into account.

The mathematical model of heat and mass transfer in the building of a non-production building looks as follows:

$$\frac{\partial}{\partial t}\rho E + \nabla\vec{w}\rho E = \nabla(\lambda_{eff}\nabla T - \sum_{i=1}^{n} h_i \vec{J_i}) + q_v. \quad (1)$$

$$\frac{dI(\vec{r},\vec{s})}{ds} + (A+\sigma_s)I(\vec{r},\vec{s}) = A\frac{\sigma_0 T^4}{\pi} + \frac{\sigma_s}{4\pi}\int_0^{4\pi} I(\vec{r},\vec{s}\,')d\Omega'. \quad (2)$$

$$\frac{\partial}{\partial t}\rho + \nabla(\rho\vec{w}) = 0; \quad (3)$$

$$\frac{\partial}{\partial t}\rho\vec{w} + \nabla(\rho\vec{w}\vec{w}) = \nabla\rho\nu\nabla\vec{w} + \rho\vec{g}; \quad (4)$$

$$\frac{\partial}{\partial t}\rho C_i + \nabla(\rho\vec{w}C_i) = \nabla\left((\rho D_{i,m} + \rho D_t)\nabla C_i\right) + S_i. \quad (5)$$

In equations (1) – (5): t – time; ρ – density of the gas mixture; $\vec{w}$ – speed of air movement; E – specific internal energy of the gas mixture; λ_{eff} – effective coefficient of thermal conductivity; h_i – specific enthalpy of the i-th component of the mixture; $\vec{r}$ – radius-vector; $\vec{s}$ – unit vector in the direction of scattering; $\vec{s'}$ – unit vector in the direction of the incident radiation; A – absorption coefficient; σ_s – scattering coefficient (for pure air $\sigma_s = 0$); σ_0 – Stefan-Boltzmann constant; I – brightness of radiation at a given point ($\vec{r}$)in a given direction ($\vec{s}$); T – local temperature; Ω – solid angle; q_v – capacity of internal sources of heat; ν – kinematic viscosity coefficient; g – acceleration of gravity; $\vec{J_i}$

– density of the diffusion flux of the i-th component of the mixture; C_i – local mass concentration of the i-th component of the mixture; $D_{i,m}$ – diffusion coefficient of the i-th component of the mixture; D_t – coefficient of turbulent diffusion; S_i – mass source of the i-th component of the mixture.

A detailed description of the mathematical model is given in [1, 2, 3].

Thermophysical and radiation properties of objects that participate in the process of heat and mass transfer are determined from reference data [4], and the radiative properties of the air mixture are calculated by specifying the volume fractions of the individual components. Sources of thermal energy, moisture and carbon dioxide in the premises of residential, public and administrative buildings are people, heating appliances and household appliances. It is known that, on average, a person exhales in a second $1{,}2\cdot10^{-4}$ кг air, whose composition 78,5 % N_2, 14,2 % O_2, 3,6% H_2O, 3,7 % CO_2, and emits 105 Wt of thermal energy [5].

Starting conditions for the solution of the system of equations (1) - (5) (temperature of surfaces of the room, air temperature, the relative humidity and speed and speed of indoor and outdoor air) are determined from the experimental data.

The power of the heating devices is determined experimentally, and the heat dissipation of electrical appliances is set according to the equipment's passport data, depending on the mode of operation of the equipment. Temperature of surfaces of the room (T_w) and sources of thermal energy is accepted by results of the experiment.

The air velocity is calculated from the experimental data or the normative value of the air exchange [6].

The k-ε model of turbulence was chosen to model turbulence from the analysis of literature [7] and analysis of the results of computational experiments [3, 7].

The geometric and grid models of the experimental room of the university's educational audience in the ANSYS Fluent SCC and the features of the computing process of microclimate parameters are described in detail in the paper [8].

Figure 1 shows the results of mathematical modeling and experimental measurement of air temperature at four points in the experimental room: in the center of the room (point 1 on the figure 1), at a distance of 0.5 m from the left wall (point 2 on the figure 1) and at a distance of 0.5 m from the front wall (point 3 on the figure 1).

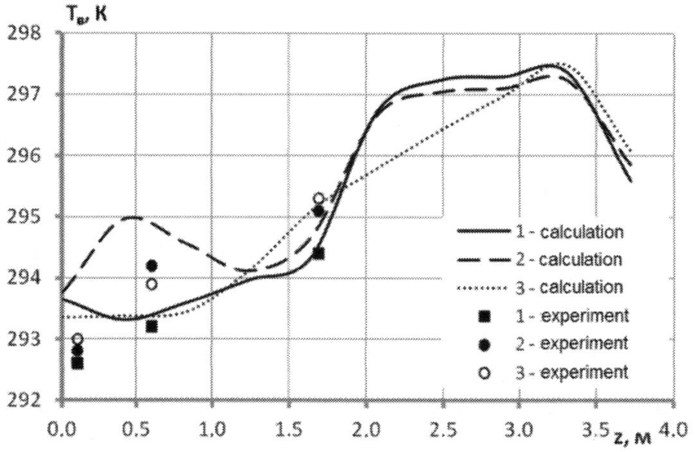

Fig. 1. Air temperature in the experimental room.

3 Results and discussion

Analysis of figure 1 shows a satisfactory accuracy in the calculation of air temperature in the experimental room.

Calculation of the relative humidity of air and air velocity in ANSYS Fluent showed a difference in calculated and experimental data of not more than 9.5%.

4 Conclusions

A mathematical model of heat and mass transfer is developed that describes the formation of a microclimate in a non-production building that takes into account all the main features of the object (people in the room, reducing the radiative heat exchange of air and the actual air exchange). The adequacy of the mathematical model of the microclimate was proved by comparing the calculation results in the ANSYS Fluent with experimental data.

References

1. V.A. Arutyunov, V.V. Bukhmirov, S.A. Krupennikov, *Mathematical modeling of thermal operation of industrial furnaces* (Metallurgy, Moscow, 1990) [In Russian]
2. Yu.A. Tabunshchikov, M.M. Brodach, *Mathematical modeling and optimization of thermal efficiency of buildings* (ABOK-PRESS, Moscow, 2002) [In Russian]
3. ANSYS Fluent User's Guide. Release 15.0 (2013)
4. V.V. Bakhmirov, *Reference materials for solving problems on the course "FHeat and Mass Transfer"* (ISPU, Ivanovo, 2009) [In Russian]
5. A.P. Laptev, S.A. Polievsky, *Hygiene* (Physical Culture and Sports, Moscow, 1990) [In Russian]
6. SP 118.13330.2012 *Public buildings and facilities* (FAU Federal State Administration, Moscow, 2014) [In Russian]
7. A.A. Yun, B.A. Krylov, *Calculation and modeling of turbulent flows with heat exchange, mixing, chemical reactions and two-phase flows in the FASTEST-3D software package* (Publishing house MAI, Moscow, 2007) [In Russian]
8. V.V. Bukhmirov, MATEC Web Conf. **92** (2017)

Investigation of thin film energy-saving coatings

Vyacheslav Bukhmirov[1,*], *Alexander* Gaskov[1]

[1]Ivanovo State Power Engineering University, 153003 Ivanovo, Russia

Abstract. The report presents the results of an experimental study of the thermophysical properties and energy efficiency of thin-film energy-saving coatings consisting of hollow microspheres and a binder material from styrene-acrylic dispersion. The value of the thermal conductivity coefficient of the energy-saving paint is estimated depending on its composition and temperature, and the thermal diffusivity coefficient is determined. Experimental results of energy efficiency of using thin-film coatings for insulation of facades of buildings and as thermal insulation for pipelines with a hot coolant are presented.

1 Introduction

Reduction of heat losses is an urgent task of increasing the energy efficiency of heat networks and objects of housing and communal services. To reduce the loss of thermal energy use a variety of thermal insulation materials, which must have a number of special qualities, the main of which is a low value of the coefficient of thermal conductivity. The main share of the market (~97%) of thermal insulation materials is occupied by inorganic materials - mineral wool insulation and polyfoams [1]. At present, thin-film coatings (energy-saving paints), consisting of hollow microspheres and bonding materials that have the properties of paints, but with a lowered value of the coefficient of thermal conductivity, have become widely used to save energy.

2 Statement of the Problem

The analysis of the scientific literature has shown that the known data on the consumer and thermophysical properties of energy-saving paints differ significantly from each other, that is why the study of the properties of thin-film coatings on the basis of hollow microspheres is an actual scientific task.

3 Experiment and processing

3.1 Investigation of thermophysical properties

* Corresponding author: buhmirov@tot.ispu.ru

This section presents the results of the experimental determination of the thermal conductivity coefficient of energy-saving paint as a function of temperature and mass content of MC-V2L hollow glass microspheres in a binder of styrene-acrylic dispersion “Acrylan 101” and experimental values of the thermal diffusivity of energy-saving paint, Produced at a chemical enterprise in the city of Ivanovo.

An estimate of the thermal conductivity was made experimentally for energy-saving paint samples with a mass content of microspheres of 8%, 25% and 32.6% in acrylic binder and for a sample of acrylic coating without addition of microspheres. The thickness of the test samples was: 1.5 mm for a binder with the addition of microspheres and 1 mm for pure acrylic. The coefficient of thermal conductivity of the samples was obtained on a laboratory bench by the method of a cylindrical layer in the steady-state regime of thermal conductivity. A series of experiments was performed at different values of the heat flux passing through the samples under study.

The coefficient of thermal diffusivity of energy-saving paint was found by the regular regime method using a modified air “α-calorimeter”, in which the test sample was asymmetrically heated longitudinally by the flow of hot air around it. The air flow velocity was chosen so that the condition of the thermally thick body Bi> 100 was satisfied for the heated sample. The test sample are made in the form of parallelepiped with dimensions of 45x60x75 mm, made of energy-saving paint with a mass content of microspheres of ~ 32.6% in a binder of acrylic. The sample, insulated with mineral wool on all sides except the working surface, was blown by a stream of hot air from the supercharger with a built-in heater. Using a supercharger control unit, it was possible to change the speed and temperature of the air flow in the interval of 0÷20 m/s and 20 ÷120°C, respectively. During the experiments, the speed and temperature of the air flow were measured using a MES-200A meteometer with a velocity measurement error of ± 1.5 m/s and a temperature of ±0.5°C. The temperature of the sample at three points (in the center, on the upper and lower faces) is determined using thermocouples of type “T” and a secondary device ADAM-4000 with an error of measurement of ±1°C. A series of experiments was performed at different air temperature values.

3.2 The study of energy efficiency

This section presents the results of a energy efficiency study of energy-saving paints used for the building facades insulation and as thermal insulation for pipelines with a hot coolant.

The study of the energy efficiency of energy-saving paints for the warming of the facades of buildings was carried out at a full-scale research site installed on the wall of a residential building. The landfill is pre-plastered with cement mortar 10 mm thick, a section of the house wall divided into zones of 3.9 m high and 2 m wide with various types of thermal insulation. To study the energy efficiency, the following types of coatings were used:

A) At the plot No. 1 of the polygon, the thermal insulation is made using the technology of a “wet” facade made of mineral wool (δ=100 mm, λ=0.038÷0,042 W/(m·K), ρ=131÷135 kg/m^3);

B) In plots 2 and 3, the thermal insulation is made by applying an energy-saving paint layer of 1 mm and 3 mm, respectively. To ensure the same thickness of the layers, the energy-saving paint was applied to the wall by the machine using a spray gun.

C) Plot No. 4 of the landfill was covered only with 10 mm thick cement plaster without applying energy-saving coatings.

During the experiments, the heat fluxes and temperatures on the surfaces of the experimental sections were measured by the ITP-MG4.03 / X (I) “Potok” instrument with a measurement error of the heat flux of ± 6% and a temperature of ± 0.2°C; Temperature between the layers of coatings and the temperature of the external surfaces of each experimental area measured by thermocouples type "T" with the secondary device ADAM-4000 which had an error of measurement of ± 1 ° C; The outside temperature with the help of the weather station Davis AdvantagePro2 with a measurement error of 0.5°C.

To investigate the effectiveness of energy-saving paints with the reducing heat losses from hot water pipelines, a series of experiments was carried out at a thermal station in which there were two water supply lines with a diameter of 219 mm and a length of 2300 mm are located. Due to the high water cut of the heat point, the service personnel use the pipelines as gangways to pass to the stop valves, which leads to mechanical damage to the thermal insulation and its waterproofing layer. Therefore, the thermal insulation gets wet, and high humidity leads to corrosion of the metal pipelines. To reduce corrosion of the metal and reduce the heat loss of the coolant, it was proposed to cover the pipelines with energy-saving paint with a thickness of δ = 2.5±0.2 mm. The efficiency of using energy-saving paint was determined by comparing the outer wall of pipelines temperature and heat flows from their surface "before" and "after" the application of a thin-coating coating. Between the measurements, the heat point and the painted pipelines were flooded with hot water due to an accident on the heating main.

4 Results and Discussion

As a result of the research:

1) The average value of the coefficient of thermal conductivity in the temperature range 20÷100°C was: for the binder (acrylic) - ~0.028 W/(m·K); At 8% content of microspheres in energy-saving paint by weight - ~0.025 W/(m·K); At 25% microsphere content - ~0.022 W/(m·K); At a 32.6% content of microspheres - ~0.019 W/(m·K). The coefficient of thermal conductivity of the energy-saving paint in the temperature range 20 ÷ 100°C with an error of no more than 12% can be approximated by the formula, W/(m·K):

$$\lambda = \frac{2.24}{10^4}T - \left(\frac{2.44}{10^6}T + \frac{1.27}{10^4}\right)C + \frac{1.61}{10^2} \tag{1}$$

where C is the mass concentration of microspheres,%; T is the temperature, °C.

2) The average thermal diffusivity of the energy-saving paint is in the range $2.7\cdot10^{-8}\div3.1\cdot10^{-8}$ m^2/s.

3) It is established that the thermal losses of the building during the application of heat insulation by the technology of a "wet" facade made of mineral wool slabs with a thickness of 100 mm are reduced by 62%, when the energy-saving paint is applied in layers of 1 mm and 3 mm, the heat losses are reduced by 8.3% and 11.8 % Respectively. At the same time, the thermal losses from the surface were taken as the base value of heat losses without applying energy-saving coatings (plot No. 4).

4) Data are obtained on the value of the average integral degree of blackness of the insulating materials being studied. In the temperature range -20 ÷ 0 ° C, the average integral degree of blackness in all thermal protection coatings studied is 0.96÷0.98.

5) The effect of hot water with a temperature of ~60°C for 5 days did not lead to the destruction of the layer of energy-saving paint, which indicates the resistance of this type of thin-film coating to the action of hot water with a temperature of ~60°C. It was experimentally established that after deposition of a thin-film coating, first, the heat losses decreased from the direct pipeline by 42.5%, and from the return pipeline by 63.4% and, secondly, the surface temperature of the direct pipeline decreased by 17°C, and the surface temperature of the return pipeline is 23.1°C.

5 Conclusions

1. The main thermophysical properties of thin-film energy-saving coatings based on hollow microspheres are determined. The obtained data can be used in engineering calculations and mathematical modeling of heat transfer processes.

2. At values of thermal conductivity coefficient for energy-saving paints and mineral wool insulation of the same order to achieve the same energy-saving effect, the thickness of energy-saving paint and mineral wool insulation should be comparable, which is economically unprofitable. However, in conditions of high humidity, the thermal insulation properties of mineral-wool heat-insulating materials deteriorate sharply and the use of thin-film coatings from energy-saving paint becomes an economically viable energy-saving measure.

References

1. O.A. Ignatova, *Technology of insulating building materials and products. Part 2. Heat- and hydro-insulating materials and products: training.* (Publishing Center Academy, Moscow, 2012) [In Russian]

Experimental study of liquid evaporation rate from coniferous biomass

E.E. Bulba[1,*], *N.A.* Ivanova[1]

[1]National Research Tomsk Polytechnic University, 634050 Tomsk, Russia

Abstract. The results of experimental studies of moisture evaporation from coniferous wood (spruce, pine) are presented. The dependences of the mass evaporation rate on temperature and time are obtained. The calculation of the accommodation coefficient for the corresponding temperature ranges has been performed. The analysis of temperature regimes of drying of two typical coniferous wood species is carried out.

1 Introduction

Many regions of the planet have an important ecological significance for the world community, since they have a large reserve of renewable forest resources and, as a consequence, a sufficient amount of biomass that can be used as fuel for the generation of thermal and electrical energy. Because of the processing of wood, a large amount of waste appears. Since the forest residues are not practically used, their disposal is associated with high financial costs. Despite these problems, it is well known that waste wood processing can be used to generate thermal and electrical energy [1].

Biomass is a promising energy carrier. Its main advantages - are renewability; ecological cleanliness, in comparison with other types of fuel; neutrality with respect to greenhouse gas emissions [1].

To date, most developed countries are seeking to develop alternative energy [2,3], since classical energy sources are exhaustible. Bioenergetics has more advantages, as biomass is used as a starting material [1], the cost of which is minimal.

However, the use of woody biomass in power engineering is complicated because of the large, as a rule, moisture content of wood. The drying processes of wood are long and energy-intensive. Therefore, the analysis of the regularities of moisture removal processes is one of the main tasks, the solution of which is necessary in the development of technologies for burning woody biomass in the furnaces of steam and hot water boilers.

The removal of moisture from wood under conditions of intense heating is a very complex process, the description of which has not yet been mathematically determined at the level sufficient for carrying out experimental-design work. Even relatively simple processes of evaporation of water droplets at high temperatures have only recently been investigated experimentally [4,5]. The theory of phase transformations (evaporation,

* Corresponding author: bulba@tpu.ru

condensation, boiling) does not even have water droplets at high temperatures. Therefore, the aim of the work is an experimental study of the processes of moisture removal from coniferous wood samples and the determination of the numerical values of the process characteristics necessary for its description in mathematical modeling.

2 Experiment Procedure

The experiment was carried out in a drying chamber, type SU 32. The chamber has a corrosion-resistant stainless steel surface. The working chamber was heated by wire heaters located around it. Such heating panels have good heat transfer and provide uniform heating of working chamber surface. Drying of materials was carried out on two removable shelves.

As experimental samples, biomass of two types (spruce, pine) was used.

The experiment was carried out in several stages:

- Preparation of a sample of wood weighing 100 grams;
- weighing of the container used for drying, using laboratory scales;
- filling the container with the sample of biomass;
- switching on the drying chamber;
- the required temperature in the drying chamber was set with electric contact thermometer;
- the container with biomass was placed in the drying chamber with time interval (10 minutes).
- after the necessary time elapsed, the capacity was pulled out and its mass was measured;
- processing of experimental data;
- analysis of the results.

Each experiment under fixed conditions was carried out 3 times for the chosen temperature range in the furnace in the same time interval.

3 Results and discussion

In the course of the experiments, the rates of evaporation of two coniferous species of wood were determined as functions of the drying time and temperature.

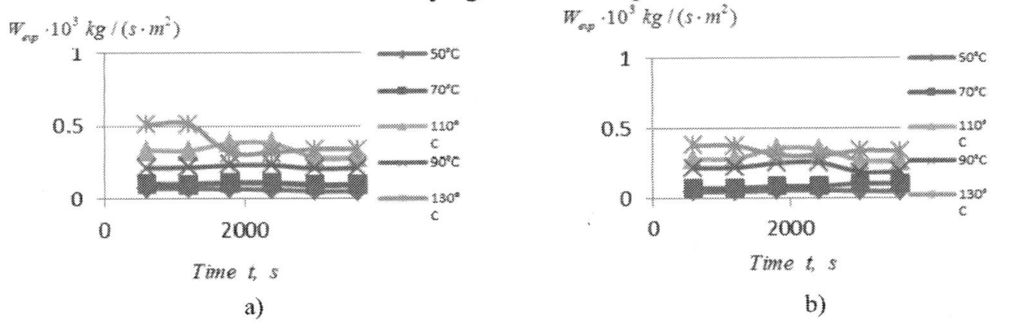

Fig. 1. Dependences of the mass evaporation rate on time (a - spruce, b - pine).

As can be seen from the obtained of dependences, the evaporation of the liquid is stationary in the temperature range 323-363 K for spruce and 323-343 K for pine. This is because at such temperatures for 3600 seconds, the removal of bound moisture does not begin. At temperatures of 383-403 K for spruce and 343-403 for pine, the rate of evaporation decreases after twenty minutes of heating-the first drying period. It corresponds to the step of removing free moisture that covers the pore surface of the material. After this stage, the binding moisture begins to be removed. The increase in the rate of evaporation with the onset of the second period is explained by the fact that the free and bound moisture evaporates, which increases the overall rate of moisture removal.

 DOI: 10.1051/matecconf/201711001020

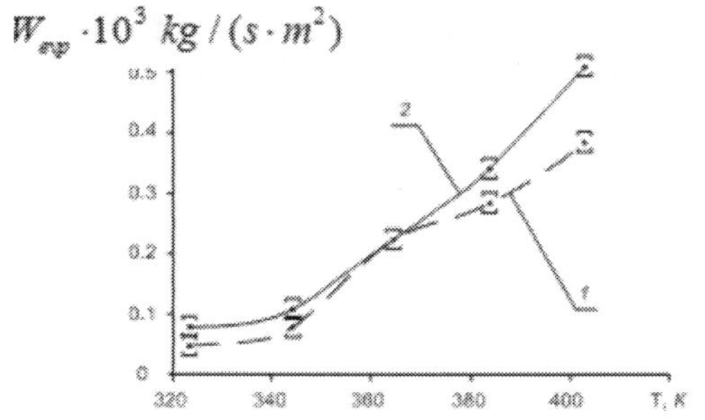

Fig. 2. Dependence of the mass evaporation rate on temperature, at t = 1200 seconds (1 - pine, 2 - spruce).

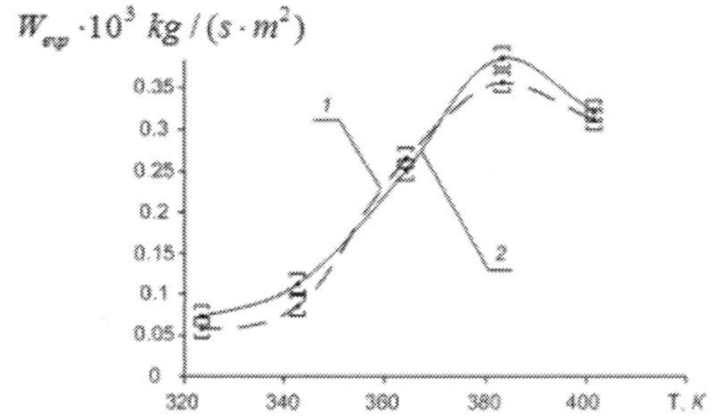

Fig. 3. Dependence of the mass evaporation rate on temperature, at t = 2400 seconds (1 - pine, 2 - spruce).

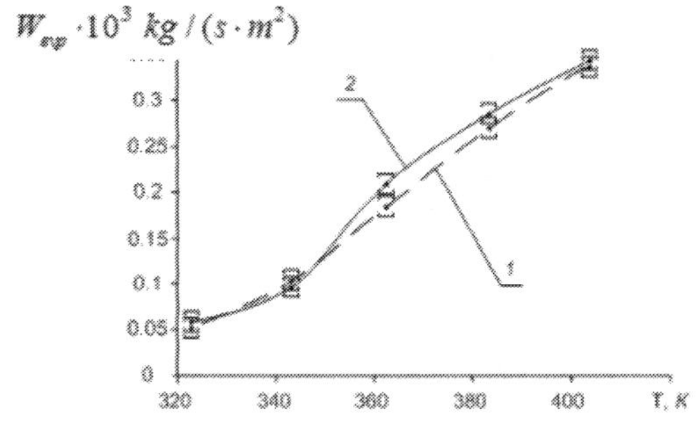

Fig. 4. Dependence of the mass evaporation rate on temperature, at t = 3600 seconds (1 - pine, 2 - spruce).

Analysis of the obtained dependences showed that the rate of moisture removal from the spruce is higher than the rate of moisture removal from pine by an average of 30%. This is explained by the fact that the density of spruce is lower than the density of pine by 10%. From which it follows that the drying of wood spruce is carried out at a lower energy cost. The temperature at which the highest evaporation rate is reached is 403 K.

For the mathematical description of evaporation and condensation processes, different approaches are currently used [6, 7]. The most common is the use of the hypothesis of the possibility of applying the Hertz-Knudsen law obtained for evaporation to vacuum for any density of the medium near the evaporation surface. The coefficient of agreement between the model and the real process is the accommodation coefficient, determination of which is quite a difficult task in any conditions.

The velocity with which gas molecules hit the surface can be calculated on the basis of the kinetic theory from the Hertz-Knudsen equation:

$$W_{evp} = \frac{A \cdot (P^S - P^*)}{\sqrt{\frac{2\pi RT}{M}}}; \qquad (3)$$

where W_{evp} - the mass evaporation rate, kg / m2 · s;

P^S - saturated vapor pressure, MPa;

P * - partial pressure, MPa;

R = 8.31 - the universal gas constant, J / mol · K;

M - molecular weight;

T - the temperature of the substance, K;

A - accommodation coefficient.

Using the obtained experimental data and the Hertz-Knudsen law, the accommodation coefficients for moisture removal from biomass were calculated in the temperature range from 323 K to 403 K in the same way as in [8,9].

Tables 1, 2 show the calculated values of the accommodation coefficient in the temperature range from 323 K to 403 K for spruce and pine.

Table 1. Coefficient of accommodation of spruce drying process.

T, K	$W_{evp}\cdot 10^3$, kg/(m²·s)	P^*, Pa	P^s, Pa	$M\cdot 10^3$, kg/mol	R, J / mol · K	Humidity,φ,%	Accommodation coefficient $\cdot 10^6$
323	0.066	6170	12344	18	8.31	50	10.3
343	0.103	15590	31176	18	8.31	50	6.6
363	0.222	35060	70117	18	8.31	50	6.5
383	0.335	71620	143240	18	8.31	50	4.9
403	0.391	135010	270020	18	8.31	50	3.1

Table 2. Coefficient of accommodation of pine drying process.

T, K	$W_{evp}\cdot 10^3$, kg/(m²·s)	P^*, Pa	P^s, Pa	$M\cdot 10^3$, kg/mol	R, J / mol · K	Humidity,φ,%	Accommodation coefficient $\cdot 10^6$
323	0.051	6170	12344	18	8.31	50	8
343	0.087	15590	31176	18	8.31	50	5.6
363	0.22	35060	70117	18	8.31	50	6.4
383	0.301	71620	14324	18	8.31	50	4.4
403	0.34	135010	27002	18	8.31	50	2.7

4 Conclusion

In the course of the experimental studies, the following parameters were determined: the mass evaporation rates of liquid from coniferous wood, the dependence of the mass evaporation rate on temperature was obtained, the accommodation coefficient for the respective temperature ranges was calculated. The temperature regimes of drying two typical coniferous species are analyzed.

The results of the experimental studies carried out are the basis for further refinement of the mechanism of moisture removal from the porous structure of the moisture-containing material and development of the general theory of drying. The dependence of the mass evaporation rate of the wood samples under study on temperature makes it possible to determine the accommodation coefficient in the mathematical expression of the Hertz-Knudsen law.

References

1. Y.P. Semenov, *Forest bioenergetics* (Tutorial), (Publishing house of Moscow State University of Forestry, Moscow, 2008) [In Russian]
2. M.V.Vasilevsky, E.G.Zykov, A.S.Razva, Theoretical Foundations of Chemical Engineering, **45**, 304 (2011)
3. A. Razva, M. Vasilevsky, V. Rykov, MATEC Web Conf. **23**, p. 01035(2015)
4. G.V. Kuznetsov, P.A. Strizhak, Technical Physics Letters, **40**, 499-502 (2014)
5. R.S. Volkov, G.V. Kuznetsov, P.A. Strizhak, Technical Physics, **59,** 959-967 (2014)
6. G.V. Kuznetsov, P.A. Kuybin, P.A. Strizhak, High Temperature, **53**, 254-258 (2015)
7. G.V. Kuznetsov, P.A. Strizhak, High Temperature, **52**, 568-575 (2014)
8. E.E. Bulba, A. A. Malinovsky, MATEC Web Conf. **23**, 01053 (2015)
9. E.E. Bulba, EPJ Web Conf. **82,** 01024 (2015)

The Implementation of the Feedback Principle in Science and Technics

Anna L. Buran[1,*] , *Vladislav* E. Kozlov[1], and *Andrey* V. Filyukov[1]

[1]National Research Tomsk Polytechnic University, 634050 Tomsk, Russia

Abstract. In this study positive and negative influence of the feedback have been examined. A feedback principle in the alternating current amplifier and in the speed regulator of the turbine rotation was described. The main conclusion of the positive and negative effect was obtained. Negative feedback worsens the properties of an object, reducing a strengthening factor, delaying regulatory action. Positive feedback considerably raises device work stability. This research will enable to identify the importance and effectiveness of the feedback principle in science and technics.

1 Introduction

Scientists and researchers refer a feedback principle to one of the fundamental and universal principles. Some scientists present robust overlapping load frequency output feedback control of multi-area interconnected power systems [1], others develop an integrated guidance and feedback control scheme for steering an underactuated vehicle through desired waypoints in three-dimensional space [2] and investigate active disturbance attenuation control for permanent magnet synchronous motor via feedback domination and disturbance observer [3]. The aim of this work is to study the effect of feedback principle in science and technics, using the examples of various technical devices, such as an operational amplifier and a regulator of turbine speed. It is assumed that no system exists without feedback. The system, which doesn't have feedback degrades and collapses. Feedback can be positive and negative, in both cases a feedback problem is to return the system to an optimum condition. To solve a problem with control means to support an operating object into a desirable condition. The analysis of control processes in various areas, allows to establish that control, irrespective of kinds of objects, is always carried out according to a feedback principle. There can be control processes in technics, in society and in nature and in education. Some researchers refer to potential of case technology in technics [4], others prove that control of different objects is carried out according to one principle, that's why all control systems must have common features.

2 A feedback principle in the alternating current amplifier

* Corresponding author: aburan@tpu.ru

Regarding the above confirmed theory, the characteristics of an operational amplifier should be studied carefully. At an amplifier output the divider of voltage R1–R2 is put, and the part of target voltage moves back to an amplifier input. There is an input voltage u_{input} at an amplifier input, which is necessary to enhance, and feedback voltage U f.b. If these types of voltage are added up, feedback is called positive if they are subtracted, it's negative.

For the purpose of this research, we should know about negative feedback. At an amplifier input the difference of voltage u_{input} – uo.c is put. This difference amplifies in Ko times. The achieved result is an amplifier target voltage.

The results indicated that negative feedback lowers proportional-action coefficient. It will be provided after considering a numerical example. A proportional-action coefficient without feedback K_o is 100 and the feedback factor β =50. Then to the amplifier with feedback we will get: K=100/1+0.5·100=100/51≈1.96

The strengthening factor has decreased exactly in 51 times. There is a reasonable question: what such feedback, which worsens amplifier characteristics is necessary for. But, losing in strengthening, we can get another advantage. It's negative feedback, which raises the stability of amplifier work. Regarding this issue we assume that for any reasons the factor of strengthening K_o has increased by 20 % and K_o=120, instead of 100, as we thought. The results indicated that the amplifier without feedback output voltage will also increase by 20 %. Now we will see what will happen, if the amplifier is captured by negative feedback: K=120/1+0.5·120=120/61=1.97. We achieved the result that gain constant has changed very little.

So, negative feedback reduces amplifier coefficient, but it increases stability of amplifier work. We will come back to strengthening factor with negative feedback: $K=K_o/1+\beta{\cdot}K_o$. At that time we assumed that product $\beta{\cdot}K_o$ is much more than one. Then it is possible to consider approximately: $K=K_o/\beta{\cdot}K_o=1/\beta$.

We see that proportional-action coefficient K_o is absent in our formula. It follows that if values β and K_o are chosen correctly, changes of transistor parameters or other elements of the amplifier don't influence on the strengthening value. Work stability of the scheme depends only on the feedback circuit stability. In feedback circuits there are no transistors or any other elements, which change their parameters considerably. That's why the factor β can be supported with great accuracy and consequently the whole amplifier will work precisely.

There are also other important advantages which have the amplifiers with negative feedback. One of them influences on the size of input resistance of the amplifier. Good amplifier should have high resistance. Such amplifier allows to intensify lower power signals very much and doesn't deform operating mode of electronic devices.

So negative feedback increases the amplifier input resistance at $(1+\beta{\cdot}K_o)$ times. These are not the only advantages of negative feedback. It's found that in the amplifiers captured by the feedback input resistance reduces with $(1+\beta{\cdot}K_o)$ time. The amplifier with small input resistance can work at any loading. Thus, the size of load resistance doesn't render the influence on amplifier work.

3 A feedback principle in speed regulator of the turbine rotation

The energy resources are frequently wasted in Russia [5]. Regarding this issue, we will find out the technology of getting electrical energy at the Russian thermal power station using the coal. Large pieces of coal are transformed into coal dust. This dust is blown by a stream of hot air into a furnace through special burners. The flame heats up water in thin tubes which cover furnaces' walls and the heated top internal gases volatilize through the chimney. At first they meet water tubes, transforming warm water into steam. Then gases are supplied to a special device – economizer, warming cold water in it and refilling water in the furnace, and, at last, – in air heater. The fuel burns, having transferred heat to water. The water in a furnace has been heated up and has vaporized, but this steam is not hot enough, cooling down, it will turn to water drops quickly. That's why, first of all, steam gets into superheater coil pipes where steam is heated up to very high temperature. This steam goes into turbines through steam pipes. Fresh "sharp" steam from a furnace is under the pressure of 150–200 atm. It passes through stationary blades. Shovels of the first row of a rotor start moving. Then steam gets to channels of the first diaphragm. And again the part of its pressure transforms into speed, and this speed works well together on the second blade tip. This repeats 10–12 times till steam pressure becomes moderate. So, initial pressure is used up smoothly and sparingly.

After that the steam goes to the condenser where cold water circulates continuously through thin brass tubes. It cools steam and transforms it into water, which is called condensate. A lot of water is provided by the circulating pump. So, steam is formed, having turned to condensate – very pure water which does not contain harmful chemical or mechanical impurities. Such clear water comes back from the condenser into a copper with the help of special pumps. The results indicated that the powerful steam turbine has a speed of 3000 rpm and it rotates a rotor of the electric generator and due to it in a winding stator the generator (a motionless part) according to the law of electromagnetic induction forms the alternate current with a frequency of 50 Hz (3000 rpm equals 50 about/c) and pressure from 6 to 20 sq. Regarding this issue, it can be concluded from research that electric energy is the main "product" of a power station.

Regarding the process of frequency regulation, it's assumed that speed regulators consist of the following basic parts: instrument movement centrifugal pendulum, having a role of instrument movement; a sliding piston; the oil engine and the wave of rotary axis speed.

The speed regulator works in the following way: pendulum spheres move to each other after frequency falling, that's why the instrument movement descents and occupies position A*. As B and G are fixed, so the lever arm A–B rotates about point B and lever arm G–E rotates about point G. Consequently, buckets of sliding piston go down and open the windows, following which pressure oil comes to lower bucket separation of oil engine and it goes up. This bucket is connected to valves. When the engine bucket moves up, admission of propellant factor and rotor speed is rising, so point A of the instrument movement goes up too.

Simultaneously with moving of the oil-engine piston lever arm A–B turns round the point A and lifts up hinged points B, D, E and sliding piston together. This action is called as disengagement. Due to this, slide windows close earlier than rising of propellant factor inlet attacks. If there were no disengagement, sliding piston would rotate around the neutral location and value of rotor speed would change periodically. Since control process ends, when points B, D, E return to starting point, so the new location of the oil engine bucket fills with a new point A position. The more inlet of the moving factor we have the lower is the new position of the point A, the less is turbine speed rotation.

The results indicated that, the peculiarity of regulators of turbine frequency rotation is the necessity of maintenance of rather considerable efforts for changing the position of steam turbines regulating valves or the hydraulic turbines directing device. And moving on these regulating elements should be carried out with great accuracy. Now the hydraulic engines often named in the systems of regulation as servo motors are used everywhere to solve the problem mentioned above. For the elements which help form the certain control mechanical, hydraulic and electric elements can be used. For example, we take the hydraulic servomotor operated by valve. Oil is brought in a valve's axle box under the pressure of P_H. In a neutral starting position valve's pistons block the channels connecting valve's axle box with the top and bottom cavities of the hydraulic engine. These cavities are filled with oil. At displacement of valve's pistons, for example downwards, the bottom cavity of the engine cylinder connects with a pressure line (pressure PH), and the top one with a line of drain (pressure PC). The engine piston will move upwards until valve's pistons don't have neutral position or till the engine piston doesn't get its ultimate position. For replacement of the hydraulic engine by a link in the block diagram of the regulatory system and giving a definition of its transfer function, we will take the management connecting the expense of oil and moving of the engine piston during a short period of time Δt: $S\Delta\mu = nK_v(P_I - P_N)\Delta t$ where $\Delta\mu$ – moving of the piston of the engine, n – open section of valve's windows, Kv between the pressure and speed of oil flowing, S – the area of the engine piston, accepted for simplification of the identical one from the top and bottom cavities. For a differential short period of time: $S{\cdot}d\mu/dt{=}n{\cdot}K_v\ (P_n{-}P_c)$We will come to relative units in (1), having accepted as basic values: μmax – the maximum working piston stroke of the engine, nmax – the area of cross-section section of completely open valve's windows. Then $T_d{\cdot}(d\mu^*/dt) = n^*(2)$, where $T_d{=}S\ {\cdot}\mu_{max}/n_{max}{\cdot}K_v\ {\cdot}\ (P_n{-}P_c)$ – the time constant of a hydraulic engine. During T_d the engine piston moves from one ultimate position to another with fully opened valve's windows. According to the transferring function of the hydraulic engine is W_d (p) $=1/T_d{*}p$ – that means the hydraulic engine can be presented as an integrating link. The time constant of T_d defines the speed of the regulatory system. The decrease at (T_a) action of a regulator towards power decrease is especially important. For example, it is required for steam turbines, tend towards closing to be about 0.2 seconds. For getting such short amount of time it is required the pump to be very productive to deliver liquid into a servo motor. As the great speed of the piston movement is reached mainly by closing of valves it is practically provided, not with the pump productivity increases, but using a unilateral servo motor which has oil pressure only while opening, and by force of a spring when closes.

Deng, W. &Yao, J. analyze in their research adaptive integral robust control and application to electromechanical servo systems, giving experimental results, proved the of the effectiveness of the proposed controllers [6].

Yang, WH. considers that computer simulation technology can play an important role in the teaching reform of power system protective relaying course [7].

Depending on particular conditions and requirements of rotation frequency regulator a servo motor can be supplemented with rigid, flexible or combined feedback. Feedback can be executed with the help of different devices. As the ways of technical realization of feedback aren't principal while taking into consideration the processes of frequency and capacity regulation, we will take only the influence of kinds and parameters of feedback elements on hydraulic servo motor transferring function [8].

Regarding the servo motor with rigid negative feedback it can be concluded that at any transferring function of feedback link is [W_d (p) $=1/T_d{*}$p$+W_{oc}$ (p)]. Transferring function of a feedback link is [W_{oc}(p) $=K_{oc}$], and transferring function of a servo motor with such feedback is [W_d (p) $=1/K_{oc}$ * (T_d/K_{oc}*p+1)]. With rigid negative feedback the servo motor is an inertial link of the first order with time constant: [T_d/K_{oc}] and strengthening factor: [$K{=}1/K_{oc}$].

4 Conclusion

The paper has described the importance of the feedback principle in science and technics. Giving examples of two different technical devices the basic regularities for feedback have been revealed. Negative feedback worsens the properties of an object, reducing a strengthening factor, delaying regulatory action a bit. Positive feedback considerably raises device work stability. This research will enable to identify the importance and effectiveness of the feedback principle.

References

1. A. Ahmadi, M. Aldeen, Int. J. Electrical Power Energy Systems, **89**, 156 (2017)
2. S.P. Viswanathan, A.K. Sanyal, E. Samiei, J. Intell. Rob. Systems: Theory Appl. (2017)
3. X. Niu, C. Zhang, H. Li, IET Control Theory and Applications, **11** (6), 807 (2017)
4. Z. Fedorinova, V. Vorobeva, M. Malyanova, XV International Conference Linguistic And Cultural Studies: Traditions And Innovations. Procedia - Social and Behavioral Sciences, **206,** 247 (2015)
5. A. Buran, G. Nizkodubov, I. Pryahin, MATEC Web Conf. **91,** 01054 (2016)
6. W. Deng, J.Yao, ISA Transactions **67,** 256 (2017)
7. WH.Yang, 2016 International Symposium On Computer, Consumer And Control (Is3c), (2016)
8. S.A. Stavitsky, N.E. Palukhin, Y.V. Kobenko, E.S. Ryabova, EPJ Web Conf. **110**, 01059 (2016)

Experimental study of the boron-containing mixtures burning rate

Yana Dubkova[1,*], *Alexander* Zhukov[1], *Ilya* Zhukov[1, 2], and *Vladimir* Promakhov[2]

[1]National Research Tomsk State University, 634050 Tomsk, Russia
[2]Institute for Problems of Chemical and Energetic Technologies, 659300 Biysk, Russia

Abstract. The results of the experimental determination of the burning rate of boron-containing fuel mixtures in the pressure range (0.1 ÷ 6) MPa are presented. The results of a comparative analysis of the burning rate of samples with component different compositions are given.

1 Introduction

The design and development of advanced solid propellant systems used in the propulsion plants for various purposes, requires a detailed study of the basic properties of the new compositions. In particular, the use of a new generation of solid-fuel systems can provide increased specific impulse thrust propulsion to (15 ÷ 20) %. One of the main energy characteristics of fuel systems, along with the heat of combustion, heat capacity and thermal conductivity, is the burning rate [1, 2]. Despite the development of technologies for obtaining new materials currently, there are no experimental data on the basic functional properties of the new fuel systems depending on external dynamic and thermal effects. In this regard, actually creating a database on the properties of the new solid systems on the basis of experimental studies simulating real nature conditions [3].

This paper presents the results of experimental determination of the burning rate of the samples of the two groups of boron-containing fuel mixtures based on titanium, depending on external conditions.

2 Materials and methods of the experimental research

To determine the burning rate experimentally investigated two groups of boron-containing fuel mixtures – P1 and P2.

Examples of group P1 were made from a stoichiometric mixture of powders of titanium and boron amorphous (69 wt. % Ti+31 wt. % B) with inserted high-energy additives – fine aluminium powder ($d_p \sim 100$ μm). The data of raster electron microscopy (REM-images) of the original powders is shown in Fig. 1.

* Corresponding author: y.a.dubkova@niipmm.tsu.ru

DOI: 10.1051/matecconf/201711001022

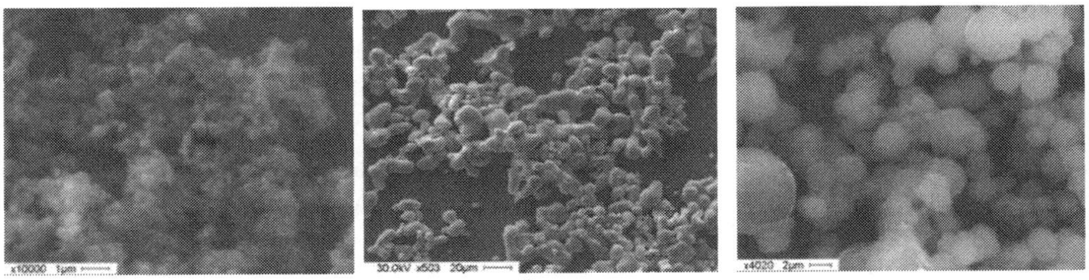

a b c

Fig. 1. REM-images of amorphous boron powder (a), of titanium powder type PTOM, 0 – 40 μm (b) and of aluminium powder type ASD.

The content of the input aluminum powder, the minimum required to ensure sustainable combustion, was determined by the results of preliminary experimental studies [4] and was (1.0 ÷ 5.0) wt. % in excess of the mass of a stoichiometric mixture of Ti + B. The tests were subjected to cylindrical specimens made by compacting mixtures of the initial powders. As a binder used was ultra-high molecular weight polyethylene (UHMWPE) [5] in the solid state in the form of spherical granules with a size of from 40 to 80 microns and polyvinyl alcohol (PVA) $(C_2H_4O)_x$ [6] in the form of a 3 % aqueous solution.

Samples of group P2 were made by through passage pressing the mixture of powders with different content of oxidizer, metal fuel, catalysts, technological additives with a polymeric fuel-binder. The components of the compositions of boron-containing fuel mixtures of group P2 are given in Table. 1.

Table 1. The components of the compositions of boron-containing fuel mixtures of group P2.

Components, wt. %	The fuel mixtures				
	1	2	3	4	5
SKDM-80	14.8	13.6	11.0	11.8	16.0
Ammonium perchlorate	70.2	71.4	74.0	73.2	69.0
TiB_2	15.0	–	–	–	–
AlB_2	–	15.0	–	–	–
B	–	–	15.0	–	–
AlB_{12}	–	–	–	15.0	–
ASD-4	–	–	–	–	15.0

Samples of boron-containing fuel mixtures of groups P1 and P2 were burned in the chamber at a constant pressure (Crawford Bomb) in an inert environment (argon). A series of experiments conducted while changing the pressure in the chamber in the range (0.1 ÷ 6) MPa.

The burning rate of samples of boron-containing fuel mixtures of groups P1 and P2 were determined using high-speed digital video camera. Obtained in the combustion process of samples of the video files were processed using special software processing of video files by the "TimeLapse". The video frames of the typical experiments for samples of fuel mixtures of groups P1 and P2 are given in Fig. 2, 3.

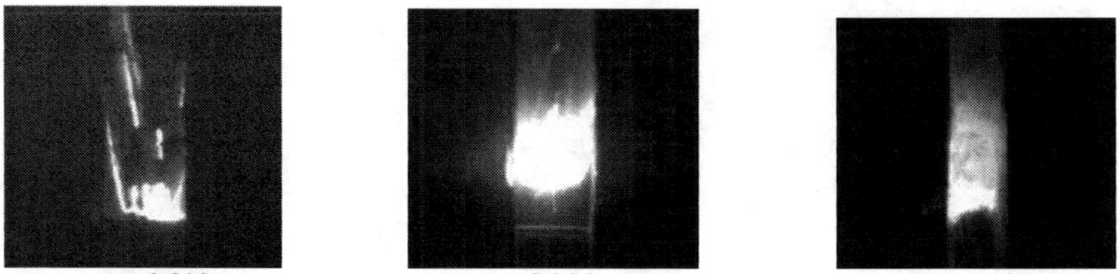

t = 0.010 s a t = 0.055 s b t = 0.108 s c

Fig. 2. Video frames of the combustion process of the sample of group P1.

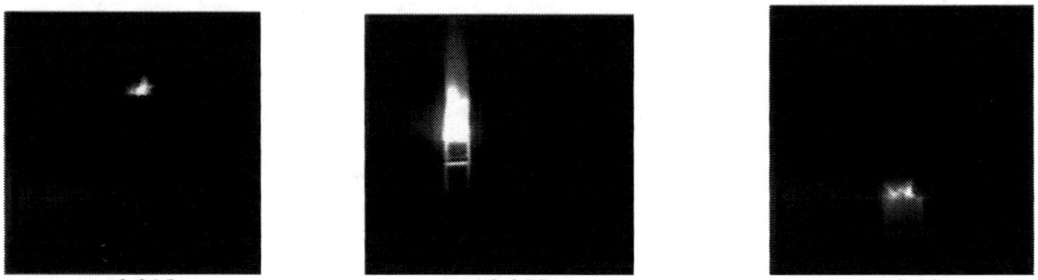

t = 0.015 s a t = 0.055 s b t = 0.18 s c

Fig. 3. Video frames of the combustion process of the sample of group P2.

The resulted frames illustrate the characteristic moments of the process, corresponding to the instant of ignition (Fig. 2, 3, a), to burning of the samples (Fig. 2, 3, b) and to extinction of the fuel mixtures (Fig. 2, 3, c).

For each composition were conducted at least three duplicate experiments.

The typical data of measurements of the burning rate of samples of group P2 in argon pressure 2 MPa, obtained on the basis of a series of experiments, are given in Table. 2.

Table 2. The data of measurements of the burning rate of samples of group P2.

Experiment number	**The burning rate, mm/s**				
	1	2	3	4	5
1	10.0 ± 0.2	24.7 ±0.2	13.0 ± 0.2	30.0 ± 0.2	7.1 ± 0.2
2	10.0 ±0.2	24.4 ±0.2	16.9 ± 0.2	26.8 ± 0.2	7.4 ± 0.2
3	10.2 ±0.2	24.5 ±0.2	20.1 ± 0.2	26.0 ± 0.2	7.0 ± 0.2

3 Results of experimental study

The visualization of the combustion process showed that the samples of boron-containing fuel mixtures with component different compositions are burning steadily in the pressure range (0.1 ÷ 6) MPa with the intense glow of the flame.

1. The burning rate of the samples of fuel mixtures of both groups in the pressure range under consideration does not depend on the pressure.
2. The burning rate of the samples of group P1 was (100 ÷ 120) mm/s in the pressure range under consideration when aluminum powder content in the fuel mixture was at least 2.5 wt. %.
3. The burning rate of samples of group P2 in the pressure range under consideration varied within (6 ÷ 30) mm/s depending on the material composition of the fast-burning component in the fuel mixture (Table 1).

4. It is obtained that the maximum of the burning rate of samples of group P2 is reached in the pressure range under consideration when the fuel mixture contains amorphous boron or aluminum polyboride (Table 2).

4 Conclusions

Burning rate of the two groups of fuel compositions was obtained experimentally. Maximum burning rate for samples of group P2 was observed on compositions containing amorphous boron. Burning rate of samples of group P1 amount to (100 ÷ 120) mm/s and in the considered range of pressures (0.1 ÷ 6) MPa does not depend on external conditions. It can be concluded that the powder composition of group P1 has a higher burning rate and can be used to reinforce fuel pellets. Powder compositions of group P2 have high performance, and the burning rate does not depend on pressure, which gives reason to talk about the possibility of their use as high-energy materials.

The work was financially supported by the RFBR, research project No. 16-38-60028 mol_a_dk and with the financial support of the grant of the President of MK-837.2017.8 agreement N 14.Y30.17.837-MK.

References

1. H. Yaman, V. Celik, E. Degirmenci, Fuel, **115** (2014)
2. A. Dey, RSC Advances, **5** (2015)
3. N. Kubota, *Propellants and explosives: thermochemical aspects of combustion* (John Wiley & Sons, London, 2015).
4. Y. Dubkova, A. Zhukov, MATEC Web Conf. **92**, 01068 (2017)
5. S.S. Gabaliev, R.Z. Hairulin, V.P. Arhireev, Journal of Kazan University of Technology, **50** (2008)
6. I.P. Losev, E.B. Trostanskaya, *The Chemistry synthetic polymer* (Chemistry, Moscow, 1971) [In Russian]

Analysis of the Irtysh surface water as natural sources of water supply and steam production in the heat and power industry

Galina I. Egorova[1], *Andrey* N. Egorov[1], *Yuriy* V. Kobenko[2,*], *Tatiana* A. Martseva[2], and *Elena* S. Riabova[3]

[1]Tyumen Industrial University, 626150 Tobolsk, Russia
[2]National Research Tomsk Polytechnic University, 634050 Tomsk, Russia
[3]Samara University, 443086 Samara, Russia

Abstract. The article demonstrates the results of experimental analysis of the Irtysh and Tobol underground and surface water as natural water supply sources and steam production in the heat and power industry; priority locations for water sampling have been identified and substantiated; a qualitative analysis of the Irtysh and Tobol samples has been proved; the reasons for the deviation from the water quality and the required degree of treatment for bringing it to the required indicies are revealed; the mineral composition of water is determined taking into account the peculiarities of the hydrological layers through which the river flows; it is proved that the organic composition is formed due to compounds that are washed out of the reservoir's soil as a result of biological processes; the mineral and organic substances that are brought into the Irtysh, Tobol by waste waters of populated areas and industrial enterprises are revealed; it is proved that the water of the Irtysh River is fresh, belongs to hydrocarbonate class and is suitable for steam production in the heat and power industry.

1 Introduction

The aim of the research is to conduct a quantitative and qualitative analysis of underground and surface water samples of the Irtysh and Tobol river as natural sources of water supply and steam production in the heat and power industry. The methods of the research are chemical and microbiological analysis.

Nowadays water purification is the most important task of water treatment engineers at thermal power plants. For long-term and trouble-free operation of thermal and power equipment, it is necessary to purify water from suspended solids and aggressive factors that cause metal corrosion. In order to complete this task heat power companies have a chemical water treatment section, the equipment of which depends on the source water quality and the purpose of purified water consumption [1].

* Corresponding author: serpentis@list.ru

DOI: 10.1051/matecconf/201711001023

Tobolsk CHP was designed and built to meet the needs of the Tobolsk industrial site in the live steam and electric power, and also to provide the city of Tobolsk with thermal energy [3].

In the heat power industry the raw material for steam production and the coolant is water. After proper processing, it is used for the following purposes: a) as a source material for steam production in boilers, steam generators, boiling water reactors, evaporators, steam converters; b) for the condensation of spent vapor in steam turbines; c) in order to cool various items and systems at TPP and NPS; d) as a coolant in the primary circuit of NPS and WWER, in heat networks and hot water systems. Unique properties and water indices are known (T_{cr}=100.0°C, T_m=0.0°C, T_b=374.12°C, p_{cr}=221.2 bar, q_{cr}=317.8 kg/m^3). In comparison with all the liquid and solid bodies, water has the greatest heat capacity. Having heated up during the summer period, the water cools down in winter, rejecting the heat to the atmosphere. Water has the greatest surface tension in comparison with all known liquids, with the exception of mercury, which is of great importance for the life of the plant world. Water surface tension and density determine the height at which it can rise in the capillary systems. This property provides the water rise in plant stems and tree trunks [4, 5].

It is important to highlight that one of the most important physicochemical properties of water is its ability to form solutions. Water is the best solvent, incomparable with any other liquid, it dissolves solid compounds, liquids and gases. Water is a reactive substance. Oxides of many metals and metalloids combine with water, forming bases and acids; a large number of salts form crystalline hydrates with water; the active metals react with water to liberate hydrogen. When water interacts with salts of weak acids or weak bases, hydrolysis of these salts takes place, that is, there is an interaction of salt ions with water ions [6].

The listed water characteristics and, first of all, its exclusive ability to dissolve various substances have caused the presence of several natural waters on the Earth, which often stand in marked contrast with each other in their qualitative characteristics. We have developed a program for monitoring the quality of underground and natural water of the river Irtysh, Tobol. Priority sampling points in the area of Tobolsk city and its district are the following: the ferry opposite the village of Bekerevo, village Abalak, settlement Sumkino. The periodicity of analyses was carried out on a quarterly basis [7]. The analysis of the samples quality was carried out according to the current strategies and methods of analysis (SanPiN 2.1.4.1074-01 "Hygienic requirements for water quality of centralized drinking water supply systems, GN 2.1.5.1315-03 "Maximum permissible concentration of chemicals", etc.). The qualitative characteristics of the water samples of the Irtysh and Tobol river were also compared with GOST (All-Union State Standard) 2761-84. The composition of surface water sources met the following requirements:

1) dry residue not more than 1000 mg/dm^3 (according to the sanitary-epidemiological service it is allowed up to 1500 mg/dm^3);

2) the concentration of chlorides and sulfates is not more than 350 and 500 mg/dm^3, respectively, the total hardness is not more than 7 mmol/m^3 (according to the sanitary-epidemiological service it is allowed up to 10 mmol/m^3),

3) concentration of chemicals (Table 1) should not exceed the maximum allowable concentration for household, cultural and social water use, as well as radiation safety standards approved by the Ministry of Health.

Table 1. Quality indicators of underground water supply sources according to GOST (classes).

Indicators	1	2	3
Turbidity, mg/dm^3, not more than	1.5	1.5	10.0
Colour, degree, not more than	20	20	50
Hydrogen index (pH)	6-9	6-9	6-9
Iron, (Fe), mg/dm^3, not more than	0.3	10	20

Manganese (Mn), mg/dm^3, not more than	0.1	1	2
Hydrogen sulphide (H_2S), mg/dm, not more than	Absence	3	10
Fluorine (F), mg/dm^3, not more than	1.5-0.7	1.5-0.7	5
Permanganate oxidizability, mgO/dm^3, not more than	2	5	15
The number of coliform bacteria (CGB), in 1 dm^3, not more than	3	100	1000

2 The results of experimental analysis

The qualitative analysis of the Irtysh and Tobol samples allowed us to divide them into three classes according to the composition of the samples, to identify the reasons for the deviation from the water quality and the required degree of processing in order to bring it to GOST indices that are suitable as water supply sources.

Table 2. Underground water quality indicators of water supply sources Bekerevo village, Abalak village, Sumkino settlement.

Quality indicators	Bekerevo village	Abalak village	Sumkino settlement
Turbidity, mg/dm^3, not more than	1.5	1.5	10.0
Colour, degree, not more than	20	20	47
Hydrogen index (pH)	7. 1	6. 9	6. 8
Iron, (Fe), mg/dm^3, not more than	5	10	20
Manganese (Mn), mg/dm^3, not more than	1	0. 2	1. 5
Hydrogen sulphide (H_2S), mg/dm, not more than	2	0. 2	10
Fluorine (F), mg/dm^3, not more than	1.5-0.7	0.7	1. 5
Permanganate oxidizability, mgO/dm^3, not more than	3	2	8
The number of coliform bacteria (CGB), in 1 dm^3, not more than	100	8	400

The quality indicators of surface water sources – water supply sources of CHP in the districts of villages: Bekerevo, Abalak, Sumkino, were studied on a quarterly basis (table 2).

Table 3. Indicators of surface water sources quality in the village of Bekerevo, village Abalak, settlement Sumkino.

	Bekerevo village	Abalak village	Sumkino settlement
Turbidity, mg/dm^3, not more than	1400	1500	10000
Colour, degree, not more than	35	120	200
Odour at 20 and 60 °C, points, not more than	2	3	4
Hydrogen index (pH)	6.9	6. 8	7.5
Iron, (Fe), mg/dm^3, not more than	1	2. 5	3. 5
Manganese (Mn), mg/dm^3, not more than	0.1	1.0	1. 5
Phytoplankton, mg/dm^3, not more than	1	5	8
Permanganate oxidizability, mgO/dm^3, not more than	13	15	18
Ultimate BOD, mgO_2/dm^3, not more than	3	3. 5	3. 8
The number of lactose-positive E. coli in 1 dm^3 of water, not more than	1000	10000	20000

The results analysis of the research of the water taken from the surface of the Irtysh, Tobol river shows the concentration excess of total iron, manganese, high oxidation in the water compared with the maximum permissible concentration that is several times higher in the Sumkino area. The hydrogen index is within the maximum permissible concentration. Moreover, the analysis of surface water for the content of ions was carried out. The content of sodium ions, ion sulfate, ion nitrate, ion silicate (mg/dm^3), as shown by the analysis, is contained in all surface and underground water at the indicated locations of the samples (Table 4).

Table 4. Water quality of a surface source (the Irtysh River).

Sampling place	PM, mg/dm^3	Content of ions, mg/dm^3					Solid residue, mg/dm^3	Total hardness, mg – eq/dm^3	Alkalinity, mg-eq/dm^3
		Na^++ K^+	SO_4^{2-}	Cl^-	NO_3^-	SiO_2^++ SiO_3^{2-}			
Abalak village	172	36.1	40.5	17	1.8	4.7	244	2.8	2.7
Bekerevo village	165	37.1	41.5	19	2.4	8.7	344	3. 0	2.7
Sumkino settlement	180	35.1	42. 5	20	2.7	9.7	360	3. 5	2.7

High indicators of solid residue show a high content of salts. At the same time. the total hardness is quite low. which proves the possibility of water usage for water supply and steam production in the heat and power industry of Tobolsk and its district.

3 Conclusion

The results of the experiment have shown that the underground and surface water of the Irtysh. Tobol are qualitative natural sources of water supply. The waters of the Irtysh river contain substances of mineral and organic origin. as well as microorganisms. The mineral composition of water is determined mainly by the features of the hydrological layers through which the river flows. The organic composition is formed mainly by compounds that are washed out of the soil by the reservoir itself as the result of biological processes. Mineral and especially organic substances are brought to the Irtysh by waste water of populated areas and industrial enterprises. The Irtysh water is fresh. belongs to the hydrocarbonate class. For water supply and steam production in the heat and power industry of Tobolsk region. the source water is purified with the help of ionic desalination. Water pretreatment for desalting is carried out using mechanical filters [8].

In the experimental work on the research of the underground and surface water quality. students of chemical-engineering department were involved. The effectiveness of the experimental work was reflected in the thesis research and term projects on the request of the industrial enterprise.

References

1. A.I.Petrov. *Current problems of the natural environment changing* (Moscow, 2005) [in Russian]
2. Yu.V. Alekseev. *Heavy metals in soil and plants* (Moscow, 2014) [in Russian]
3. Ya.M.Amosova. *Soil protection from chemical pollution* (Moscow, 2015) [in Russian]
4. G.P. Bespamyatnov. *The maximum permissible concentrations of chemicals in the environment* (Saint Petersburg, 2014) [in Russian]
5. V.D. Shantarin. *Health Safety and Industrial Safety* (Tyumen, 2014) [in Russian]

6. I.I. Borodavchenko. *Integrated use and protection of water resources* (Moscow, 2016) [in Russian]
7. G.I. Egorova. *Current problems of chemistry. chemical technology, ecology* (Tyumen, 2011) [in Russian]
8. D.S. Luzhkovoy. O.Yu. Romashova. A.A. Tubolev. Yu.V. Kobenko and E.S. Riabova. MATEC Web Conf. **92**. 01054 (2017)

Finite element analysis of radiant heating systems based on gas-fired infrared heat emitters

Anton N. Ermolaev[1], *Sergey* A. Khaustov[2,*]

[1]Tyumen industrial University, 625000 Tyumen, Russia
[2]National Research Tomsk Polytechnic University, 634050 Tomsk, Russia

Abstract. The article presents a finite element model for simulating a gas-fired IR radiation system. Simulation of gaseous combustion and discrete ordinates radiation model were used to solve a number of heat-transfer problems in ventilated rooms with radiant heating. We used Ansys Multiphysics software and Fluent CFD solver for implementing finite element analysis. To solve differential equations of heating and gas dynamics, the following boundary conditions were considered. Dry methane was used as the fuel and air with 21% of oxygen, as oxidizer. Fuel consumption was 0.5 m3/hour; the gas pressure before the nozzle was 1270 Pa. The air–fuel ratio was 9.996.

1 Formulation of the problem

Large-scale production buildings can only be effectively heated by modern radiant heating systems based on gas-fired infrared heat emitters [1, 2]. However, high operating temperatures of emitter ceramic surfaces create a number of shortcomings [3, 4].

We used Ansys Multiphysics software and Fluent CFD solver to implement finite element analysis [3]. The goal of this analysis was to establish a correlation between experimental and finite element temperatures in reference points of gas-fired infrared heat emitters. A discrete ordinates radiation model simulating gaseous combustion was used to solve a number of heat-transfer problems in ventilated rooms with radiant heating.

$$\frac{dI(\vec{r},\vec{s})}{ds}+(a+\sigma_s)I(\vec{r},\vec{s})=an^2\frac{\sigma T^4}{\pi}+\frac{\sigma_s}{4\pi}\int_0^{4\pi}I(\vec{r},\vec{s}')\Phi(\vec{s}\cdot\vec{s}')d\Omega' \quad (1)$$

where ŕ = position vector; ŝ = direction vector; ŝ' = scattering direction vector; s = path length; a = absorption coefficient; n = refractive index; σ_s = scattering coefficient; σ = Stefan-Boltzmann constant; I = radiation intensity; T = local temperature; Φ = phase function; Ω = solid angle.

To solve differential equations of heating and gas dynamics, the following boundary conditions were considered. Dry methane was used as the fuel and air with 21% oxygen, as

[*] Corresponding author: khaustovsa@tpu.ru

 DOI: 10.1051/matecconf/201711001024

oxidizer. Gas-fired infrared heat emitters with heat output of 5, 10, 15, 20, 30, 40 kW were studied. The gas pressure before the inlet nozzle was 1270 Pa. The air-fuel ratio (AFR) was 9.996.

A gas-fired infrared heat emitter is made of steel 12Cr18Ni10Ti of 2 mm with thermal conductivity coefficient, $\lambda=0.018\cdot T+9$ W/(m·K); specific heat capacitance is $c=-2\cdot10^{-4}\cdot T^2+0.45\cdot T+324.5$ J/(kgK)

The finite element model with grid and partitions is shown on Fig. 1. Tetrahedron is adopted as a finite element with 287403 cells. The gas-fired infrared heat emitter and computational domain of the heat transfer are CFD cell-zones with double walls at the contact points. On the walls bounding computational domain the third order heat transfer conditions are set. The internal air temperature was 20 °C.

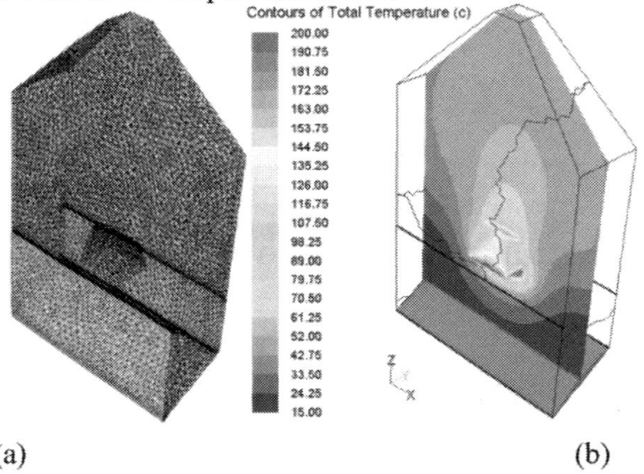

Fig. 1. The finite element model with grid (a) and partitions (b) colored by temperature (°C).

2 Results of the study

Along with the standard gas-fired infrared heat emitter (Fig. 2, a) with a reflector made of polished stainless steel the following technical solutions were considered. A gas-fired infrared heat emitter with a low thermal conductivity coefficient of the reflector insulation (Fig. 2, b); a gas-fired infrared heat emitter with water cooling of the reflector (Fig. 2, c).

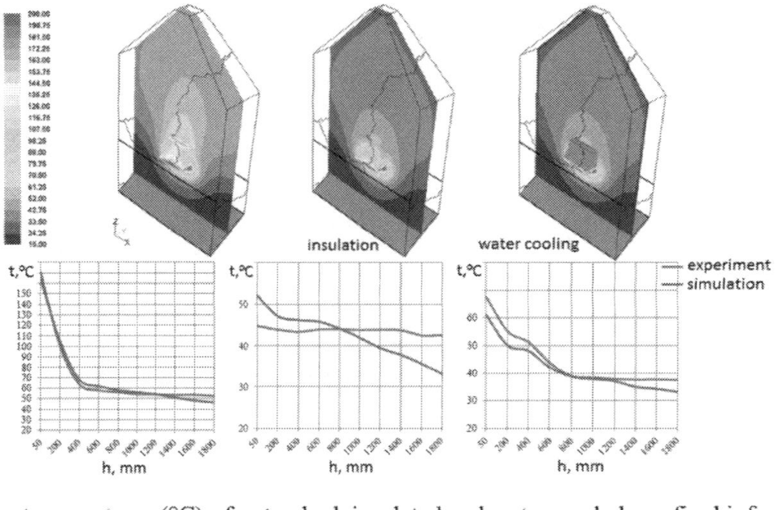

Fig. 2. Temperature contours (°C) of a standard, insulated and water-cooled gas-fired infrared heat emitter with thermal power of 5 kW and its comparison with the experimental data (charts).

 DOI: 10.1051/matecconf/201711001024

The results of the mathematical modeling of heat and mass transfer in a turbulent reaction medium with a combustion reproduce the experimental data produced by a measurement in real operating conditions of the gas-fired infrared heat emitter (see Fig. 2). According to the model, the temperature of the radiating ceramic tiles for all investigated emitters did not depend on the performance of the reflector and was 900 °C (Fig. 3). This leads to the conclusion that the performance of the reflector does not affect chemical processes and combustion.

Detailed temperature contours of the ceramic tiles (Fig. 3) enable estimation of the diffusion and reaction zone dimensions. The diffusion of the fuel jet into the air stream occurs in a special mixing chamber over the ceramic tiles of the gas-fired infrared heat emitter. Therefore, fuel is mixed with air before entering the holes in the ceramic tile. Air-fuel mixture moves in these cylindrical holes, heats and ignites forming the narrow laminar combustion front. The maximal combustion rate is observed at the outlet of the cylindrical holes.

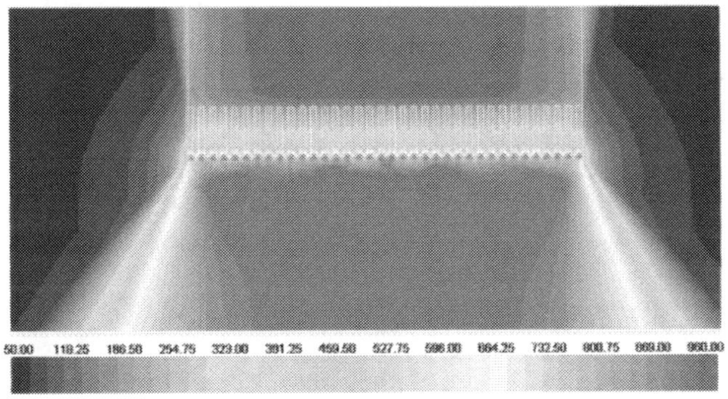

Fig. 3. Temperature contours (°C) in the longitudinal section of the ceramic tiles of the gas-fired infrared heat emitter.

To verify the fulfillment of thermal comfort conditions, the distribution of the intensity of thermal irradiation of the working area was calculated. The maximum local heat fluxes of infrared irradiation in the working area should not exceed 150 W/m^2. For the parameters studied, this condition corresponds to the following heights: for 5 kW height is 4 m (119.3 W/m^2); for 10 kW height is 6 m (106.1 W/m^2); for 15 kW height is 7 m (116.9 W/m^2); for 20 kW height is 8 m (119.3 W/m^2); for 30 kW height is 9 m (114.6 W/m^2); for 40 kW - 10 m (148.2 W/m^2). The most uniform intensity of irradiation of the working area is ensured by installing a number of gas-fired infrared heat emitters with an equal thermal power at the distance mentioned above. The value of heat flux for such a radiant heating system is 25% higher than a single gas-fired infrared heat emitter (Fig. 4). It has been established that the higher gas-fired infrared heat emitter location causes the more uniformly distribution of local heat fluxes in the working zone and its lesser magnitude in the epicenter of irradiation.

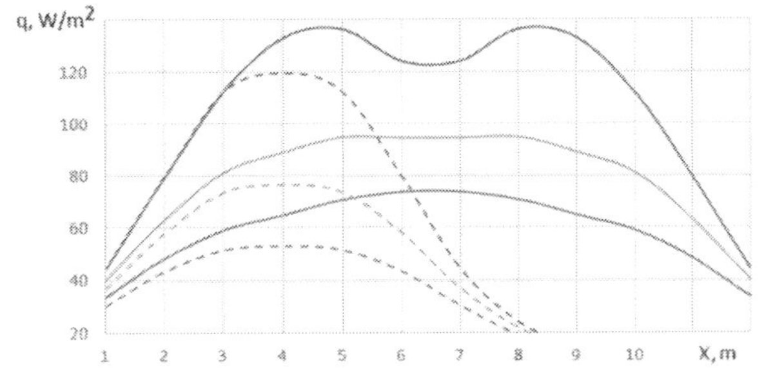

Figure 4. Local heat fluxes q (W/m^2) of thermal irradiation in the working area.

4. Conclusion

The results of studies based on bench tests and mathematical modeling:
- discrete ordinates radiation and gaseous combustion mathematical simulation are suitable for description of high-temperature gas-fired infrared heat emitters elements interaction;
- the temperature of the radiating ceramic tiles for all investigated emitters was 900 °C and did not depend on the performance because the reflector does not affect chemical processes and combustion;
- the most energy efficient way to heat large-scale production buildings is to use modern radiant heating systems based on gas-fired infrared heat emitters with an equal thermal power at the distances 4, 5, 7,.8, 9, 10 for 5, 10, 15, 20, 30, 40 kW of heat output respectively;
- virtual prototypes of high-temperature gas-fired infrared heat emitters based on discrete ordinates radiation model enable assessment of its efficiency; the value of heat flux for proposed radiant heating system is 25% higher than a single gas-fired infrared heat emitter.

References

1. Y. Luo, L. Chevalier, F. Utheza, X. Nicolas, International Polymer Processing. Journal of the Polymer Processing Society, **30**, 5,554 (2015)
2. M. Ostafin, M. Balkenhol, J. Erlemeyer, M. Kohler, Materialwissenschaft und Werkstofftechnik, **41**, 11 (2010)
3. S. Dolgov, E. Savchenko, S. Khaustov, R. Tabakaev, A. Zavorin, EPJ Web Conf. **110**, 01074 (2016)
4. K.A. Al-attab, J.C. Ho, Z.A. Zainal, Exp. Therm Fluid Sci. **62**, 1 (2015)
5. D. Patino, B. Crespo, J. Porteiro, J.L. Miguez. Appl. Therm. Eng. **100**, 849 (2016)

MATEC Web of Conferences **110**, 01025 (2017) DOI: 10.1051/matecconf/201711001025
HMTTSC-2017

Permeability of a triangular packing of open nanotubes

Aleksy M. Bubenchikov[1], *Igor* A. Evseev[1], and *Egor* A. Tarasov[1,*]

[1]National Research Tomsk State University, 634050 Tomsk, Russia

Abstract. We consider a layer of open nanotubes identically oriented in space. We investigate permeability of a nanoporous carbon system for a regular triangular packing of tubes. Curves defining layer permeability depending on the degree of spatial proximity of tubes are constructed.

1 Introduction

Problems of nanoporous structures permeability are of a great importance in solving tasks of gas separation, water treatment, as well as for the study of life support conditions in biological systems. Graphenes, fullerenes and nanotubes are ideal surface carbon systems. If to use a regular spatial packing for these atomic systems, the resulting structures will also be ideal and convenient to study using mathematical methods. The resulting spatial structures containing molecule-impenetrable clusters of carbon atoms and nanoscale pores are interesting in terms of gas passage through the constructed layers. Despite the obvious idealization of matrix structures of carbon atoms with respect to the actually existing nanoporous systems, conclusions on permeability of ideal objects are significant for development of practical gas separation technologies.

If the Lennard-Jones potential is modified in a certain way, performing integration over the surface of a 2D-material (nanotubes, graphenes) it is possible to find the interaction energy of long nanotubes or relatively large graphene plates. This could allow defining an equivalent homogeneous layer and determining its permeability. Previously, the integral impact of volume nanoparticles using the method of defining particularities was found in the papers [1-7]. Continuous integration on the basis of a modified LJ-potential was tested in [8-14]. This approach can provide significant advances in theoretical studies on membrane transport of neutral particles, as well as in solving numerous problems of surface phenomena physics [15-21].

According to the data in [22], the average longitudinal dimension of tubes is of the order of 10^3 nm. Such tubes contain about 10^5 carbon atoms. Therefore, studying interactions between these structures and molecules and atoms of gas components should be conducted using continual models of energy distribution over the tube surface.

The classical Lennard-Jones potential has a too strong singularity at zero and can not be integrated neither over the surface nor the volume. Without changing the nature of the Van der Waals forces action in the determining area of intermolecular distances, we reduced the value of singularity to $1/r$. Therefore, the resulting modified potential, in this respect,

[*]Corresponding author: diomedis@mail.ru

became equivalent to the potentials of Newton or Coulomb which, as known, are integrable both over the surface and the volume.

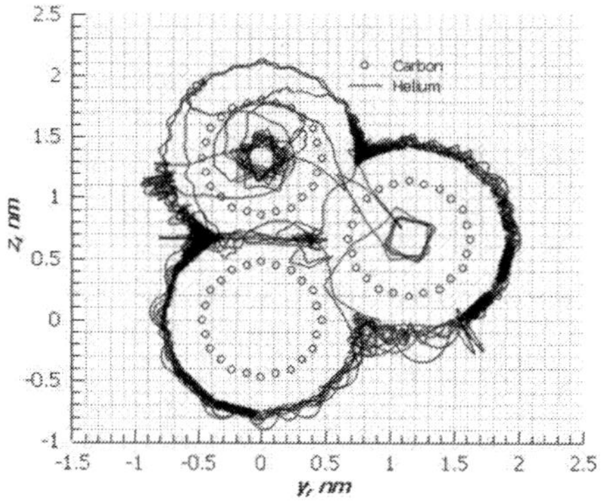

Fig. 1 Trajectory of low-energy helium atoms **O** – carbon atoms; - - helium trajectory.

However, if to take a short piece of a tube, such as of 12 rows containing carbon atoms located in the axial direction for which it appears that its length is equal to 1.136nm, it is quite acceptable to study permeability on the basis of the model of discrete-positioning of carbon atoms. At the same time, we have to remember that the problem of a single molecule moving through the structure is six-parameter (three parameters of position and three components of velocity) and requires too many launches of molecules for the purpose of determining the permeability coefficient of the nanoporous structure. Fast molecules either pass through the structure, or get reflected from it having an uninformative trajectory. Low-energy molecules move around the structure until they are knocked out of it by faster molecules. Within a limit, such spatial orbital motion is infinite and it clearly shows the presence of permeability zones in the structure. With the help of the discrete model it is quite appropriate to consider a packing of a small number of tubes.

Fig. 1 illustrates several trajectories of low-energy molecules having an initial velocity of the range [50, 100] m/s. The molecules were launched through a packing along the nanotubes axes, however, they did not leave the layer composed of the tubes but got involved in the sorption motion. Therefore, we further took into account their presence in the barrier of the packing. The contribution of existing sorption molecules into the value of the energy barrier was approximately 15% (compared with the contribution made by the energy carbon structure atoms). Calculations show that permeability of the tube with the length of 1.136 nm is the same as that of the tube of 10^3nm.

Let us consider a carbon nanotube of an arbitrary length and find its interaction energy by integration. Thus, in case of continuous integration over a 2D-material surface, we should have a convergent value of the integral action. The above mentioned modification has the form:

$$dU(\rho)=4q\varepsilon\left(\frac{\sigma}{\rho}\right)\text{th}\left[\left(\frac{\sigma}{\rho}\right)^{11}-\left(\frac{\sigma}{\rho}\right)^{5}\right]ds\ , \tag{1}$$

where dU is the energy of the selected fragment; ε, σ are the LJ- potential parameters; ρ is the distance between the centre of the test molecule and the centre of the elementary area on the surface of the 2D-material; q is the distribution density of energy sources on the surface; ds is the elementary area on the tube surface. For a flat or a cylindrical graphene structure it is easy to find that $q\approx 28$ nm^{-2}; and for a spherical - $q\approx 20$nm^{-2}. Naturally, introducing cylindrical coordinates and integrating (1) over the side surface of the open nanotube we obtain

$$U(r,z)=4\varepsilon aq\int_0^{2\pi}\int_0^{l}\left(\frac{\sigma}{\rho}\right)\text{th}\left[\left(\frac{\sigma}{\rho}\right)^{11}-\left(\frac{\sigma}{\rho}\right)^{5}\right]d\varphi' dz' \tag{2}$$

Moreover, $\rho=\sqrt{r^2+a^2-2ar\cos\varphi'+(z-z')^2}$; a is the radius of the nanotube; l is its length; r,z are the cylindrical coordinates; φ', z' are the integration variables. The double integration in (2) can be performed numerically, using the trapezoidal rule consistently for each of the integrals.

When the integration in (2) is completed, we can find the radius-averaged value of the potential. In this case, the integrated potential will depend only on the coordinate z. Let us denote it as $W_1(z)$. To assess permeability of the rectangular packing of nanotubes we will use the method of molecular penetration through an equivalent homogeneous layer. With regard to the case of packing shown in Fig. 2, the energy of this layer is determined as follows. Potential (2) is averaged by the radius equal to the side of the shaded fragment, with the help of which the whole area under consideration is parqueted. The obtained average value of the potential $W_1(z)$ is determined by the influence of only one tube. Since in the considered case the nearest neighbours of the selected elementary fragment are three tubes, due to the symmetry of the influences, the average value of the potential in the shaded area will be $W(z) = 3W_1(z)$. The influence of other tubes is neglected because their contributions are really small. On defining the average potential, we can launch a test molecule perpendicular to the shaded area, i.e. along the axes of the tubes. Passing through the uniform layer, such molecule will not experience any effects diverting it from the straight path. Therefore, the equation of motion for such molecule is one-dimensional:

$$m\frac{dw}{dt}=-gradW(z)\cdot \tag{3}$$

Here m is the mass of the molecule penetrating through the layer; w is the velocity component with respect to the nanoporous layer.

This equation, when applied to the passage through the averaged barrier, encompasses all the statistics related to the location of the molecule's starting point with respect to the tubes which make up the parqueting element. We will try to take into account the statistics on the velocity values using the Maxwell distribution, which is as follows:

$$f(v)=\frac{dN}{Ndv}=\frac{4v^2}{\sqrt{\pi}}\lambda^{3/2}e^{-\lambda v^2}. \tag{4}$$

Here f is the distribution function; v is the velocity module of the molecule; $\lambda=\frac{m}{2kT}$, k is the Boltzmann constant; T is the absolute temperature; N is the number of molecules per unit volume; dN is the fraction of molecules whose velocities are in the range of $[v, v + dv]$.

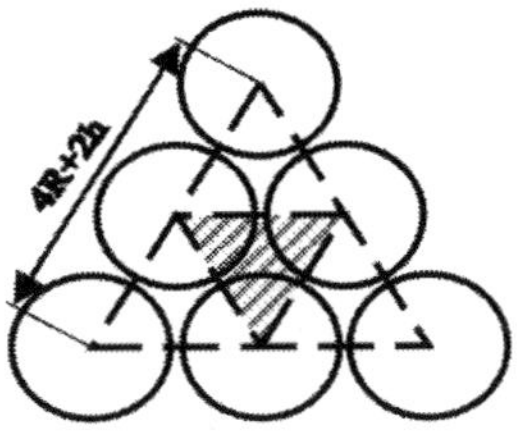

Fig. 2 Triangular packing of nanotubes.

DOI: 10.1051/matecconf/201711001025

Practical calculations show that, if at a certain velocity a molecule passes through the packing of nanotubes, it can pass through it at a greater value of velocity as well. So, finding the minimum speed of passage and further integrating the Maxwell distribution from this value to infinity, we can define the fraction of molecules which pass through the packing. The minimum velocity values for helium atoms and methane molecules for the considered triangular packing of tubes are shown in Fig. 3. The next figure represents the distribution of the permeability coefficients for helium and methane as well. Fig. 4 shows that at $0.3 \leq h \leq 1.0$ nm the structure provides good separation basing on the permeability coefficient, i.e. it functions as a separator.

To integrate equation (4) it is quite useful to apply the standard version of the Runge-Kutta scheme of the fourth-order accuracy.

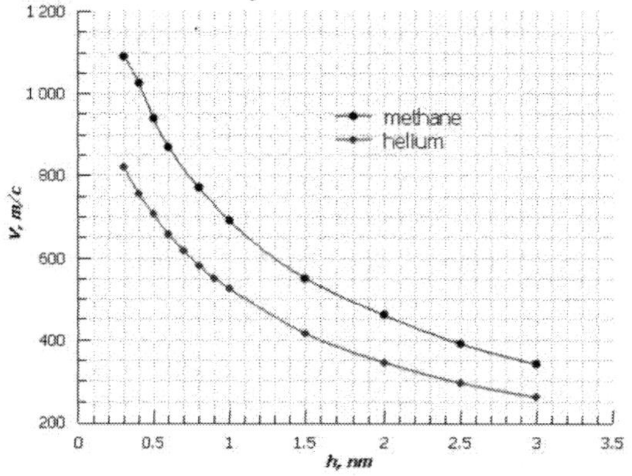

Fig. 3 Limit permeability rate for triangular packing. Black – methane; blue – helium.

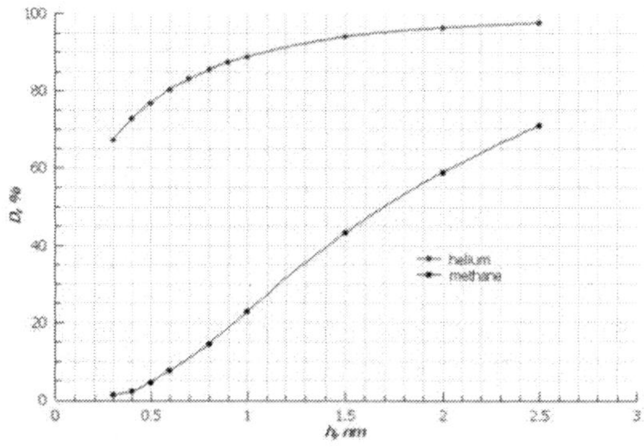

Fig. 4 Permeability curve for triangular packing of nanotubes in respect of helium atoms (blue) and methane molecules (black).

2 Summary

The calculations made it possible to easily find the minimum limit velocity of passage through the homogeneous layer for a given type of molecules and to assess by its value the degree of molecular passage through this layer. Comparing the obtained permeability values for various components of the initial gas mixture, we can draw conclusions about selective properties of a layer of parallel nanotubes or other structures created on the basis of 2D carbon materials. In the present study we found that, at least for a triangular packing of identically oriented tubes, the layer functions as a separator at the tubes convergence values h equal to 0.3 to 1 nm.

References

1. V.Ya. Rudyak, S.L. Krasnolutskiy, D.A. Ivanov, J. Doc. Phys. **57** (2012)
2. V.Ya. Rudyak, A.A. Belkin, V.V. Egorov, D.A. Ivanov, Int. J. Multiph. **5** (2011)
3. V.Ya. Rudyak, S.L. Krasnolutskiy, D.A. Ivanov, J. Microfl. Nanofl. **11** (2011)
4. V.Ya. Rudyak, S.N. Dubtsov, A.M. Baklanov, Tech. Phys. Lett. **34** (2008)
5. V.Ya. Rudyak, S.L. Krasnolutskiy, E.N. Ivashchenko, J. Eng. Phys. Thermophs. **81** (2008)
6. V.Ya. Rudyak, J. VDI-Berichte, **1803** (2003)
7. V.Ya. Rudyak, A.A. Belkin, Tech. Phys. Lett. **29** (2003)
8. M.A. Bubenchikov, A.I. Potekaev, A.M. Bubenchikov, Russ. Phys. J. **56** (2013)
9. M.A. Bubenchikov, A.I. Potekaev, A.M. Bubenchikov, Russ. Phys. J. **56** (2013)
10. A.I. Potekaev, A.M. Bubenchikov, M.A. Bubenchikov, Russ. Phys. J. **55** (2013)
11. M.A. Bubenchikov,A.I. Potekaev, A.M. Bubenchikov, O.V. Usenko, A.V. Ukolov, IOP Conf. Series: Mat. Sc. Eng. **87**, 1 (2015)
12. A. M. Bubenchikov, M.A. Bubenchikov, O.V. Matvienko, E.A. Tarasov, O.V. Usenko, AIP Conf. Proc. **1698** (2016)
13. M. A. Bubenchikov, A. M. Bubenchikov, O.V. Usenko, V.A. Poteryaeva, S. Jambaa, EPJ Web Conf. **110** (2016)
14. A.M. Bubenchikov, M.A. Bubenchikov, A.I. Potekaev, E.E. Libin, Yu.P. Khudobina, Russ. Phys. J. **57** (2014)
15. R. Sazonov, G. Kholodnaya, D. Ponomarev, G. Remnev, O. Razumeko, J. of the Korean Physical Society, **59** (61)
16. G.E. Remnev, V.V. Uglov, V.I. Shymanski, S.K. Pavlov, N.M. Lapchuk, Vacuum, **89** (1) (2013)
17. V.V. Uglov, G.E. Remnev, N.T. Kvasov, I.V. Safronov, J. of Surface Investigation, **8** (4) (2014)
18. G.E. Remnev, V.V. Uglov, V.I. Shymanski, S.K. Pavlov, A.K. Kuleshov, Applied Surface Science (2014)
19. A. Poloskov, I. Egorov, V. Ezhov, G. Remnev, Key Engineering Materials, **685**
20. N.E. Aktaev, G.E. Remnev, Surface and Coatings Technology (2016)
21. N.E. Aktaev, G.E. Remnev, IOP Conf. Series, **110,** 012054 (1),
22. R. Jansen, P. Wallis, South West Nanotechnologies, Inc. 2501 Technology Place Norman, OK 73071 (2017)

Measurement of free surface energy of laser treated aluminum substrate

Kseniya Batishcheva[1,*], *Dmitriy* Feoktistov[1], *Anastasiya* Islamova[1], and *Ivan* Shanenkov[1]

[1]National Research Tomsk Polytechnic University, 634050 Tomsk, Russia

Abstract. The change in the surface energy of the aluminum substrate after laser treatment has been determined experimentally. Surface energy is determined using the Advance KRUSS software. The laser treatment of the surface is found to increase its free surface energy. It changes the ratio of the polar and the dispersion components. Analysis of the crystalline planes of aluminum surface after laser treatment was provided by XRD X-ray diffractometry using a Shimadzu XRD 7000S diffractometer.

1 Introduction

Hydrophobic and hydrophilic materials are in demand in many spheres of human activity. The most of known methods both improving and deteriorating wettability of various materials [1-3] have a disadvantage. It is a limited period of use after which the surface acquires the initial wettability properties. Such changes occur under the influence of the surrounding aggressive environment or a simple mechanical abrasion of the surface. The development of a method for the surface properties changes due to transformations in the surface texture [4] is the most crucial topic nowadays. In addition, the period of use for such materials will significantly increase. One of these methods is laser radiation treatment.

It is known [5] that the behavior of a liquid drop on a solid surface depends on surface properties, such as surface and interfacial tension, surface free energy (SFE). The surface free energy can be found experimentally in determining the wetting angle of the surface by liquids having different polar and dispersion components. There are several scientific approaches for the definition of SFE. The most common are the following:

1. Zisman plot method [6]. From the experimental values of the contact angle θ, a dependence is plotted in the axes $cos\theta$ (Y axis) – θ (X axis), then extrapolation is used to determine the SFE. This method is used for wetting of non-polar surfaces since Zisman method does not take into account the intermolecular interaction between a liquid and a solid surface. In addition, it fails for high-energy samples where no linear trend in the plots was observed [7].

2. The Owens–Wendt–Rabel–Kaelble method (OVRK) [8, 9] differentiates polar and dispersion components of the surface energy and uses a geometric mean of these in the expression for it according to the following equation:

* Corresponding author: bka1801@mail.ru

$$\frac{\sigma_L(\cos\theta+1)}{2\sqrt{\sigma_L^D}}=\frac{\sqrt{\sigma_S^P}\cdot\sqrt{\sigma_L^P}}{2\sqrt{\sigma_L^D}}+\sqrt{\sigma_S^D} \quad (1)$$

where σ_L is surface tension of liquid; σ_S surface free energy of a solid; σ_L^D dispersion component of liquid; σ_S^D dispersion component of surface; σ_L^P polar component of liquid; σ_S^P polar component of surface.

It is assumed that the van der Waals force has the greatest influence on the dispersion component of the solid, the interaction forces and hydrogen bonds – on the polar component.

3. Fowkes method [10] is based on theory of liquid adhesion and wetting of solid and used for determination of the SFE of polar surfaces.

4. Wu method [11]. The method is based on the equation:

$$\sigma_L(\cos\theta+1)=4\left(\frac{\sigma_S^D\cdot\sigma_L^D}{\sigma_S^D+\sigma_L^D}+\frac{\sigma_S^P\cdot\sigma_L^P}{\sigma_S^P+\sigma_L^P}\right) \quad (2)$$

It is used for solids with relatively large surface free energy [5].

5. van Oss and Good method [12] is based on equation:

$$(1-\cos\theta)\cdot\sigma_L=\sigma_S+\sigma_L-2\cdot(\sqrt{\sigma_S^D\cdot\sigma_L^D}+\sqrt{\sigma_S^+\cdot\sigma_L^-}+\sqrt{\sigma_S^-\cdot\sigma_L^+}) \quad (3)$$

The difference from the above mentioned methods is in the assumption that the polar component of the SFE consists of acid and base parts.

An analysis of the use of the methods described above was made in [7]. The results of the SFE determination are found to depend on the test liquids chosen. The most accurate results were obtained using water and methylene diiodide. The OVRK method is recommended as the most accurate.

The aim of the work is to determine the effect of laser radiation on the change in the SFE of an aluminum surface.

2 Materials and methods

Experimental studies were carried out using a shadow optical system on a setup presented in Fig. 1 [13].

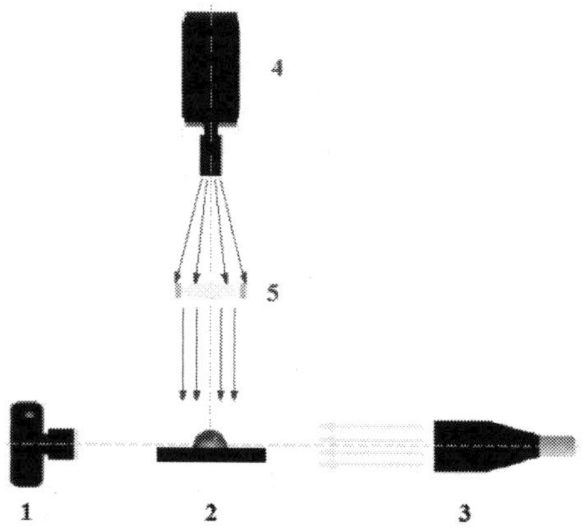

Fig. 1. Schematic of the experimental setup: 1 – photographic camera, 2 – substrate with a drop, 3 – source of plane-parallel light, 4 – high-speed video camera, 5 – condensing lens.

A drop of distilled water (methylene diiodide) of 10 μL was placed on an aluminum substrate (2) by high-precision dispenser and illuminated by a plane-parallel light (3), wherein a part of the light was dissipated by a drop, and non-dissipated light was fixed by the camera. Dash-dotted line shows the optical axis of the shadow system. Video camera (4) and condensing lens (5) were placed above the liquid drop and allowed visualization of evaporation process from the top and control of drop symmetry.

The experiment was carried out using four aluminum alloy (AMG-6) substrates. The surfaces were polished. After polishing, three substrates were treated with laser radiation, one was not processed. The texture is treated by a laser system "Mini Marker 2 M 20". Technical characteristics of the laser system used for surface treatment of aluminum experimental samples are listed in Table 1.

Table 1. Technical characteristics of laser system used for surface treatment.

Frequency of the laser beam, Hz	99
Travel speed of the laser beam, 10^{-3} m/sec	800
Power of the laser radiation, W	6, 14, 18

Surface energy is determined using the Advance KRUSS software.

The surfaces after laser treatment were studied by XRD X-ray diffractometry using a Shimadzu XRD 7000S diffractometer (Cukα – radiation, λ=1.540598, with focusing on the secondary beam monochromator). Interpretation of the phase composition was carried out by processing the obtained diffractograms in the Search-Match software package using the PDF2 database.

3 Results and discussion

Figure 2 provides the diffractograms of samples treated by laser beam with different powers in comparison with the diffractogram of polished substrate. The dominant phase is expected to be pure aluminum. The intensity of maxima reduced with increasing the power of laser radiation (Fig. 2). It can be explained by etching a part of the samples surface and, correspondingly, by reducing the useful area of the sample which interacts with X-rays. This affects the reduction of integrated statistical information which causes a decrease in the intensity. The position of maxima on the 2 θ -axis refers to alumina phase (Fig. 2). The change of the preferential orientation of sample maxima obtained at the radiation power of 6 W is the most interesting. The explanation for this fact can be a change in the preferential

DOI: 10.1051/matecconf/201711001026

orientation of the planes in the investigated subsurface layer. However, for the other two samples, the ratio of the maxima remained unchanged.

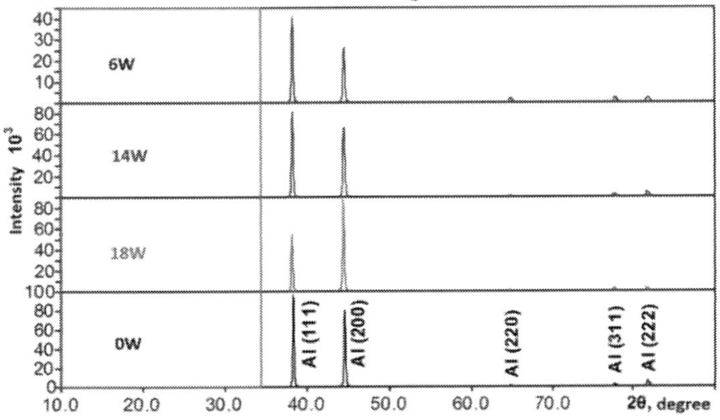

Fig. 2. Diffractograms of surfaces after laser radiation treating.

Figure 3 shows the diffractograms of surfaces in the region of small angles 2θ up to 35°.

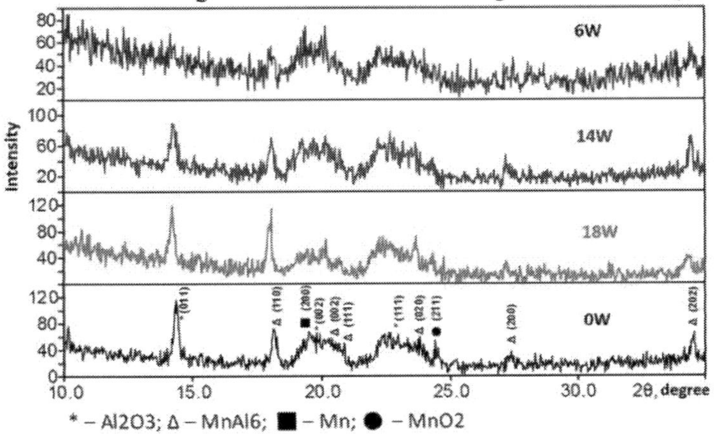

Fig. 3. Diffractograms of surfaces in the region of small angles.

According to the results presented in Fig. 3 it is found that there are reflections of crystal planes of the following phases: aluminum oxide Al_2O_3, aluminum and manganese alloy $MnAl_6$, pure manganese Mn, and manganese oxide MnO_2. The presence of these impurities containing manganese is also explained by its presence in alloy composition. The presence of oxide phases is quite natural, such as aluminum is chemically active element and is easily oxidized in air. Based on diffractograms (Fig. 3) it is seen that with increasing laser beam power, the intensity and the number of maxima of the impurity phases decrease. Traces of other elements which are present in the alloy in large quantities were not found on diffractograms.

Figure 4 shows the results of determining the change of surface energy.

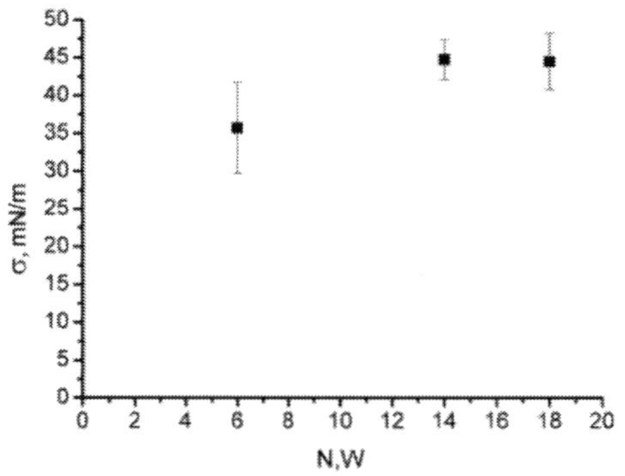

Fig. 4. Dependence of the surface energy on the laser beam power.

4 Conclusions

The influence of laser radiation on the change in the SFE of the aluminum surface is determined experimentally. The SFE is found to increase with increasing laser power. It changes the ratio of the polar and dispersion components. The proportion of the dispersion component increases. Analyses of the crystalline planes of aluminum surface after laser treatment was provided by XRD X-ray diffractometry using a Shimadzu XRD 7000S diffractometer (Cukα – radiation, λ=1.540598, with focusing on the secondary beam monochromator). It is found that the laser surface treatment changes the orientation of the crystal planes. It should be noted that this result is obtained only at a relatively low power of the laser beam (6 W). The orientation of the crystalline planes of the aluminum surface did not change at high power of the laser.

The reported study was supported by the grant of President of Russian Federation for the government support of young Russian scientists (MK-6810.2016.8).

References

1. D. Zaitsev, D. Kirichenko, O. Kabov, Technical Physics Letters, **41**, 6 (2015)
2. E.G.Orlova, E.R.Malyhina, D.V.Feoktistov, L.S.Zhidkova, MATEC Web Conf. **72**, 01062 (2016)
3. Ya. Misyura, Appl. Surf. Sci. **414**, 188 (2017)
4. D. V. Ta, A. Dunn, T. J. Wasley, R. W. Kay, J. Stringer, P. J. Smith, C. Connaughton, J. D. Shephar, Appl. Surf. Sci. **357**, 248-254 (2015)
5. M. Tryznowski, J. Izdebska-Podsiadly, Z. Zolek-Tryznowska, Prog. Org. Coat. **109**, 55-60 (2017)
6. H. Fox, W. Zisman, J. Colloid Sci. **5**, 514-531, (1950)
7. F. Hejda, P. Solar, J. Kousal, WDS'10 Proceedings of Contributed Papers, Part III, 25 (2010)
8. D. Owens, R. Wendt, J. Appl. Polym. Sci. **13**, 1741 (1969)
9. D. Kaelble, J. Adhes. **2**, 66 (1970)
10. S. Fowkes, J. Phys. Chem. **67**, 2538 (1963)
11. S. Wu, J. Polym. Sci. C **34**, 19 (1971)
12. C. van Oss, M. Chadhury, R. Good, Adv. Colloid Interface Sci. **28**, 35 (1987)
13. G. Kuznetsov, D. Feoktistov, E. Orlova, K. Batishcheva, Colloid J. **78**, 3 (2016)

MATEC Web of Conferences **110**, 01027 (2017) DOI: 10.1051/matecconf/201711001027
HMTTSC-2017

Efficiency of utilization of heat of moisture from exhaust gases of heat HRSG of CCGT

Nikolay Galashov[1,*], *Svyatoslav* Tsibulskiy[1], *Denis* Mel'nikov[1], *Alexandr* Kiselev[1], and *Al'bina* Gabdullina[1]

[1]National Research Tomsk Polytechnic University, 634050 Tomsk, Russia

Abstract. The paper discusses the technology of utilizing the heat of exhaust gas moisture from heat recovery steam gases (HRSG) of combined-cycle gas turbine (CCGT). Particular attention focused on the influence of the excess air factor on the trapping of the moisture of the exhaust gases, as in the HRSG of the CCGT its value varies over a wider range than in the steam boilers of the TPP. For the research, has been developed a mathematical model that allows to determine the volumes of combustion products and the amount of water vapor produced according to a given composition of the burned gas and determine the amount of moisture that will be obtained as a result of condensation at a given temperature of the flue gases at the outlet of the condensation heat exchanger (CHE). To calculate the efficiency of the HRSG taking into account the heat of condensation of moisture in the CHE an equation is derived.

1 Introduction

CCGT of the utilization type currently has the highest net efficiency in the generation of electricity, which at the most powerful and perfect CCGT 9.02.02 of GE Power has reached 63.7%. The working medium of the steam turbine unit of this CCGT is water. The ways to further increase the efficiency of the CCGT are given in following work [1]. This is an increase in the initial temperature in front of the gas turbine and a decrease in heat losses in the steam turbine cycle and HRSG. At the same time, a decrease in heat loss in HRSG is the simplest way.

The combustion of a hydrocarbon fuel, depending on the type of fuel in the exhaust gases is formed about 10-15% moisture vapor. To prevent condensation of this moisture in the flues and the chimney so that there is no corrosion and destruction of the equipment, the exhaust gases are withdrawn at a temperature above the dew point. In this case, the heat of condensation of water vapor is lost and the atmosphere is contaminated with them.

The technology of cooling off gases in boilers below the dew point using condensation heat recovery units is now recognized as one of the most energy efficient [2-6].

This technology is especially important for HRSG of CCGT, their efficiency is not more than 90% due to the low temperature of the gases entering the boiler. At the same

* Corresponding author : gal@tpu.ru

time, HRSG of CCGT operate with an excess air factor α greater than 2, and the articles on this topic consider the operation of boilers with α <2. For cooling of flue gases in CHE, can be used low-boiling substances and can be obtained additional generation of electricity based on the Rankine organic cycle with an air condenser [7–10].

To evaluate the effect on the amount of moisture captured and on the efficiency of the HRSG fuel characteristics, the operating parameters of the flue gases and the air excess factor, a mathematical model is worked out for the operation of the HRSG with a CCGT.

2 Mathematical model of calculating the HRSG with CHE

The mathematical model of the HRSG with CHE represents a system of equations for calculating the parameters and indicators of its operation.

The initial data:
– gas composition;
– air moisture content d_{air}, kg/kg.dry;
– temperature t_0,°C and pressure P_g, Pa exhaust gases at the input of the recovery boiler t_1 and t_2;
– coefficient of excess air in the HRSG α.

The following constants are used in the model:

Molar mass (kg/mol): water μ_{H2O} = 18.016; nitrogen μ_{N2} = 28.013; oxygen – μ_{O2} = 32.0; argon μ_{Ar} = 39.95; carbon dioxide – μ_{CO2} = 44.0.

R = 8314.41 J/(mol·K) – universal gas constant;

Take volume fractions of elements of dry air: nitrogen r_{dryN2}=0.7812; oxygen r_{dryO2}=0.2096; argon r_{dryAr}=0.0092. Because of the low content, don't take into account the fraction of carbon dioxide. Molar mass of dry air: $\mu_{dry.air}=\mu_{N2}r_{dryN2}+\mu_{O2}r_{dryO2}+\mu_{Ar}r_{dryAr}$=28.96 kg/mol.

Ratio $\mu_{H2O}/\mu_{dry.air}$=0.622.

Volume fraction of moist air: moisture $r_{mH2O}=d_{air}/(0.622+d_{air})$; nitrogen $r_{mN2}=0.622r_{dryN2}/(0.622+d_{air})$; oxygen $r_{mO2}=0.622r_{dryO2}/(0.622+d_{air})$; argon $r_{mAr}=0.622r_{dryAr}/(0.622+d_{air})$.

2.1 Calculation of the composition of combustion products and it enthalpies at the input of the HRSG

Table 1. Equations for properties of combustion products at the input of the HRSG.

Index	Determined rate	Equation
1	Theoretical volume of air required for complete combustion of fuel under normal conditions	V_0=0.0476(0.5CO+0.5H_2+1.5H_2S+2CH_4+3.5C_2H_6+5C_3H_8 +6.5C_4H_{10}–O_2), nm³/nm³
2	Theoretical volume of nitrogen in the combustion of 1 nm³ of fuel	$V_{0.N2}$ = 0.79V_0+0.01O_2, nm³/nm³
3	Theoretical volume of nitrogen in the combustion of 1 nm³ of fuel	$V_{0.RO2}$ = 0.01(CO+H_2S+CH_4+2C_2H_6+3C_3H_8+4C_4H_{10}), nm³/nm³
4	Theoretical volume of H_2O in the combustion of 1 nm³ of fuel	$V_{0.H2O}$=0.01·(H_2S+H_2+2CH_4+3C_2H_6+4C_3H_8+5C_4H_{10})+1.61 V_0, nm³/nm³
5	Theoretical volume of gases during combustion of 1 nm³ of fuel	$V_g = V_{0.N2} + V_{0.RO2} + V_{0.H2O} + (\alpha-1)V_0$, nm³/nm³
6	Actual volume of H_2O in the combustion of 1 nm³ of fuel	$V_{H2O} = V_{0.H2O} + 0.0161(\alpha-1)V_0$, nm³/nm³
7	Volume of dry gases during combus-	$V_{dry.g.} = V_g - V_{H2O}$, nm³/nm³

	tion of 1 nm^3 of fuel	
8	Volume fractions of the combustion products at the entrance HRSG exclude their content in excess air	$r_{0.N2} = V_{0.N2}/V_g$; $r_{0.RO2} = V_{0.RO2}/V_g$; $r_{0.H2O} = V_{0.H2O}/V_g$
9	Volume fraction of air in the combustion products at the input of the HRSG	$r_{0.Air} = 1 - r_{0.N2} - r_{0.RO2} - r_{0.H2O}$
10	Volume fractions of the combustion products at the entrance with the HRSG content of excess air:	$r_{1.N2} = r_{0.N2} + r_{mN2}r_{0.Air}$; $r_{1.RO2} = r_{0.RO2}$; $r_{1.O2} = r_{mO2}r_{0.Air}$; $r_{1.Ar} = r_{mAr}r_{0.m}$; $r_{1.H2O} = r_{0.H2O} + r_{mH2O}r_{0.Air}$

The enthalpy of flue gases at the input of the HRSG is determined from the proportions of combustion products obtained in (10) and the set temperature t_0 and the pressure of the P_g using the function of the REFPROP program [11]

h_0 = enthalpy("nitrogen";";";$r_{1.N2}$;";";"CO2";";";$r_{1.RO2}$;";";"oxygen";";";$r_{1.O2}$;";"; "argon";";";$r_{1.Ar}$;";";"water";";";$r_{1.H2O}$;"TP";"SI"; t_0+273,15; P_g). (11)

2.2 Calculation of the composition of the combustion products and it enthalpies at the input and outlet CHE

Table 2. Equations for properties of combustion products at the input and outlet of CHE.

Index	Determined rate	Equation
12	Saturation pressure of water vapor, Pa	$P_{S1} = f(t_1)$, $P_{S2} = f(t_2)$
13	Partial water vapor pressure, Pa	$P_{1.H2O} = P_g r_{1.H2O}$, $P_{2.H2O} = P_g r_{2.H2O}$
14	Dewpoint temperature, °C	$t_{dp1} = f(P_{1.H2O})$, $t_{dp2} = f(P_{2.H2O})$
15	Relative humidity water vapor, %	$\varphi_1 = 100P_{1.H2O}/P_{S1}$, $\varphi_2 = 100P_{2.H2O}/P_{S2}$.
16	Volume fraction of dry gases	$r_{1.dry.g} = 1 - r_{1.H2O}$, $r_{2.dry.g} = 1 - r_{2.H2O}$
17	Partial pressure of dry gases, Pa	$P_{1.dry.g} = P_g r_{1.dry.g}$, $P_{2.dry.g} = P_g r_{2.dry.g}$
18	Molar mass of dry flue gases, kg/mol	$\mu_{1.dry.g} = \mu_{N2}r_{1.N2} + \mu_{O2}r_{1.O2} + \mu_{CO2}r_{1.CO2} + \mu_{Ar}r_{1.Ar}$, $\mu_{2.dry.g} = \mu_{N2}r_{2.N2} + \mu_{O2}r_{2.O2} + \mu_{CO2}r_{2.CO2} + \mu_{Ar}r_{2.Ar}$.
19	Gas constant of dry flue gases, J/(kg·K)	$R_{1.dry.g} = R/\mu_{1.dry.g}$, $R_{2.dry.g} = R/\mu_{2.dry.g}$.
20	Coefficient	$K_{d1} = R_{1.dry.g}/R_{H2O}$, $K_{d2} = R_{2.dry.g}/R_{H2O}$
21	Flue gas moisture content, kg/kg.dry	$d_1 = K_{d1}P_{1.H2O}/(P_g - P_{1.H2O})$, $d_2 = K_d P_{2.H2O}/(P_g - P_{2.H2O})$.
22	Volume fraction of water vapor at the outlet of the CHE	$r_{2.H2O\cdot} = r_{1.H2O} - \Delta d/(K_d + \Delta d)$.
23	Volumetric fractions of elements of dry flue gases at the outlet of CHE	$r_{2.i} = r_{1.i}K_d/(K_d + \Delta d)$.
24	Enthalpy of flue gases at the inlet of the CHE,kJ/kg	h_1 = enthapy("nitrogen";";";$r_{1.N2}$;";";"CO2";";"; $r_{1.RO2}$;";";"oxygen";";";$r_{1.O2}$;";";"argon";";"; $r_{1.Ar}$;";";"water";";";$r_{1.H2O}$;"TP";"SI"; t_1+273,15; P_g)
25	Drop of moisture content in the CHE, kg/kg.dry	$\Delta d = d_2 - d_1$
26	Enthalpy of flue gases at the outlet of the CHE,kJ/kg	h_2 = enthalpy("nitrogen";";";$r_{2.N2}$;";";"CO2";";"; $r_{2.RO2}$;";";"oxygen";";";$r_{2.O2}$;";";"argon";";"; $r_{2.Ar}$;";";"water";";";$r_{2.H2O}$;"TP";"SI"; t_2+273,15; P_g).
27	Enthalpy of the exhaust gases at	$h_{15.1}$ = enthalpy("nitrogen";";";$r_{1.N2}$;";";"CO2";";";

	a temperature of 15°C in the HRSG without CHE,kJ/kg	$r_{1.RO2}$;";";"oxygen";";";$r_{1.O2}$;";";"argon";";"; $r_{1.Ar}$;";";"water";";";$r_{1.H2O}$;"TP";"SI"; 15+273,15;P_g).
28	Enthalpy of the exhaust gases at a temperature of 15°C in the HRSG with CHE,kJ/kg	$h_{15.2}$ = enthalpy("nitrogen";";";$r_{2.N2}$;";";"CO2";";"; $r_{2.RO2}$;";";"oxygen";";";$r_{2.O2}$;";";"argon";";"; $r_{2.Ar}$;";";"water";";";$r_{2.H2O}$;"TP";"SI"; 15+273,15;P_g).
29	Efficiency HRSG without CHE	$\eta_{HRSG.1} = (h_0-h_1)/(h_0-h_{15.1})$
30	Efficiency HRSG with CHE	$\eta_{HRSG.2} = (h_0-h_2)/(h_0-h_{15.2})$
31	Incrementation of efficiency HRSG	$\Delta\eta_{HRSG} = \eta_{HRSG.2} - \eta_{HRSG.1}$

The model is implemented by the program in the Excel package with the calculation of the parameters of water, fuel and flue gases according to the functions of the REFPROP base [9].

3 Results and analysis

With this program, the calculations were carried out with the following input data:
Composition of gas fuel, %: $CH_4 = 98.9$; $C_2H_6 = 0.94$; $C_3H_8 = 0.08$; $C_4H_{10} = 0.08$.
Moisture of air d_{air} =0.01 kg/kg.dry.
The temperature of the gases at the input of the HRSG $t_0 = 550$°C.
The gas pressure in HRSG $P_g = 10100$ Pa.
The temperature of flue gas at the input of the CHE $t_1 = 100$°C.

Table 3. Summary.

Temperature,°C; r, kJ/kg	Properties	α						
		1	1.5	2	2.5	3	3.5	4
t_0 = 550	h_0, kJ/kg	1258.6	1145.9	1087.4	1051.6	1027.4	1010.0	996.9
t_1 = 100	h_1, kJ/kg	718.2	623.8	574.8	544.8	524.6	510.0	499.0
	d_1, g/kg.dry	151.5	106.3	84.3	71.3	62.7	56.5	52.0
	t_{dp1}, °C	60.4	54.0	49.8	46.8	44.5	42.7	41.2
	φ, %	20.0	14.8	12.1	10.4	9.2	8.4	7.7
	$h_{15.1}$, kJ/kg	622.9	531.1	483.5	454.4	434.7	420.5	409.8
	$\eta_{HRSG.1}$, %	85.0	84.9	84.9	84.9	84.8	84.8	84.8
t_2 = 50; r = 2382	h_2, kJ/kg	509.9	517.4	522.7	526.5	529.2	531.3	532.9
	Δd, g/kg.dry	67.7	21.6	-0.8	-14.1	-22.9	-29.2	-33.9
	$h_{15.2}$, kJ/kg	472.6	479.9	485.1	488.7	491.4	493.4	494.9
	$\eta_{HRSG.2}$, %	95.3	94.4	93.7	93.3	92.9	92.7	92.4
	$\Delta\eta_{HRSG}$, %	10.3	9.5	8.9	8.4	8.1	7.8	7.6
t_2 = 40; r = 2406	h_2, kJ/kg	432.5	428.9	428.0	427.7	427.7	427.7	427.8
	Δd, g/kg.dry	104.0	58.4	36.1	22.9	14.2	8.0	3.4
	$h_{15.2}$, kJ/kg	406.6	402.8	401.9	401.6	401.5	401.6	401.6
	$\eta_{HRSG.2}$, %	96.96	96.50	96.19	95.98	95.82	95.70	95.60
	$\Delta\eta_{HRSG}$, %	11.9	11.6	11.3	11.1	11.0	10.9	10.8

$t_2 = 30$; $r = 2430$	h_2, kJ/kg	387.7	378.5	374.4	372.2	370.7	369.8	369.1
	Δd, g/kg.dry	125.1	79.6	57.5	44.4	35.7	29.5	24.9
	$h_{15.2}$, kJ/kg	372.4	363.1	359.0	356.8	355.3	354.4	353.6
	$\eta_{HRSG.2}$, %	98.3	98.0	97.9	97.8	97.7	97.6	97.6
	$\Delta\eta_{HRSG}$, %	13.3	13.1	13.0	12.9	12.9	12.8	12.8
$t_2 = 20$; $r = 2454$	h_2, kJ/kg	359.0	346.8	341.0	337.7	335.6	334.1	332.9
	Δd, g/kg.dry	137.2	91.9	69.8	56.7	48.1	41.9	37.4
	$h_{15.2}$, kJ/kg	353.9	341.7	336.0	332.6	330.5	329.0	327.8
	$\eta_{HRSG.2}$, %	99.4	99.4	99.3	99.3	99.3	99.3	99.2
	$\Delta\eta_{HRSG}$, %	14.4	14.4	14.4	14.4	14.4	14.4	14.4

Figure 1 shows the dependence of the increase in the efficiency of the HRSG on α and t_2,, and in Fig. 2 - dependence of the decrease in moisture content in the CHE from α and t_2.

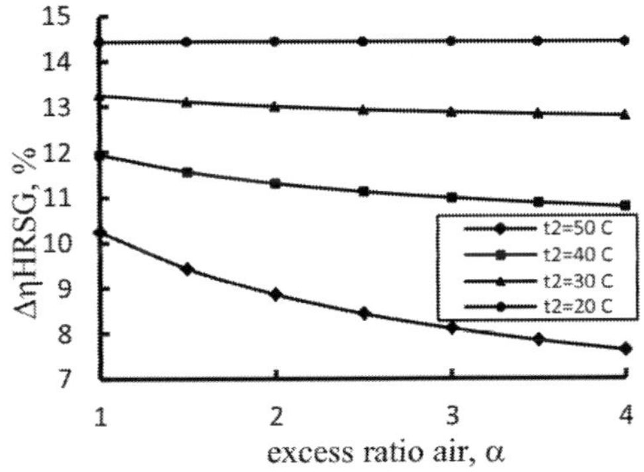

Fig.1. Dependence of the efficiency gains of HRSG.

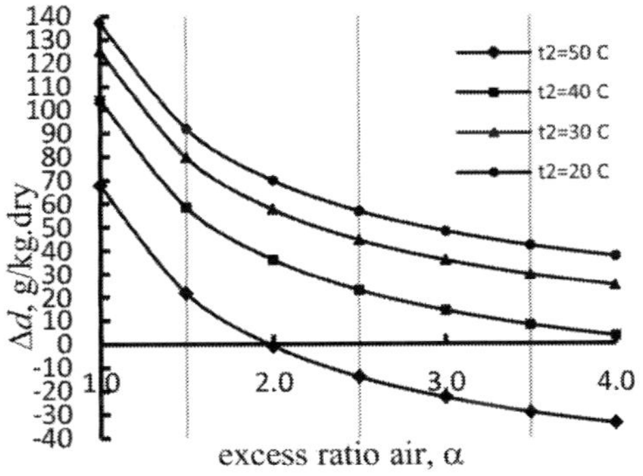

Fig. 2 –Dependence of decrease in moisture content in CHE.

The calculations based on the model showed that, depending on the temperature of the gases emerging from the CHE, the increase in the efficiency of the HRSG at the gas temperature at the outlet of the CHE 20 °C is 14.4% and is independent of the air excess ratio α. At a temperature of 30°C, the efficiency gain drops slightly with an increase in the excess air coefficient from 13.3% at $\alpha = 1$ to 12.8% for $\alpha = 4$. At a temperature of 40°C, the efficiency gain decreases from 11.8% at $\alpha = 1$ to 10.8% for $\alpha = 4$. At a temperature of 50°C, the increase in efficiency gain decreases from 10.2% at $\alpha = 1$ to 7.6% at $\alpha = 4$. A strong drop in efficiency at a temperature of 50 °C is due to the fact that for $\alpha > 1.9$ the temperature of the outgoing gases becomes higher than the dew point, so if the contact point is of the contact type, then there is no condensation of moisture, but evaporation.

References

1. A. Gabdullina, N. Galashov, S. Tsbuskiy, D. Melnikov, I. Asanov, A. Kiselev, Promising Direction of Perfection of the Utilization Combine Cycle Gas Turbine Units, **91**, 01004 (2016)
2. D.V. Gvozdyakov, V.E. Gubin, Chemical and Petroleum Engineering, **51**, 201 (2015)
3. S.A. Yankovsky, V.E. Gubin, A.A. Tolokolnikov, A.V. Zenkov, Strategic Technology (IFOST), 299 (2016)
4. S. Xiaojun, C. Defu, A. Brian, G. Jianmin, Int. J. Heat Mass Transfer, **54**, 606 (2011)
5. K. Hwang, C. Song, K. Saito, S. Kawai. Applied Thermal Enjineering, **30**, 2730 (2010)
6. N. Galashov, S. Tsibulskii, A. Matveev, V. Masjuk, EPJ Web Conf. **110,** 01019 (2016)
7. Hazell D., *Modeling and Optimization of Condensing Heat Exchangers for Cooling Boiler Flue Gas* (Ph.D. Dissertation, Lehigh University, 2011)
8. N. Galashov, S. Tsibulskiy, A. Kiselev, MATEC Web Conf. **72**, 01029 (2016)
9. N. Galashov, S. Tsibulskiy, T. Serova, EPJ Web Conf. **110,** 01068 (2016)
10. E. Lemmon, M. Huber, M. McLinden, REFPROP, standard reference database 23, version 8.0, National Institute of Standard and Technology (2007)

MATEC Web of Conferences **110**, 01028 (2017) DOI: 10.1051/matecconf/201711001028
HMTTSC-2017

Experimental study of thermal processing of solid waste

Olga Kolibaba[1,*], *Ramil* Gabitov[1]

[1]Ivanovo State Power Engineering University, 153003 Ivanovo, Russia

Abstract. In order to develop ways to improve the efficiency of an energy use in thermal reactors, a thermal analysis of the municipal solid waste (MSW) mixture average morphological structure and its components was spent. Experimental studies performed on synchronous termoanalizatore NETZSCH STA 449 F3 Jupiter, combined with a quadrupole mass spectrometer QMC 403. According to the results of the experiments, the temperature ranges of the pyrolysis process and the speed of the mass loss of the sample during solid waste drying and oxidative pyrolysis, thermal effects that accompany these processes, as well as the composition and volume of gases produced in the oxidative pyrolysis of solid waste and its components in the environment oxygen content of 1% and 10% were determined.

1 Introduction

One of the main directions of Russian energy policy is the use of renewable energy sources. These sources include solid organic waste, including household (MSW). The energy use of this raw material allows us to solve a number of problems, such as receiving a gaseous fuel, reducing harmful effects on the environment, involvement in economic turnover of land released from landfills and other.

Currently, there is the possibility of cost-effective, environmentally safe and technically available to recycle MSW without pre-screening using thermal methods, including pyrolysis.

The MSW recycling technologies use different schemes of the pyrolysis installations [1], for example, shaft furnaces in which the pyrolysis is accompanied by a small air leak (oxidative pyrolysis). Oxidative pyrolysis method with subsequent burning of the pyrolysis gases is universal and does not depend on the fractional composition of the waste and a phase condition, their moisture and ash content. At the same time, the products of combustion of pyrolysis gases is less contaminated with fly ash and soot as compared to the direct burning of solid waste, which greatly simplifies the scheme of gas purification.

To improve the efficiency of recycling MSW by oxidative pyrolysis the important task is the experimental study of this process in order to evaluate the influence of several parameters (moisture content of the feedstock, the composition of solid waste, oxygen

* Corresponding author: koli-baba@mail.ru

increase in the atmosphere, and others.) on the gas discharge and composition of the pyrolysis gases [2].

2 Statement of the Problem

Carrying out the thermal analysis of the mixture of MSW average morphological structure and its components for the development of ways to improve the efficiency of its energy use in heat technologies.

As methods of thermal analysis to study MSW oxidative pyrolysis are used:
- thermal gravimetric analysis (TG - analysis) [3], during which the change in mass of the test sample at a change in temperature was recorded continuously;
- differential scanning calorimetry (DSC), during which the thermal effects, which were accompanied by chemical and phase transformations in the sample heating were recorded continuously;
- mass spectrometry gas mixtures (MS - analysis).

Studies oxidative pyrolysis were conducted for samples of the MSW mixture the average morphological composition (paper - 43% food waste - 42%, textile waste - 7.5%, and polyethylene - 7.5%) [4], and the individual components of the mixture in an atmosphere having an oxygen content of 1% and 10%. Experimental studies were carried out using synchronous thermal analysis complex NETZSCH. Measuring complex consists of a block of thermal analysis STA 449 F3 Jupiter, Quadrupole Mass Spectrometer QMS 403 C Aeolos unit and pulse thermal analysis PulseTA.

Synchronous TG - termoanalizator DSC analysis allows to simultaneously measure heat flow and change in weight of the sample depending on the temperature or time. Simultaneous analysis allows not only process a large number of measurements, but also simplifies the interpretation of the results, by enabling to separate endothermic and exothermic processes, which is not accompanied by a change in mass, from processes where mass change takes place.

As a result of the thermal analysis, so-called thermogravimetric (TG) curves - the dependence of the sample mass on the temperature or time are obtained. Based on the value of the derivative of the TG curve (differential thermal gravimetric analysis - DTG), you can find the temperature or the time interval at which the maximum reduction in sample mass takes place.

3 Experiment and processing

By the differential scanning calorimetry (DSC) the amount of heat which is released or absorbed in the chemical reactions was determined. The composition of gas released in pyrolysis is determined by mass - spectrometer QMC 403.

Experimental studies of oxidative pyrolysis was carried out in the following sequence. MSW sample was charged in a platinum crucible, which was mounted on a microbalance in the first cell of the differential scanning DSC sensor. The second comparison sample cell was placed - the crucible without sample (reference). Etalon (crucible without sample) was placed in the second slot. The presence of a double measuring cell allows you to compensate for outside influences (processes associated with heat or heat absorption) on the measurement result. Heat was applied by electric heater in chamber in which the samples were installed. Sample weight ranged 3 ÷ 15 mg. A test sample temperature was measured by differential thermocouples placed on the bottom surface of the crucible, which were connected to the secondary device. Heating was performed in an inert gas (argon) atmosphere, an oxygen-enriched (10% or 1 vol.) of 20°C to 900°C at 5°C / min.

During the experiment, the STA 449 F3 Jupiter continuously recorded decline MSW mass of the sample in the process of heating (TG curve), the temperature of the sample and the DSC signal. Maximum sensitivity of the device by weight was 0.1 g and at a temperature of - 0.001°C.

The qualitative composition of the pyrolysis products was determined with the help of the Quadrupole Mass Spectrometer QMS 403C Aëolos by finding the partial pressure of the individual components of the gas mixture. Separation of the mixture components was produced by its ionization current. A beam of gas ions by using an electric field accelerates to high speed and passes through a quadrupole mass analyzer, which consists of four metal rods, to which a voltage is applied so that at any one time through the analyzer flying ions with only one ratio of mass to charge ratio - M / z.

The process of data collection and processing was automated by means of the device software NETZSCH Proteus Thermal Analysis.

The quantitative composition of the pyrolysis gas is determined by internal standard method [5], according to which test compounds are compared with the signal intensity of the standard substance signal (calibration gas), and to calculate the mass fractions of pyrolysis gas components, the methodology proposed by Kozlov A.N. and Keiko A.V. was used [6].

According to this procedure, the amount of substance of i - th the pyrolysis gas component is determined by the formula:

$$n_i = \frac{S_i}{k_i^0} \tag{1}$$

where: S_i - the peak area under the ion current curve of i - th the pyrolysis gas component (CO, CH_4, H_2, CO_2, H_2O) (Fig. 3), A • s; k_i^0 - calibration factor i - th the pyrolysis gas component.

The area under the peak of the ion current curve i - th component of pyrolysis gas by the formula:

$$S_i = \int_{\tau 1}^{\tau 2} I_i d\tau \tag{2}$$

where: I - ion current, A; τ - time, s.

The calibration factor i - th the pyrolysis gas component is calculated by the formula [9]:

$$k_i^0 = \frac{S_i^0 \cdot 10^8 \cdot R_\mu \cdot T^0}{P^0 \cdot V_i^0 \cdot x_i^0} \tag{3}$$

where i - index of the component (CO, CH4 , H2 , CO2, H2O); S_i^0 - peak area under the curve of the ion current calibration gas, A • s; R_μ - universal gas constant, J / mol • K; V_i^0 - volume of calibration gas, m3; P_i^0 - gas pressure at the inlet of the device, Pa; T^0 - gas temperature at the inlet of the device, °C; x_i^0 - concentration of a component in calibration gas,%.

The calculation of the calorific value of the pyrolysis gas is formed on the principle of additive:

$$Q_n^r = Q_{CH4} \cdot g_{CH4} + Q_{CO} \cdot g_{CO} + Q_{H2} \cdot g_{H2} \tag{4}$$

DOI: 10.1051/matecconf/201711001028

where: Q_{CH4}, Q_{CO}, Q_{H2} - calorific value of methane, carbon monoxide and hydrogen, respectively, J / kg; g_{CH4}, g_{CO}, g_{H2} - mass fractions of methane, carbon monoxide and hydrogen, respectively.

4 Results and Discussion

The thermal analysis results of the sample SHW mixture consisting of paper, textile, food waste, and a polymeric film is shown in Fig.1 as DSC, TG and DTG curves.

The results of TG analysis (TG curve) showed that the curve of a sample mass change during heating may be divided into several distinguishing intervals. The first temperature interval from 20°C ÷ 220°C is characterized by a slight change in mass of the sample (0.5 to 8%) due to evaporation of physical and bound moisture from the surface of the porous structure of the material. With further increase in temperature to 320-330°C, a sufficiently rapid decrease in mass takes place due to thermal decomposition of the substance and volatile compounds formation. When the temperature reached 330°C the mass change curve (Line 1 in Fig. 1) has a distinct bend, which demonstrates decreasing release rate of volatile compounds. At a temperature of 490°C, a sharp drop in the mass loss rate of the substance, slowing of the thermal decomposition of solid waste, takes place. At a temperature of 540°C volatile content ends.

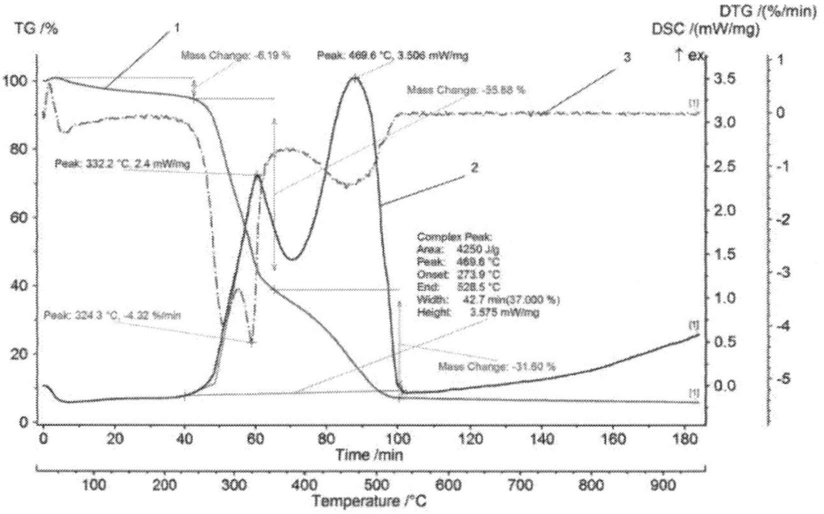

Fig. 1. Curves 1 TG 2 - DSC and 3 - DTG for MSW samples at an oxygen content in the atmosphere of 1%.

Results of differential scanning calorimetry (line 2 in Fig. 1) show that at temperatures up to 220 ° C endothermic processes associated with evaporation of moisture occur in the samples. With further increase of the sample temperature exothermic processes occur.

The DSC curve (line 2 in Fig.1) has two evident exothermic peak at temperatures of 330°C and 470°C. The function then begins to decrease sharply. This indicates that the intensity of exothermic processes is reduced.

The results of differential thermogravimetric analysis (lane 3 in Fig. 1) showed that the maximum decrease in sample mass takes place at temperatures of 330 ° C and 470 ° C, which corresponds to the maximum output of the gas phase MSW pyrolysis.

A similar experiment was performed in an environment with oxygen content of 10%, that intensifies pyrolysis process. The thermal analysis results of the pyrolysis MSW mixture and its individual components in an environment with an oxygen content of 1% and 10% are given in Table 1.

Table 1. Results of thermal analysis of solid waste and its component.

N	Temperatures of the beginning and end of the pyrolysis process, °C		Energy release peak temperature, °C		Residual weight, %		Maximum rate of weight loss, %/min		Thermal effect of pyrolysis, J/g	
O_2	1%	10%	1%	10%	1%	10%	1%	10%	1%	10%
MSW	225÷540	215÷550	469	454	6.48	6.49	4.27	5.89	4250	6685
Paper	245÷492	245÷488	477	444	10.8	9.69	6.06	6.61	4951	6624
Potatoes	225÷540	215÷550	511	511	3.8	4.13	5.29	6.75	3672	6696
Polyethylene	340÷540	280÷550	395	375	10.01	9.96	6.67	5.23	3954	8685
Textile	240÷500	240÷490	345	336	14.34	13.43	10.4	8.68	4376	5925

The qualitative composition of the pyrolysis gas is obtained by treatment of the mass spectrometer signals (ion current as a function of temperature) distributed in the mass numbers (Fig.2). Figure 2 shows the results of measurements in the MSW pyrolysis in an atmosphere with oxygen content of 1%.

Similar experiments were performed for MSW pyrolysis in an atmosphere with oxygen content of 10%

The experimental results are presented graphically in Fig. 2.

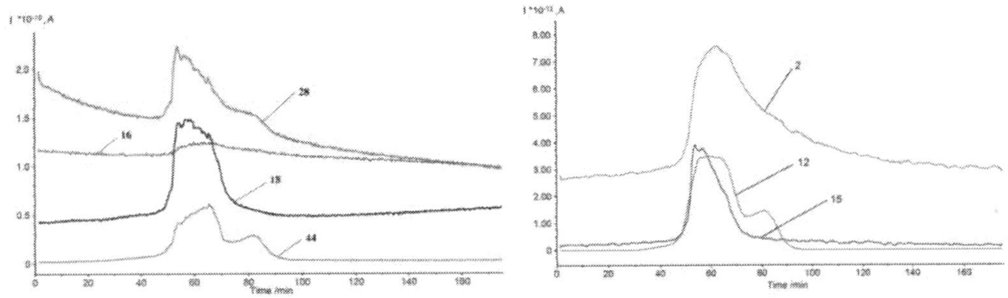

Fig. 2. Mass spectrum of gas formed during MSW pyrolysis in an atmosphere with oxygen content of 1%: a) gases with a high value mass - spectrometer signal, b) gases with a low value of the mass – spectrometer signal (The numbers on the curves indicate the mass number).

To identify the components of the pyrolysis gas, base mass spectrometers NIST data [7], containing information on the mass spectrum of substances were used.

Analysis of Fig.2 shows that the main components of the pyrolysis gas are: H_2 (2), CH4 (16), H_2O (18), CO (28), CO_2 (44). The contribution of other components with mass numbers 12, 13, 14, 15, 17 and 46 in a mass component of the main components is determined by comparison with the results presented in the databases NIST.

Table 2 shows the mass spectrum data of the main components of the gas mixture in the form: Experiment / mass basic value - NIST spectra.

Table 2. Comparison of experimental data with NIST database.

m.ch.	2	12	13	14	15	16	17	18	28	44
H_2	100	-	-	-	-	-	-	-	-	-
CH_4	-	2.8/3	6.8/8	17/20	80/87	100	-	-	-	-
CO	-	4.5/5	-	-	-	0.8/1.1	-	-	100	-
H_2O	-	7.9/9	-	-	-	0.7/1	20/23	100	-	-
CO_2	-	-	-	-	-	8.8/9	-	-	10/13	100

Table 3 shows the qualitative and quantitative composition of gas released in the process of oxidative pyrolysis with oxygen in the atmosphere of 1% and 10%.

Table 3. Mass fractions of the main components of the pyrolysis gas, %.

N	CH_4		CO		H_2O		CO_2		H_2	
O_2	1%	10%	1%	10%	1%	10%	1%	10%	1%	10%
MSW	22.5	11.93	41.25	32.91	18.9	26.66	17.21	28.47	0.14	0.03
Paper	20.7	11	52.51	44.59	13.84	19.64	12.9	24.75	0.05	0.02
Potatoes	22.94	14	32.52	22.55	22.34	32.34	22.11	31.08	0.09	0.03
Polyethylene	16.24	7.2	41.89	35.82	20.58	26.98	20.48	29.94	0.81	0.06
Textile	21.82	9.32	45.05	29.93	20.28	29.75	12.78	30.99	0.07	0.01

The calculation results of the combustible gas calorific value produced in the oxidative pyrolysis in an atmosphere having an oxygen content of 1% and 10% are shown in table. 4.

Table 4. The calorific value of the pyrolysis gas.

O_2	1%, Q, MJ/kg	10%, Q, MJ/kg
MSW	16.73	10.00
Paper	16.44	10.64
Potatoes	16.12	10.10
Textile	13.40	7.70
Polyethylene	15.71	8.30

The oxidative MSW pyrolysis in an atmosphere with oxygen content of 1% is characterized by a high content of carbon monoxide (up to 42% by mass). In this process gas calorific value was 16.8 MJ / kg and heat (pyrolysis effect) of 4.25 MJ / kg.

In the oxidative MSW pyrolysis of in an atmosphere with oxygen content of 10% calorific value of the pyrolysis gas was 10 MJ / kg and heat of 6.7 MJ / kg.

5 Conclusions

1. Thermal analysis of the process of oxidative pyrolysis of a SDW mixture of average morphological structure and its components with an oxygen content in the atmosphere of 1% and 10% is performed. As the result, the data on the output of the volatile compounds, the maximum rate of weight loss and temperature of heat peaks were obtained.
2. Oxidative pyrolysis in the presence of oxygen in the working zone of the reactor 1% is accompanied by a thermal effect less than pyrolysis with oxygen content of 10%, but it is formed over caloric pyrolysis gas.
3. The composition and volumes of gases produced by oxidative pyrolysis of MSW mixture average morphological structure and its individual components are experimentally determined.
4. The results of experimental studies of oxidative pyrolysis can be used to calculate the design and operational parameters of the thermal reactor for processing solid waste.

References

1. P. Basu, *Biomass gasification and pyrolysis. Practical design and Theory* (Academic Press., N.Y., 2010)
2. G.V. Kuznetsov, M.A. Polsongkram, J. Scientific - Technical Gazette Volga, **1**, 125 (2011)
3. W. Wendland, *Thermal methods of analysis* (Mir, Moscow, 1978)

4. Thermal utilization of solid household waste, Krasnoyarsk (2006)
5. A.T. Lebedev, *Mass spectrometry of organic chemistry* (BINOM, Moscow, 2003) [In Russian]
6. A.N. Kozlov, D.A. Svishchev, I.G. Donskoi, A.V. Keiko, J. Journal Thermal Analysis and Calorimetry **3**, 1311 (2012)
7. An open database of mass spectra of NIST : http: //webbook.nist.gov/chemistry

Flue gas moisture capacity calculation at the outlet of the condensation heat recovery unit

Nikolay Galashov[1,*], *Svyatoslav* Tsibulskiy[1], *Denis* Mel'nikov[1], *Alexandr* Kiselev[1], and *Al'bina* Gabdullina[1]

[1]National Research Tomsk Polytechnic University, 634050 Tomsk, Russia

Abstract. As a result, study equation has been obtained which determine the flue gas moisture capacity at the outlet of the condensation heat recovery unit with an error of less than 1%. It possible to at the temperature of the flue gas below the dew point and the known air-fuel ratio efficient. The equation can be used to calculate plants operating on products of gas combustion without Use of tables and programs for calculating the water-vapor saturation pressure.

1 Introduction

The ways to increase gas turbine unit's efficiency are an application of the condensation heat recovery unit (CHE) [1-5]. The condensation heat recovery unit is applied in regions with low air temperature, where air can be used to cool flue gas more efficiency [6]. If the CHE is applied in installations operating on the Rankine Organic Cycle [7-9], the efficiency will be increased due to using the outlet of flue gas. At the same time, by decrease the flue gas temperature and application the heat of condensation of moisture from them, increase of the boiler efficiency can by 10-15%. The most effective is application of CHE in gas burning units because condensed water liberated from flue gas almost does not contain harmful substances except dissolved CO_2 and O_2, which can be easily deaerated. Therefore, the condensate can be useful for both the installation itself and for other purposes.

2 Moisture capacity calculation at the input and output of the CHE

To determine the flue gas moisture content at the input and output of the installation is necessary for calculating CHE. Considering that moisture content determination by calculation is very time-consuming, Seminuk's equations are used for estimative computation of natural gas combustion [8]. It is assumed that at the entrance to the CHE, water vapor in gases has a temperature above the dew point, and at the outlet it is lower. The moisture content is determined in *kg* or *g* of moisture ratiu to 1 *kg* of dry gases.

The moisture content at entrance to the CHE was calculated from the equation

[*] Corresponding author : gal@tpu.ru

$$d_n = (0.13 - d_{nv}\ \alpha_{yx})/(\alpha_{fg} - 0.058), \quad (1)$$

where α_{fg} is the air-fuel ratio in the flue gas; d_{nv} - the air moisture content before burning, kg / kg (cg).

The moisture content at flue gas to the CHE when it is cooled below the dew point was calculated from the equation

$$d_{fg} = P_{wv} /(P_g - P_{wv})\ R_{dg} / R_{wv}, \quad (2)$$

where P_g static pressure in the gas boiler pass, P_{wv} - water vapor saturation partial pressure in Pa, is determined by tables [11] on the temperature of flue gas behind the CHE (t_{fg}) ; R_{dg} / R_{wv} is the ratio of the gas constant of dry gases to the gas constant of water vapor, for natural gas depending on α_{fg} was calculated from the equation [8]

$$R_{dg} / R_{wv} = (0.09924 + 0.622\alpha_{fg})/(0.199 + \alpha_{fg}). \quad (3)$$

The values of R_{dg}/R_{wv} are given in Table 1.

Table 1. Dependence of the ratio R_{dg}/R_{wv} to a α_{fg}.

α_{fg}	1	1.5	2	2.5	3	3.5	4
R_{dg} / R_{wv}	0.6015	0.6076	0.6108	0.6129	0.6143	0.6154	0.6162

In table 2 are presented the calculations of d_{fg} according to equation (2) as a function of α_{fg} and t_{fg} at P_g = 100000 Pa.

Table 2. d_{fg} value, g/kg(dry) was calculated from the equation (2).

Parameters		α_{fg}						
		1	1.5	2	2.5	3	3.5	4
t_{fg}, °C	10	7.48	7.55	7.60	7.62	7.64	7.65	7.66
	20	14.41	14.55	14.63	14.68	14.71	14.74	14.76
	30	26.68	26.95	27.09	27.18	27.25	27.29	27.33
	40	47.96	48.44	48.70	48.87	48.98	49.06	49.13
	50	84.77	85.62	86.08	86.37	86.57	86.72	86.83
	60	149.87	151.38	152.19	152.71	153.06	153.32	153.52

These equations are derived on the basis of reduced fuel characteristic.

Seminuk [8] has suggested equation for moisture content calculation that does not require saturated vapor pressure determination and is widely-used in scientific literature, kg/kg(dry)

$$d_{fg} = (0.0006382 + 0.004\alpha_{fg})/(0.199 + \alpha_{fg})\mathrm{EXP}(0.062 t_{fg}). \quad (4)$$

Table 3 shows the results d_{fg} in g/kg(sg) are presented by the equation (4), and table 4 – the ratio error are presented by equation (4) with ratio to (2).

Table 3. d_{fg} value, was calculated from the equation (3).

Parameters		α_{fg}						
		1	1.5	2	2.5	3	3.5	4
t_{fg}, °C	10	7.19	7.26	7.30	7.33	7.34	7.36	7.37
	20	13.37	13.50	13.57	13.62	13.65	13.68	13.69
	30	24.85	25.10	25.23	25.32	25.38	25.42	25.45
	40	46.19	46.66	46.91	47.07	47.18	47.26	47.32
	50	85.87	86.73	87.20	87.49	87.70	87.84	87.96
	60	159.63	161.23	162.10	162.65	163.02	163.30	163.51

DOI: 10.1051/matecconf/201711001029

Table 4. Ratio error value was calculated from the equation (3) ratio to (2), %.

Parameters		α_{fg}						
		1	1.5	2	2.5	3	3.5	4
t_{fg}, °C	10	3.86	3.86	3.86	3.86	3.86	3.86	3.86
	20	7.22	7.22	7.22	7.22	7.22	7.22	7.22
	30	6.85	6.85	6.85	6.85	6.85	6.85	6.85
	40	3.69	3.69	3.69	3.69	3.69	3.69	3.69
	50	-1.30	-1.30	-1.30	-1.30	-1.30	-1.30	-1.30
	60	-6.51	-6.51	-6.51	-6.51	-6.51	-6.51	-6.51

Table 4 shows that the ratio error calculation by equation (4) for most points more than 6-7%, so in this research, based on multidimensional approximation of the according table. 2 was receive the equation

$$d_{fg} = (4.287+0.0321\alpha_{fg})\text{EXP}(0.0597t_{fg}). \quad (5)$$

Percentage error of computation using equation (5) relative to data in table 2 does not exceed 5%. However, obvious trend is observed between the difference of calculation according to equations (5) and (2) depending on t_{fg}. This trend was approximated by the third degree polynominal with determination coefficient of 0.9992. The result the equation was obtained.

$$d_{fg} = (4.287+0.0321\alpha_{fg})\text{EXP}(0.0597t_{fg}) + (0.7068-0.208t_{fg}+0.01254t_{fg}^2-0.000175t_{fg}^3)(6)$$

Table 5 show the calculations of d_{fg} by equation (6), and in table. 6 – ratio error of calculation according to equation (6) ratio to (2). The error for all points does not exceed 1%. in calculation by equation (6).

Table 5. d_{fg} the value in g/kg (dry) to (6).

Parameters		α_{fg}						
		1	1.5	2	2.5	3	3.5	4
t_{fg}, °C	10	7.55	7.58	7.61	7.64	7.67	7.70	7.73
	20	14.42	14.47	14.52	14.58	14.63	14.68	14.73
	30	26.92	27.02	27.12	27.21	27.31	27.40	27.50
	40	48.29	48.47	48.64	48.82	48.99	49.17	49.34
	50	85.24	85.56	85.88	86.19	86.51	86.83	87.15
	60	150.82	151.40	151.98	152.55	153.13	153.71	154.28

Table 6. Ratio error of calculation by (6) and (2), %.

Parameters		α_{fg}						
		1	1.5	2	2.5	3	3.5	4
t_{fg}, °C	10	-0.97	-0.35	-0.20	-0.24	-0.39	-0.60	-0.85
	20	-0.06	0.57	0.74	0.71	0.58	0.39	0.16
	30	-0.92	-0.27	-0.09	-0.11	-0.23	-0.41	-0.63
	40	-0.69	-0.05	0.13	0.10	-0.02	-0.21	-0.44
	50	-0.56	0.07	0.23	0.20	0.07	-0.13	-0.37
	60	-0.63	-0.02	0.14	0.10	-0.04	-0.25	-0.50

3 Conclusion

As a result, study equation has been obtained which determine the flue gas moisture capacity at the outlet of the condensation heat recovery unit with an error of less than 1%. It possible to at the temperature of the flue gas below the dew point and the known air-fuel ratio efficient.

The equation can be used to calculate plants operating on products of gas combustion without Use of tables and programs for calculating the water-vapor saturation pressure.

References

1. A.A. Kudinov, Increase in the efficiency of the surface-type condensation heat. Industrial energy, **6**, 49 (2002)
2. A.A. Abramovskikh, U.A. Ulyanova, S.V. Lavrinenko, A,A. Matveeva, D. Malishev, MATEC Web Conf. **91,** 01002 (2017)
3. K. Larionov, D. Gvozdjakov, A. Zenkov, Z. Zaytsev, MATEC Web Conf. **110,** 01034 (2016)
4. D. Hazell, *Modeling and Optimization of Condensing Heat Exchangers for Cooling Boiler Flue Gas* (Ph.D. Dissertation, Lehigh University, 2011).
5. N. Galashov, S. Tsibulskiy, A. Kiselev, MATEC Web Conf. **72,** 01029 (2016)
6. N. Galashov, S. Tsibulskii, A. Matveev, V. Masjuk, EPJ Web Conf. **110,** 01019 (2016)
7. N. Galashov, S. Tsibulskiy, T. Serova, EPJ Web Conf. **110,** 01068 (2016)
8. L.G. Semenyuk, Industrial energy, **8,** 47 (1987)
9. Ya.L. Pekker, Energy (1977)
10. A.A. Alexandrov, B.A. Grigoriev, MPEI (1999)
11. E. Lemmon, M. Huber, M. McLinden, REFPROP, standard reference database 23, version 8.0, National Institute of Standard and Technology, (2007)

Possible directions for using bran in heat-power engineering

Maria Gaydabrus[1,*], *Igor* Razov[2], *Roman* Tabakaev[1], and *Vitaly* Lebedev[3]

[1]National Research Tomsk Polytechnic University, 634050 Tomsk, Russia
[2]Tyumen industrial University, 625000 Tyumen, Russia
[3]Omsk State Transport University, 644046 Omsk, Russia

Abstract. In connection with the necessity of providing of ecofriendlyness of energy the researches sent to the increase of stake of biomass in fuel and energy balance are actual direction of science and technique. The aim of work is put the study of heating engineering properties of bran and research of possibility of briquetting as it applies to the power use in the stratified fire-box devices. Research methods included standard methodologies of ISO and physical experiment. It is shown as a result of research that bran on the warmth of combustion is comparable with brown coals, but has a less ash-content considerably. However shallow factious composition of bran and caking in the process of ashing will result in high operating costs at their incineration in a natural kind. The variant of the power use of bran is considered as the preforms got the thermal pressing, and the optimal temperature of pressing, equal 150°C. On the example of pressing of shell of cedar nuts possibility of the use of bran is shown as connective, is set here.

1 Introduction

Bran is a by-product of flour-miller production as a hard shell of grain and, in fact, is departure of food industry. They found the application as feed addition for agricultural animals and bird. However, according to [1] bran serves only as a supplement and cannot completely replace the existing diet. The folded situation results in the market entry of surplus amount of bran, compelling to search the alternative ways of their use.

The aim of work is put the study of heating engineering properties of bran and research of possibility of briquetting as it applies to the power use in the stratified fire-box devices.

2 Material and methods

Bran (Fig. 1) is considered, wheats got at processing one of enterprises of the Kemerovo region. Bran had heterogeneous factious composition: the stake of bran made 14.1% sizes a more than 1 mm; 0.5-1.0 mm – 38.6%; 0.2-0.5 mm – 35.9%; less than 0.2 mm – 11.4%. Before an analysis bran is preliminary dried up at a room temperature to an air-dry state.

* Corresponding author: maria.lazy597@gmail.com

Fig. 1. Prospected bran.

Heating engineering descriptions of bran, certain concordantly ISO 5068-1:2007, ISO 1171:2010, ISO 5071-1:3013, presented in a Table. 1. The value of lower heating and ultimate analysis are set on a high-fidelity analytical equipment – calorimeter of ABK-1 (Russia) and analyzer of Vario Micro Cube (Germany) accordingly.

Table 1. Peat characteristics.

Peat sample	Moisture content W^a, %	Dry ash content A^d, %	Volatile yield V^{daf}, %	Lower heating value Q_i^r, MJ/kg	Ultimate analysis (dry ash-free basis, %)				
					C^{daf}	H^{daf}	N^{daf}	S^{daf}	O^{daf}
Bran	11.5	5.9	81.0	16.6	48.44	6.92	3.05	–	41.59

Bran has a high enough ash content (A^d=5.9%) for biomass. Lower heating value (Q_i^r=16.6 MJ/kg), comparable with the lower heating of row of brown coals [2], allows to examine bran as a fuel for caldron options. A non-flying remain appearing after determination of volatile yield had the alloyed character (Fig. 2).

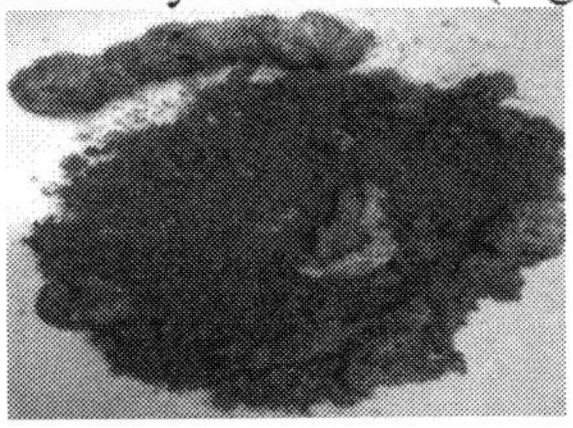

Fig. 2. Non-flying remain appearing after determination of volatile yield from bran.

3 Results and discussion

Taking into account the volumes of exit of bran as wastes (a few tons in twenty-four hours within the limits of only one enterprise), the possible sphere of their power use is see a domestic economy, objects of small energy or coverage of own power expenses of enterprise. An in this case comburent equipment as it applies to that it is needed to examine bran are hot-water boilers with the fire-box devices of the stratified type. By analogy with peat transportation of bran on distance a more than 25-30 kilometres will promote their cost considerably [3] and, probably, will do uncompetitive as compared to other types of fuel.

As it applies to this type of equipment the factious composition of bran given above will result in the high size of failure through a furnace-bar grate. What be more, simultaneously with it in the process of incineration there will be caking of chark remain, hampering work of blow fans and reducing efficiency of burning, that in totality will result in low output-input ratio options and to the high operating costs.

In connection with above-stated by one of perspective directions of processing of bran for the subsequent power use, briquetting can be examined [4].

Possibility of the cold briquetting of bran is considered at a room temperature with the receipt of preform sizes concordantly GOST 54248-2010 at pressure 5-15 MPa. But to form

 DOI: 10.1051/matecconf/201711001030

durable preforms this character not succeeded: preforms collapsed in the process of extraction from a press-form.

The thermal briquetting of bran was investigational at pressure 10 MPa (mean value from the considered range at the cold briquetting). A temperature of briquetting was 100-180°C. On a Fig. 3 the preforms got here are shown.

A preform (Fig. 4) also had a durable protective layer, stability to the abrasion and destruction.

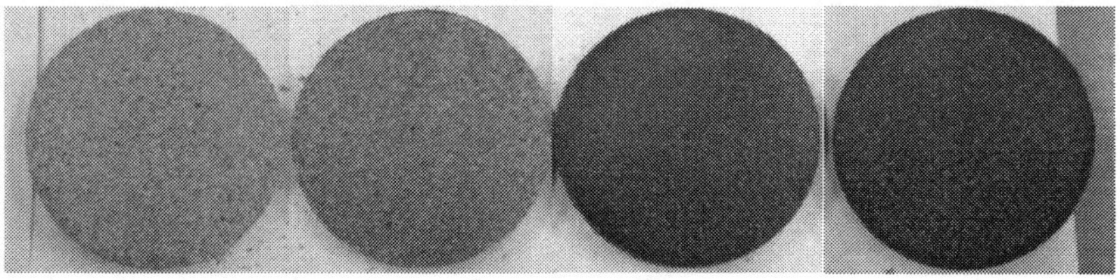

a b c d

Fig. 3. Preforms got the thermal pressing bran (a diameter is a 50 mm): a) – 100°C, b) – 125°C, c) – 150°C, d) – 180°C.

Fig. 4. Preform from bran and shell of cedar nuts, got pressing at 150°C.

4 Conclusions

Bran on the lower heating value is comparable with brown coals, but has less dry ash content considerably. However shallow factious composition of bran and caking in the process of ashing will result in high operating costs at their incineration in a natural kind.

Possibility of the power use of bran is shown as the preforms got the thermal pressing. Thus the optimal temperature of pressing is 150°C.

Bran it maybe to use as connective at making of mixed preforms.

The optimization of the combustion of fuels based on coal and oil processing waste was performed within the framework of the strategic plan for the development of National Research Tomsk Polytechnic University as one of the world-leading universities.

References

1. E.B. Etuk, A.V.Ifeduba, U.E. Okata, I. Chiaka, Okoli, C. Ifeanyi, N.J. Okeudo, B.O. Esonu, A.B.I. Udedibie, J.C. Moreki, J Anim Sci Adv. **2**, 510 (2012).
2. D.F. Umar, H. Usui, B. Daulay, Fuel Process. Technol. **87**, 1007 (2006).
3. V.I. Markov, N.I. Volkova, Ecology and industry in Russia, **1**, 58 (2014) [in Russian].
4. R. Tabakaev, I. Shanenkov, A. Kazakov, A. Zavorin, J. Anal. Appl. Pyrolysis, **124**, 94 (2017)

DOI: 10.1051/matecconf/201711001031

Numerical research of the perforation of targets by ice impactors at different temperatures

Gennady N. Bogomolov[1,*], *Victor* P. Glazyrin[1], and *Maxim* Yu. Orlov[1]

[1]National Research Tomsk State University, 634050 Tomsk, Russia

Abstract. In the present work, the process of interaction of ice impactors of spherical shape with targets of aluminum alloy and asbotextolite has been numerically investigated. The initial temperature of the impactors was set from 0 to 50^0 C. The speed of the impactors was set in the range from 500 m/s to 1500 m/s. The behavior of the medium was described from the general positions of the mechanics of continuous media and corresponded to modern concepts of the destruction of solids under load. The material was considered as isotropic, elastoplastic, compressible, porous medium. The equation of state is chosen in the Walsh form. The problem was solved in a two-dimensional formulation for the case of axial symmetry. The calculations were carried out using the Lagrangian technique of computer simulation, which takes into account the fragmentation of the material in shear and shear failure. The current configurations of the impactor-target and the velocity of the leading fragments are obtained when penetrating the targets.

1 Introduction

The relevance of research into the interaction of ice impactors of spherical shape with targets of aluminum alloy and asbestos cloth is caused by the extensive use of the targets in question when creating new models of aviation and rocket and space technology. In [1, 2] numerical results were obtained on the process of pulsed loading of ice at a certain fixed temperature. However, in terrestrial conditions [3,4] the ice can be kept at a temperature of 0^0C and up to -60^0 C. Consequently, the mathematical model of ice behavior should take into account the dependence of strength physical-mechanical characteristics on temperature.

In the present paper, the problem of the interaction of ice impactors of a spherical shape with targets of an aluminum alloy and asbotextolite is solved. Impactors are at different temperatures. The behavior of materials is described by the basic system of equations of continuum mechanics, i.e. Equations of continuity, motion and energy, written in independent Lagrangian variables [5, 6]. The problem was solved in a two-dimensional formulation for the case of axial symmetry. The current configurations of the impactor-target and the velocity of the leading fragments are obtained when penetrating the targets.

*Corresponding author: glvp@list.ru

This approach makes it possible to predict the behavior of promising impact-resistant protections in order to facilitate the search for the optimal design variant.

2 Mathematical model

The physical formulation of the problem is formulated as the shock interaction of spherical ice bodies with three types of targets equivalent in mass. The initial velocity of the impactor is directed along the axis of symmetry normal to the obstacle plane and is set equal to 500 m/s, 1000 m/s and 1500 m/s. The initial temperature of the impactors was set equal to - 15^0C and - 50^0C. The radius of the impactor is 0.5 cm, the density is 0.92 g/cm^3, the bulk modulus is 8.4 GPa. The first type of target is made of aluminum alloy AMG-6, 0.5 cm thick. The second type of target is Asbotextolite (AsT), 0.875 cm thick. The target of the third type is two-layer. The first layer is AsT, 0.438 cm thick. The second layer is an AMG-6 alloy with a thickness of 0.25 cm. The density of the AMG-6 alloy is 2.85 g/cm^3, the bulk modulus is 76.5 GPa. The density of AsT is 1.6 g/cm^3, the bulk modulus is 4.5 GPa. The remaining physicomechanical characteristics of the materials in question, which are necessary for the calculation, are taken from [3, 4, 6].

We shall assume that the strength characteristics of ice depend on temperature in the following manner [3, 4]:

$$f_i = f_{io}(1+\frac{1}{f_{io}}\frac{\partial f}{\partial t}\Delta T) = f_{io} + a_i \Delta T, \qquad (1)$$

where f_i - shear modulus μ_0, yield stress σ_{TO}, tension spall σ_{ko} and critical work of plastic deformation A_o; f_{io} - initial values of these characteristics. Constants a_i determined from experimental data [3, 4, 7] and the condition of the vanishing ice at zero temperature. The following values of strength characteristics used in the calculations:

μ_0 = 3,1 MPa; σ_{To} =1.85 MPa; σ_{ko} = 3.06 MPa; A_o=0.3 KJ/kg; a_1 = 0.035 MPa/degree; a_2= 0.030 MPa/degree; a_3 = 0.030 MPa/degree; a_4 = 0.02 KJ/ (kg×deg).

The medium under consideration is assumed to be compressible, isotropic, mass forces, internal heat sources and thermal conductivity are absent. Shock-wave phenomena, as well as the formation of detachment and shear destruction, are taken into account. The stress tensor is divided into deviator and spherical components. To determine the hydrostatic pressure, the equation of state in the Walsh form is used. The deviator components of the stress tensor are based on the model of elastoplastic flow based on the Prandtl-Reis equations associated with the Mises yield condition [5, 8, 9].

3 Numerical simulation

Numerical analysis of the processes of deformation and destruction of materials of impactors and obstacles was carried out using a software package simulating the nonstationary motion of a compressible elastoplastic, porous medium, taking into account the detachment and shear nature of the fractures, with the possibility of the appearance of new free surfaces, including dividing the body under consideration into separate fragments [1, 5].

The initial conditions in problems of the shock interaction of solids for the unperturbed state assume that the stress components, the pressure, and the internal energy are zero. The boundary conditions on the contact surfaces allow their sliding without friction. The solution of the given boundary-value problem is carried out in a two-dimensional formulation for axial symmetry by the Johnson Lagrangian method, generalized to the case of large deformations and fracture of the material into fragments by introducing the mechanisms of splitting the calculated nodes and destroying the calculated elements into the computational part of the model [5, 9].

4 Results

The calculated current configurations of the target striker in the interaction of the ice ball with the obstacle from AMG-6 with an initial velocity of 500 m / s are shown in figure 1. It is obtained that already for 4 μs the impactor is covered by a dense network of cracks, and by 100 μs it is completely destroyed into small Fragments. In this case, the target is practically not deformed. Figure 1b shows the results of calculating an analogous variant with an initial shock velocity of 1000 m / s. It can be seen that also for 4 microseconds the impactor is covered with a dense network of cracks, and for 100 microseconds it is completely destroyed into small fragments. The barrier is considerably deformed and destroyed both from the front and from the rear. However, there is no through penetration either at a temperature of -15^{0}C or -50^{0}C.

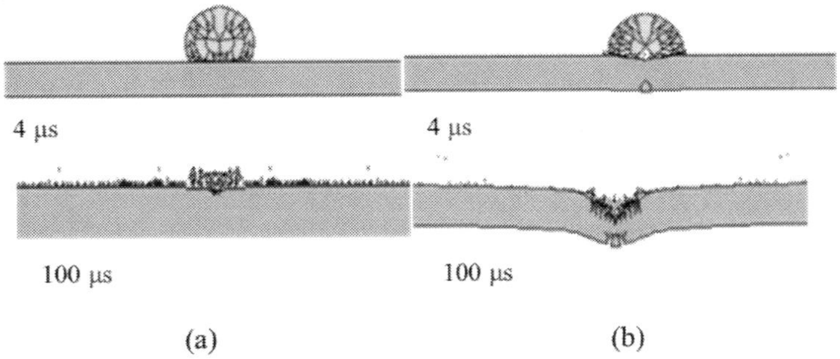

Fig. 1. Results of calculation of the interaction with targets projectiles at T = -15^{0}C; a – V_o = 500 m/s, b - Vo = 1000 m/s.

The calculated current configurations of the impactor -target for obstacles from AMG-6 with an initial velocity of 1500 m / s are shown in figure 2a. On the 4th microsecond, the impactor is practically destroyed. There are also disruptions in the target. At the time of 10 microseconds, the collapsed impactor penetrated the target by half its thickness. At a time of 30 microseconds, a through hole exists in the target, and the leading fragments move with a constant velocity equal to 510 m / s for the variant with the impactor temperature T = -15^{0}C. For the variant with the impactor temperature T = -50^{0}C, the leading fragments move at a velocity equal to 565m/s.

Figure 2b shows the results of calculating the variant with an obstacle from AsT and an initial strike velocity of 500 m/s. It is seen that for 6 μs the impactor is rapidly destroyed, and by 30 μs it is completely destroyed into small fragments. The barrier has fractures of a cross-cutting nature. At 100 μs, a small fragment of the rear surface of the obstacle is observed moving at a speed of 100 m/s. For a variant with a impactor temperature T = -50^{0}C, the velocity of such a fragment is 112 m/s.

 DOI: 10.1051/matecconf/201711001031

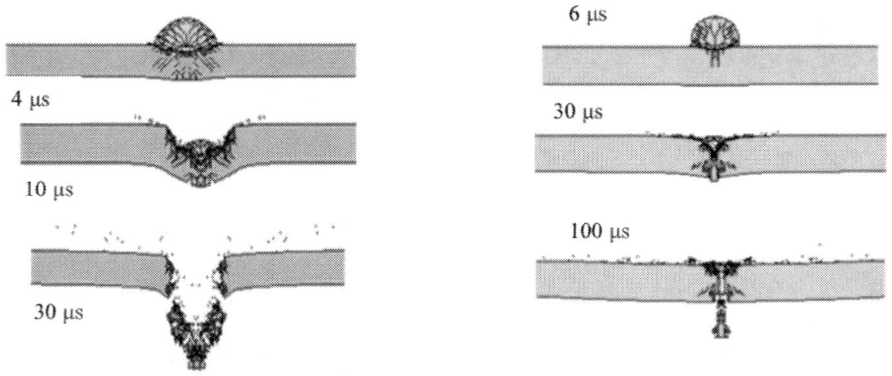

(a) (b)

Fig. 2. Results of calculating the process of interaction of impactors with targets at T = -15^{0}C; a - targets AMG-6, Vo = 1500 m/s, b - AsT targets, Vo = 500 m/s.

Figure 3a shows the results of calculating an analogous variant with an initial impactor velocity of 1000 m/s. It can be seen that in the process of interaction the impactor completely collapsed. At 100 µs, the barrier has a through hole, and the leading fragments move at a constant velocity of 345 m/s. At the impactor temperature T = -50° C, the leading fragments move at a velocity equal to 378 m/s.
For the variant with an initial velocity of 1500 m/s (figure 3b), it turned out that the hole diameter is somewhat larger than in the previous case, and the leading fragments move with a constant velocity equal to 600 m / s. At the impactor temperature T = -50° C, the leading fragments move at a velocity equal to 655 m/s.
The results of the calculation of the process of interaction of impactors with two-layered targets from AsT (0.438 cm) and AMG-6 (0.25 cm) are shown in Fig. 4. It can be seen that at an interaction velocity of 500 m / s (figure 4a), the asbotextolite layer is destroyed in the center, the layer from AMG-6 has small deformations.

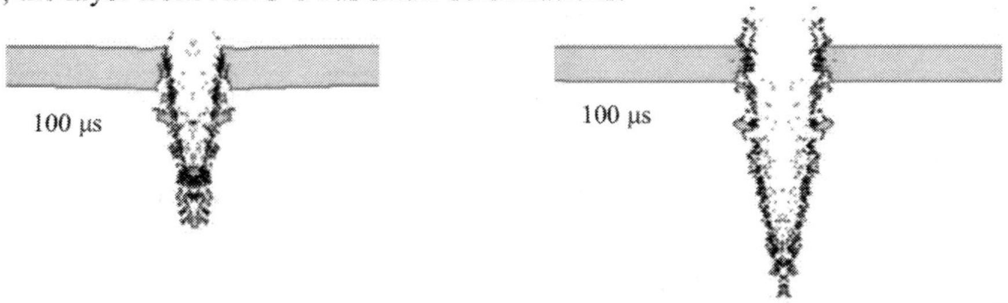

(a) (b)

Fig. 3. Results of calculating the process of interaction of impactors with targets at T= -15^{0}C with targets from AcT; a – V_0 = 1000 m/s, b – V_0 = 1500 m/s.

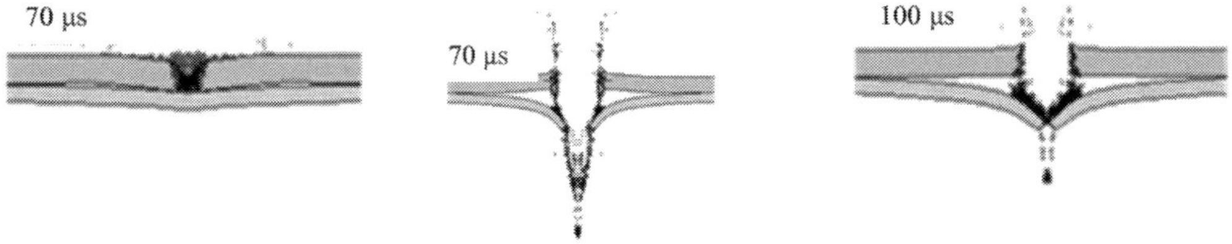

(a) (c) (b)

Fig. 4. Results of calculating the process of interaction of impactors with targets at T= -15^{0}C with two-layered targets from AcT and AMG-6; a – V_0 = 500 m/s, b – V_0 = 1000 m/s, c – V_0 = 1500 m/s.

At an initial velocity of 1000 m/s, there is a hole in the asbotextolite layer, and in the layer of AMG-6 - the destruction along the center along the entire thickness. At a coupling velocity of 1500 m/s (figure 4c), a two-layer obstruction at a time of 70 μs has a through hole, and the leading fragments move at a constant velocity equal to 585 m/s. At the impactor temperature T = -50° C, the leading fragments move at a velocity equal to 635 m/s.

5 Conclusions

Thus, with the use of the developed software complex, numerical investigation of the penetration of targets by ice impactors at various initial temperatures have been carried out. It is established that with the growth of the negative temperature of the shock workers from 15 to 50 degrees, the velocity of the burglary fragments increases from 9.5% to 11.5%. The results of the calculations can be useful in constructing approximating formulas that connect the inhibitory velocity of the leading fragment of the target with the initial velocity of the damaging element for carrying out an express analysis of the corresponding process of impact interaction of solids.

The work was carried out with the financial support of the Ministry of Education and Science of the Russian Federation.

References

1. V.P. Glazyrin, Ju.N. Orlov, Ju.N. Orlova, Izvestija vuzov. Fizika, **52**, (7) 2 (2009)
2. A. Combescure, Y. Cyuzel-Marmot, J. Fabis, Int. J. Solids Structures, **48**, 2779 (2011)
3. V.V. Bogorodskij, V.P. Gavrilo, *Led. Fizicheskie svojstva. Sovremennye metody gljaciologii* (Gidrometeoizdat, Leningrad, 1980) [In Russian]
4. K.S. Carney, Int. J. Solids Structures, **43**, 7820-7839 (2006)
5. A.V. Gerasimov, V.P. Glazyrin *Vysokoskorostnoj udar. Modelirovanie i ehksperiment* (Izdatel'stvo NTL, Tomsk, 2016) [In Russian]
6. V.P. Glazyrin, M.Ju. Orlov, Ju.N. Orlov, Izvestija vuzov. Fizika, **50**, (9) 2 (2007) [In Russian]
7. R.G. Kraus, Earth. Planet. Sci. Lett. **289**, 162 (2010)
8. G.N. Bogomolov, V.P. Glazyrin, M.Yu. Orlov, MATEC Web Conf. **72** (2016)
9. V.P. Glazyrin, Ju.N.Orlov, Research J. International Studies, **11**(42), 3 (2015)

Determination of acceptable limits of adjustment of heat consumption loading

Sergey Glushkov[1,*], *Nikolay* Kachalov[1], *Elena* Senkiv[1], *and Victoriya* Glushkova[2]

[1]National Research Tomsk Polytechnic University, 634050 Tomsk, Russia
[2]Tomsk State Pedagogical University, 634061 Tomsk, Russia

Abstract. The article describes the numerical study of the temperature fields in the living quarters of an apartment building, which made it possible to evaluate the dynamics of the air temperature changes in unsteady conditions as a result of an increase or decrease in the capacity of heating appliances in relation to the standardized value. The permissible limit for adjusting the heat consumption for panel houses is determined, where the permissible period of power variation is not more than 6 hours. The range of power change in this case reaches 40%. The obtained results can find its practical application in intelligent heat supply systems to minimize consumption of fuel and energy resources.

1 Introduction

The use of intellectual technical complexes is a promising reality, allowing at the present stage of society development to ensure the growth of its welfare. In the world practice, the tendency of using these complexes in engineering urban infrastructure was termed "smart city" [1]. It is understood as a city infrastructure of which is built on the basis of technologies allowing rational use of material, labor and energy resources and minimizes environmental impacts. It is obvious that with the implementation of the "smart city" concept, engineering infrastructure systems should have additional options compared to traditional ones. In particular, the elements of energy supply systems should be able to coordinate the production and consumption of energy according to the criterion of minimizing the consumption of fuel and energy resources by the criterion of minimization of consumption of fuel and energy resources. In this formulation, the problem of determining the acceptable limits for adjusting the loads of heat consumption is of theoretical and practical interest. It is known, that the main types of heat load in the social and domestic sphere are heating and hot water supply. The first (heating) during the heating season (for Siberia, its duration is more than 4000 h per year) varies from 20 to 100% of its calculated maximum and is determined mainly by the temperature of the outside air. The second load (hot water supply) on average for residential development is 20-25% of the calculated maximum of the heating load and varies within a day from 0 to 100%, having two pronounced peaks on weekdays and a specific structure on Sundays and holidays, depending on the season of the year.

[*] Corresponding author: sglu2000@mail.ru

According to the control technology in a heating system that is not equipped with automated process control systems, heat release with a qualitative control method is carried out according to the average temperature of the previous day. In the presence of an intelligent superstructure with automated control systems of technological processes in control systems, the optimal temperature averaging period is determined by a number of factors characterizing the structure of the heat network connected to the heat source and the structure of the connected load [2, 3].

In this case, when planning the operation of a heat source, forecasted meteorological data is taken into account. Naturally, in these conditions, there are problems of optimizing the operation of heat sources (equalizing the daily load for boiler houses, working on electric diagrams for thermal power plants, etc.). In addition, it is important to know the limits of heat shortage in emergency situations. To solve these problems, numerical studies were carried out to determine the change in temperature fields in residential premises in case of a decrease (increase) in the consumption of heat by heating systems.

Heating devices are implemented in the model as distributed in the space heat sources, according to the location and other characteristics similar to the corresponding real heating devices.

2 Methods of research

The purpose of conducting numerical experiments is to determine the effect of the type of heater on the dynamics of changes in the temperature of the indoor air in a heated room due to a change in the heat input to the heating system. For a numerical experiment was taken a room located on the ground floor: height H = 2.5 m, length L = 6.0 m. The calculated area in size and shape was a vertical section along the window of the enclosing constructions and the air environment of the room. The thermophysical characteristics of the materials of the enclosing constructions and the initial conditions are identical for all calculations. The outer wall was a three-layer panel. The material of the outer and inner layers is reinforced concrete; the heat-insulating layer is made of expanded polystyrene. The material of the window frame and window sill is wood. Glass units are made of the material with equivalent thermal characteristics. The wall is reinforced concrete. In all calculations, the material of the ceilings separating the room in question from other rooms with an air temperature of 20°C takes into account the properties of the reinforced concrete floor with a floor covering. The peculiarity of the floor construction above the basement, in which the air temperature was 5°C, was that on the reinforced concrete floor an additional layer of thermal insulation was fixed from below (in order to provide a normalized value of the resistance to the heat transfer of the ceiling above the basement). In the calculations, radiant heat exchange of the surfaces inside the room, window panes and the environment was taken into account.

The previous studies have shown that with diurnal fluctuations of the outdoor air temperature of ± 12°C, the amplitude of fluctuations of the temperature of the indoor air in a similar room is not more than ± 0.4°C for different types of heating appliances. Therefore, in this study, the temperature of the outside air was set constant in each of the calculations.

We considered systems with three types of heating devices: 1) radiator, 2) convector, 3) floor heating. The first two types of heating devices were located under the window. In accordance with the features of the calculation algorithm, the power of the Qtot heating devices (W) was set at 1 m of the room width. Heat transfer from the convector and radiator consisted of radiant and convective parts [4].

Calculations were carried out for different values of the characteristic temperatures of the outside air. Each of the calculations consisted of two stages. The first is the starting calculation, which is carried out for reaching the stationary mode. For this purpose, the heat loads of Qtot devices were set in accordance with the heat losses of the room, calculated according to the normative method, from the condition that the internal air temperature is 20 ° C. The second stage is the basic calculation performed with the changed capacity of the heater for a calculation period of 24 hours. During this calculation, the change in the average room temperature was observed.

In the case of using a radiator and a convector the similar patterns of air movement are observed. It is connected with the compactness of these appliances and similar location - under the window. A completely different nature of the emerging pattern of air movement in conditions of using floor heating is caused by the distribution of heating capacity almost across the whole floor area of the room.

With an increase in the input power of up to 30% the temperature increases within 2°C per day, its visible growth occurs in the first 12 hours, and the intensive one - in the first 6 hours. Thus, an increase in power up to 30% does not lead to significant "overheating" of premises.

We can assume that the allowable reduction in the input power is about 40%, at which the temperature in the room decreases by 1.5-2.2°C for a period of 6 hours for the systems with radiators and convectors. For the floor heating, the period of decrease in room temperature by 2°C is about 24 hours. For the remaining outside air temperatures, the decrease in the room temperature over the corresponding time periods is even less [5].

Consequently, it can be considered that the acceptable regulation of the input power to the heating systems is within 40% of the calculated ambient air for a given temperature and a time period of up to 6 hours in case of compliance with the daily balance of the supplied thermal energy, taking into account fluctuations in the outside air temperature. It is crucially permissible to carry out a deeper unloading of the heating system, but in such case the variants when the duration of the cycle "decrease / increase in power" can exceed 12 hours or more are possible. Then it will be difficult to adjust the balance of the energy supplied within a day.

3 Resume

1. The performed numerical study of the thermal fields of living quarters as a result of the reduction (increase) in the capacity of the heating devices made it possible to determine the permissible limits of the regulation of the power of the heat supplied to the targets. The possibilities of accumulating thermal energy by residential buildings without significant fluctuations in the air temperature of the living quarters due to their warming up are limited to a time range of up to 6 hours. For the same period, it is permissible to reduce the heat power supplied to heating systems of residential buildings. The range of power variation can be up to 40%.

2. The obtained results are expedient for use in automated control systems of technological processes of heat supply for the coordination of modes of generation and consumption of energy by the criterion of minimization of fuel costs.

References

1. E. Paes, Rise of the Wise: Next Steps for Smart Cities // Clinton Foundation. URL: https://www.clintonfoundation.org/blog/2013/12/09/rise-wise-next-steps-smart-cities
2. S. Glushkov, N. Kachalov, E. Senkiv, V. Glushkova, MATEC Web Conf. **92**, 01072 (2017)
3. V.A. Vtyurin, *Automated control systems of technological processes: teaching manual* (Publishing house SPbGLA, St. Petersburg, 2006)

4. A.I. Eremkin, T.I. Koroleva, *Thermal conditions of buildings* (Phoenix, Rostov-on-Don, 2008)
5. D.S. Luzhkovoy, O.Yu. Romashova, A.A. Tubolev, Yu.V. Kobenko, E.S. Riabova, MATEC Web Conf. **92**, 01054 (2017)

Bubble size analysis generated on the submerged orifice, depending on the gas supply mode

Sergey Goldaev[1], *Sergey* Basalaev[2,*], and *Kirill* Afanasyev[3]

[1]National Research Tomsk Polytechnic University, 634050 Tomsk, Russia
[2]Tomsk State University, 634050 Tomsk, Russia
[3]Transneft Central Siberia, 634050 Tomsk, Russia

Abstract. A numerical analysis of an algebraic equation of the fifth degree describing the achievement of the diameter of a gas bubble at an aperture is made, the results obtained earlier for limiting cases by simplifying the initial equation and solving it by an analytical method are refined and generalized. With the help of the ratio of the Froude criterion to the Weber criterion, the boundaries of realization of the static, transient and dynamic regimes are established. Approximating formulas are proposed for these regimes, which, with an allowable error in the performance of applied calculations, make it possible to find the value of the detachment diameter of the bubble.

1 Introduction

Process of formation of bubbles of gas on opening cut at its expiration in liquid accompanies aeration processing [1], when heating water with devices of submersible burning [2], stage of cooling of combustion products before filling of covers of buoyancy [3].

The surface area of gazo-liquid structure having impact on efficiency of the considered processes represents total surface of bubbles of the different sizes [2].

It is known that with reduction of diameter of bubbles, the size of phase boundary at identical gas content increases in bubbling layer. Therefore decrease in the sizes of bubbles is one of ways of increase of efficiency of processes of gas-liquid contact [2].

Dynamics of education and separation of bubbles at the expiration of gas in liquid through small opening is described by diameter of d_0 in work [4]. Eighteen publications in which the question by determination of detachable diameter of bubble of D was considered are mentioned in its survey part, is noted that there is no uniform model of this process. Some semi-empirical techniques are given in the monograph [2] according to D which with different accuracy describe the considered process, however results from work [4] are not mentioned.

* Corresponding author: tarm@niipmm.tsu.ru

Fig. 1 shows the scheme for the formation of a gas bubble of diameter D through a small hole d_0 located at a depth h, a gas flow with velocity v_0.

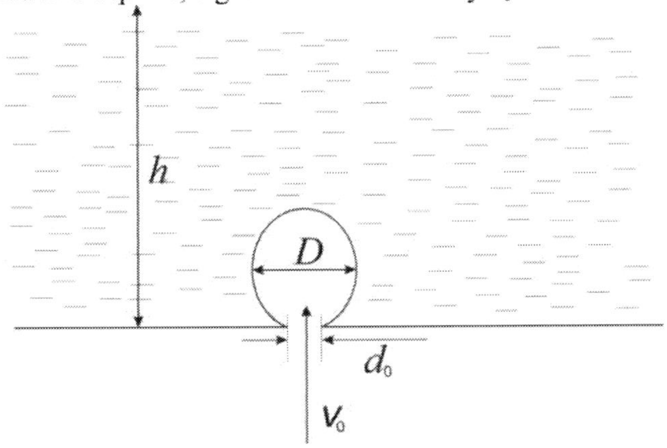

Fig. 1. Scheme of formation of the gas.

The purpose of the real work consists in quantitative analysis of the size of detachable diameter of bubble depending on parameters of the considered process by the numerical solution of the equation proposed in [4] and receiving the approximating expressions allowing to carry out evaluation calculations at preliminary stage of creation of bubblers.

2 Analysis results calculation bubble diameter

In work [4] on the basis of use of equilibrium condition of the gas bubble which is at the time of separation across small hole under the influence of four forces: lifting, surface tension, hydrodynamic pressure of gas and inertial force of liquid, the equation for calculation of detachable diameter of gas bubble is received.

$$L^5 + \left(1.5K_g \cdot \mathrm{Fr} - 6 \cdot \mathrm{We}\right)L^2 = 4 \cdot K_w \cdot \mathrm{Fr}, \quad (1)$$

where $\mathrm{We} = \sigma / \left[\left(\rho_w - \rho_g\right) g d_0^2\right]$ Weber's number; $\mathrm{Fr} = v_0^2 / g d_0$ Froude number; dimensionless parameters $K_g = \rho_g / \left(\rho_w - \rho_g\right); K_w = \rho_w / \left(\rho_w - \rho_g\right)$; $\rho_g, \rho_w -$ gas densities and liquids; σ coefficient of surface tension; g acceleration due to gravity; v_0 gas exhaust speed in liquid; D, d_0 bubble and hole diameters.

Analysis of the values of the ratio $\delta_{FW} = \left(K_g / 4\right) \cdot \mathrm{Fr/We}$ allows us to clarify the boundaries of the implementation of a particular regime.

Quantitative results were obtained in [4] for limiting cases when it was possible to simplify the initial equation (1) and solve it analytically. Specific values of the criteria We and Fr for the respective modes are not given.

In this paper we present the results of a detailed analysis of the effect of the components under consideration on L.

The equation (1) has been solved by method of bisektion [5]. During calculations $d_0 = 1 \ldots 6$ mm, $v_0 = 0.25 \ldots 20$ m/s which corresponded to the experimental data [4] varied.

A comparison of the calculated data with the experimental data from [6] showed a satisfactory agreement.

Taking into account the possible use of bubbling devices in high-pressure plants or gas generators to fill hard and soft pontoons located at a depth of several hundred meters [3], the influence of this factor, which affects the density of the gas filling the bubble, was estimated.

To calculate the bubble diameters in the case of the action of a large hydrostatic pressure, an approximate formula

$$L_{ph} = k_h \left(4 \cdot K_w \cdot \mathrm{Fr}\right)^{0.2}, \tag{2}$$

where $k_h = \begin{cases} 1, h < 250; \\ \mathrm{We} + \left(0.005h / h_1\right)^{-0.7}. \end{cases}$

Calculations have shown that with depth $h = 500$ м value $L_{ph} = 5.62$, which for 5% exceeds diameter of the bubble found the numerical solution of the equation (1), and at $h = 1000$ м – $L_{ph} = 4.49$, and it is 2% more than value of L.

Simulation of the static regime of bubble formation was accomplished by setting low rates of gas outflow into water.

It is established that at d_0 = 1 mm and exhaust speeds of v_0 <0.75 m/s, are carried out inequality $\mathrm{Fr} \ll K_{L3}$, and the error of calculation for formula (2) is in tolerance limits. Therefore it is possible to say that the static mode is implemented. However for diameters of d_0 = 3 … 5 mm and small exhaust speeds the mentioned inequality is not carried out that leads to big error of calculation of L for formula (2).

In order to generalize the calculated data for the static and transient regimes by the method of selected points [4], an interpolation formula was obtained containing the Froude and Weber criteria:

$$L_a = \left(1.296 + 0.493\,\mathrm{Fr}^{0.318}\right) k_c, \tag{3}$$

where $k_c = \begin{cases} 1/\left(1 + 0.25\,\mathrm{We}\right), & \mathrm{We} < 0.85 \\ \mathrm{We}/\left(1 + 0.65\,\mathrm{We}\right), & \mathrm{We} > 0.85 \end{cases}$.

Table 1 shows the results of the calculation of the dimensionless diameter of the L_a bubble according to this formula, and also the relative error in comparison with the values of L.

Table 1. Diameter of the bubble at different flow rates and hole diameters.

v_0, m/s	0.2	0.4	0.6	0.8	1.0
	d_0 = 0.003 m; We =0.85				
Fr	1.36	5.44	12.2	21.8	34
$\delta_{FW} \cdot 10^3$	0.55	2.2	4.9	8	14
L	1.88	2.14	2.39	2.61	2.81
L_a	1.52	1.77	1.98	2.16	3.07
	d_0 = 0.005 m; We =0.31				
Fr	0.8	3.26	7.34	13.04	20.4
$\delta_{FW} \cdot 10^3$	0.9	3.7	8.2	15	22.9
L	1.49	1.80	2.06	2.28	2.47
L_a	1.64	4.20	4.94	5.54	6.06

It is established that the proposed approximation formula describes, with an error not exceeding 10%, the results of solving equation (1) in the static and transition regions.

Since the volume flow of gas through the hole, the diameter of the bubble and the frequency of its formation are related by the relation [2]

$$Q_0 = f\pi D^3 / 6,$$

where f the frequency of forming of gas bubble in liquid layer calculated on formula

$$f = 1.5(v_0 / D)(d_0 / D)^2,$$

Thus, the calculation of the diameter of the bubble with greater accuracy makes it possible to reasonably choose the parameters of the bubbler.

3 Conclusion

By numerical solution of the algebraic equation of the fifth degree describing the achievement of the diameter of the gas bubble on the hole, the results obtained earlier for limiting cases by simplifying the initial equation and solving it by an analytical method are refined and generalized. It is established that at values of parameter $\delta_{FW} \approx 10^{-4}...10^{-3}$ the static mode takes place, at $\delta_{FW} \approx 10^{-3}...7\cdot10^{-2}$ transitional, at $\delta_{FW} > 7\cdot10^{-2}$ the dynamic. Approximating formulas for the listed modes which with admissible error of performance of applied calculations allow to find value of detachable diameter of bubble are offered.

Research is funded by Russian Science Foundation grant (project No 15-19-10014).

References

1. S.S. Kutateladze, M.A. Styrikovich, *Gidrodinamik of gas-liquid systems* (Energy, Moscow, 1976) [in Russian]
2. A.N. Alabovsky, P.G. Udyma, *Devices of submersible burning* (Publishing house of MEI, Moscow, 1994) [in Russian]
3. V.D. Barsukov, S.V. Goldaev, N.P. Min'kova S.A. Basalaev, *Underwater burning solid fuels dibasic* (Publishing house of Tom., Tomsk, 2016) [in Russian]
4. A.A. Voloshko, A.V. Vurgaft, V. N. Frolov, J. Engin. Phys. Thermophys. **35** (2), 213 (1978)
5. L.I. Turchak, P.V. Plotnikov. *Basics of numerical methods* (Fizmatlit, Moscow, 2005) [in Russian]
6. V.P. Myasnikov, A.P. Mitronov, N.A. Kochergin, V.V. Dil'man, Dokl. Academy Sci. of the USSR, **269** (4), 827 (1983) [in Russian]

MATEC Web of Conferences **110**, 01034 (2017) DOI: 10.1051/matecconf/201711001034
HMTTSC-2017

Power engineering challenges in Zambia

Elena Gorkaltseva[1, *], *Uchivyo* Sinkala[1]

[1]National Research Tomsk Polytechnic University, 634050 Tomsk, Russia

Abstract. The paper presents a review of current power engineering challenges in Zambia in relevance to the required potential necessary for the country's modern economic development. An outline of the energy sector is provided. The driving need for enhancing the development of the energy sector is highlighted, accounted by the country's annual increase of GDP, leading to an increase in electricity demand. Possible mitigations are suggested, with the development of Zambia's rich potential in renewable energy, as well as upgrading the operating power plants and constructing more off-grid and on-grid generating facilities.

1 Introduction

Generating electricity can be used to represent the reality of human intellect. Constituting a physical phenomenon of moving electrons, it is empowered in energy as a thermodynamic quantity equivalent to the capacity of a physical system to do work. As the human effort is slowly being replaced by machines that require scientific energy (power), so the need for a country to develop or do some adjustments in the energy sector increases. Power generation has an enormous impact on the people around, or a country in general. Unfortunately, in some countries electrical energy is still an Olympic gold medal.

Zambia, a Southern African country, has the size of 752.681 square kilometers, a population close to 15 million. These facts result in a conflicting statement, since the vast territory for the specified number of people could be viewed as a favorable factor for the country's development. Nonetheless, a vast majority of land is still virgin, and only segmented portions of it are densely populated. Zambian economy is predominantly determined by mining, agriculture, manufacturing, and tourism.

Zambia, which won its independence in 1964, has the heritage of a few decades of one-party socialist government. The country is currently a multi-party democracy state and has been going through a period of economic and political reforms, many of which have affected the energy sector. Back in 1991 the government started privatizing state-owned enterprises, trying to increase their performance.

However, its feats for modernizing its economy have not been as effective as they should be, due to its inadequate supply of power. The country's GDP has been increasing at the rate above 6% [1], which technically has shown that each year there is a 3% increase in electricity demand [2].

* Corresponding author: elena_gork@mail.ru

According to the estimates done in 2014, electrification rate in Zambia was extremely low - at 22% with 45% city dwellers and only 3% of people in rural areas [1]. It has slightly improved since then. Still, only about 25% of the total population have access to electricity, while many people have to rely on kerosene and candles almost every single day. The current situation in power supply in Zambia is described as energy crisis.

2 Energy sector review

Zambia's energy resource potential includes bio-energy resources, hydro, solar, wind and geothermal power. In spite of its richness, the potential in power generation and supply needs to be enhanced and balanced.

The primary energy source for the majority of people is biomass (firewood and char coal), which builds up 80% of supply. It is relied on to cover people's needs for cooking and heating, and accounts for over 70% of energy consumption (with 57% of urban populations and 97% in rural communities) [1].

The major source of electrical energy in Zambia is hydro power, which cumulatively account for about 95.9% of the total installed electricity generation in the country [3, 4]. Hydro power only covers 12% of the energy needs [1].

The generation of electricity in the country is predominantly carried out by three hydro power stations: Victoria Falls Power Station, Kariba North Bank, Kafue Gorge. The Victoria Falls Power Station, comprising 3 power stations, has the total capacity of 108 MW. The Kariba North Bank, one of the world's biggest man-made reservoirs, currently supplies about 720 MW of power; The Kafue Gorge generates 900 MW of electricity. However, the installed capacity of some of these power plants is higher [2].

Table 1. Major Power Stations in Zambia.

Power station	Installed Capacity
The Victoria Falls Power Station	108 MW
The Kariba North Bank	1080 MW
The Kafue Gorge	990 MW

All these three major suppliers of the country's power are owned by ZESCO (Zambia Electricity Supply Corporation) [3], the largest power producer. So far, it has experienced considerable difficulties in meeting the demands of the country's growing economy. Hydro power generation potential in Zambia is theoretically estimated to be 6000 megawatts [2]. However, it has not been employed to its full, with only 1700 MW currently exploited [1]. A considerable impediment in sustainable reliance on hydropower is dependence on rainfall [5, 6], among other factors, the major one is the lack of investments for modernizing the plants.

The remaining 4.1% of the electrical energy is obtained from diesel sources and mainly covers the needs for transport, agriculture, and mining. However, the country depends heavily on export in obtaining diesel fuel.

3 Measures to resolve energy crisis

To alleviate the energy deficit the government has implemented the procurement programs [4]. They are known as "load shedding program" and can be regarded as one of the most unwelcomed programs in modern Zambia. The programs involve electricity supply according to revised load shedding schedules of between 5 and 6 hours for the domestic, industrial and commercial customers on a rolling basis through a 24-hour period. Only mines are the unaffected entities in this crisis, due to their vital role in the healthy state of the economy. Industrial produce has drastically decreased due to reduced power supply.

This incapability of ZESCO and others players in the energy sector of Zambia to meet the modern energy needs has raised concerned citizens and the government to develop measures to make the power deficit less impactful. The government has implemented policies like the" feed-in tariff policy" which involves collaboration with different stakeholders in order to increase investments in the energy sector. The construction of new and upgrading of the hydro-electric power stations are highly demanded [2].

Currently an important direction in which measures are intended and attempted to be taken is furthering the development of renewable energy sources. Zambia's potential in this area is believed to be significant [1, 2].

Biomass, being the major source of renewable energy, has been holding an important position in supplying energy to cover population's living needs. This energy source is rather varied and includes agricultural residues (maize, rice, wheat), forest residues, livestock waste, municipal waste [7, 8]. Moreover, according to the conducted research, this energy source even holds favorable prospects in terms of addressing the problem of Zambia's dependence on imports of petroleum products [9].

Nonetheless, more attention should be granted to more advanced technologies, the major of which is furthering the initiated procurement process for the deployment of up to 300 MW from solar power [10]. Solar power has a high potential in Zambia, since the country experiences about 3000 sunshine hours per year.

Another field of mitigations is increasing investment in geothermal energy, with the prospect to use the power of more than 80 hot springs. Exploring and exploiting geothermal energy is demanding in terms of cost and resources. Nowadays there is only one geothermal plant, built in late 1980s, operating and is planned to be upgraded for the production of 2 MW of electricity [2].

The strongest suggestion in terms of the energy crisis in Zambia is the application of decentralized energy. It means creating power-supply networks with energy sources built near the place of consumption [11]. One of the major tasks in coping with energy deficit is building mini-hydro power stations [1], especially in the North and the North-West of the country. Solutions for both on-grid and off-grid applications are supported.

The other measures are the importation of power from the neighboring countries. Zambia cooperates with a number of countries (Zimbabwe, Namibia, Tanzania, Botswana) in an interconnected grid.

4 Conclusions

The struggle for a modern country can best be fought with the country's electricity abundance. The following steps can be proposed to overcome the current power engineering challenges in Zambia:

- the construction of new on-grid and off-grid hydro-electric power stations, locating them near the place of consumption is highly advisable;
- upgrading the existing power stations;
- focusing on alternative sources of power, specifically solar, bio-energy sources, wind and geothermal power.

Zambia remains a remarkable investment destination in the energy sector. Its location and the development of new interconnectors coupled with power deficit in the region builds up a good export market potential.

References

1. *Zambia renewables readiness assessment.* (2013) Available: http://www.irena.org/DocumentDownloads/Publications/RRA_Zambia.pdf [Accessed April; 2017]
2. *Energy sector profile* (by Zambia Development Agency) (2014) Available: www.zda.org.zm/?q=download/file/fid/55 [Accessed April; 2017]
3. J. Kbaki. *Case study in Corporate Governance: Industry Focus – Electricity.* (2009) Available: https://www.oecd.org/daf/ca/corporategovernanceofstate-ownedenterprises/44200010.pdf [Accessed April; 2017]
4. D. Mulongoti, G. Mugala, B.Kumwenda, H. Louie, GHTC 2016 - IEEE Global Humanitarian Technology Conference: Technology for the Benefit of Humanity. Conference Proceedings, **7857324**, 488 (2017)
5. B. Tembo, B. Merven, J. Energy Southern Africa, **24**(2), 16 (2013)
6. K. Pilli-Sihvola, S. Väätäinen-Chimpuku, Int. J. Disaster Risk Reduction, **19**, 461 (2016)
7. A. Shane, Sh.H. Gheewala, B. Fungtammasan, Th. Silalertruksa, S. Bonnet, S. Phiri, Bioenergy resource assessment for Zambia. Renewable and Sustainable Energy Reviews , **53** (2016)
8. L. German, G. Schoneveld, Rev. of Policy Research, **29** (4), 467 (2012)
9. D. Drabik, H. de Gorter, G.Timilsina, Food Policy, **59**, 103 (2016)
10. K. Kornbluth, B.Pon, P. Erickson, Renewable and Sustainable Energy Reviews, **16,** 6737 (2012)
11. N.L. Behnke, T. Klug, R. Cronk, K.F. Shields, K. Lee, E.R. Kelly, G. Allgood, J. Bartram. Resource mobilization for community-managed rural water systems: Evidence from Ghana, Kenya, and Zambia. J. of Cleaner Production (2017)

Mathematical modeling of the formation of sedimentary acid precipitation in the atmosphere in view of the evaporation of moisture from their surface

Dmitry Gvozdyakov[1,*], *Vladimir* Gubin[1], *Svetlana* Shvab[1], Anastasia Sidorova[1], and *Nikolay* Semenov[1]

[1]National Research Tomsk Polytechnic University, 634050 Tomsk, Russia

Abstract. The article presents the results of numeric simulation of the formation of sedimentary acid precipitation in the atmosphere taking into account the evaporation of moisture from their surfaces. It is established that the joint condensation of vapors of sulfuric anhydride and water vapor, given the flow of solar energy and the evaporation process significantly slows the growth of drops. The possibility of achieving the underlying surface by the formed sediments is analyzed.

1 Introduction

The Earth's atmosphere is a composite of many gaseous compounds of natural [1] (oxygen, nitrogen, etc.) and anthropogenic (oxides of nitrogen, sulfuric anhydride, etc.) origin [1–4].

It is known [5] that the geometric characteristics of atmospheric droplets repeatedly changed in the process of movement. When the droplet diameter (d) from 10^{-6} m to $100 \cdot 10^{-6}$ m, typically, the droplets have a spherical shape [6], if $100 \cdot 10^{-6} < d < 1 \cdot 10^{-3}$ m of the ellipsoid. Based on the results of [6] we can assume that the geometric parameters of the droplets will affect the area of their deposition on the Earth's surface, as the motion of droplets of non-spherical shape different from the spherical movement not only quantitative characteristics (higher values of coefficient of drag – C_D), but the mechanism of movement.

Previous investigations [7-8] illustrate the mechanism of formation of sedimentary acid precipitation in the atmosphere in areas of thermal power stations location without considering the process of evaporation of moisture from their surface.

The aim of this work is mathematical modeling of the formation of sedimentary acid precipitation of spherical shape in the atmosphere in view of the "moderate" evaporation and the major significant factors.

We used the iterative algorithm developed for solving nonlinear heat conduction problems in a local heat exposure [9] method of finite differences.

* Corresponding author: dim2003@.tpu.ru

According to [6], the influence of non-sphericity coefficient drops characterized non-sphericity G, as determined ratio of the surface area of a real drop in the area of similar size sphere.

2 Mathematical problem

Heat equations drops and movement within the accepted physical model of the test process is as follows:

$$\left.\begin{aligned} \rho_1 \cdot C_P \cdot \frac{\partial T_1}{\partial t} &= \lambda \left(\frac{\partial^2 T_1}{\partial r^2} + \frac{2}{r} \cdot \frac{\partial T_1}{\partial r} \right) \\ \frac{dU_X}{dt} &= \frac{3 \cdot G \cdot C_D \cdot \rho_1 \cdot (U_X - U_0) \cdot (U_X - U_0)}{8 \cdot \pi \cdot \delta \cdot \rho_2} \\ \frac{dV_Y}{dt} &= g + \frac{3 \cdot G \cdot C_D \cdot \rho_1 \cdot (V_Y - V_0) \cdot (V_Y - V_0)}{8 \cdot \pi \cdot \delta \cdot \rho_2} \end{aligned}\right\} \qquad (1)$$

where $\rho_{1,2}$ – density, kg/m^3; C_P – isobaric heat capacity, $J/(kg \cdot K)$; λ – coefficient of thermal conductivity, $W/(m \cdot K)$; T – temperature, K; t – time, s; V_X, V_Y – drop velocity in the direction of the axes x and y, respectively, m/s; C_D – drag coefficient; g – acceleration of gravity, m/s^2; U_0, V_0 – velocity components ambient, m/s.

The initial and boundary conditions are:

$$t = 0: T = T_0, 0 \le r \le \delta, \qquad (2)$$

$$r = \frac{\delta}{2}: \quad -\lambda_1 \cdot \frac{\partial T_1}{\partial r} + Q \cdot \left(W_K^{SO_3} + W_K^{steam} \right) - P = -\lambda_2 \cdot \frac{\partial T_2}{\partial r}, \quad T_1 = T_2; \quad r = 0: \quad \frac{\partial T_1}{\partial r} = 0, \qquad (3)$$

where T_1, T_2 – the temperature drops and a gaseous medium, respectively, K; $W_K^{SO_3}$, W_K^{steam} – mass condensation rate, $kg/(m^2 \cdot s)$; Q – the heat of reaction, J/kg; δ and r – the characteristic dimensions of the drops, m; P – density of solar energy flux, W/m^2.

Mathematical modeling performed at the following values of the parameters characterizing the initial conditions of the considered process: U_0=5 m/s; V_0=0.5 m/s; temperature drops at the initial moment of time: summer – 294 K; winter – 262 K. Investigations of the influence of geometrical characteristics of emerging acid drops are made for range of time from 6000 s to the surface of the condensation nuclei sized from $5.00 \cdot 10^{-6}$ m.

In Fig. 1, 2 shows the trajectory of the droplets in the air space in winter and summer.

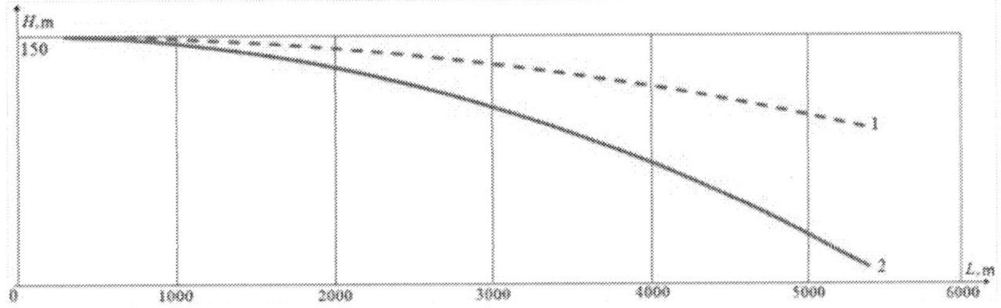

Fig. 1. The trajectory of the drops during the summer: 1 – taking into account the process of evaporation; 2 – without taking into account the process of evaporation.

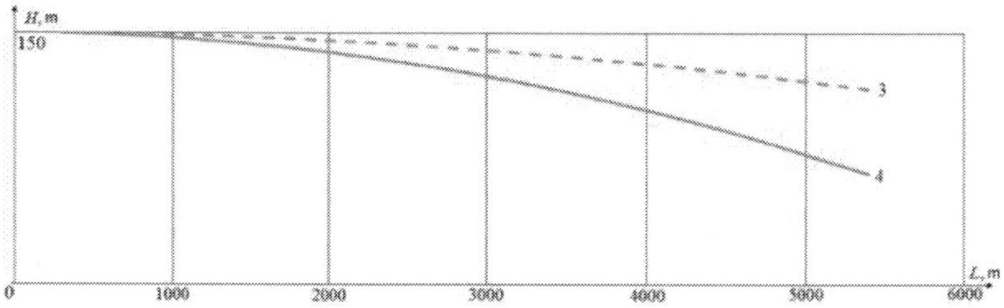

Fig. 2. The trajectory of the drops in winter: 3 – taking into account the process of evaporation; 4 – without taking into account the process of evaporation.

Two thermal periods are analyzed, typical for regions with sharply continental climate given the mechanism of moisture evaporation from the surface of the forming precipitation. The value of the flux density of solar energy was taken equal to the value of the solar constant.

The simulation results show that the trajectory of the particles change significantly given the mechanism of moisture evaporation from the surface of the forming precipitation. It is established that in summer, during the time interval, precipitation are formed that are unable to reach the underlying surface in a given radius (6000 m from the emission source, located at the height of 150 m). Atmospheric drops in this period of the year under given settings and conditions are to be at the height of 120-150 m from the underlying surface. A similar pattern is observed in the winter (see Fig. 2, feature 3).

3 Conclusion

Analysis of the results of mathematical modeling of process of formation of sedimentary acid precipitation leads to the conclusion that co-condensation of vapor of sulfuric anhydride and steam with account for the flow of solar energy and evaporation of moisture from the surface drops significantly (up 40%) retards the growth drops. The resulting precipitation in the studied radius with given parameters are not reached the underlying surface.

The research has been accomplished with financial support of the Russian Federation Ministry of Education and Science within implementation of the project №13.7644.2017/BCH.

References

1. G.E. Zaikov, S.A. Maslov, V.L. Robalo, *Acid rain and the environment* (Chemistry, Moscow, 1991)
2. E.P. Volkov, E.I. Gavrilov, F.P. Dugih, *Flue pipe TPP and NPP* (Energoatomizdat, Moscow, 1987)
3. K.V. Slyusarskiy, K.B. Larionov, V.I. Osipov, S.A. Yankovsky, V.E. Gubin, A.A. Gromov, Fuel, **191**, 383 (2017)
4. K.V. Slyusarskiy, A.V. Zenkov, S.A. Yankovsky, A.A. Matveeva, S.V. Lavrinenko, A.A. Gromov, MATEC Web Conf. **72** (2016)
5. Y.E. Geguzin, *Drop* (Science, Moscow, 1973)
6. V.I. Terekhov, M.A. Pakhomov, *Heat and mass transfer and hydrodynamics in gas-drop flows*, (Novosibirsk, NSTU, 2008)
7. D.V. Gvozdyakov, V.E. Gubin, Chem. Pet. Eng. **51** (3), 201 (2015)
8. D.V. Gvozdyakov, V.E. Gubin, S.A. Shvab, A.O. Tanishev, EPG Web Conf. **72** (2016)
9. G.V. Kuznetsov, M.A. Sheremet, Russ. Microelectron. **37** (2), 131 (2008)

Modeling of the polymerization process of the insulating layer of cable thermocouples

Iashutina Olga[1,*], *Asadullina* Liliya[1], and *Chernova* Galina[1]

[1]National Research Tomsk Polytechnic University, 634050 Tomsk, Russia

Abstract. A mathematical model describing the main physical processes occurring during thermal processing of cable thermocouples (CT) during their production is presented. The polymerization of the insulating layer is taken into account in the conditions of its intense heating. The developed mathematical model allows to carry out calculations of heat transfer processes under the conditions of substantial heterogeneity of cable thermocouple structure. The computational experiments carried out have established significant differences in the characteristics of heat transfer in CT in the presence of thermoelectrodes with differing thermophysical characteristics.

1 Introduction

One of the most important controlled parameters of technological processes in chemical [1], petrochemical [2], energy [3], fuel, and other industries is temperature. The latter is the most informative characteristic that determines the entire physics of the technological process.

At present, two main methods for measuring temperature have been developed: contact (with the help of thermocouples, resistance thermometers, liquid thermometers, etc.) [4] and non-contact (using a pyrometer) [5]. The most preferred method of temperature measurement is the contact method. The latter is due to the cheapness and simplicity of the corresponding equipment of contact methods in comparison with non-contact methods. The most common of contact methods for measuring temperature is thermocouple [6]. The thermocouple is a sensor that consists of two interconnected dissimilar electroconductive elements, usually metallic conductors or sometimes semiconductors [7].

Unlike conventional industrial thermocouples in a sturdy ceramic housing, the cable thermocouple has such advantages as the flexibility and tightness of the housing [8]. Cable thermocouples are used in a very wide temperature range (from 23K to 2800K) [9] in all industries (from cryogenic technology to high-temperature heat engineering installations).

At present, in spite of the large-scale experimental [10] and theoretical [11] studies of heat and mass transfer processes that take place in the manufacture of thermocouple products, the physical theory and the mathematical basis for the thermal processes that occur in the manufacture of cable thermocouples have not been developed yet. The latter is

* Corresponding author: osyl@tpu.ru

DOI: 10.1051/matecconf/201711001036

due to the complexity of calculating the characteristics of heat transfer processes occurring under conditions of intense thermochemical (polymerization) transformations in a small (up to $1 \cdot 10^{-3}$ m) one.

The purpose of the work: mathematical modeling of heat transfer processes proceeding together under conditions of intense radiation-convective heat transfer in the manufacture of cable thermocouples.

2 Problem statement

In the formulation of the problem, it was assumed that at the initial instant of time the cable thermocouple enters the high-temperature medium (intracameral space of the polymerization furnace). Under the influence of intense radiation-convective heat transfer, its heating occurs. As a result, the process of polymerization of the internal insulation of the cable thermocouple is initiated. It should be noted that thermoelectrodes with different thermophysical characteristics (thermal conductivity λ, heat capacity C, density ρ) are symmetrical in relation to the thermocouple axis. Accordingly, it can be reasonably assumed that the location of the thermal center of symmetry will differ from the localization of the geometric center of the thermocouple. In mathematical modeling, the system in Figure 1 was considered.

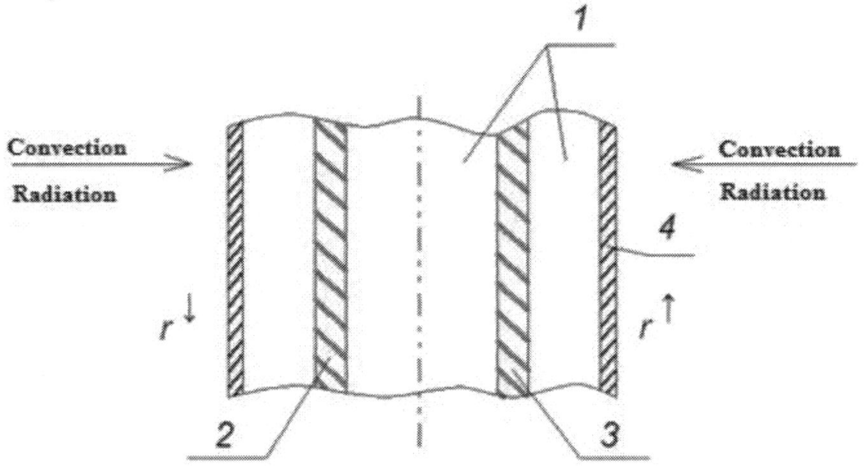

Fig. 1. Scheme of the problem solution area: 1 - insulation powder (MgO); 2, 3 - thermoelectrodes; 4 - cable thermocouple sheath.

3 Results and discussion

In order to verify the mathematical model, a test calculation of heat transfer was carried out with the thermophysical characteristics of thermoelectrodes being equal. The results of the computational experiments are shown in Fig. 2. It can be noted that the temperature curve is a parabola with a center of symmetry on the axis of the thermocouple. The latter testifies to the physical nature of the calculation results. Accordingly, we can say that the developed mathematical model conducts a prognostic analysis of heat transfer processes under conditions of intense heating of CT with a sufficiently high degree of accuracy.

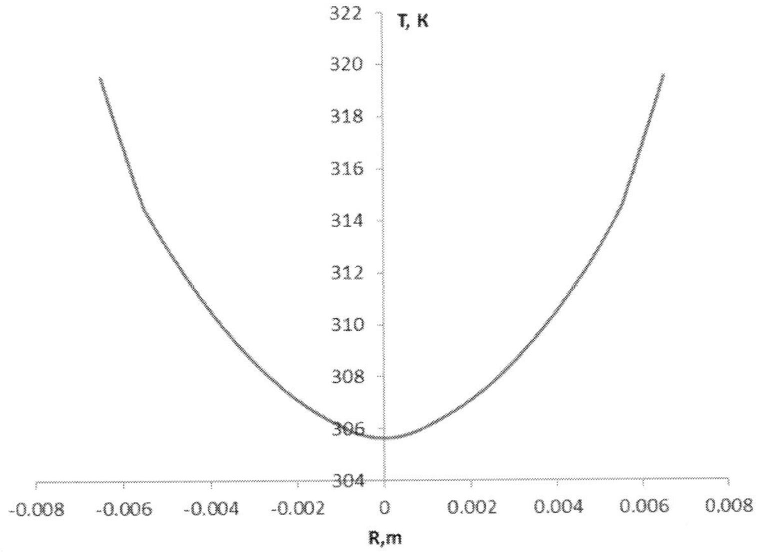

Fig. 2. The parabolic temperature curve.

Figure 3 shows the temperature dependence on the material of the cable thermocouple thermoelectrodes.

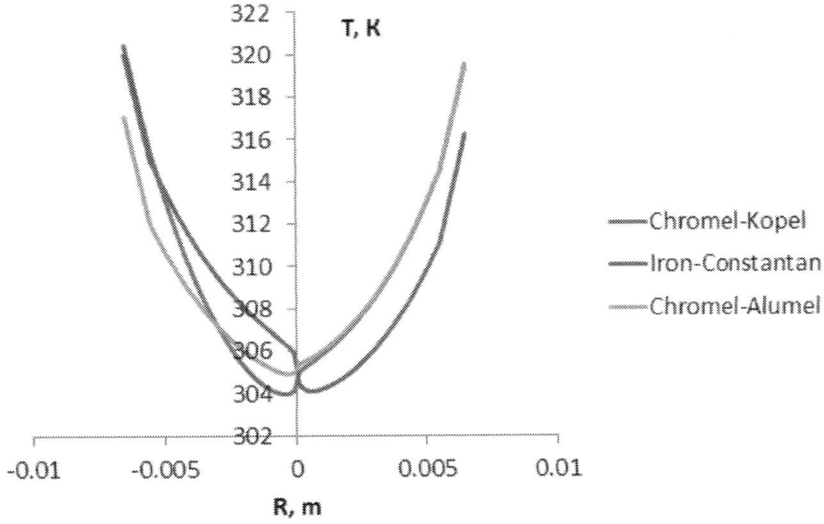

Fig. 3. Temperature distribution for different thermoelectrode materials.

Analyzing the obtained dependences, it can be noted that the temperatures at the boundaries of the "cable thermocouple-external environment" system ($r_0^{\downarrow} < r < r_0^{\uparrow}$) are different. The latter indicates significant differences in the heat transfer conditions in both parts relating to the thermocouple axis.

Figure 4 shows the distribution of the polymerization parameters along the radius of the system under consideration.

The highest degree of polymerization is observed on the surface of a cable thermocouple. This is due to the high heat flows at the outer boundary and the high thermal conductivity ($\lambda = 250$ W/m·K) of the aluminum shell.

 DOI: 10.1051/matecconf/201711001036

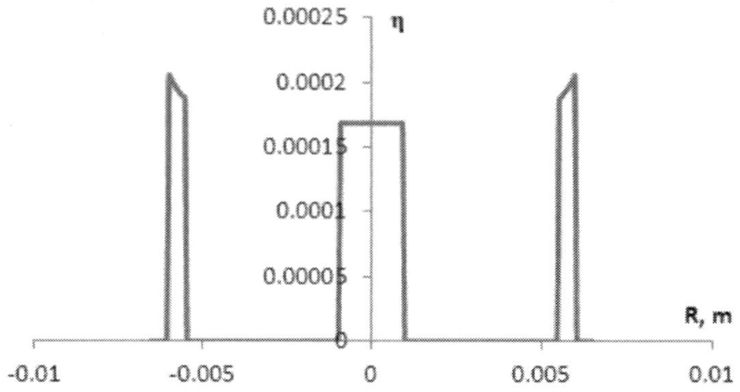

Fig. 4. Distribution of the polymerization parameters along the radius of the system.

3 Conclusions

For the first time, a mathematical model of the processes of heat and mass transfer proceeding together under conditions of intense thermochemical transformations (polymerization) when heating the cable thermocouples is presented. It should be noted that the displacement of the thermal center of symmetry relative to the axis of the thermocouple has been taken into account.

Based on the results of numerical modeling of heat transfer under conditions of radiation-convective heating, a significant inhomogeneity of the temperature field has been established, due to the difference in the thermophysical characteristics of thermoelectrodes. It was also shown that the degree of polymerization depends significantly on the location of the insulating layer relative to the center of the thermocouple.

References

1. M.V. Miroshkin, Exposition Oil Gas, **34** (2014)
2. I.S. Sheleba, Photon-Express, **6** (2015)
3. V.I. Nazarov, A.L. Burov, E.N. Kriskina, *Thermotechnical measurements and devices* (Tutorial, 2012) [in Russian]
4. E.V. Khaliulin, A.V. Kozlov, Bulletin of the Ufa State Aviation Technical University, **3** (2014) [in Russian]
5. A.V. Frunze, Adv. Powder. Metall. Part. Mater., **5**, 40-42 (2009)
6. Yu.K. Atroshenko, I.P. Ozerova, P.A. Strizhak, Scientific and technical statements of SPbSPU. Electrical engineering, **214** (2015) [in Russian]
7. Yu.K. Atroshenko, P.A. Strizhak, Sensors and systems, **192** (2015)
8. V.A. Karzhavin, A.V. Karzhavin, Instrum Sci. Technol., **12**, 12-22 (2010)
9. O.A. Geraschenko, A.N. Gordov, A.K. Eremina, *Temperature measurements*, (1989)
10. Y.K. Atroshenko, P.A. Strizhak, O.S. Yashutina, EPJ Web Conf., **82** (2015)
11. Y.K. Atroshenko, P.A. Strizhak, EPJ Web Conf., **76** (2014)

MATEC Web of Conferences **110**, 01037 (2017) DOI: 10.1051/matecconf/201711001037
HMTTSC-2017

Evaluation of the laser-flash method and its errors for determining of materials thermal diffusivity at high temperatures

Mark D. Katz[1,*] , *Ilija* M. Katz[1]

[1]National Research Tomsk Polytechnic University, 634050 Tomsk, Russia

Abstract. The article deals with evaluation of determining constructional materials thermal characteristics systematic errors by laser-pulse method under high temperature conditions. The article gives a detailed analysis of the influence of heat exchange with surrounding air on systematic errors. It is shown that laser pulse method can be used for calculation of thermal diffusivity and thermal capacity of the metals and alloys with high thermal diffusivity for temperatures not exceeding 1173 K.

1 Introduction

Applying advanced materials for engineering systems requires knowledge of their thermal characteristics [1]. For this purpose, nowadays, different methods of thermal characteristics (TC) determination are used. But for TC determination at high temperature these methods become more complicated. More expensive experimental facilities and measurement systems should be used in this case. Also, a preparation time and duration of experiments are increasing too [2, 3]. In light of this, more simple methods of thermal characteristics determination should be used. One of these methods is a laser flash method (LFA) [4–8]. According to LFA, the maximum temperature (T_{max}) of a sample heating and heating time (τ_{05}) of the sample to half of T_{max} are defined using experimental time-based temperature data of the reverse ("cold") surface of the sample.

The thermal diffusivity a of the material can be found through approximate solution of the one-dimensional heat conduction problem for an infinite plate [4]:

$$a = 0.1388L^2 / \tau_{0.5}, \tag{1}$$

where L is trickiness of the sample.
For equation (1) solution values T_{max} and τ_{05} must be known.

Conditions for TC reliable determination at low temperature band and estimation errors of it are described in a fairly large number of papers [4-10]. TC at medium and high temperatures (no greater than 1173 K) can be determined only in experimental way. But

* Corresponding authors: riki@tpu.ru

these experiments are carried out under conditions of active heating of a sample. Under such conditions TC of the sample is very hard to detect [1, 5, 6]. So, at the first stage of the TC determination process it might be useful to carry out theoretical experiment. Results of this experiment could be used for planning of physical experiment. In this case, the determination of TC by theoretical way is based on mathematical modelling of thermal conductivity process in plane sample, under relevant conditions for LFA [9,10].

Comparison of calculated TC by mathematical modelling with TC assumed during this simulation in [9,10] is allowed to evaluate influence various factors to errors of TC determination. The approach [9, 10] is used for determination of TC of highly heat-conductive materials at medium and high temperatures (no greater than 1173 K).

The aim of the present work is theoretical analysis of the errors in determining the TC of materials for engineering systems at medium and high temperatures by the LFA.

2 Statement of the problem

During the LFA development it was supposed, that there is not abstraction of heat form the sample surface to the outdoor environment [4, 5]. As was shown in [9, 10], such assumption doesn't influence the TC of the sample at low temperature. In the case under consideration, when temperature is over 1000 K, heat-gravitational convection will significantly affect the heat flow pattern of a sample during the experiment. This phenomenon also has an impact on the further results of measurement of T_{max} and $\tau_{05.}$ Therefore, it is necessary to take into account abstraction of heat from the sample surface, which caused by natural convection. Different approaches to the solution of this task are used.

In particular, this process can be described by a conjugate convective heat transfer problem [11, 12]. However, implementation of algorithms [11, 12] requires the significant time resources. To simplify them, integral parameters (surface heat-transfer coefficients) should be used.

Simulation of heating processes during laser pulse exposure with thermal flux density q and pulse time Δt were made for case of horizontal placed thin plate. Abstraction of heat from the sample surface, which caused of the natural convection, was taken into account during this simulation.

Simulation was made for typical construction materials: stainless steel 12H18N10T, titanium and silicone carbide at temperature band 673–1173 K.

Heat transfer was modeled under the following assumptions: 1) the materials are gray and their self-radiation is small in comparison with the external radiant flux; 2) thermophysical parameters of materials at the temperature band of heating by laser pulse correspond to the temperature of sample preheating; 3) cooling-down of the sample is owing to abstraction of heat from the front and back surface of the sample, which caused the natural convection; 4) linear dilatation of the sample during temperature rise was not taken into account.

The first assumption doesn't restrict the problem statements and real conditions of heat interchange through the material.

The mathematical model that corresponds to the adopted formulation of the problem involves one-dimensional nonstationary differential energy equations for thin plate with constant thermal flow q directed to the heated surface during laser pulse time Δt and convective heat transfer to the outdoor environment (fig. 1).

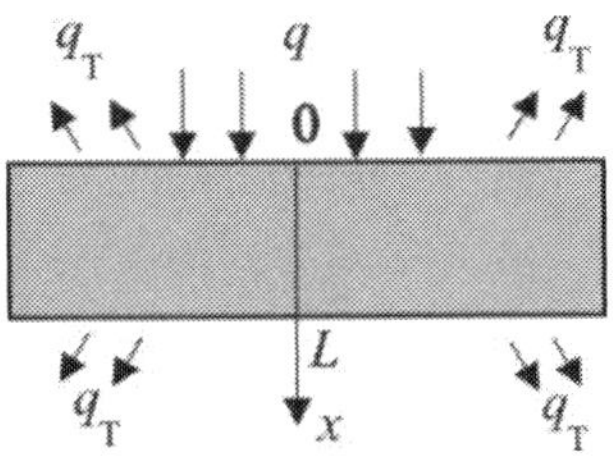

Fig. 1. Scheme of the area of solution.

The mathematical model includes nonstationary partial differential heat-transfer equation (2) with boundary conditions (3)–(5) and initial condition (6):

$$c\rho\frac{\partial T}{\partial t}=\lambda\left(\frac{\partial^2 T}{\partial x^2}\right),\quad 0<x<L \tag{2}$$

$$x=0:\qquad -\lambda\frac{\partial T}{\partial x}=q+\alpha(T_e-T),\quad \alpha>0,\quad \Delta t>t>0; \tag{3}$$

$$x=0:\qquad -\lambda\frac{\partial T}{\partial x}=\alpha(T_e-T),\quad \alpha>0,\quad t>\Delta t; \tag{4}$$

$$x=L:\qquad \lambda\frac{\partial T}{\partial x}=\alpha(T_e-T),\quad t>0; \tag{5}$$

$$t=0:\qquad T=T_0, \tag{6}$$

where c - heat capacity, J/ (kg·K); ρ- the density, kg/m^3; T – temperature, K; t – time, s; Δt – the duration of the laser pulse, s.; q – radiation flux density directed to the heated sample surface, W/m^2; T_0 – initial value of the current temperature, K; T_e – outdoor environment temperature, K; α – heat-exchange to outdoor environment coefficient, W/(m^2○K).

The above mentioned nonlinear boundary value problem was solved by the method of finite differences using an iteration algorithm, developed for the numerical solution of high-temperature heat-transfer problems under conditions of active physicochemical transitions of high-energy condensate liquid substances [13]. The reliability of the results of the numerical modeling was checked by comparison them with experimental data [14]. Also conservatism of used difference scheme according to procedure discussed in [13] was analyzed.

3 Results and discussion

Simulation of heat transfer process was made for stainless steel 12H18N10T, titanium and silicone carbide at temperature band 673–1173 K under the following conditions: $q=3.5\cdot10^7\ldots6.6\cdot10^7$ W/m^2, $\Delta t=1.0$ ms, $T_0=300$ K. Numerical values of TC were adopted in accordance with the papers [15, 16]. Samples thickness was chosen 5 mm according to conditions in [9] to minimize error of TC determination method.

DOI: 10.1051/matecconf/201711001037

During the simulation a time step was chosen 1 ms. The calculations were performed with moves in space to 2.5 microns. Heat-exchange to outdoor environment coefficient α varied within the limits from 0 to 5 W/m^{2}K.

The values of the T_{max} and τ_{05} were found based on the results of the numerical solution of the problem (2–7). The thermal diffusivity was calculated by (1).

Errors of TC determination by LFA method under heat-exchange to outdoor environment and convective heat transfer in the sample conditions were defined by comparing reference value of thermal diffusivity [16, 17] with calculated one.

The calculated thermal diffusivity for steel 12H18N10T was also weighed against experimental data [14] at high temperatures T_0. Results of this comparison showed, that the difference between these values didn't exceed 5 %. That's why we can conclude that mathematical model (2)–(7) is valid for simulation of heat transfer in a plate, heated by short laser pulse at high-temperature region (no greater than 1173 K). Also this mathematical model can be used for TC determination of construction materials at temperature band 673–1173 K.

During calculation of sample temperature pattern, heated to high temperature, phenomenon of heat abstraction from the sample surface at plate borders should be taken into account, because heat losses are increasing with rising of heated and "cold" surfaces temperature. That, in turn, led to additional errors of T_{max} and τ_{05} determination.

Figure 2 presents the dependences of the errors in the determination of the thermal diffusivity δ_a on the heat-exchange to outdoor environment coefficient α under conditions of natural convection. It was established that δ_a depends not only on level of heat abstraction, but also it depends on material thermal conductivity.

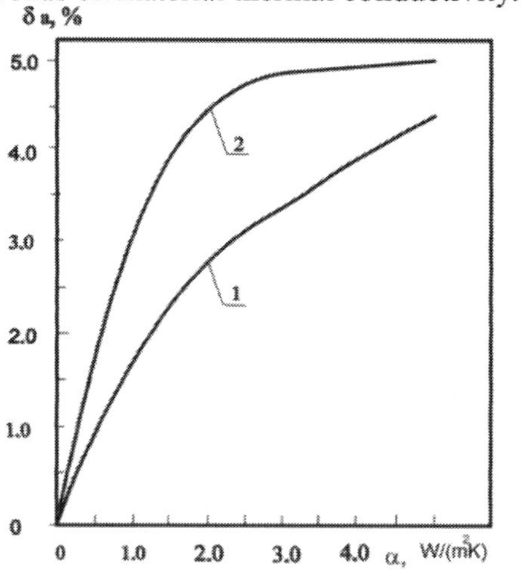

Fig. 2. Errors in determination of the thermal diffusivity δ_a the convective heat transfer coefficient at heat temperature 873 K: 1 – titanium, 2 – silicone carbide.

This fact is shown in fig.2, the error in the determination of the thermal diffusivity δ_a of silicone carbide exceeds errors of other materials. It mainly owing to relatively low thermal conductivity of silicone carbide as against titan thermal conductivity. As shown in fig. 2, under conditions of short heating pulse, steel sample doesn't warm fast. Temperature of it surface doesn't exceed 899.5 K. The heat abstraction process form the sample surface to the outdoor environment (air) occurs at heat difference about 600 K. During the heating of silicone carbide the surface temperature is higher (939 K) and the heat difference is about 639 K. Stored during the pulse duration heat at thin surficial region of front sample surface doesn't transfer to the "cold" surface, which peak temperature is used as governing characteristic for a calculation. This heat is transferred to the outdoor environment.

The results of the simulation give good reason to draw conclusions that the laser pulse method [4] can be used for determination of TC of materials (metals and alloys) for engineering systems at medium and high temperatures (no greater than 1173 K). For experimental determination TC of materials with low thermal conductivity measurements should be done in special closed boxes. The temperature in these boxes must be close to the sample heating temperature. It's necessary to provide the conditions of low heat losses of the sample.

Conclusion

An analysis of the systematic errors in the determination of the thermophysical characteristics of materials (metals and alloys) for engineering systems at medium and high temperatures (no greater than 1173 K) has been made by of the mathematical simulation of heat transfer processes.

Applying of LFA for TC determination under condition of convective transfer to environment and its errors were evaluated.

It was established, that LFA method can be used for TC determination of highly heat-conductive materials under high temperature conditions.

References

1. E.E. Son, High Temp. **51**, 3 (2013)
2. M.V. Ilichev, V.B. Mordinskii, D.V. Tereshenok, High Temp. **53**, 2 (2015)
3. Z.A. Musaeva, V.E. Peleckii, High Temp. **40**, 6 (2002)
4. W.J. Parker, R.J. Jenkins, C.P. Butler, J. Appl. Phys. **32**, 9 (1961)
5. M. Akoshima, T. Baba, Int. J. Thermophys. **27**, 4 (2006)
6. O. Altun, B. Erhan, A. Kalemtas, J. Achievements Mater. Manuf. Eng. **30**, 2 (2008)
7. D. Moskal, J. Martan, V. Lang, Int. J. Heat Mass Transf. **102**, 1 (2016)
8. S. Lee, D.H. Kim, High Temp. High Pressures, **45**, 5 (2016)
9. G.V. Kuznetsov, M.D. Kats, Meas. Techn. **55**, 4 (2012)
10. G.V. Kuznetsov, M.D. Kats, Meas. Techn. **53**, 6 (2010)
11. G.V. Kuznetsov, V.A. Sheremet, Computational Thermal Sciences, **1**, 3 (2009)
12. G.V. Kuznetsov, V.A. Sheremet, Fluid Dynamics, **41,** 6 (2006)
13. O.V. Vysokomornaya, G.V. Kuznetsov, P.A. Strizhak, Chem. Phys. **30**, 8 (2011)
14. S.V. Stankus, I.V. Savchenko, A.V. Baginskii, High Temp. **46**, 5 (2008)
15. V.S. Chirkin, *Thermal-physical properties of materials for nuclear engineering: Manual* (Atomizdat, Moscow, 1968) [in Russian]
16. A.S. Zubchenko, *Steels and alloys grade guide* (Machinostroenie, Moscow, 2003) [in Russian]

DOI: 10.1051/matecconf/201711001038

Research of thermal conditions over high-temperature gas-fired infrared emitters

Anton N. Ermolaev [1], *Sergey* A. Khaustov [2,*]

[1]Tyumen industrial University, 625000 Tyumen, Russia
[2]National Research Tomsk Polytechnic University, 634050 Tomsk, Russia

Abstract. The paper presents the study results of the thermal conditions in the area above high-temperature gas-fired infrared emitter. A number of bench tests and experiments were made on the basis of production facilities to control the distribution of temperatures above emitter in different heating system operating modes. Impact of the thermal characteristics in the area above high-temperature gas-fired infrared emitter on the heating system performance was estimated. Comparison of the bench tests results with existing experimental data has shown a good result convergence for both efficiency and accuracy. The obtained results can be used in the emitter development phase and in the construction phase of modern gas-fired radiant heating systems.

1 Introduction

Recently, there are increasingly used radiant heating systems based on high-temperature gas-fired infrared emitters with the goal of energy resource saving [1, 2]. This heating method has significant benefits in comparison with traditional convection heating systems [3, 4]. Modern radiant heating systems based on gas-fired infrared heat emitters lead to high energy efficiency when heating large-scale production buildings [1, 5].

Despite the fact that contemporary high-temperature gas-fired infrared emitters have high energy efficiency, during their operation for heating large-scale production buildings a number of drawbacks was discovered [5, 6]: upper zone was overheated; smut appeared on walling surface resulting in wall subsequent destruction. Listed drawbacks are caused by considerable heat losses with stack gases (24-30 %) and by emitter design (9-18 %). Strauch-Sürken, Michael Wendt. Tierärztliche Praxis suggest that heat-insulating properties of reflector material effects the radiant heating system, and propos a self-brazing aluminum clad steel [7] for producing gas-fired infrared emitter's reflecting surfaces.

We carried out a series of experiments on specially designed experimental setup to obtain temperature fields formed in the area above high-temperature gas-fired infrared emitter's installation. Object of the research were high-temperature gas-fired infrared emitters with thermal capacity of 5, 10, 15, 20, 30, 40 kW.

* Corresponding author: khaustovsa@tpu.ru

DOI: 10.1051/matecconf/201711001038

2 Experimental setup

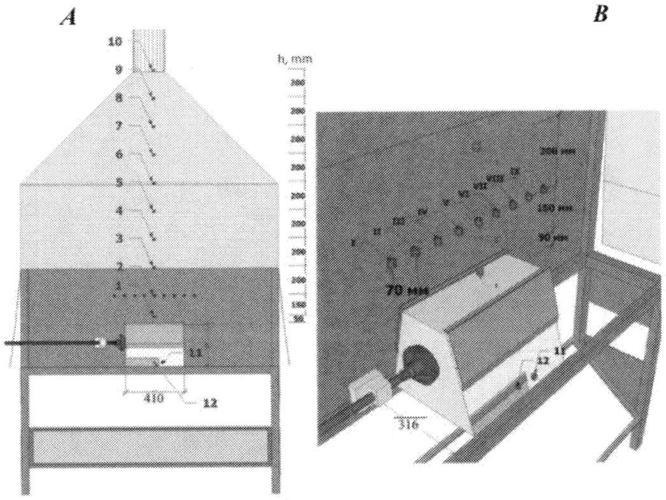

Fig. 1. Longitudinal section (A) and isometric view (B) of experimental setup: numbers from 1 to 10 indicate vertical positions of thermocouples, roman numbers from I to IX indicate horizontal positions of thermocouples, sensors numbered with 11 and 12 are spring-loaded probes Testo-480 for measuring the casing surface temperatures.

A gaseous medium inside the gas-fired infrared heat emitter was carried out on an experimental setup shown in Fig.1. The heating control was implemented with certified instrumentation. Thermoelectric converters THC-0006 were used for measuring the temperature of the gaseous medium and a spring-loaded probe Testo-480, for measuring the casing surface temperatures.

During the experiment, a vertical pitch between the sensors remained unchanged, x = 100 mm. The horizontal pitch between the sensors was adjusted in proportion to the heat output of the emitter. The heat output at different horizontal pitches is shown in table 1.

Table 1. The heat output at different horizontal pitches.

Heat output	Horizontal pitch
5 kW	70 mm
10 kW	115 mm
15 kW	160 mm
20 kW	205 mm
30 kW	180 mm
40 kW	225 mm

Experimental data from the sensors was transferred to PC by an analog-digital converter for Finite Element Comparison.

Gas-fired infrared heat emitters with heat output of 5, 10, 15, 20, 30, 40 kW were studied. The gas pressure before the inlet nozzle was 1270 Pa. The air-fuel ratio (AFR) was 9.996. Tests of each high-temperature gas-fired infrared emitter's thermal capacity were carried out for 40 minutes with temperature recording for every 40 seconds after launching. This allowed to study the thermal conditions change above the emitter before and after stationary conditions reaching. All high-temperature gas-fired infrared emitter's tech specs are taken according to nominal data.

3 Results of the study

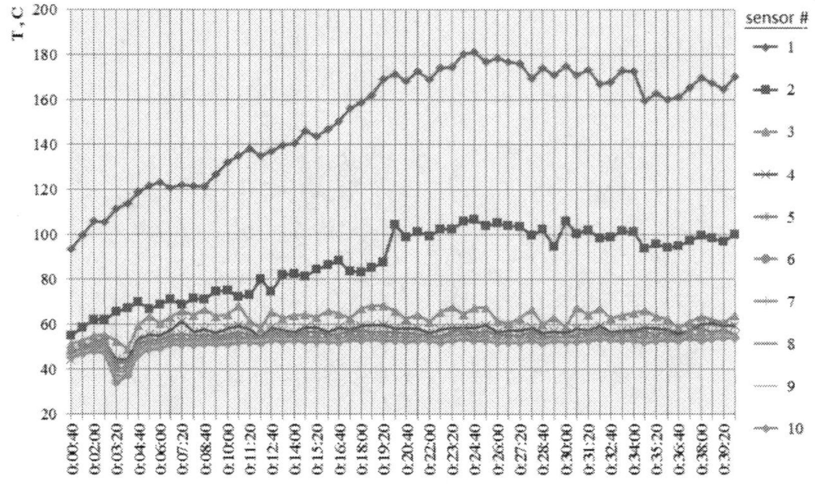

Fig. 2. Temperature distribution in locations of the sensors 1-10 over 5 kW high-temperature gas-fired infrared emitters before and after stationary conditions reaching.

Fig. 2 shows that temperature pulses in time. This pulsation range characterizes turbulent flows appearing above high-temperature gas-fired infrared emitter. The free turbulent flow is caused by gas pressure difference at the exit of region bounded by reflector (excess) and above high-temperature gas-fired infrared emitter (vacuum). These pulsations of temperature in time are peculiar to all thermal capacities. Transition to sustainable mode was reached after 20 minutes of continuous operation under full thermal charge.

It should be noticed that thermal capacity of high-temperature gas-fired infrared emitters is defined by a number of perforated ceramic tiles, unit capacity (1.5 kW at an average). The tile surface temperature at an average is 900 °C and depends on gas pressure in front of the gas-infrared burner nozzle and the ceramic tile composition.

The maximum value of the temperature is recorded at a distance of 50 mm above the gas-fired infrared heat emitter at the location of the sensor #1. With the distance from the radiator structure (sensors 1-10) increasing the shape of the curve turns into hyperbola. First 600 mm temperature falls significantly. The difference between the temperatures of the first 600 mm and at higher distance increases with heat output. The temperature above gas-fired infrared heat emitter 5 kW is reduced by three times, while for 40 kW, barely 1.5 times.

The temperature distribution along the gas-fired infrared heat emitter (sensors I-IX) is uneven. If heat output increases, the spread in temperature values reduces. The results of temperature measurement on outer (sensor No11) and inner (sensor No12) sides of the reflector surface are shown in table 2.

Table 2. The results of gas-fired infrared emitter's casing temperature measurement.

High-temperature gas-fired infrared emitter thermal capacity, kW	Casing temperature on outer side in location of the sensor No11, °C	Casing temperature on inner side in location of the sensor No12, °C
5	148.08	202.81
10	151.04	206.86
15	144.93	205.06
20	140.59	204.85
30	130.80	125.18
40	139.15	142.40

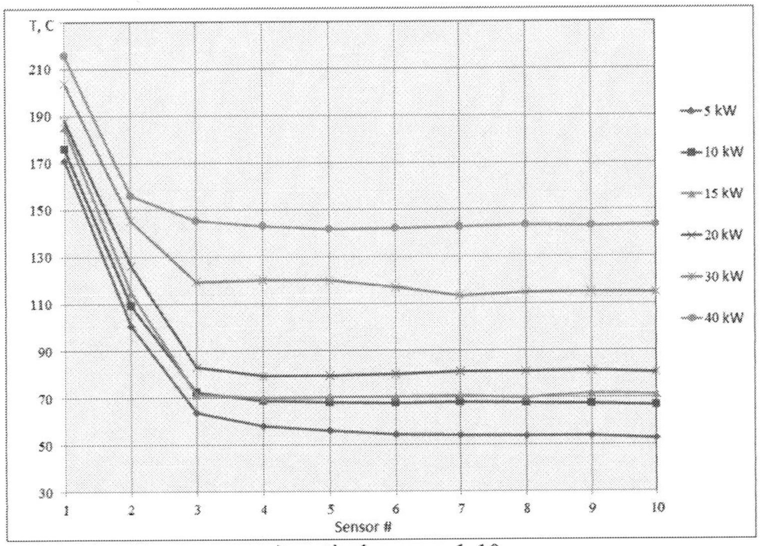

a) vertical sensors 1-10

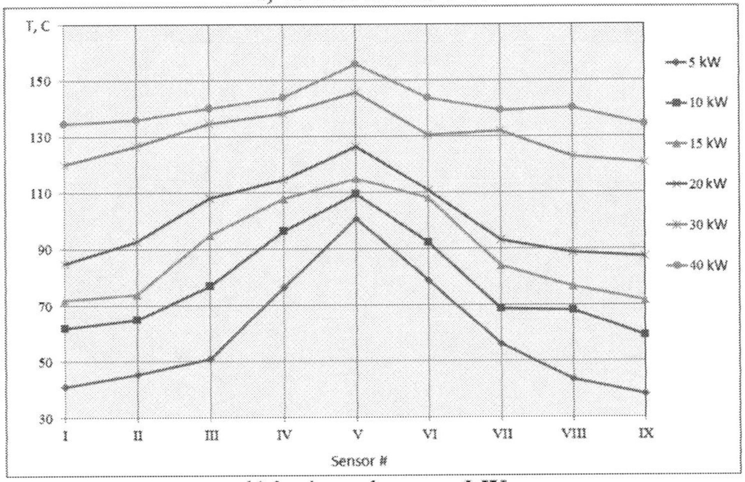

b) horizontal sensors I-IX

Fig. 2. Experimental temperature distribution above 5-40 kW high-temperature gas-fired infrared emitters.

Analysis of the obtained data indicates a feature of high-temperature gas-fired infrared heat emitter. It produces significant thermal radiation in the upper zone of the experimental setup by the heated surfaces with simultaneous flow of convective heat and mass transfer processes.

4 Conclusion

According to the obtained results, it can be concluded: reflector temperature on outer and inner sides does not significantly change provided that reflector is made of the same material; temperature change more than 50 °C inside of 30-40 kW high-temperature gas-fired infrared emitters caused by the use of alternative reflector material. Structural design of 5-20 kW high-temperature gas-fired infrared emitters assumes the use of self-brazing aluminum clad steel with trademark feran ®, and for 30-40 kW high-temperature gas-fired infrared emitters – polished stainless steel 12Cr18Ni10Ti. Thus, the use of feran ®

increases thermal-insulating properties of the reflector, increasing the heat flow intensity into working area.

A high thermal conductivity of the feran ® forming the reflector of gas-fired infrared emitter causes an intensive heat exchange of combustion medium with the outer air. Thus, the temperature distribution at the measuring points over gas-fired infrared emitter becomes uneven in height and time. The thermal conditions above the high-temperature gas-fired infrared emitters depends on its thermal power. The higher thermal output of the burner causes the higher temperatures at each sensor installed. This dependence is explained by the fact that as the thermal power of the gas-fired infrared emitter increases, the proportion of combustion products also increases with respect to the air volume in the experimental setup. As a result, this reduces the intensity of dilution of the combustion products and increases the temperature at each point.

References

1. Y.M. Luo, L. Chevalier, F. Utheza, X. Nicolas, International Polymer Processing. Journal of the Polymer Processing Society, **30** (5), 554 (2015)
2. M. Ostafin, M. Balkenhol, J. Erlemeyer, M. Köhler, Materialwiss. Werkstofftech. **41**, 11 (2010)
3. D. Patino, B. Crespo, J. Porteiro, J.L. Miguez, Appl. Therm. Eng. **100**, 849 (2016)
4. S.A. Khaustov, A.V. Kazakov, G.A. Cherkashina, L.A. Sobinova, EPJ Web Conf. **110**, 01028 (2016)
5. S. Dolgov, E. Savchenko, S. Khaustov, R. Tabakaev, A. Zavorin, EPJ Web Conf. **110**, 01074 (2016)
6. K.A. Al-attab, J.C. Ho, Z.A. Zainal, Exp. Therm. Fluid Sci. **62**, 1 (2015)
7. L. Strauch-Sürken, M. Wendt, Tierärztliche Praxis. Ausgabe G, Grosstiere/Nutztiere, **43**, 4 (2015)

MATEC Web of Conferences **110**, 01039 (2017) DOI: 10.1051/matecconf/201711001039
HMTTSC-2017

Preserved boreal zone forest massif mass estimation during fire extinguishing by liquid aerosol

Nikita E. Shlegel, *Galina* S. Nyashina, and *Alyona* O. Zhdanova*

National Research Tomsk Polytechnic University, 634050 Tomsk, Russia

Abstract. This study contains results of experimental studies into establishing the principles of decreasing the model of fire source mass when extinguishing the ground cover of deciduous and mixed forests by liquid aerosol. The experiments were carried out with typical forest fuels - birch leaves and mixed non-living components of temperate forest. Densities of forest fuel samples in a model of fire source were variated in the corresponding real practice ranges: 20.26–54.70 kg/m^3 for birch leaves and 27.54–72.18 kg/m^3 for a mixed forest fuel. Dimensions of droplets generated by the nozzle amounted to 0.01–0.12 mm. It is consistent with modern aerosol fire extinguishing systems. The dependence of the initial forest fuel sample mass on the remaining mass after ending of the pyrolysis reaction was established.

1. Introduction

Forest fires are considered to be spontaneous fire spread in forest zones with homogeneous and mixed vegetation [1]. As a result, after forest fires in Russia thousands hectares of forest are damaged and killed every year, forest animals and birds die, huge quantities of carbon dioxide and smoke are emitted into the atmosphere, residential settlements are destroyed, millions of rubles are spent from the state budget to extinguish forest fires and restore territories suffered from fire. The optimal way to eliminate the fire zone is a local discharge of quenching liquid from the side of the aircraft into the combustion zone [2]. Studies have shown [3] that with this approach to fire extinguishing most of the discharged liquid is absorbed into the ground. As a rule, it does not suppress the pyrolysis reaction in the soil covering of boreal zone. Only 5–7% of the total liquid volume intended for reducing the temperature during the combustion of forest fuel (FF) evaporates due to the termination of combustion and soil covering pyrolysis [3]. It is important to estimate the mass of FF, which can be preserved in the eliminating the bottom ignition process in the boreal zone.

* Corresponding author: zhdanovaao@tpu.ru

The aim of this study is to quantify the preserved forest massif during extinguishing the forest zone by liquid aerosol.

2. Experimental setup and procedure

Figure 1 shows the experimental setup used in carrying out experimental researches.

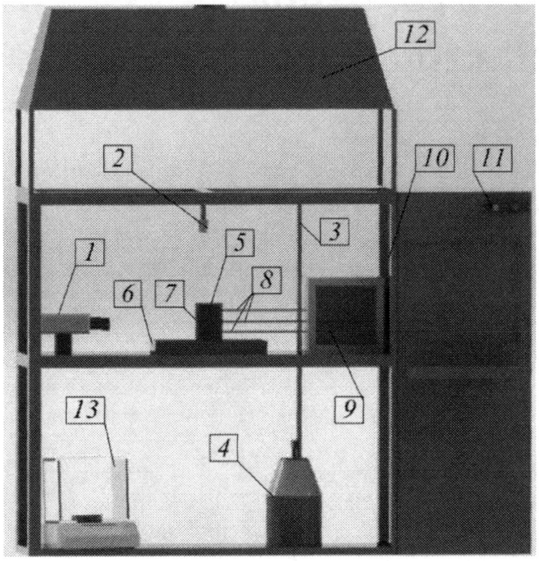

Fig.1. A scheme of experimental setup: *1* – high-speed video camera; *2* – spraying nozzle; *3* – water supply channel; *4* – water tank; *5* – FF; *6* – metal tray; *7* – cylinder with a FF; *8* – thermocouples; *9* – multichannel temperature recorder; *10* – workstation; *11* – control panel of turning on/off ventilation system; *12* – ventilation system; *13* – analytical microbalance.

Type-K needle thermocouples *8* (temperature measurement range 223–1473 K, accuracy ± 3 K, heat retention no more than 0.1 s) were used to measure temperature of the pervasive combustion process. Temperature readings were fixed on a multichannel recorder *9*. We realized quenching using spray nozzle *2* (droplets' size R_d=0.01–0.12 mm; irrigation density ξ_f=0.014–0.016 l/m^2·c; water consumption μ_w=0.00063 l/c). Liquid was supplied from the tank *4*. The speed of the droplet flow was determined using the high-speed video camera *1* and the "Actual Flow" software for processing the results of the experiments using optical methods of "Particle Image Velocimetry" (PIV) diagnostics. Dimensions of dispersed stream droplets were calculated by the IPI method [5].

Two groups of samples were used during researches. The first group included only birch leaves; the second group consisted of mixed non-living component of temperate forests with a following mass ratio: 25% birch leaves; 15% pine needles; 60% branches of hardwoods. We used distilled water (GOST 6709-72) for extinguishing the standardized fire. At the first stage of the experiment, FF sample was weighed on the analytical microbalance ViBRA HT 84RCE 13 with an accuracy of 10^{-5} g (mass m_{f0} was determined). After that, forest fuel was placed to the bottom of the cylinder. The initial weights and

DOI: 10.1051/matecconf/201711001039

densities of samples are shown in figure 2. The initial weight of forest fuel was chosen so that its density varied from one experiment to another in a narrow range.

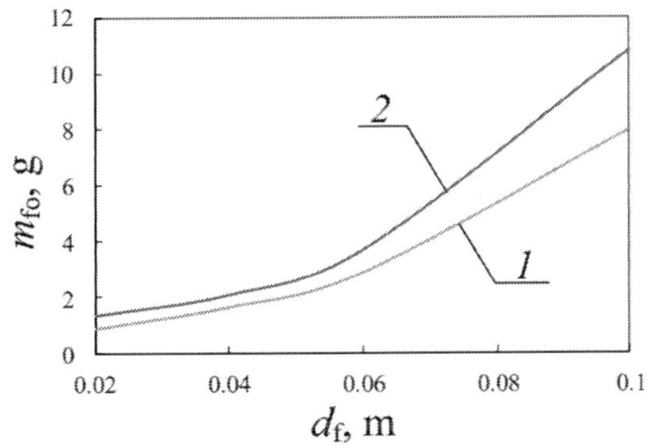

Fig. 2 Dependence of average FF weight (m_{f0}) on the original brazier diameter (d_f), with a constant height of forest fuel bed (h_f=0.04 м): *1* – birch leaves, *2* – mixed FF.

Ignition of the samples was carried out uniformly over the entire area of the open FF surface. We with used three piezoelectric gas burners simultaneously. For each model of fire source we conducted from 15 to 20 experiments. The standardized fires were extinguished by a liquid aerosol with a droplets' radius R_d=0.1–0.12 mm. Spraying continued until the complete suppression of thermal decomposition process Ending of this process was recorded visually and using thermocouple readings. After suppression of the pyrolysis reaction, the fraction (residue) of unreacted FF (m_r) upon extinguishing was dried for 24 hours at room temperature (fig. 3) and weighed on laboratory microbalance.

a

DOI: 10.1051/matecconf/201711001039

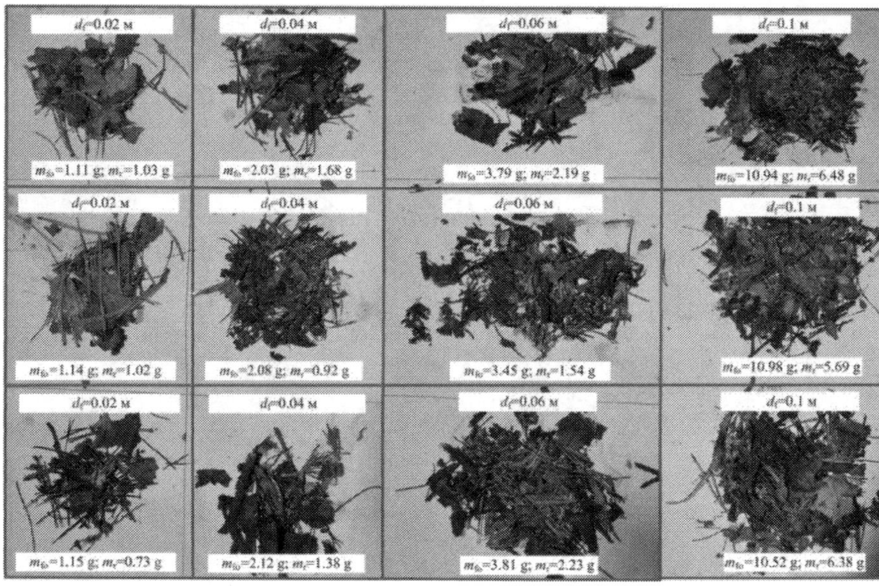

b

Fig. 3. Typical video records of the residual FF after extinguishing by liquid aerosol (R_d=0.1–0.12 mm): *a* – birch leaves, *b* – mixed FF.

3. Results and discussions

Figure 4 shows dependencies of the mass loss ($\Delta m = m_{f0} - m_r$) on initial mass of the sample in extinguishing.

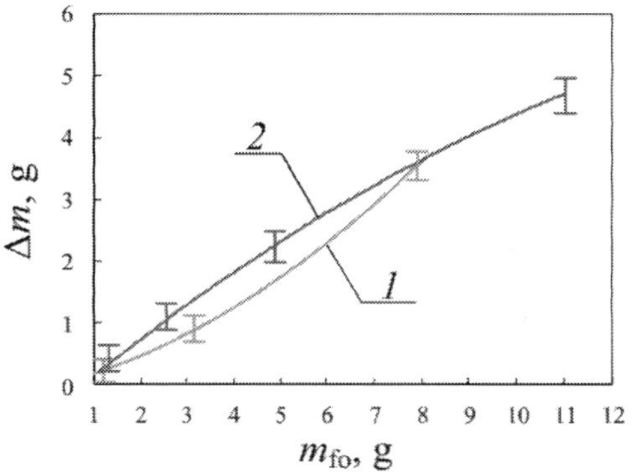

Fig. 4. Dependence of the parameter Δm ($\Delta m = m_{f0} - m_r$) on initial sample mass m_{f0} (Δm is parameter characterized FF mass loss during extinguishing by liquid aerosol): *1* – birch leaves, *2* – mixed FF.

An analysis of the location of the approximation curves of fig. 4 enable to conclude that the mass loss when extinguishing mixed FF is greater than for birch leaves (on 0.21–0.54 g). This effect may be due to the forest fuels covering structure. The area of birch leaves is larger than the area of needles and twigs. During extinguishing by a liquid aerosol, we observed water droplets accumulation on the birch leaves surface over a period of time. As a consequence, two extinguishing schemes were realized. These were blocking the oxidant entering the reaction zone and directly reducing the temperature of the decomposing FF bed – birch leaves. Because the differences in structure of samples with a mixed FF and birch leaves, water drops did not accumulate on the sample surface, but penetrated through the sample bed. The temperature of mixed forest fuel bed was reduced because of the water droplets evaporation. Penetrated into the pores of the sample water drops and the evaporation during the interaction of the products of thermal decomposition and the dropping stream – a cloud of water vapor was formed and displaced pyrolysis products.

Figure 5 shows dependencies of the remaining weights of FF (m_r) on the initial ones (m_{f0}). During comparing the initial mass of the sample and the remainder when extinguishing with a liquid aerosol (figure 4), it was established that a significant portion of the sample remained:

- for birch leaves - 53–83% of the initial mass of the sample (m_{f0});
- for a mixed FF- 54–85% of the initial mass of the sample (m_{f0}).

Figure 5 shows, that remaining mass increases with increasing of initial mass of FF. For birch leaves and mixed forest fuel there is an increase in m_r with a variation in the initial mass by 3.7 and 5.2 g, respectively. This effect may be explained by the heat content of the FF sample. During the experiments, the times of suppression the thermal decomposition reaction of forest fuel also increased with an increasing braziers' area.

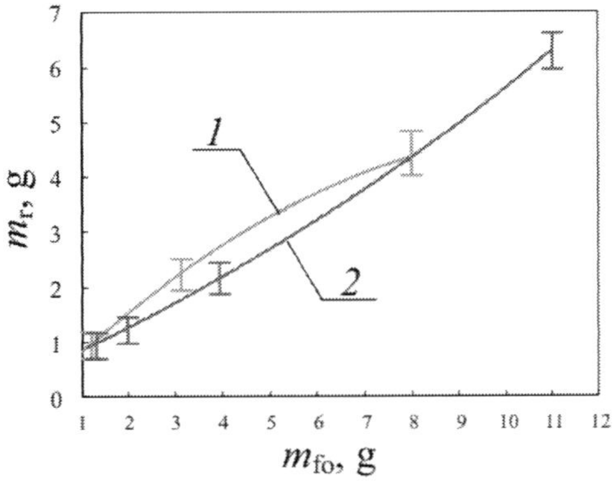

Fig. 5. Dependence of the remaining FF mass (m_r) on initial one (m_{f0}): *1* – birch leaves, *2* – mixed FF.

4. Conclusion

The obtained result leads to conclusion about the effectiveness of extinguishing forest fire by a liquid aerosol. Experimental studies allow us to establish that up to 80% of the initial mass of the soil covering in the boreal zone were preserved. The established principles reflect the positive outcome of forest fire extinguishing

The investigations of heat and mass transfer processes in the suppression of forest fire were supported by Russian Science Foundation (project 14–39–00003). Experiments to determine the specific differences between the processes of thermal decomposition and combustion of various types of forest fuel were carried out within the framework of the President of the Russian Federation grant MK-1684.2017.8.

References

1. A. Dimitrakopoulos, C. Gogi, G. Stamatelos, I. Mitsopoulos, J. Environ. **20**, 327 (2011)
2. N.P. Kopylov, I.R. Khasanov, A.E. Kuznetsov, D.V. Fedotkin, E.A. Moskvilin, P.A. Strizhak, V.N. Karpov, Fire safety **2**, 49 (2015)
3. *Atlas of fires risk in the territory of the Russian Federation* ("Design. Information. Cartography" 2010)
4. R.S. Volkov, G.V. Kuznetsov, P.A. Strizhak, Journal of Engineering Physics and Thermophysics **87**, 450 (2014)
5. R.S. Volkov, A.O. Zhdanova, G.V. Kuznetsov, P.A. Strizhak, Thermal Engineering **62**, 586 (2015)

Experimental study of the effect of the substrate roughness on the thermocapillary breakdown of a horizontal liquid film

Dmitry Kochkin[1,2,*]

[1]Kutateladze Institute of Thermophysics, 630090 Novosibirsk, Russia
[2]Novosibirsk State Technical University, 630073 Novosibirsk, Russia

Abstract. The paper describes experiments on study of influence of the substrate roughness on the phenomenon of thermocapillary breakdown in a horizontal nonisothermal film of liuqid placed upon a stainless steel substrates. The critical heat flux for fim breakdown is almost independent of roughness. However, the substrate roughness is signifcant for dynamics of film breakdown. The velocity of contact line propagation decreases with the roughness growth. The final size of a dry spot decreases for higher roughness.

1 Introduction

The modern industry often uses apparatus with liquid film flow. Thin liquid film is a promising approach for developing of cooling of devices with high local heat release, in particular, for high-end electronic chips [1,2]. The industry creates a need for cooling of high local heat fluxes from electronics components like computer chips, and power electronics (transistors and thyristors).

Film-based cooling systems are highly efficient since they provide a high rate of heat transfer at moderate flow rates of the coolant. The reduction in the film thickness enhances heat transfer; however, thin films are prone to film breaking and this increases drastically the temperature of the cooled element and may cause failure.

The effect of substrate wetting and liquid properties on the thermocapillary breakdown of a flowing film was studied in [3-5]. The paper [6] studied the effect of substrate structuring on phenomenon of liquid film breakdown. The paper [7] deals with the effect of inclination angle of the working section. The papers [8, 9] were pioneering in study of thermocapillary breakdown of a steady horizontal liquid layer. The papers [4, 10] demonstrated that the critical heat flux for the case of horizontal liquid films is by several times higher than the critical level for gravity-driven liquid films. This explains the relevance of presented study.

The key object of this study is the effect of substrate roughness on the threshold heat flux which causes film breakdown and also the dynamics of this breakdown.

* Corresponding author: kochkin1995@mail.ru

2 Methods

Experiments were conducted with locally heated substrates with different size of roughness made of stainless steel with the diameter of 50.8 mm and thickness of 1 mm. The substrate was mounted on a textolite basement with an embedded copper rod with the diameter of 12.7 mm (Fig. 1), and thermal paste was applied for improvement of heat contact between the substrate and basement. The copper rod has a thermal contact with a ceramic heater. The heat flux was calculated from the temperature drop over the copper rod for the situation of steady process. It was also controlled by the known electric power consumed by the heater. The working liquid was distilled super-pure Milli-Q water with the initial temperature of 25°C. The proper volume of water was fed to the substrate via a syringe pump. The substrate perimeter has cooling arrangement. The water temperature in the cooling circuit was sustained at 5°C.

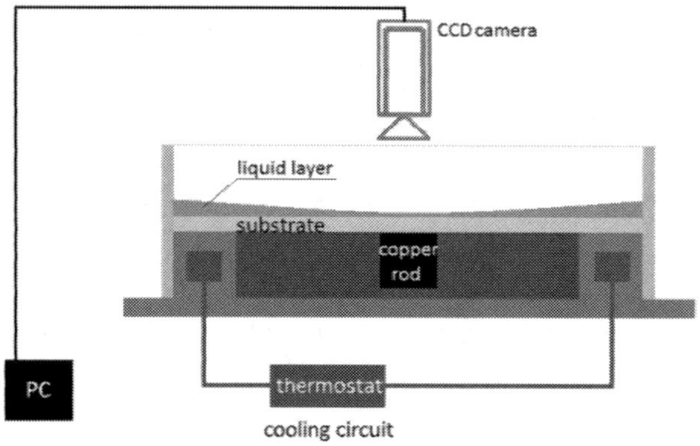

Fig. 1. Diagram of the experimental setup (the heater's power source is not shown).

The using of removable substrate in this setup ensures study of breaking a horizontal liquid layer on substrates with different roughness. The different scale of roughness was achieved by surface polishing. The morphology of the surface was analyzed using a scanning electron microscope (HITACHI S3400N) and an atomic force microscope (Solver Pro NT MDT).

The liquid film thickness was measured with a probe moved with a motor-driven microdisplacement tool. As the film surface is toughed by the probe tip, this creates a liquid meniscus recorded with a CCD camera. The total film thickness was controlled by he known volume of liquid fed using the syringe pump. Experiments were carried out for films with the initial thickness 350 µm (+/- 5 µm). The heat flux was incremented with small steps up to the threshold flux (which ensures film breakdown). The steadiness of process was controlled using thermocouples installed in the working section and heater. The film breakdown initiates a drastic growth of heater temperature.

The film breakdown was recorded with the CCD camera DMK 23GP031 placed above the working section.

3 Results

Experiments were conducted for stainless steel substrates with different levels of roughness (Ra = 0.05 µm, Ra = 1.30 µm, Ra = 2.5 µm). It was found that the threshold heat flux (initiating the film breakdown) is almost independent of roughness and it was about 10-11 W/cm^2. Experiments were carried out for the initial film thickness 350 µm (+/-5 µm), and this thickness is dictated by the fact that a moderate heat flux is required to break down this

DOI: 10.1051/matecconf/201711001040

film. In the process of heating, the contribution from water evaporation was low and the final film thickness was also about 330 μm.

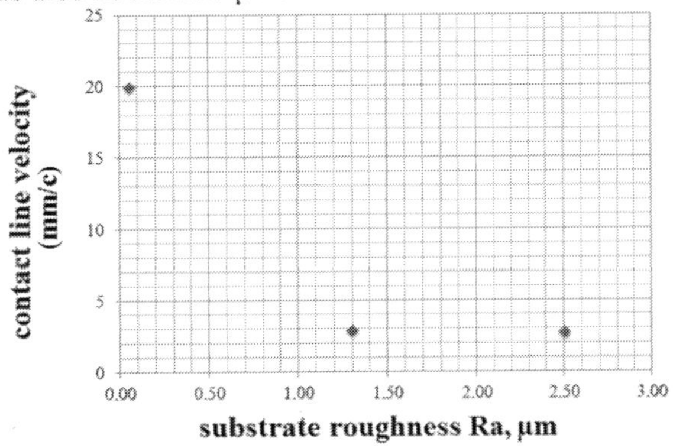

Fig. 2. Graphs of contact line velocity vs. substrate roughness.

Our study demonstrated that the speed of dry spot expansion (Fig. 2,3.) and the final dry spot size (Fig. 4.) depend significantly on the surface roughness. For example, for a surface with roughness Ra = 0.05 μm the velocity of contact line is by order higher than for the same kind of substrate with the roughness of Ra = 1.30 μm. But in experiments with a substrate with a higher roughness (Ra = 2.5 μm) the further increase in line velocity is insignificant. The experimental data are explained by a concept that the contact line after film breakdown is caught by the substrate roughness and this reduces the breakdown velocity.

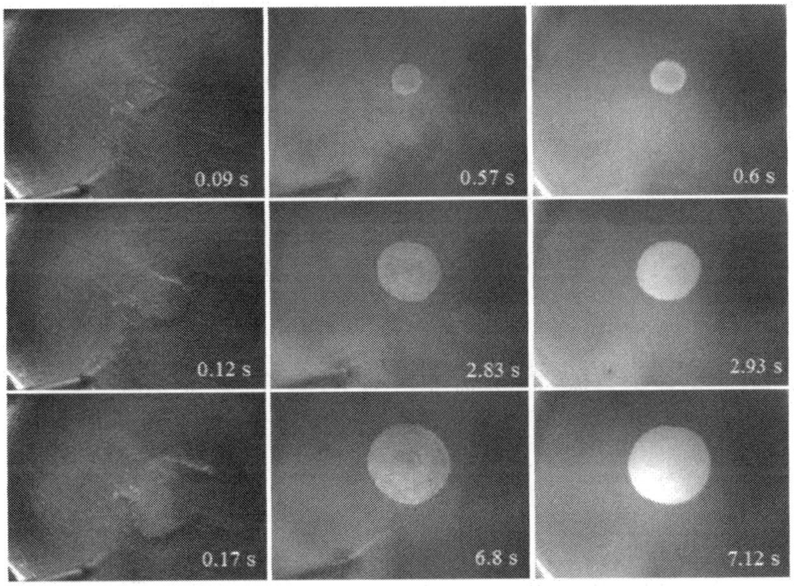

Fig. 3. The dynamics for dry spot expansion on the stainless steel substrate with the thickness of 1 mm at different scales of roughness (for initial film thickness 350 μm). The time is measured since the breakdown startup.

The smaller final size of dry spot at higher roughness is explained by counteraction of roughness to dry spot expansion.

DOI: 10.1051/matecconf/201711001040

Ra = 0.05 μm
q = 10.3 W/cm²

Ra = 1.30 μm
q = 11.07 W/cm²

Ra = 2.5 μm
q = 10.64 W/cm²

Fig. 4. Picture of the final size of a dry spot for substrates with different roughness.

4 Conclusion

The tests demonstrated that the substrate roughness has almost no impact on the critical heat flux needed for breakdown of a locally heated liquid film. However, roughness is significant for breakdown dynamics and the final size of the dry spot. For a higher roughness of substrate, the contact line velocity and the final size of dry spot become lower.

The research was supported by the Russian Science Foundation, Project No. 1419-01755.

References

1. A. Bar-Cohen, P. Wang, ASME. J. Heat Transfer, **134**, 051017 (2012)
2. D.V. Zaitsev, D.A. Rodionov, O.A. Kabov, Technical Physics Letters, **35**, 680 (2009)
3. D.V. Zaitsev, O.A. Kabov, V.V. Cheverda, N.S. Bufetov, High Temp. **42**, 450 (2004)
4. D.V. Zaitsev, D.P. Kirichenko, O.A. Kabov, Technical Physics Letters, **41**, 551 (2015)
5. D.V. Zaitsev, A.A. Semenov, O.A. Kabov, Thermophys. Aeromech., **23**, 625 (2016)
6. D.V. Zaitsev, A.M. Lozano, H. Auracher, O.A. Kabov, Microgravity sci. techn., **19** (3-4), 71 (2007)
7. E.A. Chinnov, O.A. Kabov, A.V. Muzykantov D.V. Zaitsev, Intern. J. Heat Technol. **19**, 31 (2001)
8. A. Orell, S.G. Bankoff, Int. J. Heat Mass Transf., **14**, 1835 (1971)
9. J.P. Burelbach, S.G. Bankoff, S.H. Davis, Phys. Fluids A, **2**, 321 (1990)
10. D.V. Zaitsev, O.A. Kabov, Microgravity sci. techn., **19** (3-4), 174 (2007)

MATEC Web of Conferences **110**, 01041 (2017) DOI: 10.1051/matecconf/201711001041
HMTTSC-2017

3D CFD simulation of aerodynamics of a 406 MW_t CFB boiler

Oleg M. Koksharev[1,*], *Andrey* V. Gil[1], and *Lebedev* V.M.[2]

[1]National Research Tomsk Polytechnic University, 634050 Tomsk, Russia
[2]Omsk State Transport University, 644046, Omsk, Russia

Abstract. Modeling of various processes in CFB boilers (combustion, aerodynamics, hydrodynamics, etc.) is connected with the solution of systems of differential equations with a large number of unknowns that leads to their simplification and, accordingly, affect the quality of the calculation. The work consists of several parts and aims to assess the applicability of some or other mathematical algorithms to obtain a quantitative simulation of processes in power boilers with circulating fluidized bed. In part, presented in this article, the simulation of CFB boiler aerodynamics using ANSYS FLUENT is considered.

1 Introduction

The modeling methods for CFB furnaces can be classified as fundamentals-oriented and practice-oriented models. In the fundamentals-oriented models, the modeling of the fluid dynamics is attempted with the currently available fundamental theories, such as the Eulerian–Eulerian multiphase models applying the kinetic theory of granular flow. In the practice-oriented models (or engineering models), the theories are simplified and adjusted by empirical correlations to improve the calculation speed and fit the model results with the measured data [1, 2]. In this paper, the 3-D model simulation of the aerodynamic structure of CFB boiler has been resolved by applying ANSYS FLUENT 12.1.

2 Problem statement

It is worth noting that a proper understanding of heat and mass transfer in furnaces with circulating fluidized bed is important in order to make the best design with efficient and reliable options for this type of system. During the individual review of the physical phenomena such as gas flow with the particles, turbulence, chemical reactions and heat transfer, are quite difficult to forecast and study. This study becomes more complicated when these physical processes are interconnected in a large-scale system, such as boilers with CFB.

The results of studies that have been carried out on the basis of the geometrical model presented in [3, 4] (Fig. 1), showed the ambiguity in terms of layout decisions of secondary

* Corresponding author: omk2@tpu.ru

air nozzles, burners and return leg for particles flow from a cyclone arrangements that were accepted based on empirical dependences (Fig. 1).

Therefore it was decided to create a 3-D model, where the calculation takes into account not only the volume of combustion chamber as well as the processes in the cyclone and standpipe with loop seal (Fig. 2).

a) b)

Fig. 1. The initial object of study: a - 3-D model; b – mesh.

The object of study is a CFB boiler with a capacity of 406 MW_t that is equipped with two cyclones for particulate control. The height of the combustion chamber is 24.5 meters. The furnace is equipped with 5 burners on the rear wall in 1 row and 10 nozzles of secondary air at the front wall in 2 rows. At the exit of the furnace the particles are separated into 2 streams – one is separated and through the standpipe and loop seal returns to the furnace bed layer, another flow leaves the cyclone and directed to the electrostatic separator. It should be noted that the boiler before the simulation was originally calculated on thermal design basis for boilers with CFB in accordance with [5]. 3-D model and computational grid (mesh) of the boiler are shown in Fig. 2.

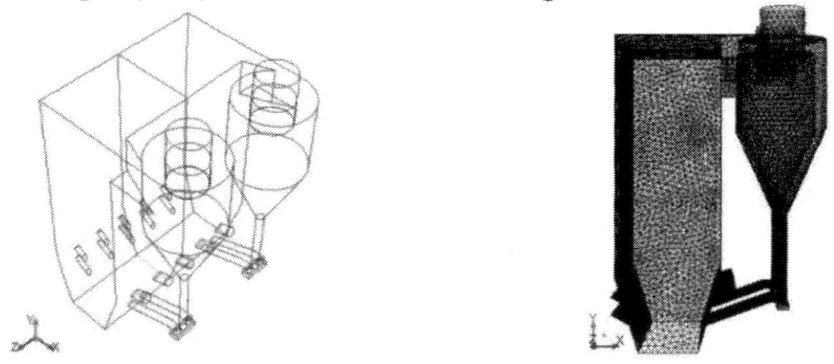

a) b)

Fig. 2. The research object: a - 3-D model; b – mesh.

3 Results and discussion

Fig. 3–4 show the contours of static pressure in horizontal and vertical sections of the boiler. The obtained results are in the agreement with the results obtained by other authors [6]. Accordingly, the highest values are observed in the lower and upper part of the combustion chamber and comprise about 17000 Pa, the average value through the volume of the furnace is 15000 Pa. The vacuum increases as it approaches the exit of the cyclone.

 DOI: 10.1051/matecconf/201711001041

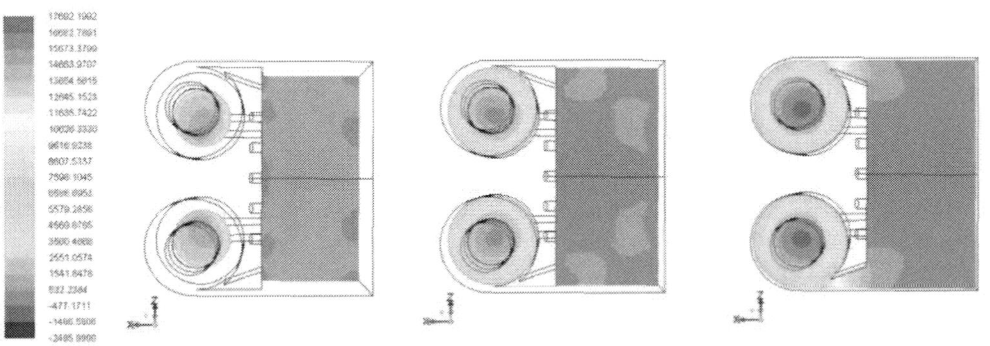

a) b) c)

Fig. 3. Contours of static pressure in horizontal sections of the CFB boiler (Pa): a – at 15 meters height; b – at 20 meters height; c – at 23 meters height.

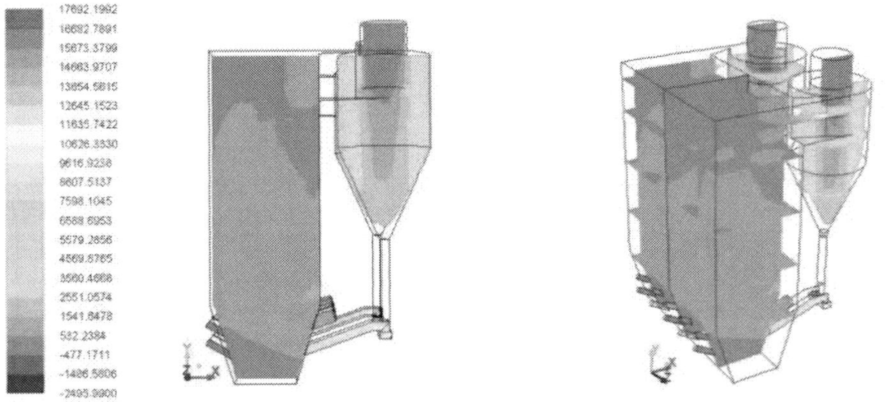

a) b)

Fig. 4. Contours of static pressure in vertical sections of the CFB boiler (Pa): a – vertical section at a distance of 2.75 meters from the side wall; b – isometric projection with horizontal and vertical sections.

According to the results, the blower fan should be selected in such a way as to create the necessary pressure for the air supply. For the loop seal under the pressure of 7000 Pa and 16500 Pa for the primary air.

Fig. 5 shows graphs of static pressure distribution over the furnace chamber height. In the case of Fig. 5a the pressure drops to 8000 Pa in the area of the recirculation pipelines inlet from the cyclone and the burners embrasure at a height of 2.5 meters in the furnace. Further, the pressure is equalized in the volume of the furnace. In the case of Fig. 5b, the pressure varies from 15200 Pa to 16400 Pa.

 DOI: 10.1051/matecconf/201711001041

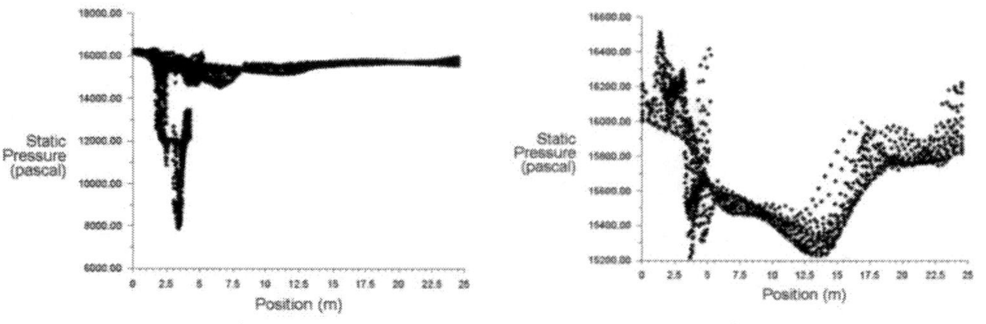

a) b)

Fig. 5. Graphs of static pressure distribution over the furnace chamber height: a – at a distance of 2.75 meters from the side wall; b – at the center of the furnace.

Fig. 6 presents the results of velocities numerical simulation in the volume of the combustion chamber and cyclone. High-speed flow in the furnace volume fairly uniform and is 6-7 m/s. According to Fig. 6c – f it may be noted a more rapid upward flow in the central zone of the combustion chamber caused by the arrangement of the burners on the back wall of the furnace at the bottom. Then, along the furnace height speed becomes higher in the peripheral parts, where the cyclones are located. Thus in this area can be more intensive abrasive damage of the heating surfaces.

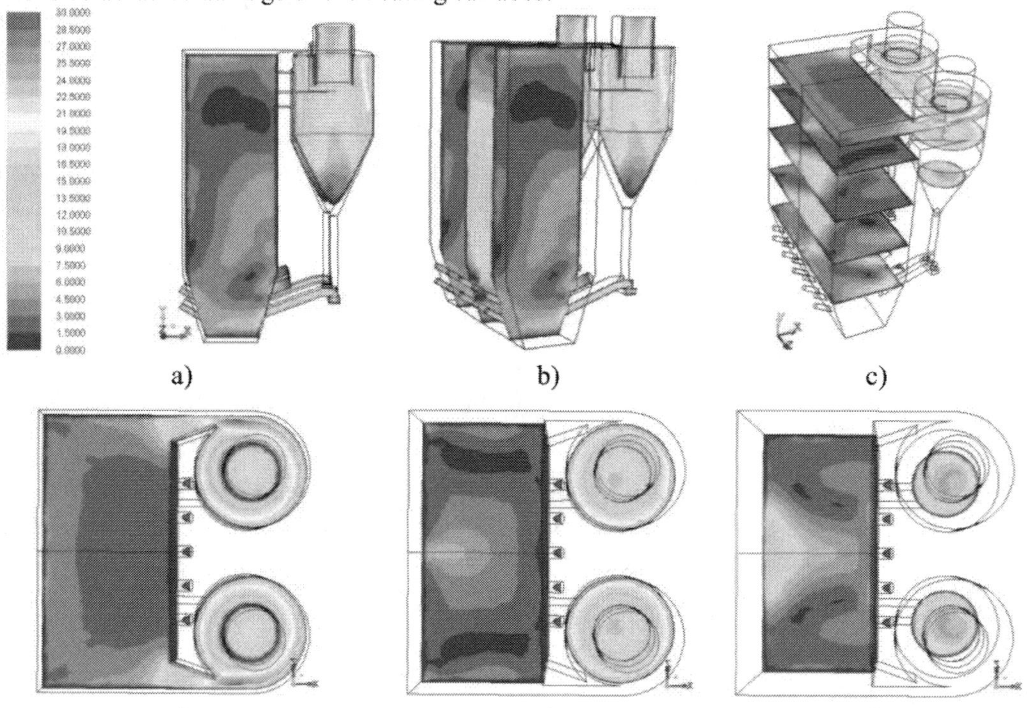

a) b) c)

d) e) f)

Fig. 6. Contours of velocity distribution in the volume of the boiler with CFB (m/s): a, b – distribution in vertical planes at a distance of 2.75 meters from the side wall; c – distribution in the horizontal planes (5, 10, 15, 20 and 23 m); d, e, f are the velocity distribution at the level of 23, 20 and 15 meters, respectively.

A further increase in speed (Fig. 6, d) corresponds to the desired value for the effective separation of solid particles in cyclones. At the entrance to the cyclone, speed reaches values of 27 m/s.

The speed of fluidization at 850 °C is 4.6 m/s, but the maximum speed in the furnace according to the calculations accepted with a 50% reserve, equal to 7 m/s. As can be seen from Fig. 6, the average velocity over the cross section of the furnace is 7 m/s. At the inlet to the cyclone, the velocity of the particles is 20 m/s. In both cases, the calculated values by the method of mathematical simulation coincide with empirical data.

Consider the trajectory of the particles inside the volume of the boiler. Fig. 7 shows a graph of the particle trajectory colored by the particles ID. For clarity, a part of the particles was not included in graph.

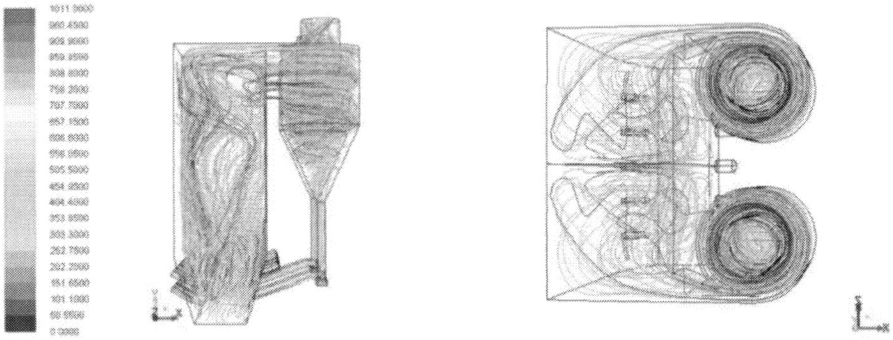

a) b)

Fig. 7. Pathlines colored by particle ID: a – side view; b – top view.

4 Conclusions

In this work the analysis of the numerical simulation results of the aerodynamic structure of furnace volume, the cyclone and the elements for particles return in CFB boiler with the help of ANSYS FLUENT was made. Presented plots of the velocity distribution and static pressure are correlated with the empirical data. It is worth noting that the adopted design of the boiler, the arrangement of the burners and secondary air nozzles provide a reliable operation and maintenance of the boiler with a stable supply of the fuel and the organization of the external circulation.

References

1. M.H. Bordbar, K. Myöhänen, Appl. Therm. Eng. **76**, 344 (2015)
2. A.V. Gil, A.Y. Gil, EPJ Web Conf. **82**, 01044 (2015)
3. A.V. Gil, D.A. Baturin, MATEC Web Conf. **72**, 01009 (2016)
4. A.V. Gil, D.A. Baturin, MATEC Web Conf. 23, 01009 (2015)
5. Pr. Basu, *Circulating Fluidized Bed Boilers: Design, Operation and Maintenance* (Springer, Switzerland, 2005)
6. N. Zhang, B. Lu, W. Wang, J. Li, Chemical Engineering Journal, **162**, 821 (2010)

Laser ignition of high-energy materials containing AlB_2 and AlB_{12} powders

Alexander Korotkikh [1,2,*], *Konstantin* Slyusarskiy [1], *Konstantin* Monogarov [1], and *Ekaterina* Selikhova [1]

[1]National Research Tomsk Polytechnic University, 634050 Tomsk, Russia
[2]Tomsk State University, 634050 Tomsk, Russia

Abstract. Boron-containing substances are known to have high gravimetric and volumetric heats of oxidation in comparison with any metals that make them promising for using in high-energy materials. This work is aimed to study ignition characteristics of ammonium perchlorate and nitrate-based composite solid propellants containing aluminium borides by means of radiative heating by CO_2-laser. It was found that the effect upon laser ignition (ignitability) at full replacement of ASD-4 aluminum powder by aluminium diboride and dodecaboride is close. The ignition time of HEM is reduced by 2.0–2.5 times compared with ASD-4 HEM sample in the heat flux density range of 90–200 W/cm^2.

1 Introduction

Boron and its compounds are considered to be fuel components with high efficiency for use in solid propellants for solid fuel rocket and ramjet engines [1–3]. Boron hydrides and organic derivatives have the high burning rates in a wide range of oxidizing medium compositions and pressure values. The specific heat of complete boron oxidation in the pure oxygen is more than 1.9 times higher compared to aviation kerosene and aluminum.

The most widespread solid propellant contents 22 wt. % of micro-sized aluminium powder. According to [4, 5] the composite fuel combustion is strongly influenced by presence of aluminium oxide and carbonaceous residues in refractory layer of aluminium powder during oxidation. Aluminium oxide melting point is significantly higher compared to pure aluminium. According to the study [4, 5] the aluminum oxidation during the combustion of composite solid propellants strongly influence by the presence on the burning surface the refractory layer containing the aluminum oxide particles and carbonaceous resudues. Melting point of alumina significantly above the melting point of aluminum. The aluminum particles combustion possible at the high temperature gradient of high-energy materials (HEMs) reaction layer near the burning surface with occurrence of cracks and destruction of the oxide layer resulting in the oxidation of active metal. It is well known [6] that the destruction of the aluminum oxide layer on the particle surface may react with the carbon particles to form aluminum carbide.

* Corresponding author: korotkikh@tpu.ru

However, HEMs, containing aluminum micropowders, have the long ignition time and the low burning rate compared to HEMs, containing ultrafine powders of aluminum and other metals [7–10].

This paper presents the experimental dependences of the ignition time and temperatures on heat flux density, the activation energy values for the model HEMs based on ammonium perchlorate and nitrate, energetic binder, containing micron powders of aluminum, aluminium diboride and aluminium dodecaboride.

2 Experimental

2.1 HEM samples

Three HEM samples were used in current study. The first sample (marked as ASD-4) contain ammonium perchlorate (fraction with particle size 160-315 μm), ammonium nitrate (fraction <50 μm), energetic binder MPVT-ASP and 30 wt. % of aluminum micron powder ASD-4. Two other samples ASD-4 aluminum was completely substituted by aluminum diboride and aluminium dodecaboride micron powders for samples AlB_2 and AlB_{12}, respectively.

2.2 Ignition of HEM

The ignition process was studied via setup for the radiant heating based on CO_2-laser with the 10.6 μm wavelength and 200 W power (Fig. 1). Before measurements all samples were cut into tablets of 5 mm in height.

The test HEM sample (6) was attached to the sample holder (8). After opening of the shutter (4) radiation was focused by the sodium chloride lens (4) on the HEM sample (6). Signals from photodiodes (7) were transmitted to the L-card-E 14-440 ADC (9) and recorded in the personal computer (10), and then processed with the software application LGraph2. The ignition time t_{ign} of HEMs was determined by difference between the moments of signals from two photodiodes (7), one of which registered the appearance of flame near the end surface of HEM sample. The ignition delay time of HEM samples were determined in atmospheric conditions. Thermo imaging was realized via thermal imager Jade J530 SB.

Video and thermal-imaging was manually started before opening the shutter and stopped after closing the shutter. Files of video and thermal imaging were synchronized with photodiode data by characteristic sound. The relative error of delay times measuring of t_{ign} was equal 5–12 % at the value of confidence probability 0.9.

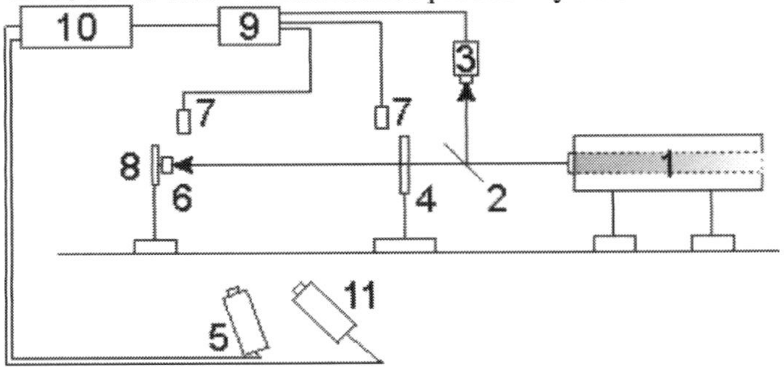

Fig. 1. The scheme of experimental setup based on CO_2-laser: 1 – CO_2-laser; 2 – beam-splitting mirror; 3 – thermoelectric sensor of radiation power; 4 – shutter; 5 – videoimaging camera; 6 – HEM sample; 7 – photodiodes; 8 – sample holder; 9 – ADCs; 10 – PC; 11 – thermal imager.

The radiation power and heat flux density of CO_2-laser beam was measured by the thermoelectric sensor of radiation power (3). The maximum radiation power was defined in the center of the laser beam using a diaphragm with diameter 2 mm. The diameter of the laser beam incident on a HEM sample was 10 mm.

3 Results and discussion

The dependences of HEM ignition time vs. the heat flux density value (in the centre of the laser beam) were determined (Fig. 2).

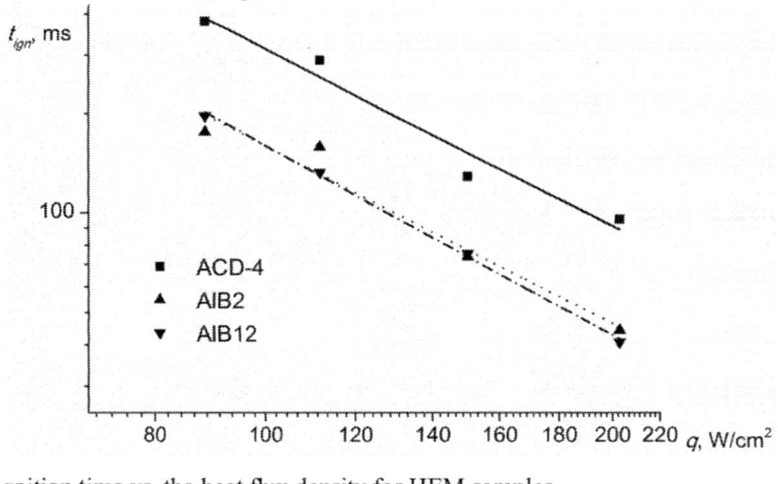

Fig. 2. The ignition time vs. the heat flux density for HEM samples.

It was found that a complete replacement of aluminium by aluminium boride powder in HEM leads to a decrease the ignition time by 2–2.5 times in the heat flux density range of 90–200 W/cm^2. It is worth mentioning that no significant difference were observed for AlB_2 and AlB_{12} samples.

Surface average and maximal temperature dependences on time for all samples are given in fig. 3.

Based on presented in Fig. 3 curve slope it could be concluded that AlB_{12} and AlB_2 samples have close values of heat capacity, while for ASD-4 sample it is greater. It is worth mentioning that gaseous gasification products has strong influence on measured surface temperature value.

The kinetic constants of HEM ignition were determined using following equation [11]:

$$t_{\text{ign}} = 0.359\left(1-\frac{T_0}{T_{\text{ign}}}\right)^2 \frac{E_{\text{a}} c_{\text{p}}}{RQz} \exp\left(\frac{E_{\text{a}}}{RT_{\text{ign}}}\right),$$

where E_a is the activation energy, R is the universal gas constant, Q is the specific (per unit weight) heat of reaction, z is a pre-exponential factor, T_{ign} is the quasi-steady ignition temperature that is determined from the condition of equality between the rate of heat supply from reaction layer of chemical interactions and the rate of heat removal into the depth of condensed phase

$$T_{\text{ign}} = T_0 + 1.2q\sqrt{\frac{t_{\text{ign}}}{\lambda c_{\text{p}} \rho}},$$

where λ, c_p, ρ are factors of the thermal conductivity, the specific heat capacity, and the density for condensed HEM.

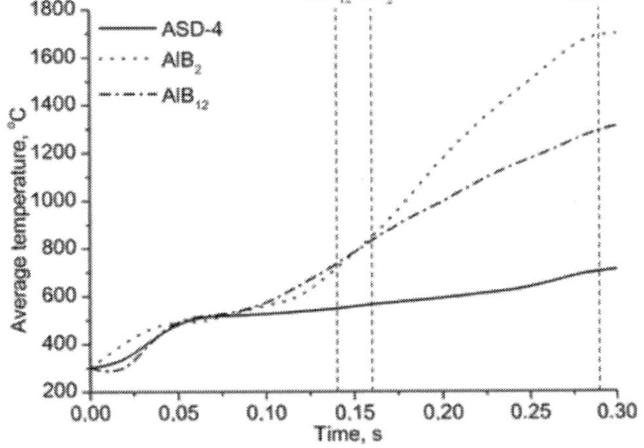

a)

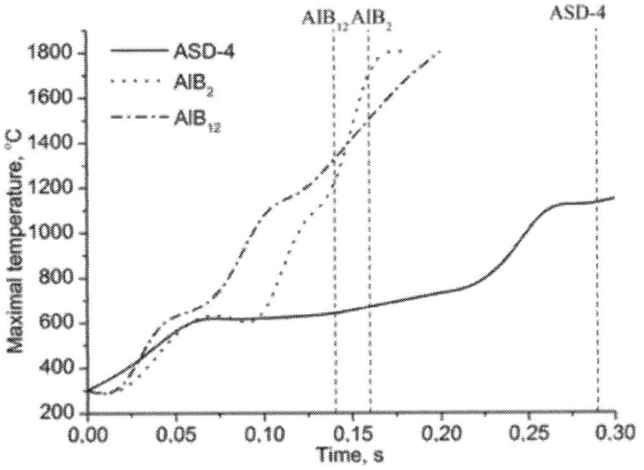

b)

Fig. 3. Dependence of average (a) and maximal (b) surface temperature for HEM samples on time.

The results of calculation of the kinetic constants of ignition for tested HEM samples are presented in Table 1. Here there are values of the average $\overline{T}$ and maximum $T_{\max}$ surface temperature on the HEM reaction layer at the ignition moment and appearance of visible flame obtained using a thermal imager Jade J530 SB at q = 114 W/cm^2. The characteristic parameters of ignition process from Table 1 reveals that there are little difference between average surface temperatures in the moment of ignition while maximal temperatures differ significantly. Despite lower ignition delay times for samples with aluminium dodecaboride the activation energy for this sample was calculated to be lower than for ASD-4 sample due to high Qz value. Opposite could be said for AlB_2 sample.

Table 1. The activation energy, specific heat of reaction, ignition temperature and surface temperature of the reaction layer for HEM samples.

Parameter	HEM sample, containing metal powder		
	ASD-4	AlB_2	AlB_{12}
E_a, kJ/mol	219	98	307
Qz, W/g	$8.768\cdot10^{10}$	$5.136\cdot10^{6}$	$4.811\cdot10^{18}$

T^*, K	1684	1114	1744
$\bar{T}$, K	983±75	990±75	1008±75
T_{max}, K	1421±120	1973±120	1600±120

4 Conclusions

The ignition parameters for the model HEM samples based on ammonium perchlorate and nitrate, energetic binder, containing micron powder of aluminium, aluminium diboride and aluminium dodecaboride are presented. It was found that the effect upon laser ignition (ignitability) at full replacement of ASD-4 by both aluminium diboride and dodecaboride is close. The ignition time of HEM is reduced by 2.0–2.5 times in the heat flux density range of 90–200 W/cm^2.

Characteristic temperatures were obtained as well. Average surface temperature at the moment of ignition was found to be nearly same for all studied HEM samples (close to 1000 K) while maximal temperatures differ significantly – they were found to be in range of 1400–2000 K. When replacing the ASD-4 aluminum powder by aluminum dodecaboride in HEM composition the activation energy is increased by 1.4 times and is decreased by 2.25 times for aluminium diboride HEM.

The reported study was partially supported by Russian Scientific Foundation, Grant 16-19-10316.

References

1. A. Gany, Y.M. Timnat, Acta Astronautica, **29**, 3 (1993)
2. T.L. Connell, G.A. Risha, R.A. Yetter, C.W. Roberts, G. Young, Journal of Propulsion and Power, **31**, 1 (2015)
3. K.-L. Chintersingh, M. Schoenitz, E.L. Dreizin, Combustion and Flame, **173** (2016)
4. P.F. Pokhil, A.F. Belyaev, Yu.V. Frolov, V.S. Logachev, A.I. Korotkov, *Combustion of Powdered Metals in Active Media* (Science, Moscow, 1972) [in Russian]
5. V.E. Zarko, O.G. Glotov, Science and Technology of Energetic Materials, **74** (2013)
6. A.A. Gromov, A.G. Korotkikh, A.P. Il'in, L.T. DeLuca, V.A. Arkhipov, K.A. Monogarov, U. Teipel, *Nanometals: Synthesis and Application in Energetic Systems, Energetic Nanomaterials: Synthesis, Characterization, and Application* (Elsevier Inc. 2016)
7. A.G. Korotkikh, O.G. Glotov, V.A. Arkhipov, V.E. Zarko, A.B. Kiskin, Combust. Flame, **178** (2017)
8. A. Abraham, H. Nie, M. Schoenitz, A.B. Vorozhtsov, M.I. Lerner, A. Pervikov, N. Rodkevich, E.L. Dreizin, Combust. Flame, **173** (2016)
9. V.A. Arkhipov, A.G. Korotkikh, V.T. Kuznetsov, A.A. Razdobreev, I.A. Evseenko, Russ. J. Phys. Chem. B, **5**, 4 (2011)
10. T.R. Sippel, S.F. Son, L.J. Groven, S. Zhang, E.L. Dreizin, Combust. Flame, **162** (2015)
11. V.A. Arkhipov, A.G. Korotkikh, Combust. Flame, **159** (2012)

Research of heat and mass transfer processes in conditions of microwave heating of wet wood

Anastasya A. Kostoreva[1], *Zhanna* A. Kostoreva[1], *Lidiya* V. Rogovaya[1], and *Vladimir* S. Loginov[1,*]

[1]National Research Tomsk Polytechnic University, 634050 Tomsk, Russia

Abstract. The results of experimental and theoretical studies of the processes of heat and mass transfer occurring during drying of wood under microwave heating are presented. The times of moisture removal are set. It is established that the dynamics of dehydration depends on the type of wood. Based on the results of experimental studies, a mathematical model of the heat transfer processes taking place under conditions of intensive phase transformations (evaporation of water) is formulated together during the dehydration of the wood. A comparative analysis of the theoretical and experimental values of the drying times showed their satisfactory consistency.

1 Introduction

Woodworking is one of the promising industries both in the Russian Federation and abroad. This is due in many respects to the geographical position of the country, the area of forests and their ability to self-renew. In recent years, the growth of the woodworking industry has begun in Russia, which requires blanks from coniferous and hardwood species. Such industries need high-quality processed products [1]. Therefore, wood is one of the most popular materials at the present time [2].

The growth of national economies is forcing the introduction of high technologies in the industry to meet the rapidly changing demands of consumers for wood products. The latter must be fulfilled as soon as possible [3].

As a rule, wet wood is brought to the production process that has not undergone preliminary processing (trimming bark, sawing, drying, etc.). At the same time, the finished products must be dry, because products made from raw wood, over time, change their properties (strength, shape, density, etc.). The main stage of wood processing is the removal of moisture. The drying process is one of the most energy-intensive processes in the woodworking industry, significantly increasing the cost of the final product by 20-30%. This is due to the high duration of the process and large losses to the environment (infiltration through the openings of the outer walls of the chamber) [4]. One of the most promising technological solutions at the present time is the drying of wood in the field of microwave radiation (microwave drying). However, to date, the technology of microwave drying has not received sufficient propagation in industry due to the lack of elaboration of

* Corresponding author: loginovvs@tpu.ru

the physical theory of moisture removal in the microwave heating of wood. Full-scale research (experiment) of heat and mass transfer processes occurring during microwave drying of wood is very expensive and complex, and sometimes it is simply impossible. This is due to the numerous group of factors (type of wood, size of billet, type of sawing, etc.) [5]. Accordingly, it is necessary to exclude a large number of expensive and long-term field experiments [6]. Therefore, we can say that the use of the apparatus of mathematical modeling. However, the mathematical models developed at the present time are based on a number of significant assumptions that do not take into account the entire complex of thermophysical and electrophysical processes. So, for example, in [7], the results of mathematical modeling of microwave heating of a dielectric are presented, it is assumed that the distribution of the electric field vector over the sample is uniform, respectively, and the volume heat release is also unevenly distributed. However, as is well known from the works of Maxwell [8], the distribution of the electric field strength in space is not substantially uniform. In [9], the results of a numerical solution of the problem of heating a wood sample in a microwave field are presented, taking into account the nonuniformity of the EM field. However, this does not solve the problem of drying the samples. It is known [10] that when dehydrating porous materials a phase transition front separating the water-saturated and dry layers of wood is formed. Due to internal heating, it is possible to form an area of dehydrated wood inside the sample. These are multi-front tasks of drying. The latter are some of the most complex in mathematical physics.

Summarizing, it can be said that the development of an "adequate" physical theory and the mathematical basis for the microwave drying of wood that provides it is an urgent and unresolved task to date. The latter will contribute to the development of optimal drying modes for wet wood.

Objective: Experimental and theoretical investigation of the processes of heat and mass transfer proceeding together under conditions of intense phase (evaporation of water) transformations with intense convective-microwave heating of moist wood.

2 Experiment

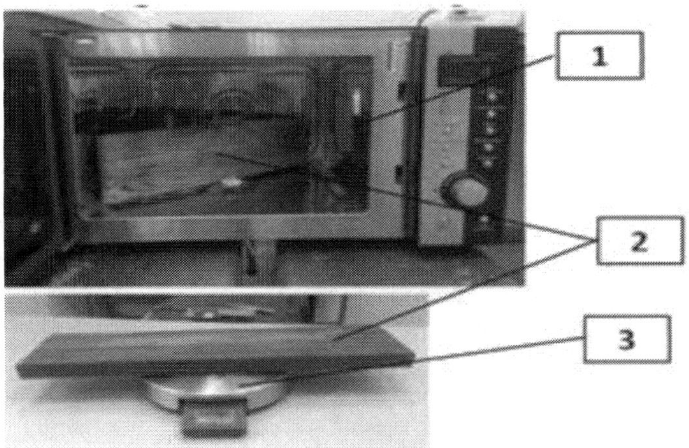

Fig. 1. Experimental setup: 1 - microwave oven, 2 - wood sample, 3 - scales.

Before the experiments, the wood blanks were brought to the moisture content φ of the freshly cut tree φ = 30-40%. After that, the samples were placed inside the microwave drying chamber. Then, the wood was removed and the control weighing was carried out (the holding time of the wet billet in the microwave oven was 60 seconds, the weighing period was not more than $0.1 \cdot t_{dry}$). After weighing, the blank was again placed inside the microwave chamber and further heating was carried out. The experiments were repeated until the humidity change in the wood in 60 seconds did not exceed 0.01%.

3 Formulation of the problem

Based on the results of the experiments, the physical and mathematical models of the investigated process are formulated.

Physical setting

At the initial time ($\tau = 0$), the wood preform enters the drying chamber and is heated by the combined action of microwave radiation. As a result of microwave exposure, internal heat generation occurs and the evaporation of moisture is initiated. Accordingly, a layer of dry wood with a low coefficient of thermal conductivity is formed inside the wood sample. The time interval from the beginning of heating to the moment of hollow drying (complete removal of moisture) is considered the drying time t.

Accordingly, it is possible to form a dry layer within the sample and the appearance of a second evaporation front that moves from the depth of the workpiece to the surface.

Mathematical statement

Mathematically, the drying process within the framework of the formulation corresponding to the above physical model is described by the energy equation for wet wood, taking into account the endothermic evaporation effect:

$$c(\mathrm{T})\rho(\mathrm{T})\frac{\partial T}{\partial \tau} = \frac{\partial}{\partial x}(\lambda\frac{\partial T}{\partial x}) - \frac{Q_{eva} \cdot W_{eva} \cdot \delta(x_V)}{h_d} + q_{UHF} \cdot \theta(x_{eva} - x) \quad (1)$$

W_{eva} — the evaporation rate (kg/(m^2·s)) have been calculated according to the formula [11]:

$$W_{eva} = \frac{\mathrm{V_0} \cdot \exp[-\mathrm{Q}_{eva} \cdot \mu_0 \cdot (\mathrm{T} - 273)]}{\mathrm{R} \cdot \mathrm{T} \cdot 273} \quad (2)$$

q_{uhf} — thermal effect (W/m^3) UHF impact:

$$q_{UHF} = \varepsilon 1 \cdot f \cdot tg\delta \cdot |E|^2 \quad (3)$$

where: c — heat capacity, j/(kg · K); ρ — density, kg/m^3; λ — thermal conductivity, W/(m·K); T — temperature, K; θ — the Heaviside function; δ — a discrete function of Dirk; Q_{eva} — heat of the endothermic effect of evaporation, j/kg; h_d — ratio to smearing of the evaporation front, m; ε_1 — the dielectric penetrability; f — frequency of electromagnetic field, Hz; $tg\delta$ — the angle; $|E^2|$— the potential stress of the electromagnetic field, V/m, μ_0 — the molar mass of water, kg/mol.

- equation (1) has been solved under the following boundary conditions:

$$\tau = 0 : T = T_0$$

$$x = 0 : \frac{\partial T}{\partial x} = 0 ; \quad (4)$$

$$x = l : \lambda \frac{\partial T}{\partial x} = \alpha(T_e - T); \tag{5}$$

The problem (1) - (5) was solved by the method of finite differences using the algorithm similarly to [12]. The rate of "dehydration" was calculated by the method of implicit separation of the evaporation front [13].

4 The results

Table 1 shows the experimentally and theoretically obtained drying times for wood. It can be noted that the times of complete drying of the wood sample from the cedar practically coincide. But for the spruce specimen, the deviation of the theoretical and experimental drying times is about 30%. The latter can be explained by the fact that the initial data used in mathematical modeling of the thermophysical and electrophysical characteristics of spruce sith differ from the analogous parameters of the common spruce used in the experiment.

Table 1. Experimental and theoretical times of wood drying.

Type of wood	t_{dry}, s. Theoretical values.	t_{dry}, s. Experimental values.	Relative deviation of drying times%
fir	568	645	11.94
spruce	564	810	29.13
pine	579	510	13.53
cedar	560	660	1.52

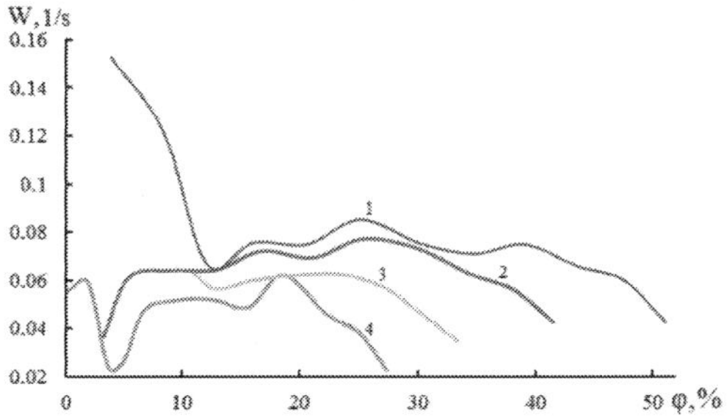

Fig. 2. Dependence of the relative rate of moisture removal W = Δφ / s on absolute humidity φ for samples made of wood: 1 - spruce; 2 - fir; 3 - pine; 4 - cedar.

Figure 2 shows the dependence of the moisture removal rates on absolute humidity for samples made from different types of wood. It can be noted that the longest dehydration period corresponds to the drying of spruce, while the lowest rate of moisture removal is recorded in cedar. This is due to the significant differences in the thermal and electrical properties of wood. The shape of the W (φ) curves for all types of wood is the same, however, for the samples from spruce and cedar in the final stage (for spruce $\tau = 720$ seconds, for cedar $\tau = 630$ seconds) drying, the dewatering rates increase abruptly.

Figure 3 shows the dependence of the absolute humidity change φ on time for samples made from different types of wood. It is possible to note the absolutely identical character of the curves. Accordingly, it can be said that the mechanism of moisture removal for all investigated samples of wood is the same. It can also be noted that the curves φ (t) are practically linear. Accordingly, it can be assumed that the drying process can be described by simple polynomial expressions.

Figure 4 shows the dependence of the change in the mass of the wood billet on the drying time for samples made from different types of wood. The characteristics of the change in the mass of the bar for spruce and cedar are almost identical, accordingly, it can be reasonably concluded that such types of wood can be dried in one drying chamber. The dehydration characteristics of fir significantly differ from those of pine, spruce and cedar. Accordingly, wood samples made of fir, it is necessary to dry separately from the rest.

4 Conclusion

The results of experimental and numerical studies of the processes of heat and mass transfer

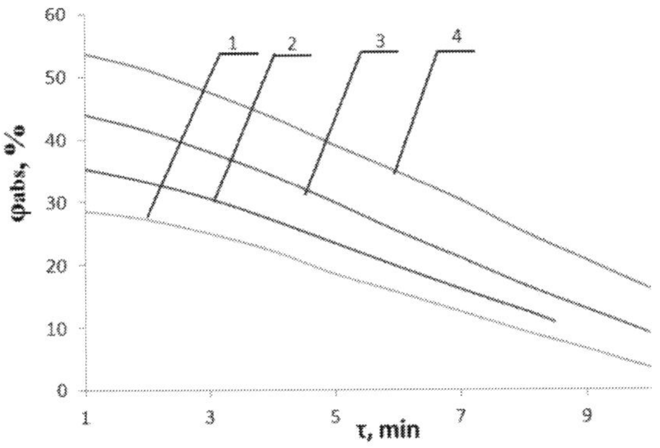

Fig. 3. Dependence of absolute humidity φ_{abs} from the drying time τ for samples made of wood:
1 – cedar; 2 – pine; 3 – fir; 4 – spruce.

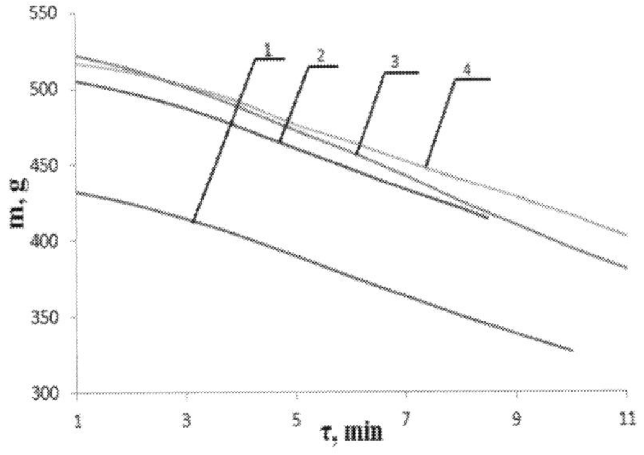

Fig. 4. Dependences of the change in sample mass (m) from the drying time τ for samples made of wood: 1 – fir; 2 – pine; 3 - tree; 4 – cedar.

in the conditions of microwave heating of moist wood are presented. Based on the results of the experimental study, the rate of moisture removal from the absolute moisture content of the wood sample was found to be significant, and it was noted that this dependence is essentially nonlinear. So in the initial drying period, the rate of moisture removal increases, however, after the time $\tau = 0.74 \cdot \tau$ drying, the dehydration rate begins to decrease. A comparative analysis of the experimental data obtained and the theoretical dependences shown showed satisfactory agreement. Based on the results of experimental studies, a mathematical model was developed for the processes of heat and mass transfer occurring during microwave drying. According to the results of mathematical modeling, it was found that the integral characteristics of the drying process depend substantially on the type of wood. For example, the time for drying pine, spruce and cedar is approximately the same, while tusch fir is significantly different. Accordingly, it can be concluded that wood blanks of pine, spruce and cedar can be dewatered simultaneously in one drying chamber.

References

1. R.R. Safin, J. of Tomsk State University **11**, 4 (2006)
2. O.A. Kizina, A.L. Adamovich, J. of Polotsk State University **3**, (2011).
3. Yu.G. Grazenberg, A.L. Adamovich, Yu.V. Matieniuk, J. of Polotsk State University. **3**, 114 (2012)
4. A.I. Raseev, *Drying of wood*, (EDP Lan', St. Petersburg, 2014)
5. G.A. Morozov, Antennas release **1**, 40 (1998)
6. G.V. Kuznetsov, M.A. Sheremet, *Difference methods for solving heat conduction problems* (Publishing house TPU, Tomsk, 2007).

7. V.V. Salomatov, S.O. Sladkov, S.E. Pashchenko, J. Eng. Phys. Thermophys. **85**, 3 (2012)
8. J.C. Maxwell, Phil. Trans. R. Soc. Lond. **155**, (1865)
9. F. Erchiqui, Z. Annasabi, M. Souli, F. Slaoui-Hasnaoui. 3D numerical analysis of the thermal effect and dielectric anisotropy on thawing frozen wood using microwave energy. Int. J. of Thermal Sciences **89**, (2015)
10. A.V. Lykov, *Theory of drying*, (Energia, Moscow, 1968)
11. A.A. Khashchenko, O.V. Vecher, E.I. Diskkaeva, J. of Altai State University. **1**, 89 (2016)
12. S.V. Syrodoy, N.Yu. Gutareva, R.I. Taburchinov. MATEC Web of Conferences **72**, 01109 (2016)
13. Syrodoy S.V., Gutareva N.Y., Bugaeva K. A. MATEC Web of Conferences **72**, 01035 (2016)

MATEC Web of Conferences **110**, 01044 (2017) DOI: 10.1051/matecconf/201711001044
HMTTSC-2017

Creation of two-level liquid cooling system of PC

Sergey V. Koynov[1], *Egor* A. Tarasov[1,*]

[1]Tomsk State University, 634050 Tomsk, Russia

Abstract. This paper devoted to creation of two-level liquid cooling system of PC. We found the solution to change a traditional air cooling system for the liquid cooling system with ethylene glycol as a coolant. The cooling system managed to reduce the temperature by up to 4-10 °C while down time and ~11 °C while load time.

1 Introduction

Nowadays technologies of PC components creation, such as *CUDA (Compute Unified Device Architecture)* hardware architecture or *APP (Accelerated Parallel Processing)* technology let us use computational models of physical processes we could never use before due to the limitation of computing power. Unfortunately, increase in this capability is accompanied by both energy consumption and heating power growth, later influencing the temperature inside the PC system unit case. Below is the table, showing energy consumption values of a numbers of modern graphics cards and CPUs:

Table 1. Energy Consumption and Cooling Systems Requirements of a Number of Modern CPUs and Graphics Cards.

CPUs		
CPU Model	Thermal Design Power (TDP)	Additional Informantion
1. Intel Core i7-6800K	140 W	Maximum Computer Case Temperature - 66.8 °C
2. AMD FX-9590	220 W	Maximum Computer Case Temperature – 57 °C
3. Xeon E5-2670 v3	120 W	Maximum Computer Case Temperature – 84.5 °C
Graphic Cards		
CPU Model	Energy Consumption	Additional Informantion
1. Nvidia GeForce GTX 1080	180 W	Maximum GPU Temperature – 94 °C
2. Nvidia GeForce GTX 980	165 W	Maximum GPU Temperature – 98 °C
3. Nvidia GeForce GTX 1050	75 W	Maximum GPU Temperature – 97 °C

* Corresponding author: diomedis@mail.ru

DOI: 10.1051/matecconf/201711001044

Utilizing PC in the environment with the temperature higher than certain value may cause a premature failure of different components. Temperature exposure is the reason for PC case crack, mechanical damage of a processor chip (due to thermal growth), electrolyte leakage in condensers or delamination of PC board, where the devices are situated. In case when elements are low powered (chips, RAM) or when constraining computational capacity, cooling system can turn the cooler only without forced air cooling, and it will manage timely heat transfer.

For example, when CPU works at the temperature above maximum for a computer case (see Table 1), it can not only damage the surrounding elements on the motherboard, but also starts throttling. Cooling systems and ways traditionally used in PCs [2,3,4], such as active and passive air-cooling (fans, radiators of intricate geometry), thermal compound on radiators' contacts and heat causing elements (CPU, GPU, northbridge and southbridge) stop coping with the cooling task, and for this reason we need the transition to liquid cooling systems.

2 Experimental

PC liquid cooling system we created represents the following. To sink the running temperature of Intel® Core™ i7-4790K CPU on the GA-Z97X-Gaming G1 WIFI-BK motherboard, we initially used the CBO Deepcool CAPTAIN 360 ready-made solution: water unit material - copper; pump - time per fault is 120000 hours, frequency - 3500 rpm; radiator - material: aluminium, sizes: 395x120x32 mm; fans - sizes: 3x 120x120x25 mm; rotational rate: 600±200~2200±10% rpm.

The run of the solution revealed one more problem with the cooling system of the PC - the cooling effect of CPU near the socket had vanished and we needed to set up one more liquid cooling system for the Intel Z97, PLX PEX8747 chips and for the brain of processor power microchip, RAM and GPU. It was necessary for Z97 chip to remain cool, to carry the heat away from the computer case.

After solving a number of engineering tasks we created the given system, which together with ready-made CPU solution comprises a two-level liquid cooling system of PC. In the figure 1 the following elements are presented:

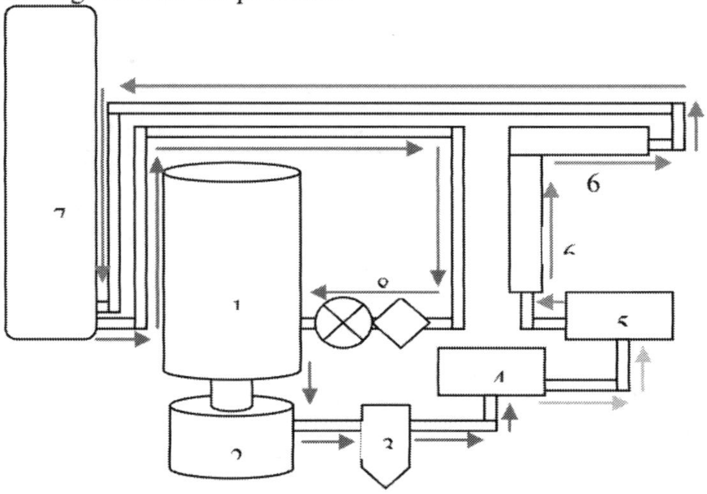

Fig. 1. The second level of liquid cooling system of PC.

1 - liquid container; 2 - MAGICOOL pump; 3 - dump; 4 - water unit (copper) MAN-kd 2013CPU on a Intel Z97 chip; 5 - water unit (copper) KD North Bridge 2015 on a PLX PEX8747 chip; 6a, 6b - big and small radiators VRM (standard radiators for GA-Z97X-

Gaming G1 WIFI-BK motherboard); 7 - STW R240P radiator (aluminium); 8 - STW flowmeter and a digital thermometer with a display. The arrows show the coolant run.

Having tested the system on a various types of cooling liquids, such as light water (tap-water), service water and service water with ethanol (6 parts water to 4 parts ethanol), we chose Thermaltake C1000 Opaque Coolant. It is ethylene glycol (1,2-dioxiethan) with anti-oxidants. Main physical and chemical characteristics are represented in the table 2.

Table 2. C1000 Opaque Coolant characteristics.

Characteristics	Value
pH level, (at 20 °C)	8.0 – 8.5
Density, (at 20 °C)	1.120 - 1.125 g/cm ^{3}
Viscosity, (kinematic, at 20 °C)	20 - 30 mm ^{2}/s
Boiling point	≥165 °C
Fusing temperature / solidifying point	≤-18 °C

To test the resulting system we compared results of VRM stress testing enumerated in [5] with the results of system load and down time:

Table 3. Comparison of experiments' results.

Data from [5]	Environmental temperature	CPU (LGA1150)	PCH (Intel Z97)	System board (VRM)
Down time	~27 °C	40.1 °C	51.1 °C	35.5 °C
Load		58.2 °C	51.6 °C	49.3 °C
Experimental data received by the authors with the help of temperature sensors, attached to ITE IT8620E chip on the motherboard				
Down time	~23 °C	32 °C	33 °C	31 °C
Load		43 °C	37 °C	34 °C

3 Summary

As we can see from the temperature sensors, the cooling system managed to sink the temperature by up to 4-10 °C while down time and ~11 °C while load time (considering difference in our environmental temperature with the one in the other experiments). The obtained results let us to draw a conclusion about successful application of the created liquid cooling PC system in the given configuration.

In the future we plan to set up computational models for the description of forced convection processes in this system, using data from previous experiments [6,7] and to cast the modelling results in our further works, and what is more, to compare them with the experimental results.

References

1. Intel® I/O *Controller Hub 10 (ICH10)* Family Thermal and Mechanical Design Guidelines. (2008). Document Number: 319975-001319975-001
2. S. Matei, Xtrem PC, **78**, 46 (2006)
3. K. Azar, B. Tavassoli, Electroics Cooling, **9** (2), 30 (2003)
4. J.R. Wilson, Military & Aerospace Electronics, 34 (2006)
5. https://3dnews.ru/903287
6. G.V. Kuznetsov, M.A. Sheremet, Thermophys. Aeromech. **16**, 119 (2009)
7. G.V. Kuznetsov, M.A. Sheremet, Russian Microelectronics, **37** (2), 131 (2008)

Analysis of influence of heat exchange conditions on the outer surface of the lithium-ion battery to electrolyte temperature under the conditions of high current loads

Alexander Krasnoshlykov[1,*]

[1]National Research Tomsk Polytechnic University, 634050 Tomsk, Russia

Abstract. Numerical analysis of thermal conditions of a lithium-ion battery using the software package ANSYS Electric and ANSYS Fluent has been carried out. Time dependence of the electrolyte temperature on the various heat exchange conditions on the outer surface has been obtained.

1 Introduction

Lithium-ion batteries play an important role as a source energy in a variety of applications (motor transport, aviation and space equipment) and have prospects for development through the low weight, high energy and power density [1, 2]. However, a major obstacle to the wide use of batteries of this type is the flammability and low performance at increasing temperature due to thermal runaway [1, 2].

Thermal runaway of the battery is the effect which occurs when electrolyte temperature increases and the current flowing through the battery can rise to values approaching the short-circuit current [2]. Thermal runaway is able to affect the safety of the use of vehicles (e.g., aircraft [1]). Thermal runaway occurs at a combination of factors: the electrolyte temperature rise due to insufficient battery heat exchange with the environment; low level of electrolyte in the battery cell; aging active electrode mass [2].

The experiments with accumulators reduce to determining the temperature of the surface under certain conditions of their use [3]. It should be noted that when designing a typical lithium-ion battery, it is necessary to take into account the heat release during the passage of an electric current through plates (Joule heat) [4]. To solve this kind of complex problems, it is most expedient to apply methods of mathematical modeling of heat transfer processes in the material of solid plates and the electrolyte volume of the battery based on solving the systems of mass transfer equations, momentum and energy [5] Joel heat.

The purpose of the work is mathematical modeling of temperature distributions in the accumulator under various heat exchange conditions on the external vertical surface.

* Corresponding author: krasnas@tpu.ru

2 Formulation of the problem

A typical segment of a lithium-ion battery, including copper and aluminum plates, a liquid layer of electrolyte between them and a solid wall is considered (Fig.1b).

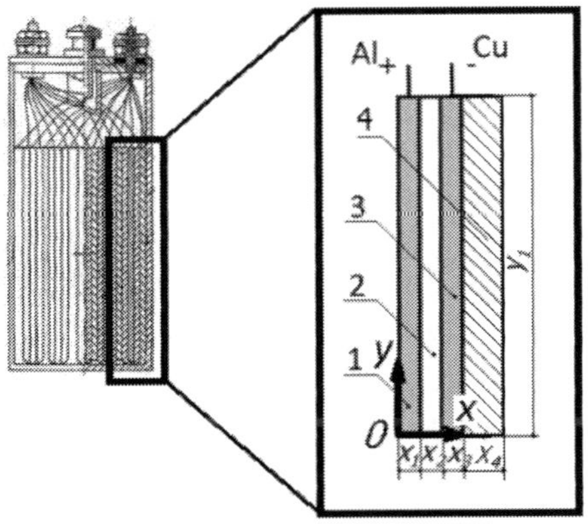

a). b).

Fig. 1. a). General view of the region of the solution; b). The considered segment of the battery. 1 - positive electrode (aluminium plate); 2 - liquid electrolyte (LiPF6); 3 - negative electrode (copper plate); 4 - battery wall (polypropylene).

In the formulation of the problem, the possible release of heat as a result of exothermic chemical reactions of the interaction of electrolyte and electrode materials was not taken into account.

Joule heat is computed using the vector potential method [6]:

$$Q^j = \frac{1}{n}\sum_{i=1}^{n}[\rho]\{J_{ti}\}\cdot\{J_{ti}\}$$

where Q^j – Joule heat per unit volume;
n – number of integration points;
$[\rho]$ – resistivity matrix;
$\{J_{ti}\}$ – total current density in the element at integration point i.

It was assumed that the heat release in the plates was distributed uniformly along the x and y directions.

The equations of continuity, momentum and energy for a layer of liquid electrolyte:

$$\frac{\partial\rho}{\partial t}+\frac{\partial(\rho u)}{\partial x}+\frac{\partial(\rho \upsilon)}{\partial y}=0; \tag{1}$$

$$\frac{\partial\rho_2 u_2}{\partial t}+\frac{\partial(\rho_2 u_2 u_2)}{\partial x}+\frac{\partial(\rho_2 \upsilon_2 u_2)}{\partial y}=\rho_2 g_x-\frac{\partial P_2}{\partial x}+\frac{\partial}{\partial x}\left(\mu_2\frac{\partial u_2}{\partial x}\right)+\frac{\partial}{\partial y}\left(\mu_2\frac{\partial u_2}{\partial y}\right); \tag{2}$$

$$\frac{\partial\rho_2 \upsilon_2}{\partial t}+\frac{\partial(\rho_2 u_2 \upsilon_2)}{\partial x}+\frac{\partial(\rho_2 \upsilon_2 \upsilon_2)}{\partial y}=\frac{\partial P_2}{\partial y}+\frac{\partial}{\partial x}\left(\mu_2\frac{\partial \upsilon_2}{\partial x}\right)+\frac{\partial}{\partial y}\left(\mu_2\frac{\partial \upsilon_2}{\partial y}\right); \tag{3}$$

$$\rho_2 C_{p2}\left(\frac{\partial T_2}{\partial t}+u_2\frac{\partial T_2}{\partial x}+\upsilon_2\frac{\partial T_2}{\partial y}\right)=\lambda_2\left(\frac{\partial^2 T_2}{\partial x^2}+\frac{\partial^2 T_2}{\partial y^2}\right). \tag{4}$$

Equation of thermal conductivity for plates and the battery wall:

$$\rho_1 C_{p1}\frac{\partial T_1}{\partial t}=\lambda_1\left(\frac{\partial^2 T_1}{\partial x^2}+\frac{\partial^2 T_1}{\partial y^2}\right); \tag{5}$$

$$\rho_3 C_{p3}\frac{\partial T_3}{\partial t}=\lambda_3\left(\frac{\partial^2 T_3}{\partial x^2}+\frac{\partial^2 T_3}{\partial y^2}\right); \tag{6}$$

$$\rho_4 C_{p4}\frac{\partial T_4}{\partial t}=\lambda_4\left(\frac{\partial^2 T_4}{\partial x^2}+\frac{\partial^2 T_4}{\partial y^2}\right). \tag{7}$$

where u, υ – the velocity components in the projection on the axis x, y, respectively; P – pressure; T – temperature; ρ – density; x, y – Cartesian coordinates; t – time; C_p – heat capacity; g – acceleration of gravity; λ – coefficient of thermal conductivity; μ – dynamic viscosity; indices 1, 2, 3, 4 – correspond to aluminum, electrolyte, copper and polypropylene.

The initial conditions for the system of equations (1-7) are as follows:
u(x,y)=0; T(x,y)=T0; P(x,y)=P0.

The boundary conditions for the equations (1-7) have the form:

$$x=0,\ 0<y<y_1\quad \frac{\partial T_1}{\partial x}=0;$$

$$x=x_1,\ 0<y<y_1\quad \begin{cases}T_1=T_2\\ \lambda_1\dfrac{\partial T_1}{\partial x}=\lambda_2\dfrac{\partial T_2}{\partial x},\end{cases}\ \begin{cases}u_2=0\\ \upsilon_2=0\end{cases};$$

$$x=x_1+x_2,\ 0<y<y_1\quad \begin{cases}T_2=T_3\\ \lambda_2\dfrac{\partial T_2}{\partial x}=\lambda_3\dfrac{\partial T_3}{\partial x},\end{cases}\ \begin{cases}u_2=0\\ \upsilon_2=0\end{cases};$$

$$x=x_1+x_2+x_3,\ 0<y<y_1\quad \begin{cases}T_3=T_4\\ \lambda_3\dfrac{\partial T_3}{\partial x}=\lambda_4\dfrac{\partial T_4}{\partial x};\end{cases}$$

$$x=x_1+x_2+x_3+x_4,\ 0<y<y_1\quad \lambda_4\frac{\partial T_4}{\partial x}=\alpha\left(T_4-T_{OC}\right);$$

$$y=0,\ 0<x<x_1\quad \lambda_1\frac{\partial T_1}{\partial y}=0; \qquad y=0,\ x_1<x<x_2\quad \lambda_2\frac{\partial T_2}{\partial y}=0,\ \begin{cases}u_2=0\\ \upsilon_2=0\end{cases};$$

$$y=0,\ x_2<x<x_3\quad \lambda_3\frac{\partial T_3}{\partial y}=0; \qquad y=0,\ x_3<x<x_4\quad \lambda_4\frac{\partial T_4}{\partial y}=0;$$

$$y = y_1,\ 0 < x < x_1\ \lambda_1 \frac{\partial T_1}{\partial y} = 0; \qquad y = y_1,\ x_1 < x < x_2\ \lambda_2 \frac{\partial T_2}{\partial y} = 0,\ \begin{cases} u_2 = 0 \\ \upsilon_2 = 0 \end{cases};$$

$$y = y_1,\ x_2 < x < x_3\ \lambda_3 \frac{\partial T_3}{\partial y} = 0; \qquad y = y_1,\ x_3 < x < x_4\ \lambda_4 \frac{\partial T_4}{\partial y} = 0.$$

3 Results and Discussion

Numerical modeling of heat transfer processes during the flow of direct current through the battery plates (Fig.1) is carried out using the ANSYS Electric and ANSYS Fluent software packages. Geometric dimensions of the solution region: the height of the plates ($0<y<y_1$) is 140 mm, the thickness of each element (aluminum, copper plates and electrolyte layer) is 0.5 mm, the wall thickness is 2 mm [7, 8].

The following variants of load current on the battery are adopted: I_n=50 A, I_n=60 A, I_n=70 A [4]. As the electrolyte is adopted $LiPF_6$ [7]. The temperature of thermal decomposition of this electrolyte is 125 °C [8], however, it should be noted that already at 60°C in the volume of electrolyte chemical reactions occur, described in [3, 4], which can lead to thermal overclock of the battery.

Figure 2 shows the dependence of the average temperature of the electrolyte on time with a heat-insulated battery case. It can be seen that in the absence of cooling from the surface of the wall of the battery case and the current load above the nominal (>50 A), the temperature reaches critical values (60 °C) in 800 s.

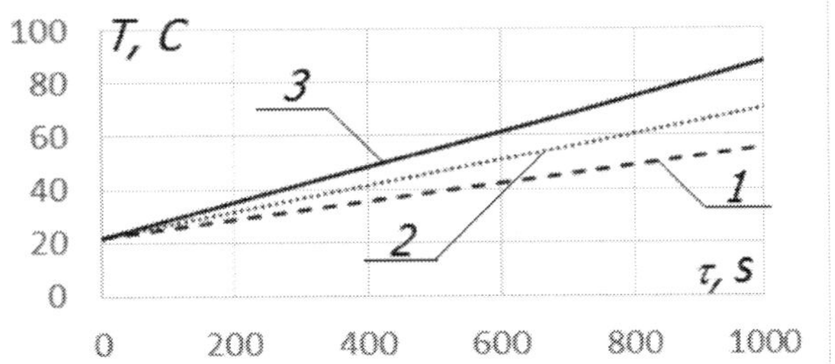

Fig. 2. Dependence of the battery electrolyte temperature on time with different current load (no cooling). 1 – I_n=50 A, 2 – I_n=60 A, 3 – I_n=70 A.

When the battery surface is cooled (α=5 W/(m^2·K) at the boundary) the temperature of the electrolyte rises slower. Up to a constant value, the temperature rises to 4000 s (Fig. 3a). With the considered current load versions of 50 A, 60 A, 70 A and a heat transfer coefficient of 5 W/(m^2·K) the electrolyte temperature in the established mode is higher than the nominal one and is 74,3 °C, 97,3 °C and 124,5 °C, respectively.

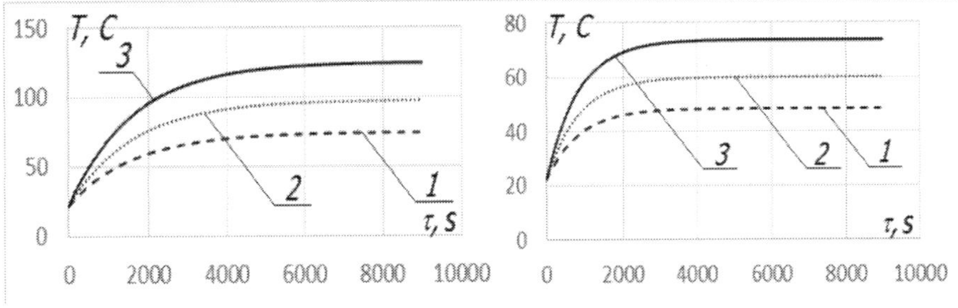

Fig. 3. Dependence of the battery electrolyte temperature on time at different current load at a). α=5 W/(m^2·K); b). α=10 W/(m^2·K)). 1 – I_n=50 A, 2 – I_n=60 A, 3 – I_n=70 A.

With intensive heat transfer from the wall surface of the battery case (α=10 W/(m^2·K)) the electrolyte temperature rises to above the regulatory values only at a load current of 70 A and is 73,7 °C (Fig.3b).

The results of the performed studies show that under conditions of sufficiently intense heat generation in electrodes plates caused by high current loads. It is possible to ensure the level of electrolyte temperatures allowed by the battery operation regulations due to the heat sink from the outer surface of the battery case. At the same time, it is enough to provide conditions for the coefficients of heat transfer from these surfaces to reach 10 W/(m^2·K).

Conclusion

Based on the results of numerical modeling of convective-conductive heat transfer, it is established that an increase in the heat transfer intensity from the wall surface of the battery cabinet in the range of variation of the coefficient α from 5 to 10 W/(m^2·K) leads to a significant decrease in the temperatures of the liquid electrolyte by 26 °C at a constant current load of 50 A compared with the thermal insulation on the outer contour of the battery case. At I_n=60 A and 70 A the temperature decreases in the cooling conditions at 37 °C and 51 °C, respectively.

With a current load of not more than 60 A and a heat transfer coefficient of 10 W/(m^2·K), the temperatures of both the battery surface and its electrolyte do not exceed the specified temperature values (60 °C).

Based on the analysis of the results of the numerical simulation, it can be concluded that even a relatively moderate heat dissipation from the surface of the outer wall has a considerable influence on the performance of the battery.

The reported research was supported by Russian Federation President Grant for state support of the Russian Federation leading scientific schools SS-7538.2016.8 (No.14.Y31.16.7538-SS).

References

1. D.E. Bruskin, *Electric power supply of aircrafts* (Higher School, Moscow,1988) [in Russian]
2. J. Van Mierlo, G. Maggetto, J. Automobile Engineering, **36**, 583 (2003)
3. S.G. Rebrov, S.V. Yanchur, V.S. Mansurov, S.A. Moskovkin, Proceedings of the MAI, **72** (2014) [in Russian]
4. A. Sambaa, N. Omar, Electrochimica Acta, **117**, 246 (2014)
5. G.V. Kuznetsov, Al-Ani, M.A. Sheremet, J. Eng. Thermophys. **201** (2011)
6. Ansys Help. FLUENT Theory Guide.
7. C. Forgez, D. Vinh Do, G. Friedrich, M. Morcrette, C. Delacourt, J. Power Sources, **195** (9) ,2961 (2010)
8. R. Younesi, G. M. Veith, P. Johansson, Energy & Environmental Science **8**, 1905 (2015)

Effective fuel temperature of wwer-1000

Alexander Kudrov[1,*], *Arian* Kuzmin[1], and *Yuri* Rakov[1]

[1]National Research Tomsk Polytechnic University, 634050 Tomsk, Russia

Abstract. The main temperature characteristics of a pressurized water reactor are distinguished, supporting its safety and reliable operation. The special role of the uranium fuel effective temperature is emphasized and the accuracy of the analytical determination of the power effect is increased. The calculation of the temperature distribution along the radius of the fuel rod was carried out taking into account the temperature dependence of the thermal conductivity UO_2. The design procedure was corrected for using the Finca-Ronchi dependence for the thermal conductivity of 95% density of the theoretical one.

1 Primary temperature characteristics of reactor

These include the following integral and differential characteristics [1, 2].

The temperature effect of reactivity (TER), which is defined by the difference of reactivity, caused by the same temperature change from the cold to the hot state of all materials in reactor core (RC). The initial temperature of the cold reactor is set to 20-40 °C. Hot reactor temperature varies with a minimum controlled power level due to external heat sources. Fuel reactivity effect is responsible for forming the neutron spectrum and its leakage.

The power reactivity effect (PRE) ρ_N is determined by the mean or effective temperature of uranium fuel T_{eff} and the actual presence an isotope ^{238}U in it. Due to the Doppler effect, there are resonance levels of the uranium isotope broadening with temperature rise, which increases the neutron absorption probability, thereby reducing reactivity. The higher the fuel temperature and the lower its concentration, the greater the effect. Also, magnitude ρ_N is important for to assess the mode extension fuel campaign using the power effect reactivity.

Differential characteristics: isothermal $\alpha_T = d\rho/dT$ and power $\alpha_N = d\rho_N/dN$ reactivity coefficients determine self-regulation, self-protection, reliability and safety of the nuclear reactor. Both coefficients to provide these nuclear reactor properties must be negative: the first is near the operative point, the second is on the entire range of power transient.

The fuel effective absolute temperature is determined by the exact integral formula

* Corresponding author: alex12kudrow@gmail.com

$$T_{эф} = \frac{\int_0^{r_T} T_T(r) 2\pi r dr}{\int_0^{r_T} 2\pi r dr} \quad (1)$$

Or approximate expression [1]

$$T_{эф} = T_c + 0.4(T_{max} - T_c), \quad (2)$$

where

$T_T(r)$ – temperature distributions over the fuel block;

T_{max} – temperatures on the axis of the fuel block;

T_c – temperatures on the surface of the fuel block.

2 Characteristic comparison on the coefficient of thermal conductivity

The method of direct explicit calculation $T_T(r)$ using the $\lambda_{UO_2}(T)$ is presented in handbooks [3, 4]. Research activity was based on the results [5], where the theoretical dependence of the thermal conductivity UO_2 was used as the equation:

$$\lambda_{UO_2} = \frac{10^2}{3.77 + 0.0258 \cdot T} + 1.1 \cdot 10^{-4} \cdot T + 1.01 \cdot 10^{-11} \cdot T^3 \cdot \exp(7.2 \cdot 10^{-4} \cdot T). \quad (3)$$

There is also the empirical formula:

$$\lambda_{UO_2} = 11.5 - 1.14 \cdot 10^{-2} \cdot T + 4.4 \cdot 10^{-6} T^2 - 5 \cdot 10^{-10} \cdot T^3, \quad (4)$$

which could be approximated by

$$\lambda_{UO_2} = \lambda_a \text{ch}[(T - a)/b], \quad (5)$$

choosing the constants: $\lambda_a = 2.3$ W/(m·K); $a = 2000$ K; $b = 870$ K.

More recent works are recommended to use the Fink-Ronchy formula to determine the thermal conductivity with density of 95%:

$$\lambda_{UO_2} = \frac{100}{7.5408 + 17.692 \cdot \tau + 3.6142 \cdot \tau^2} + \frac{6400}{\tau^{5/2}} \exp\left(-\frac{16.35}{\tau}\right), \quad (6)$$

where

$\tau = (T / 1000)$K.

The equations (3) - (6) are presented in Fig. 1 and show the necessity of the coefficients adjustment in dependence (5).

3 Updating of the design formula for the fuel block

We take the Fink-Ronchy formula (6) as the main dependence for the coefficient of the thermal conductivity UO_2 and approximate it by the trigonometric function (5)

$$\lambda_{UO_2} = \lambda^*_a \mathrm{ch}\left[\left(T - a^*\right)/ b^*\right], \quad (7)$$

which has the following constants: $\lambda^*_a = 2.1$ W/(m·K); $a^* = 2070$ K; $b^* = 980$ K.

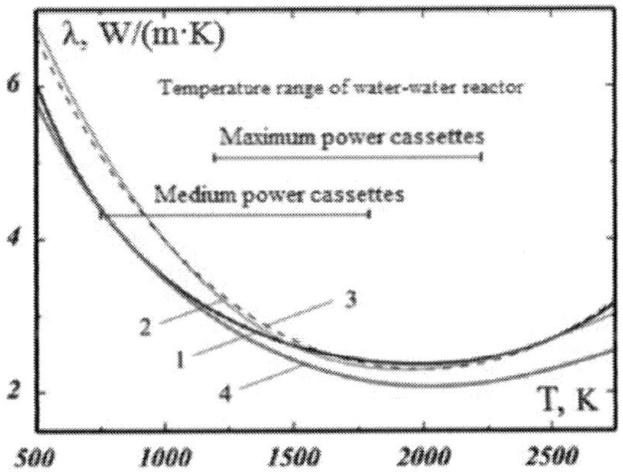

Fig. 1. Comparison of approximations using the thermal conductivity UO_2:
1 – theoretical dependence (3) of work [5];
2 – empirical formula (4), [5];
3 – formula (5);
4 – Fink-Ronchi ratio (6).

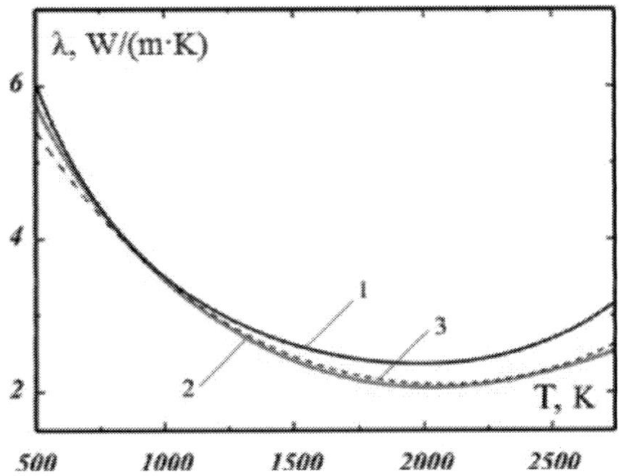

Fig. 2. Updating of approximation using the thermal conductivity UO_2:
1 – theoretical dependence (3) of work [5];
2 – Fink-Ronchi ratio (6);
3 – formula (7).

4 Analysis of results

Carrying out estimations on the proposed dependencies allows us to draw the following conclusions:

- the approximation formula for determining the effective fuel temperature (2) is in error slightly more than 3%;
- the acceptable trigonometric approximation (7) of the Fink-Ronchi formula has been received; it will allow calculating the radial temperature distribution using the temperature dependence of the uranium dioxide thermal conductivity (8);

• comparison with the known results of thermal design for the reactor fuel rods (WWER– 1000 [9]) showed slightly lower values of the temperature field; this indicates, first of all, the necessity of using the effect of density changes and deviations from stoichiometry, which requires further analysis.

References

1. G.G. Bartholomew, G.A. Bat', V.D. Bajbakov, M.S. Alhutov, *Fundamentals of the Theory and Methods for Calculating Nuclear Power Reactors* (YULAND, Yekaterinburg, 2016) [in Russian]
2. V.I. Vladimirov, *Physics of Nuclear Reactors: Practical Tasks for Their Operations* (Knizhnyy dom "LIBROKOM", Moscow, 2009) [in Russian]
3. P.L. Kirillov, Yu.S. Yur'ev, V.P. Bobkov, *Handbook of Thermohydraulic Calculations (Nuclear Reactors, Heat Exchangers, Steam Generators)* (Energoatomizdat, Moscow, 1990) [in Russian]
4. P.L. Kirillov, A.V. Zhukov, N.I. Loginov, V.M. Makhin, I.L. Pioro, Yu.S. Yur'ev, Handbook of Thermohydraulic *Calculations in Nuclear Power. Volume 2. Nuclear Reactors, Heat Exchangers, Steam Generators* (IzdAt, Moscow, 2013) [in Russian]
5. V.I. Kolyadin, E.P. Il'in, A.G. Kharlamov, V.V. Yakovlev, Atomic Energy, **36**(1), 59 (1974) [in Russian]
6. Thermophysical Properties Database of Materials for Light Water Reactors and Heavy Water Reactors: Final Report of a Coordinated Research Project 1999-2005; IAEA-TECDOC-1496. – Vienna: IAEA, 2006.
7. P.L. Kirillov, M.I. Terent'eva, N.B. Deniskina, *Thermophysical properties of nuclear materials* (IzdAt, Moscow, 2007) [in Russian]
8. Yu.G. Godin, A.V. Tenishev, V.V. Novikov, *Physical Material Science. Volume 6, Part 2. Nuclear Fuel Materials* (MIFI, Moscow, 2008) [in Russian]
9. A.M. Afrov, S.A. Andrushechko, V.F. Ukraintsev, *Pressurized Water Reactor-1000: Physical Bases of Operation, Nuclear Fuel, Safety System* (Logos, Moscow, 2006) [in Russian]

The microhardness and internal stresses allocation in superheating pipes welded node made of dissimilar steels

Aigul Ilyasova[1,*], *Roman* Kulesh[1], *Lyudmila* Lyubimova[1], *Alexander* Tashlykov[1], *Anton* Gerasimov[1], *Alexander* Zavorin[1], and *Evgenii* Puzyrev[2]

[1]National Research Tomsk Polytechnic University, 634050 Tomsk, Russia
[2]Polzunov Altai State Technical University, 656038 Barnaul, Russia

Abstract. The object of research is the welded knot made of dissimilar steels 12Cr1MoV and Di-59. Research methods are spectral analysis, microhardness measurements and X-ray analysis. It is established that the structural-phase nonuniformity, significant concentration fluctuations of alloying elements, the spread in values of microhardness and mechanical characteristics along the length of the pipe can cause significant value of internal macrostress that leads to the formation of the "flaw" defect type in the welding process. It is shown how to distinguish between flaws that have appeared on the stage of production from operating flaws.

1 Introduction

Among the broad range of tasks solved in the framework of the Russian energy development, among the first in importance is the problem of ensuring the reliability and efficiency of the thermal power plants equipment.

This applies in particular to those components where in manufacture materials different in their mechanical, chemical, strength and thermal characteristics are used. It is obvious that the most dangerous elements in this sense are welded joints which transitional zone is a kind of stress concentrators.

A characteristic feature of the weld adjacent zone (WAZ) and heat affected zones of welding (HAZ) is the pronounced structural and phase heterogeneity of the properties, which leads to the spread in values of mechanical characteristics in length and the wall thickness of the pipe. This factor in the operational conditions of frequent thermal cycles will create critical temperatures and mechanical stresses, and therefore dangerous surface areas which will impact on the performance of the design as a whole [1,2].

On the other hand, technological modes of pipes production predetermine the appearance of significant discontinuity that will cause a high level of internal structural stresses with reduced material resistance to fracture [3,4].

Control of internal stresses in welded joints is not regulated by the test operation and is not reflected in guidance documents and company standards [5]. However, the recovery

* Corresponding author: asi4@tpu.ru

heat treatment (RHT), as well as post welding heat treatment (PWHT), the ultimate goal of which is to remove residual stresses, are used as a basis for the extension of material resource and not always clearly guarantees the alignment of properties in transitional areas of the weld [6]. This is due to the fact that the real durability is not determined by design or economic life, but individual physical resource of tubular products from steel and alloys for different purposes.

In this regard, the aim of this research is to identify the distribution of the microhardness and internal structural residual stresses of the first kind associated with a complex of individual properties, ensuring operability of the design.

2 Materials. Samples. Experimental techniques

Object of research is weld knot made of dissimilar steels 12Cr1MoV and Di-59.

Low-alloy heat-resistant structural tube steel 12Cr1MoV has the following chemical composition according to GOST 20072-74, % by weight: 0.08 to 0.15 C; 0.9 to 1.2 Cr; 0.25 – 0.35 Mo; 0.15 – 0.30 V; 0.4 – 0.7 Mn; 0.17 – 0.37 Si; < 0.25 Ni; < 0.20 Cu; < 0.025 S; < P 0.030; the rest is iron.

A small study of some structural transformations aspects due to the cyclic thermal and mechanical gradients determined the relevance of studies of some materials which seem promising for supercritical pressure boilers superheaters. These include Russian austenitic chord-manganese steel 10Cr13Mg12Si2Ni2Cu2Nb (Di-59).

The chemical composition of the austenitic steel Di-59 according to GOST 5632-72 (% by weight): 0.07% C; of 1.99 % Ni; 12.45% Cr; 13.0% Mn; 0.21% Si; 0.006 % S; 0.015% P; 0.03% Al; 2.15 % Cu; 0.79 % Nb; 0.015% B; the rest is iron.

As a tool for studies X-ray diffraction, microthermometry and X-ray analysis methods were selected.

Residual stresses of the first kind (σ_I) were defined by Dron-2,0 X-ray apparatus using a radiation tube with a copper anticathode, with an average wavelength $\lambda K\alpha_{av}$ = 1,54178 Å on the basis of Hooke's law:

$$\sigma_I = (\Delta a/a)E\ ,$$

где E – modulus of elongation, МПа; a – crystal lattice parameter, Å.

Spectral analysis performed using the EDX 2800 EDXRF spectrometer based on the measurement of the spectra of the chemical elements radiation characteristic in the sample. The elements was identified by the standard perfomance curve provided as element of comparison inherent in the spectrometer program. The resulting spectrum was displayed on the computer, quantitative analysis was performed automatically according to the intensity of the experimental spectral lines and their area.

Measurement of microhardness was carried out using PMT-3 microhardness tester.

Cut-up sketch is shown in Fig. 1, 2.

 DOI: 10.1051/matecconf/201711001047

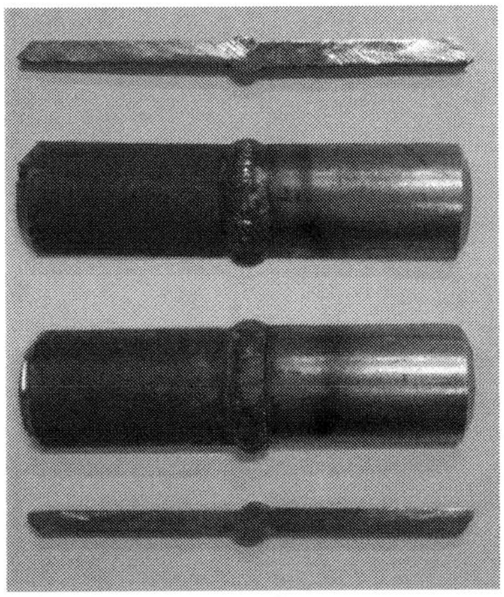

Fig. 1. Cut-up sketch of the longitudinal sample.

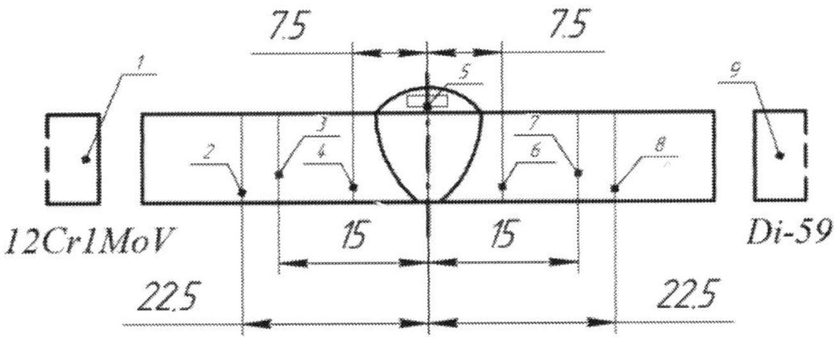

Fig. 2. Cut-up sketch from the transition zones of welded node:
1, 9 – reference samples of 12Cr1MoV and Di-59 steels, respectively;
4, 6 – samples cut from the heat-affected zones at a distance of 7.5 mm from the axis of the weld;
5 – the sample from the weld zone;
3, 7 – samples cut from the heat-affected zones of the weld at a distance of 15 mm from the weld axis;
2, 8 – samples cut from areas at a distance of 22.5 mm from the axis of the weld.

3 The experimental facts. Discussion

Fig. 3, 4 shows X-ray fragments of the weld zone (sample No. 5 in Fig. 2). It is evident that due to weld heat-affected occurs a non-uniform decomposition of the supersaturated solid solution of austenite, which is display itself in the phenomena of structural polymorphism, in the result zone of the weld becomes a mixed phase structure. One of the phases has a body-centred cubic (BCC) lattice (α -Fe), the second phase is an austenite (γ - Fe) with face-centered (FCC) lattice. The concentration of the mentioned phases are different for the outer and inner surfaces of the weld zone.

DOI: 10.1051/matecconf/201711001047

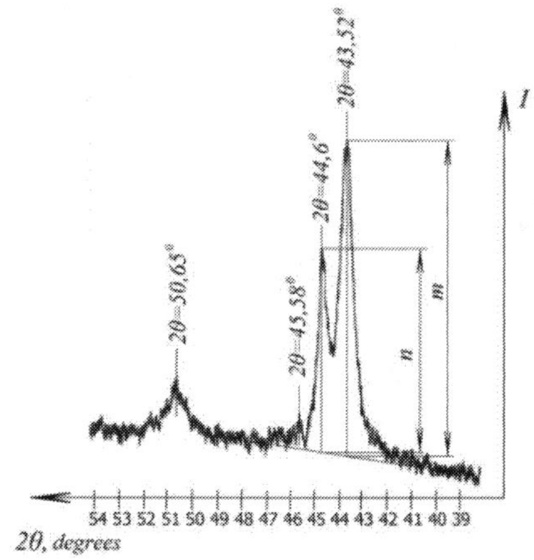

Fig. 3. The outer surface of the weld zone radiographs fragment (sample No. 5).

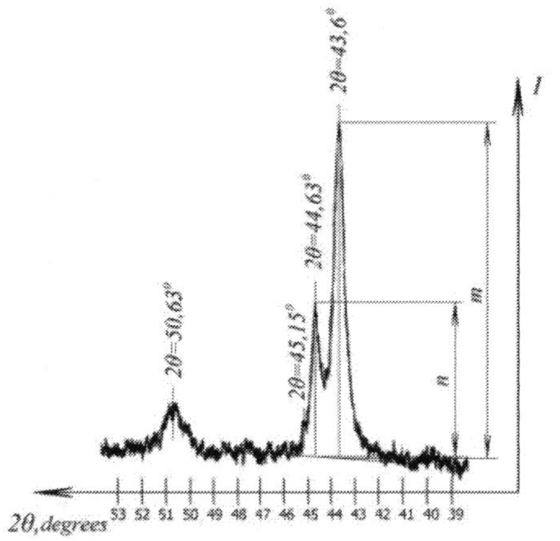

Fig. 4. The inner surface of the weld zone radiographs fragment (sample No. 5).

A consequence of the heterogeneous decomposition of solid solution is the occurrence of point defects in the form of the liberated impurities of substitution and their heterogeneous distribution in the transition and the heat affected zone of the weld (Fig. 5, 6, 7).

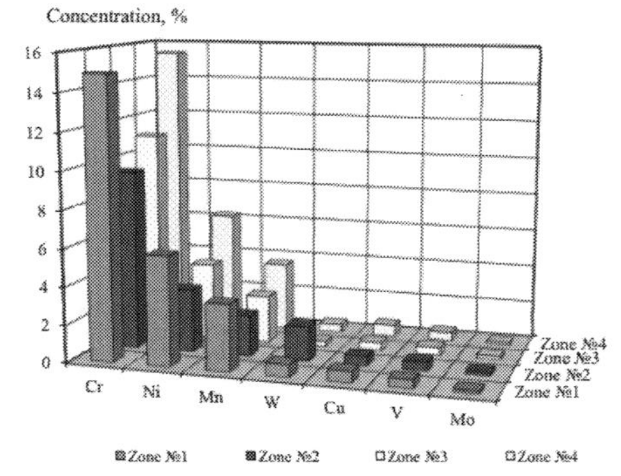

a)

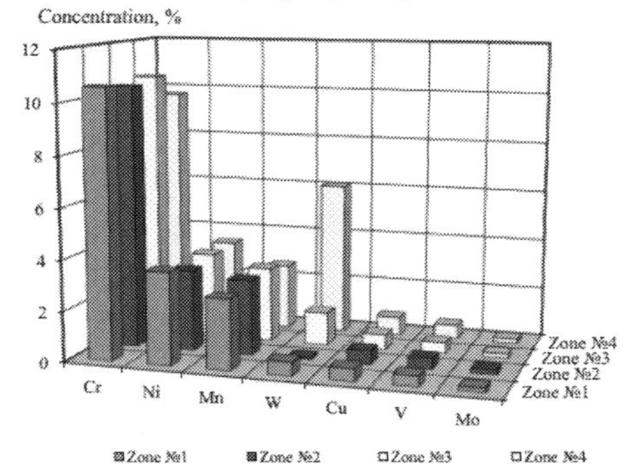

b)

Fig. 5. The results of spectral analysis of the alloying elements distribution in the sample No. 5 (1-4 – controlled area): a) outer surface; b) inner surface.

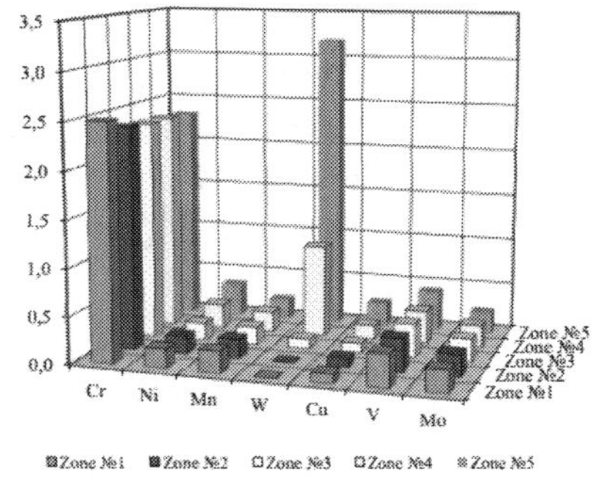

a)

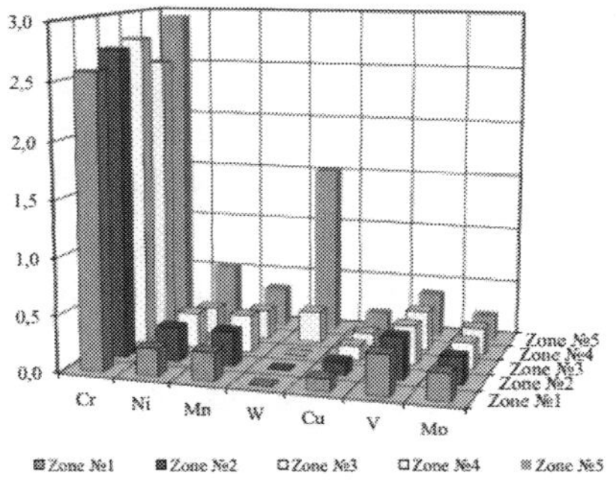

b)

Fig. 6. The results of spectral analysis of the alloying elements distribution in the sample No. 4 (1-5 – controlled area): a) outer surface; b) inner surface.

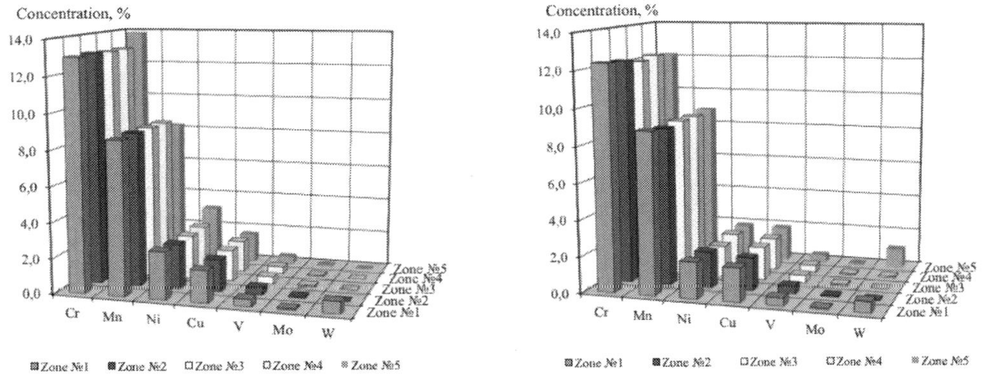

a) б)

Fig. 7. The results of spectral analysis of the alloying elements distribution in the sample No. 6 (1-5 – controlled area): a) outer surface; b) inner surface.

Fig. 8 shows the distribution of microhardness in the welded node. The microhardness was determined for the entire length of the longitudinal sample with a step of 250 μm (Fig. 1). Measurement pitch in the area of the weld was 100 μm.

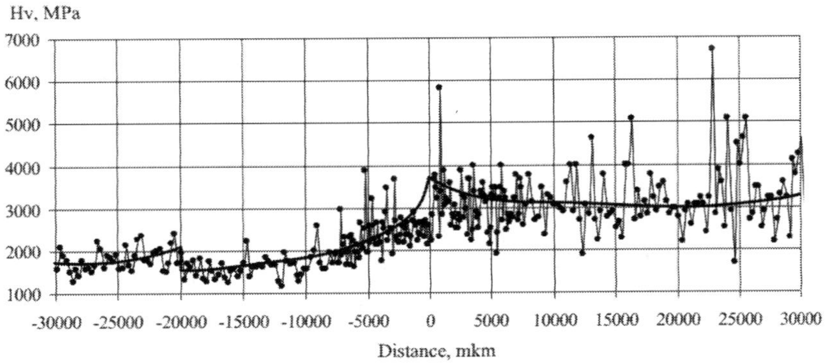

Fig. 8. The microhardness distribution in the longitudinal sample (fig. 1).

The microhardness distribution in the longitudinal sample (fig. 8) allows to conclude:

– thermal influence of welding from 12Cr1MoV steel side has led to a significant stratification of the pipe wall by the level of microhardness. In particular, at a distance of about 30,000 microns from the axis of the weld, the microhardness corresponds to the level of 2000 MPa. Further to the axis of the weld, the level of microhardness increases and reaches a maximum equal to 3500 MPa in the zone of the weld;

– the microhardness of Di-59 steel has a more sharp oscillations from an average value of 3000 MPa to 5000 – 6000 – 6500 MPa;

– in the immediate area of the weld microhardness varies from 2500 MPa to 3500 MPa in the transition from one steel to another.

The obtained results of the microhardness distribution and alloying elements in the weld zones show the existence of defects, such as delamination of metal – detected zones with high microhardness, reduced creep resistance and soft interlayer.

Structural and phase nonuniformity (Fig. 3, 4), significant fluctuations in the concentration of alloying elements (Fig. 5, 6, 7), the scatter of values of microhardness and mechanical characteristics along the length of the pipe (Fig. 8) are the cause of the significant value of internal macrostress.

Educated estimates of the stress state are shown in Fig. 9.

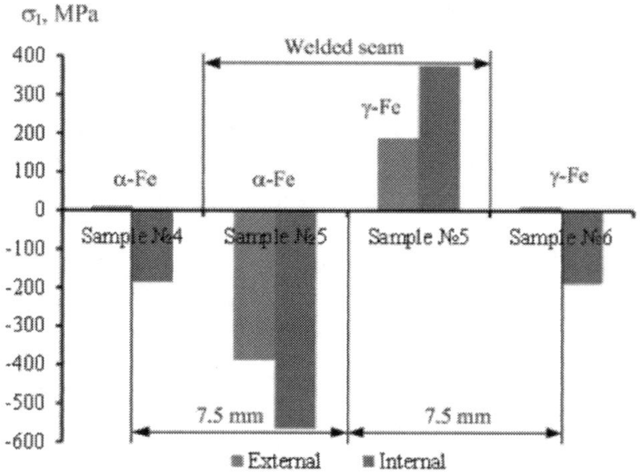

Fig. 9. Scheme of the stress state in the welded node critical zones.

The weld joint thus represents the structure with the presence of structural stresses gradients. A limit state for a typical welding assembly is brittle fracture (Fig. 10). A group of factors such as technological, shape defects, cracks, lack of penetration, contaminants, etc. has a complex influence on initiation of brittle fracture. Significant risk factor for brittle fracture include heterogeneous internal structural stresses which may reach the yield strength [7], and their uneven distribution in the fracture zone [8-12].

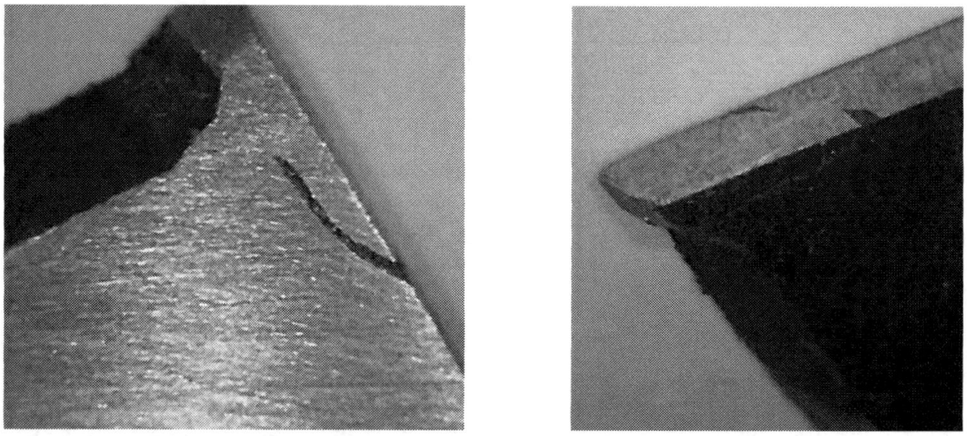

a) b)

Fig. 10. Crack (x 100) in the HAZ of welded node from the side of 12KH1MF steel: a) an exterior surface; b) crack propagation along the thickness of the wall in the zone of the weld.

Fig. 10 shows that the crack appeared in the HAZ of 12Cr1MoV steel (Fig. 10 a) and has spread to the area of the weld (Fig. 10 b). This damage may be due to operating conditions, and, most likely, due to the result of the technological process that caused the presence of welding stresses in the joint region of the product caused the formation of primary, steadily growing cracks. How to distinguish cracks appeared on the stage of production from operating cracks?

A sign of the presence of technological cracks that result from the welding process can be relax or compressive stresses in the 12Cr1MoV steel, indicating a preliminary course of relaxation processes associated with the structural damage. Relaxation is the stress relief and the formation of microcracks. The conditions formation in the zone of the weld node is associated with tensile stresses that lead to the initial crack release and its growth.

4 Conclusion

1. Revealed significant structural and phase nonuniformity of the structural element having a transition zone of heterogeneous steels in mechanical and thermophysical characteristics.
2. Non-uniform stress concentration in welded node was set.
3. The heat treatment of the welded joint and HAZ of the base metals for the purpose of leveling properties can not coincide, but may lead to a deterioration of the metal actual state and welding structural damage.

This work was supported by RFBR grant ''Fundamentals of engineering sciences'' No. 15-08-99544a in 2015.

References

1. F.A. Khromchenko, Technology of mechanical engineering, (5), 66 (2007)
2. A.A. Makeev, L.L. Lyubimova, A.S. Zavorin, A.A. Tashlykov, Steel in Translation, **39** (12), 1048 (2009)
3. L. Lyubimova, R. Fisenko, S. Breus, R. Tabakaev, MATEC Web Conf. **72**, 01059 (2016)
4. L.L. Lyubimova, R.N. Fisenko, R.B. Tabakaev, A.A. Tashlykov, A.S. Zavorin, Materials Science and Engineering: A, **682**, 248 (2017)
5. RD 153-34.1-003-01 ''Welding, heat treatment and control of pipe systems of boilers and pipelines during installation and repair of power equipment'' (RTM-1s). 01.01.2002.
6. OS 153-34.17.459-2003. A manual for regenerative heat treatment of thermal power plant equipment elements.30.06.2003.
7. N.I. Kamenskaya, Metallography and heat treatment of metals, **1** (715), 47 (2015)
8. E.A. Grin, A.V. Zelensky, Thermal Engineering, (2), 12 (2009)
9. A.B. Popov, Thermal Engineering, (2), 13 (2011)
10. S.A. Kharchenko, N.B. Trunov, N.F. Korotaev, S.L. Liakishev, Thermal Engineering, (3), 27 (2011)
11. V.V. Gerasimov, O.V. Pereverzeva, Technology of metals, (8), 26 (2008)
12. A.B. Popov Thermal Engineering, (11), 70 (2010)

Research of lignite oxidation kinetic parameters modified by $CuSO_4$ and $NaNO_3$ initiation additives

Kirill Larionov[1,*], *Ilya* Mishakov[1,2], *Alexander* Gromov[1,3], *Andrey* Zenkov[1], and *Vladimir* Glaktionov[1]

[1]National Research Tomsk Polytechnic University, 634050 Tomsk, Russia
[2]Boreskov Institute of Catalysis, 630000 Novosibirsk, Russia
[3]Nürnberg University of Technology Georg-Simon-Ohm, 90402 Nurnberg, Germany

Abstract. An experimental study and subsequent analytical assessment of activation energy change in lignite oxidation process with addition of $NaNO_3$ and $CuSO_4$ mineral salts were conducted. The results showed that injection of catalytic additives leads to reduction of coal activation energy and reaction initial temperature.

1 Introduction

Nowadays, coal is one of the most popular traditional fuel types [1]. Current trends in the development of coal-based energetics are concentrated on new technologies aimed at improving environmental, technical and economic performance of the fuel combustion installations [2]. One of the most promising coal application technologies for heat and electricity production is its promoting combustion [3]. The use of initiation additives to intensify the solid fuel combustion processes has a significant effect on the kinetics of the process. Drop of the reaction temperature range leads to reduction of sulfur and nitrogen oxides high-temperature emissions and the average burn rate increase [4]. All of these effects happen due to activation energy reduction. Activation energy is the amount of energy that needs to be provided to the system to ensure that the reaction occurs. Thus, the effectiveness of a particular catalytic additive can be assessed by the amount of activation energy.

In this paper the effect of $NaNO_3$ and $CuSO_4$ initiation additives on the change in kinetic characteristics of brown coal oxidation process is considered.

2 Experimental setup and study technique

Samples of Borodinskoe deposit lignite were used in this research. Initial fuel was grinded in a ball-mill pulverizer and sifted through a sieve with a mesh size of 80 microns. Composition of the researched fuel was determined according to standard methods [5] and is shown in Table 1.

* Corresponding author: laryk070@gmail.com

Table 1. Composition of the initial coal.

Index	Unit of measure	Value
Analytical moisture	wt./%	11.6
Volatile content		40.8
Ash content		4.0
Carbon content		43.6

The studies were conducted with the use of Netzsch STA 449 F3 Jupiter (Germany) simultaneous thermal analyzer. All experiments were performed under identical conditions with the sample heating rate of 2.5℃/min in a corundum crucible (sample weight approximately 7mg) at the temperature range 50-600℃. A mixture of air and nitrogen was used as an oxidizing medium with a flow rate of 60 ml/min and 10 ml/min respectively.

Synthesis of coal samples with initiation additive was performed by incipient wetness impregnation method by their dissolving in alcohol-water medium in the bulk concentration of 1:1. The additives weight content is 5% of the sample total mass. Coals were dried after impregnation in a drying oven at 105°C for 20 hours.

Equation, which links the reaction rate to kinetic parameters, was used to determine activation energy of oxidation reaction of pure brown coal and samples modified by promoting additives.

The reaction rate is expressed as a first time derivative of the conversion degree dα/dt. In turn, conversion degree is equal to the proportion of reacted product and is converted into equation [6]:

$$\alpha = \frac{m_i - m}{m_i - m_f}, \tag{1}$$

here m_i is initial mass; m_f is final mass; m is current mass.

The reaction rate represents multiplication of two functions [6]:

$$\frac{d\alpha}{dt} = f(T)g(\alpha), \tag{2}$$

here $f(T)$ is function depending only on temperature; $g(\alpha)$ is function depending only on conversion degree.

Function of temperature is expressed as the Arrhenius equation [6]:

$$f(T) = Aexp\left(-\frac{E}{RT}\right), \tag{3}$$

here A is pre-exponential factor; E is activation energy; R is absolute gas constant; T is absolute temperature.

In turn, a function of the conversion degree can be considered as a first order model [6]:

$$g(\alpha) = (1 - \alpha). \tag{4}$$

Thus, the activation energy can be found with the use of the experimental data and these equations. But differentiation of dα/dt is associated with a significant loss in accuracy, so different approximate solutions of this differential equation are often used in practice. Following approximate solution was used in this paper [6]:

$$ln\left[-\frac{\ln(1-\alpha)}{T^2}\right] = ln\left[\frac{AR}{\beta E}\right] - \frac{E}{RT}, \tag{5}$$

here β is sample heating rate.

It is sufficient to plot a dependence of $ln\left[-\frac{\ln(1-\alpha)}{T^2}\right]$ from 1/T using experimental data in order to use this equation to determine the activation energy. Then, it is necessary to approximate, while the angular coefficient is equal to the value of E/R, where it is easy to determine the activation energy E from.

3 Experimental results

The results of researched samples thermal analysis are presented in Figure 1 as a TG curve.

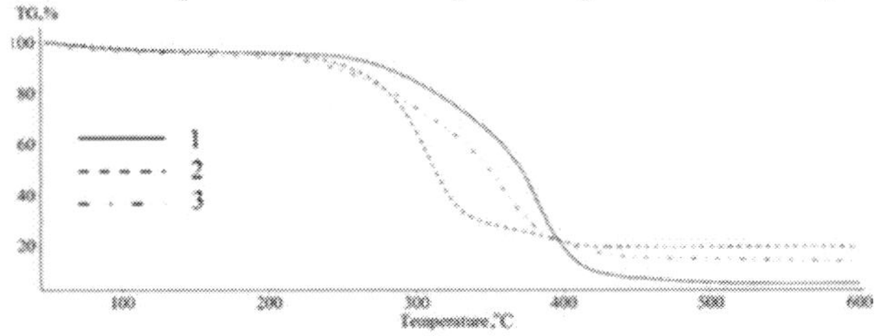

Fig 1. Thermogravimetric curves (TG) of coal samples oxidation (1 – lignite comparison sample, 2 - lignite + $NaNO_3$, 3 - lignite + $CuSO_4$).

According to figure 1, reaction start temperature of the initial sample 1 was 278°C. This temperature is 244°C for a sample with the addition of $NaNO_3$ and 255°C with the addition of $CuSO_4$. Thus, the largest shift in the direction of oxidation beginning temperature decrease is observed in the sample 2, for which the change was 34°C.

Certain activation energy numerical values of the samples are presented in Table 2.

Table 2. Activation energy of coal samples.

Sample	1	2	3
E, kJ/mol	72.0	67.4	69.6
A, 1/min	24332	19244	16314

It is seen from obtained results that the use of initiation additives reduces kinetic parameters values. The smallest value of activation energy is achieved in the case of $NaNO_3$ application and differs from the initial sample by 6.4%.

4 Conclusion

Conducted experimental and numerical studies have shown that the injection of promoting additives in the form of $NaNO_3$ and $CuSO_4$ mineral salts greatly influences the nature of the reaction. It turns out, that $NaNO_3$ is more active than $CuSO_4$, which is proved by lower value of activation energy.

The research has been accomplished with financial support of the Russian Federation Ministry of Education and Science within implementation of the project No. 13.7644.2017/BCH.

References:

1. S.A. Yankovsky, V.E. Gubin, A.A. Tolokolnikov, A.V. Zenkov, V.I. Nikolaeva, Strategic Technology (IFOST), 299 (2016)
2. S. Yankovskiy, A. Matveev, A. Tolokolnikov, A. Zenkov, MATEC Web Conf. **92**, 01036 (2017)

3. N. Galashov, S. Tsibulskiy, A. Kiselev, MATEC Web Conf. **72** (2016)
4. D.V. Gvozdyakov, V.E. Gubin, EPJ Web Conf. **82**, 01026 (2015)
5. K.V. Slyusarskiy, K.B. Larionov, V.I. Osipov, V.E. Gubin, A.A Gromov, Fuel. **191,** 383 (2017)
6. H. Wang, C. Li, Z. Peng, S. Zhang, Energy Sources, Part A: Recovery, Utilization and Environmental Effects, **37** (21), 2323 (2015)

Modelling of heat transfer process in condensing unit with titanium alloy tubes

Sergey Lavrinenko[1,*], *Pavel* Polikarpov[1], *Anastasiya* Matveeva[1], and *Vladimir* Martyshev[1]

[1]National Research Tomsk Polytechnic University, 634050 Tomsk, Russia

Abstract. One of the most important units of heat transfer equipment of a nuclear power plant is the condenser. Currently, at the nuclear stations of concern "Rosenergoatom" work is actively underway to replace the tubes of copper containing alloys to various steels or titanium alloys. Also, active work is underway on modernization of heat-exchange equipment of operating units. It is necessary to make the modelling of the parameters of condenser, to ensure that after the upgrade, the unit will continue operating normally. For this purpose, was created the model of module of tube bundle of condenser unit of K-33160 with tubes of titanium alloy. The modelling process is based on the equation of heat balance. In this work were modelled condenser of NPP of K-33160 with WWER-1000 reactor and the tubes of a titanium alloy VT0-1. The calculation was carried out for three presented methods and the error was less than 2%.

1 Introduction

At present, Rosenergoatom's nuclear power plants are actively working on the modernization of existing equipment, in particular, replacement of tubes from copper-bearing alloys to various steel or titanium alloys [1, 2]. This is due to the fact that the copper that contained in the heat-exchange surfaces washed out and is being eroded by the working environment of the second circuit, in which the hydrazine is dosed, during the technological process.

The dosage of hydrazine in the working environment is to protect internal surfaces of equipment and pipelines, because hydrazine, reacting with the iron, forms a resistant oxide film that is called magnetite (Fe_3O_4) and neutralizes the oxygen that is got in the circuit.

But, at the same time, hydrazine under high temperature (more than 100 °C) decomposes to ammonia, which dissolves the copper contained in the copper-nickel alloys ("MNZH-5").

As a result, the dissolved copper gets into the water of the second circuit and causes corrosion of the tubes of heat transfer equipment.

These circumstances forced the project organizations to seek new materials for heat transfer surfaces of the equipment. All equipment (high pressure heaters, low pressure heaters, heat exchangers, ejectors, etc.) on the power plant units that are in the process of

* Corresponding author: serg86@tpu.ru

construction, made of stainless steel, tube condensers made of titanium alloys. Also, active work is underway on modernization of heat-exchange equipment of operating units [3, 4].

2 Methods of research

One of the most important units of heat transfer equipment of a nuclear power plant is the condenser. It is necessary to make the modelling of the parameters of condenser, to ensure that after the upgrade, the unit will continue operating normally. For this purpose, was created the model of module of tube bundle of condenser unit of K-33160 with tubes of titanium alloy (Fig. 1).

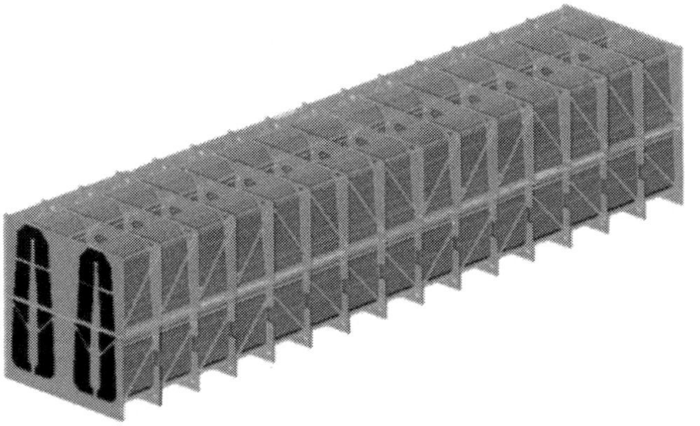

Fig. 1. The module of tube bundle of condenser.

Due to the change of material of tubes, changing the heat transfer coefficient, which would entail a change in the thermal capacity of the condenser, which is unacceptable. In order to avoid this impact, it is necessary to choose the appropriate area. The modelling process is based on the equation of heat balance [5]:

$$Q_c = G_s\,(h_c - h`_c) = G_w\,c_p\,(t_{out} - t_{in}) = k{\cdot}F{\cdot}\delta t\,, \tag{1}$$

where
Q_c – thermal capacity;
G_s – steam flow to the condenser;
h_c – enthalpy of steam;
$h`_c$ – enthalpy of water;
G_w – cooling water flow;
c_p – heat capacity at constant pressure;
t_{in}, t_{out} – temperature of the cooling water;
$k = (R_w+R_t+R_s)^{-1}$ – heat transfer coefficient;
R_w – thermal resistance of the cooling water;
R_t – thermal resistance of the tube wall;
R_s – thermal resistance of the steam;
F – heat transfer area;
δt – average logarithmic thermal pressur.

The heat transfer from the condensing steam to the cooling water through the heat exchange surface, belongs to the complex type of heat transfer. Considering the heat transfer coefficient as the reciprocal of the sum of the thermal resistances in the path of heat transfer from the steam to the cooling water.

The influence of sediments on both sides of the tube can be neglected, as the tubes are new, and unlike copper-Nickel alloy, the amount of deposits is significantly reduced, in case of long term work.

Determination of thermal resistance of water and wall of tube are not particularly hard:

$$R_t = \delta_t/\lambda_t \quad (2)$$

$$R_t = (0.023\mathrm{Re}_w^{0.8}\mathrm{Pr}_w^{0.4}(\lambda_w/d_{in}))^{-1}, \quad (3)$$

where

δ_t – tube wall thickness;

λ_t, λ_w – coefficient of thermal conductivity of tube, water.

It is necessary to know appropriate geometric and thermophysical parameters of the water and the wall.

The definition of thermal resistance from steam side of condensate is strongly complicated by the joint influence of many factors, the most important of which is the phenomenon of the flooding of the lower tubes with the condensate. It should be borne in mind that the main steam parameters will substantially change according to the movement in the tube bundle.

The heat transfer coefficient from the steam to the surface of the tube bundle:

$$\alpha_s = 28.3\Pi^{0.08}Nu^{-0.58}/\alpha_{Nu}, \quad (4)$$

where

$\Pi = \rho_s W_s^2/(\rho_c g d_{out})$ – dimensionless number;

ρ_s, ρ_c – density of steam and condensate;

W_s – flow rate of steam;

$Nu = \alpha_{Nu}d_{out}/\lambda_c$ – Nusselt number for condensation of stationary;

$\alpha_{Nu} = 0.725Ga{\cdot}Pr{\cdot}K{\cdot}\lambda_c/d_{out}$;

λ_c – coefficient of thermal conductivity of condensate;

Ga, Pr, K – number of Galileo, Prandtl and phase transition.

Also the heat transfer coefficient was determined by the simplified methods [6].

2.1 Method of "VTI"

$$k = 4070{\cdot}a{\cdot}\Phi_w\,\Phi_t\,\Phi_z\,\Phi_d, \quad (5)$$

where

a – coefficient of condition of heat exchange surface;

Φ_w – factor of rate;

Φ_t – factor of temperature;

Φ_z – factor of number of cycles of flowing of cooling water;

Φ_d – factor of steam load.

2.2 Method of "ITO"

$$k = k_{o1}\,\beta_t\,\beta_m, \quad (6)$$

where

k_{o1} – coefficient of heat transfer in case of the temperature of water at the outlet is 21°C ;

β_t – correction for temperature of output water;

β_m – coefficient that take into account the wall thickness and material.

For condensers, that were made in Russia, required one additional amendment to take account of the pollution of the tube: 0.75-0.85.

In order to be able to determine the coefficient of heat transfer, it is necessary to set the parameters of tubes:

Table 1. Parameters of tubes.

Material of the tubes	d_{in1}, mm	d_{out1}, mm	N_{t1}, pcs	d_{in2}, mm	d_{out2}, mm	N_{t1}, pcs
VT0-1	27	28	26688	26.6	28	252

3 Results

The results of determination of heat transfer coefficient of the condenser K-33160 with titanium tubes by different methods is presented in table 2.

Table 2. Heat transfer coefficient.

Material of the tubes	According to the thermal resistances	VTI	ITO
K, W/(m^{2}K)	5540	5614	5648
Δ, %	-	1.34	1.96

4 Conclusion

The process of heat exchange in the condenser is quite complicated, and its calculation is difficult. It is much complicates the process of replacing the tubes and their calculation on various parameters of the environment. In this work were modelled condenser of NPP of K-33160 with WWER-1000 reactor and the tubes of a titanium alloy "VT0-1". The calculation was carried out for three presented methods and the error was less than 2%.

References

1. E. Pavelyev, S. Lavrinenko, MATEC Web Conf. **72**, 01085 (2016)
2. V.A. Ruppel, Yu.S. Tretyakova, S.V. Lavrinenko, A.A. Matveeva, V.N. Martyshev, MATEC Web Conf. **37**, 01059 (2015)
3. N. Galashov, S. Tsibulskiy, A. Kiselev, MATEC Web Conf. **72**, 01029 (2016)
4. N. Galashov, S. Tsibulskii, A. Matveev, V. Masjuk, EPJ Web Conf. **110**, 01019 (2016)
5. Yu.M. Brody, M.A. Nirenstein, *Calculation of heat exchangers for steam turbine units* (USTU, 2001)
6. S.A. Yankovsky, V.E. Gubin, A.A. Tolokolnikov, A.V. Zenkov, IFOST, 7884252 (2017)

Simulation of the furnace of the boiler P-49 in the package of applied programs fire 3D

Evgeny Tsibizov[1,*], *Boris* Lebedev[1], and *Maria* Cherckasova[1]

[1]National Research Tomsk Polytechnic University, 634050 Tomsk, Russia

Abstract. The combustion of solid low-grade fuel in LTV-boiler furnaces is a pressing research questions currently. The aim of this work is the creation of a computational grid model LTV-furnace to calculate the package of applied programs FIRE 3D. The study created a model LTV-furnace. The model tested on brown coal from the Nazarovo Deposit. The resulting distribution of temperatures and velocities has proved the performance of the model.

1 Introduction

Low-temperature combustion technology (LTV-combustion) is one of the promising but still little-studied areas. This method of burning is interesting that in his study has identified a number of advantages: stabilization of ignition and combustion, increase thermal efficiency, reduce slagging and contamination of working surfaces, the reduction of emissions of nitrogen oxides and sulfur [1].

2 Object of study

Numerical simulation is a modern method of investigation of processes occurring in boiler furnaces.

A mathematical model of the LTV of the furnace created by using the package of applied programs FIRE 3D, which is based on modern approaches and methods of mathematical description of complex chemical and physical processes.

As the sample for the study selected boiler P-49, mounted on UDPS city Nazarovo. Boiler P-49 (Ff-1600-25-545/545) designed to receive superheated steam from the combustion of dried products Nazarovo brown coal with liquid slag removal. The boiler is designed for operation in unit with the turbine C-500-240 500 MW. Boiler P-49 is ramjet, consists of two buildings with a three-way arrangement working independently from each other. In 2015 he was transferred to NTV method of burning.

The original model was developed for boiler P-49: 12 burners, system bottom blast (SBB) is divided into five slits across the width of the combustion furnace, tertiary air of the lower tier (TL) in the amount of 24 pieces, middle tier (MT) in the amount of 12 pieces [2]

* Corresponding author: TabakaevRB@tpu.ru

3 Mathematical model

The size of the computational grid of this model 79x45x192 cells. Based on this model were carried out the first calculations. The model proved to be too cumbersome. The complete calculation took 3 to 4 days. It is not possible to quickly get the result. This model has aerodynamic ledge on the back wall of the boiler furnace. To solve these problems, the decision on the establishment of a new model. It must take into account all the flaws of the first model.

The furnace of the investigated boiler is symmetrical in width, all the burners and the air-supplying device are arranged in a plane parallel to the side wall of the furnace. Aerodynamics must be flat, therefore, can be used to model not the entire furnace volume, and only its third part.

The obtained grid has a size of 96x53x61 cells. Differences from the first model are:

– reduction of volume of the model, the number of burners and air supply four burners, SBB is a crack along the whole width of the boiler furnace, TL includes eight devices, MT and includes 4 of the device;

– the model takes into account both aerodynamic ledge;

– also considered the possibility of feeding the furnace of a dust of high concentration (DHCp) through the burner.

To do the calculation on the new model possible in a short period of time (12-20 hours). The accuracy of the calculation remains quite high.

The resulting model was test on brown coal from the Nazarovo Deposit. The station is equipped with a centralized preparation and the grinding of brown coal. Characteristics of coal dust supplied to the boiler from filesaved presented in table 1.

One of the results of mathematical simulation are the velocity field (Fig. 1) and the field of particle concentrations (Fig. 2) in cross section at the burner parallel to the side wall.

4 Characteristics of coal particles

Characteristics of Nazarovo coal deposits are presented in table 1.

Table 1. Content of produced SCF.

Name	Marking	Coal	Dust
Carbon, %	C^r	39	46.22
Hydrogen, %	H^r	2.5	3.2
Oxygen, %	O^r	12.5	15.6
Sulfur, %	S^r	0.3	0.49
Moisture, %	W^r	39.8	24
Ash, %	A^r	5.4	10
Nitrogen, %	N^r	0.5	0.49
Lowest calorific value, kJ/kg	Q_i^r	13702	22380

5 Temperature and oxygen distribution

Graphs of the distribution of temperatures and velocities obtained in the result of the calculation in the package of applied programs FIRE 3D. The results on the burners and between the burners is shown in figure 1 and figure 2.

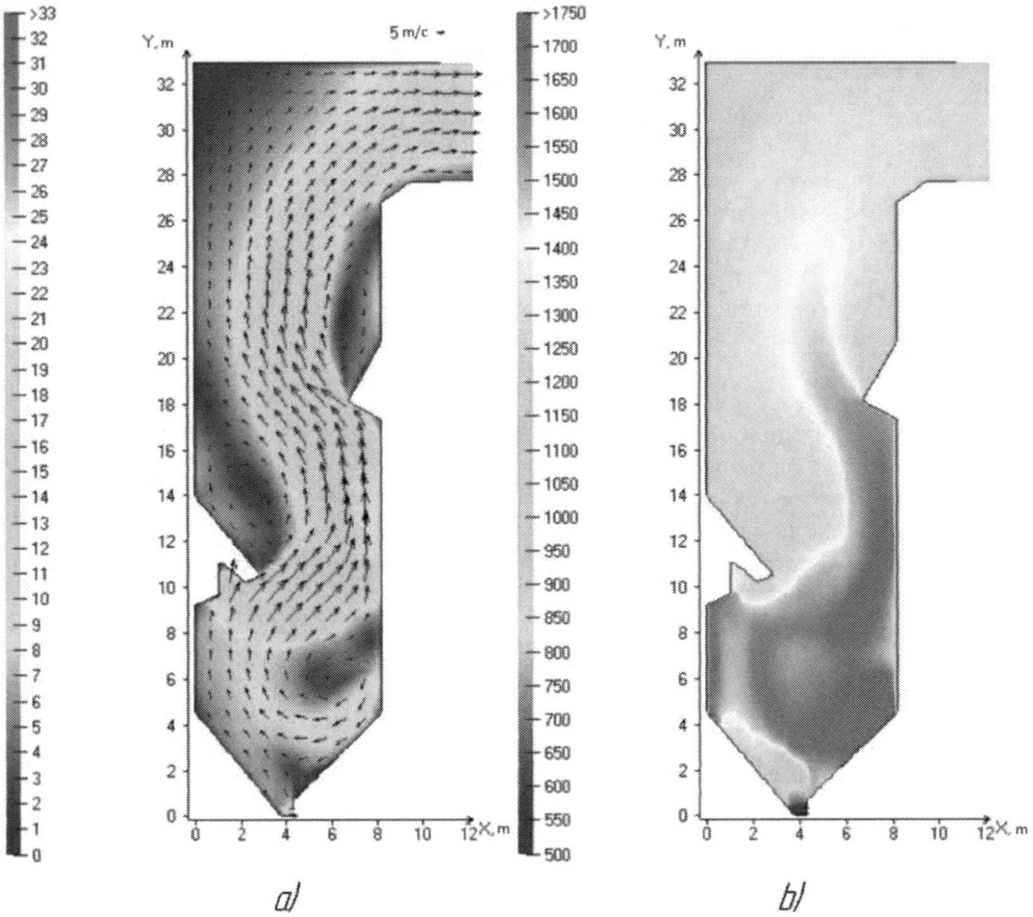

Fig. 1. The distribution of velocity (a) and temperature (b) in cross-section between the burners.

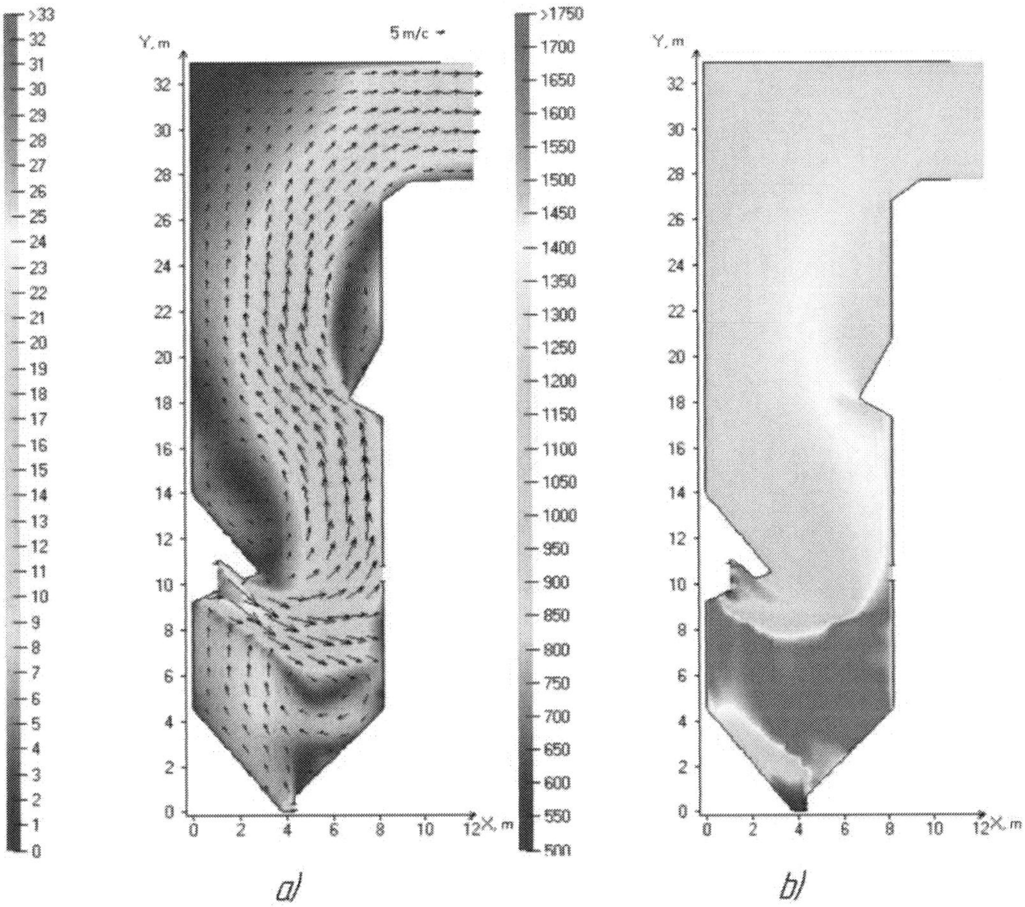

Fig. 2. The distribution of velocity and temperature in cross-section on the burners.

6 Conclusions

The mathematical model parts LTV-furnace boiler P-49 for calculations in the package of applied programs FIRE 3D. The obtained velocity field and particle concentration showed the performance of the model.

References

1. K.Y. Orlova, MATEC Web Conf. **92** (5), 01003 (2017)
2. B.V. Lebedev, A.S. Zavorin, Bulletin of Tomsk Polytechnic University, **315**, 15 (2009) [In Russian]
3. New modernized power-generating unit was put into service on Nazarovo state district power station. (Newslab.ru, 2013) [In Russian]
4. A.M. Bubenchikov, A.V. Starchenko, *Numerical modeling of aerodispersive mixings' dymanic and combustio n in channels*, (Izd-vo Tom. yn-ta, Tomsk, 1998) [In Russian]
5. A.V. Starchenko, A.M. Bubenchikov E.S. Burlutski, Proceedings of the Second International Conference on Turbulent Heat Transfer. Manchester. Great Britan. **2**, 109 (1998)

MATEC Web of Conferences **110**, 01051 (2017) DOI: 10.1051/matecconf/201711001051
HMTTSC-2017

Nonlinear dynamic model of steam boiler furnace

Mikhail Yu. Livshits[1,*], *A.* H.Elnour Mohammmad[1]

[1]Samara State Technical University, Heat and Power Engineering Department, Russia

Abstract. This paper describe a nonlinear dynamic model for furnace boiler and its addressed to be the first part of other system components which will be developed with control structure for a typical boiler, which will be used for educational purposes and research studies. This paper shows the variation of furnace dynamics (pressure, Temperature and mass flow rate) when variation in fuel mass flow rate occurs.

1 Introduction

Boiler has an unsteady state operation due to the variation in power demand. During the unsteady operation of the boiler steam, a pulsatory flow of the flue gas is created. This pulsatory flow allows the occurrence of certain pressure rhythmic variations inside the furnace leading to negative effects upon the air-flue gas circuit safety and efficiency in operation. [1]

Knowing the dynamic characteristics of the furnace is useful to develop its control system. The most practical way to know the dynamic properties of the furnace is simulation. Performing test runs is almost impossible because of the safety and economic reasons [2] the simulation model is based on the mass, heat and momentum conservation principles. The present paper is an extension of the paper [1], [2].

This paper discusses the development of a simulation toolbox for a typical furnace with standard configuration. The heart of a plant simulation is its modeling block, which for a boiler is comprised of highly nonlinear and complex algebraic, and differential equation [3]

To design a simulator, the programming language and the required software should first be selected. Modeling and programming can be run using any of the major programming languages. This method requires advanced techniques in programming or the simulator will not be economic or flexible. An alternate idea is to use simulator software to simulate and analyze the necessary calculations; this would be a better method for training and other applications.

One of the best software packages for simulation purposes is software environment MATLAB®.

* Corresponding author: entcom@samgtu.ru

2 Mathematical model of the furnace

The mathematical model was developed from the mass, energy balances and the heat transfer from hot flue gases to water and steam model and the flue gas flow through the boiler model. *[3]*

- Combustion heat balance:

$$VF\ \partial(\rho g\ hg)/\partial t = WA\ hA + WF\ CF + Wg\ hg - Weg\ heg\ RS\ (1+y/100) - Qir - Qis; \tag{1}$$

- combustion mass balance:

$$VF\ \partial \rho g/\partial t = WA + WF + Wg - Weg; \tag{2}$$

- flue gas flow through the boiler:

$$Weg = kf\ \ PG; \tag{3}$$

- furnace gas pressure:

$$Pg = Rg\ \rho g\ Tg; \tag{4}$$

-simulink model equations:

$$heg = XF\ /\ \rho eg, \tag{5}$$

$$Tg = (hEG - href / Cpg) + Tref, \tag{6}$$

$$Qir = \theta\, k\, VF\ Tg\hat{\ }4\ \ Weg/Pg, \tag{7}$$

$$Qis = (1-\theta)\, k\ VF\ Tg^4\ \ Weg/Pg, \tag{8}$$

$$Qgs = Qis + kgs\, Weg^{0,6}\ (Tg - Tst), \tag{9}$$

$$Tgr = Tg + ((Qis - Qgs))/(Cgs\, Weg), \tag{10}$$

$$Qrs = Krs\, WEG\hat{\ }0.6\ (Tgr - Trh), \tag{11}$$

$$Tge = Tgr - Qgs\ /\ (Cgs\, Weg), \tag{12}$$

$$Tgl = Tge - 1/(Cgs\ Weg)\ Qes \tag{13}$$

$$Qes = Kes\, Weg\hat{\ }0.6\ (Tge - Tet), \tag{14}$$

$$y = 100\ ((WA + Wg - WF\ Rs\ ))/(WF\ \ RS\ ) \tag{15}$$

State equations:

$$VF\ \partial XF/\partial t = 1/VF\ (CF\ \ WF + hA\ WA + \ldots$$
$$\ldots + hG\ Wg - Weg\ heg\ RS(1+y/100) - Qir - Qis, \tag{16}$$

 DOI: 10.1051/matecconf/201711001051

$$\partial \rho g / \partial t = 1/VF\,(WF + WA + Wg - Weg), \tag{17}$$

where: VF-furnace Volume [m^3]; ρg-exhaust gas density [kg/m^3]; hg-inlet gas enthalpy [kj/kg]; heg-exhaust gas enthalpy [kj/kg]; hA-air enthalpy [kj/kg]; WA-mass flow rate of the inlet air [kg/sec]; WF-mass flow rate of the inlet fuel [kg/sec]; Wg-mass flow rate of the gas to the furnace [kg/sec]; CF-fuel calorific value [J/kg]; Rs- stoichiometric air/fuel volume ratio; Qir-heat transfer to the riser [J/s]; Qis-heat transfer by radiation to the superheated [J/s]; Weg-mass flow of exhaust gas; Kf-friction coefficient (m sec); Kgs- an experimental coefficient(J/kg°K); Krs- an experimental coefficient(J/kg°K); href-reference exhaust gases enthalpy condition (J/kg); Tref-reference exhaust gases temperature condition(˚K); σ-Stefan-Boltzman constant; Pg-furnace air pressure [Pa]; ;Tg-furnace air temperture (˚K); Reg-ideal gas constant for exhaust gases; Qgs-total heat transfer to the superheater[J/sec]; Qrs-heat transfer to the reheater[J/sec]; Qes-heat transfer to the economizer(J/sec); T-gas temperature at the superheater(˚K); Tgr-gas temperature at the reheater(˚K); Tge-gas temperature at the economizer(˚K); Tgl-boiler exhaust gas temperature(˚K).

3 Simulink Model

Simulink model have been developed from the mathematical equations (1)-(19). The dynamic behavior of the furnace has been found out introducing certain variation laws of the control input which stands for the fuel mass flow rate WF and specifying the time variation of the gas pressure inside the furnace.

Conclusion

In order to illustrate the dynamic behavior of the model, computer Simulation was carried out and responses to changes in inputs were obtained. One input has been changed at a time while others were kept constant. Figures (1 to 4) illustrate the responses of the state variables and input variables for a 15% Decrease in fuel input. From the figures, the response follows the general behavior of the dynamics of the furnace as proven on literature and it concluded that the simulation model depict the dynamics of furnace quite well and it can be used further as a part of boiler simulation toolbox.

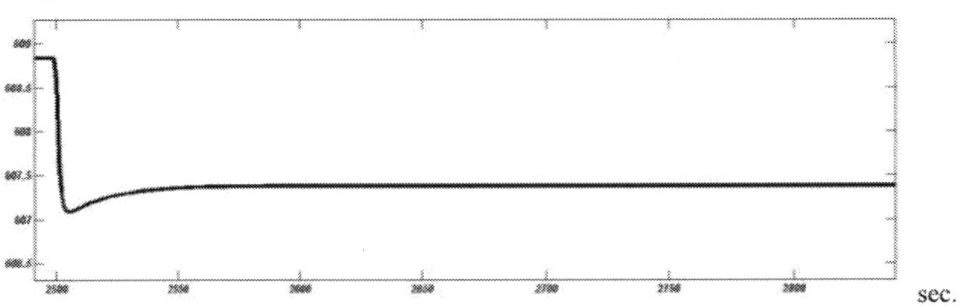

Fig. 1. The change of exhaust gas temperature Tg˚K with 15% change in fuel flow rate WF [kg/S].

DOI: 10.1051/matecconf/201711001051

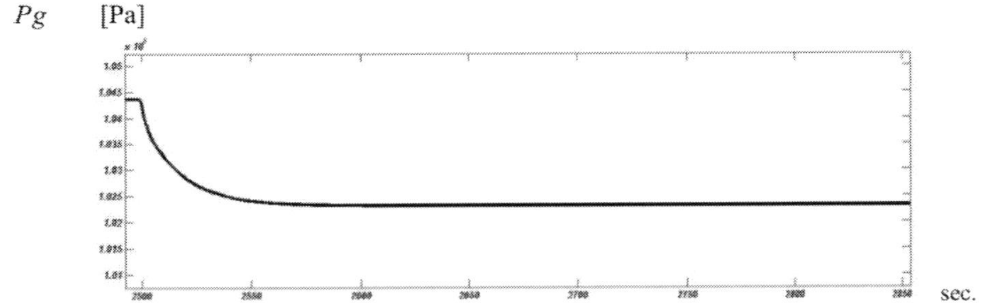

Fig. 2. The change of exhaust gas pressure Pg [Pa] with 15% change in fuel flow rate WF [kg/S].

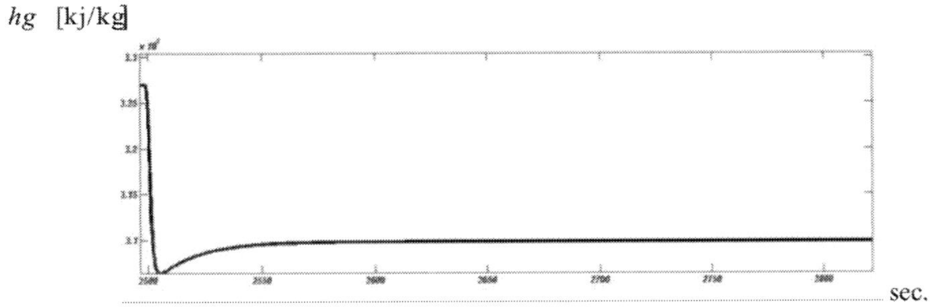

Fig. 3. The change of exhaust gas enthalpy with 15% change in fuel flow rate WF [kg/S].

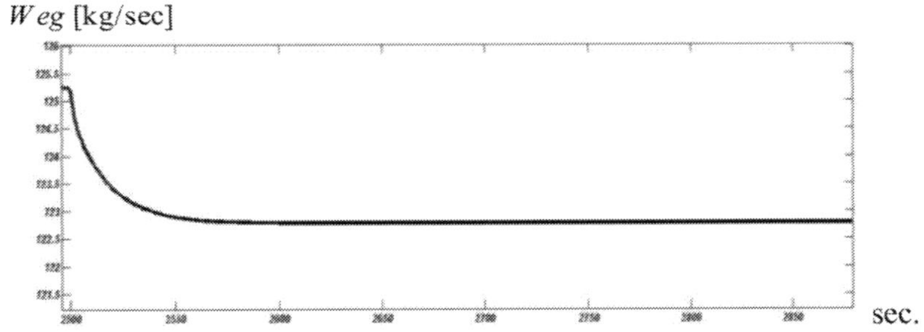

Fig. 4. The change of exhaust gas mass flow rate with 15% change in fuel flow rate WF [kg/S].

References

1. F. P. Ion, V. Ion, Technologies in Machine Building Galati, 23 (2012)
2. V.I. Ion, S. Craciun, Modelling and Optimization in the Machines Building Field MOCM-99, **2**, 143 (2003)
3. A.Salehi, A.R. Seifi.and A.A. Safavi, An Interdisciplinary Journal of Research, Theories and Observations, **9**(1), 99 (2008)
4. K.L. Chien, E.I. Ergin, C. Ling, A. Lee, Transactions of ASME, **80**, 1809 (1958)
5. Neaga C., et al., Works of the National Conference on Thermotechnics, May 23-24, 1997
6. Tuomas Kataja, Irdzhe Majanne, 14th Nordic Process Control Workshop, Sirkka-Liisa Ichia-Jounela (ed.), August 2007.

Mathematical model of the standalone electrical supply system with distributed photoelectric generation

Dmitry I. Muravyev[1,*], *Boris* V. Lukutin[1]

[1]National Research Tomsk Polytechnic University, 634050 Tomsk, Russia

Abstract. A promising way to increase the technical and economic characteristics of standalone power supply systems is to incorporate renewable energy installations in their structure. Photoelectric generators based on distributed sources are optimal technologies for the use of renewable energy sources in 0.4 kV low-voltage power grid. The most common option is a hybrid system with photoelectric power stations (PES) incorporated into the local network of the diesel power station (DPS). Photoelectric stations meet all environmental requirements and can make a significant contribution to the electrification of remote settlements, tourist and agricultural field. This paper deals with stabilization of voltage value and reduction of losses of electric energy depending on the parameters of elements of power supply systems of radial type (0,4 kV) with an installed capacity of up to 100 kW. The research has been conducted by simulating the operating modes of hybrid power systems of various configurations. To analyze the joint work of a photoelectric station with a diesel power station, a mathematical model is created in the Simulink (SimPowerSystems) application of the MatLab R2016b program. Most of the known works do not show the issues of quality and power losses in the standalone power supply system with photoelectric distributed generation. [5,6,7,9,10].

1 Object of research

A standalone system of power supply of a radial type with a voltage of 0.4 kV, which is characteristic for electrification of small settlements, was considered. The researches were carried out for the geographic conditions of the middle latitudes of the territory of the Russian Far East.

2 Research objectives

Creation of a working model of a standalone power supply system with distributed photoelectric generation, which takes into account the need for a large number of calculations associated with daily changes in insolation and power consumption. In

* Corresponding author: dim15@tpu.ru

DOI: 10.1051/matecconf/201711001052

particular, the stability of the magnitude of voltage and losses in transmission lines is of interest, depending on the ratio of the power of the DPS and PES, the points of connection of photoelectric stations to the autonomous electric network [4].

3 Results of the research

To analyze the hybrid operation of a photoelectric power station with a diesel power station, a mathematical model is created for the load in the Simulink application of the MatLab R2016b program, shown in Fig. 1. SimPowerSystems library was used for simulation of transient processes. The PES model is based on the approximated characteristics of the photoelectric generator and simplified mathematical functions [8].

The length of power transmission lines is limited by the maximum allowable voltage deviation at the load at the end of the line relative to the nominal value. The state standard establishes the maximum value of this deviation not more than 10%. Block subsystem "Power function from two parameters" describes the analytical expression of the influence of the ambient temperature and the magnitude of the intensity of solar radiation, with the arrangement of the photo panel perpendicular to the incident rays. [8,9,10].

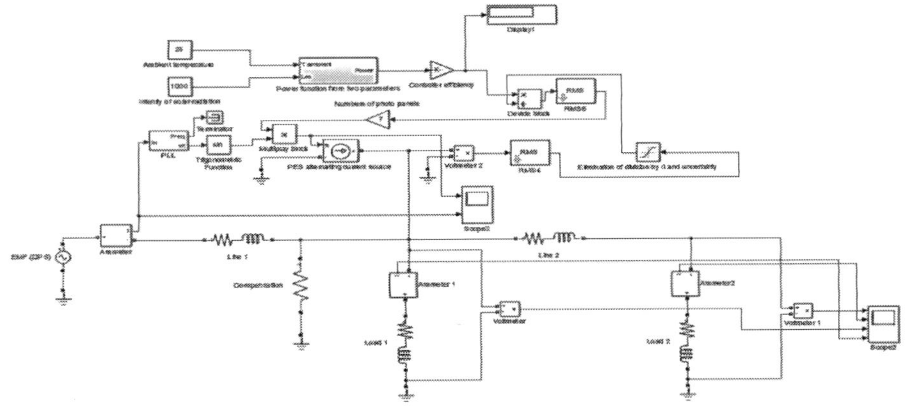

Fig.1. Mathematical model of a standalone power supply system with distributed photoelectric generation.

The equation of the solar power generation function according to the parameters of the ambient temperature and solar radiation intensity [1]:

$$(I_{SC}\cdot U_{OC}\cdot(1+(\alpha-\beta))\cdot(T_E-T_0))\cdot\lambda. \quad (1)$$

where I_{SC} - short-circuit current of the solar module, A; U_{OC} - open-circuit voltage of the solar module V; α, β - corresponding temperature coefficients; T_E - ambient temperature; T_0 is the temperature taken as the basis (20 ^{0}C); λ - is the intensity of solar radiation.

In the nominal mode of operation of the DPS, the voltage at the end of the investigated power transmission line model is reduced by 8.04% relative to the nominal. Connecting an additional source in the form of a photovoltaic station allows to redistribute the load power between the DPS and the PES. As a result, there is a change in the magnitude of the currents in various parts of the transmission line, which reduces the loss of electricity in the wires and provides greater voltage stability in the points of connection of electricity consumers [5].

The obtained data reflect the power consumption by the objects under consideration and the generation of currents in the nodes of the electric grid from the main source of the DPS and the auxiliary PES [5,9,10].

The results of the simulation showed that the corresponding power of the photoelectric station, depending on the amount of load consumed by the power supply system, makes it possible to reduce the generation of power from the main source of DPS, which entails a reduction in fuel consumption and consumption, diesel segment unloading in summer. With significantly uneven power loads of the power distribution line, it is advisable to connect the power station at the point of connection of the maximum power of the electric consumer [3].

3 Conclusions

1. The effectiveness of PES in standalone power supply systems is substantially limited by the conditions of solar insolation and temperature changes. There is a need to use power accumulators.

2. For geographic regions in the mid-latitudes of the north-eastern hemisphere, the large difference in the magnitude of insolation in summer and winter determines the expediency of maximum use of PES in the summer, which contradicts the reduction of summer power consumption.

3. To stabilize the voltage in the power supply system and reduce power losses, it is possible to use the PES closer to the end of the distribution line or to the point of connection of the most powerful consumer [5].

4. This model of a power station with distributed photoelectric generation showed satisfactory simulation results. The model makes it possible to test the investigation of various modes and configurations of hybrid photo-diesel electric power system for the purpose of stabilization and reduction of voltage losses at the investigated time intervals.

References

1. S.N. Udalov, *Renewable energy sources* (Publishing House of the National Technical University, Novosibirsk, 2007) [in Russian]
2. B.V. Lukutin, *Renewable sources of electricity* (Publishing house of Tomsk Polytechnic University, Tomsk, 2008) [in Russian]
3. N.N. Duraev, S.G. Obukhov, I.A. Plotnikov, Izvestiya Tomsk Polytechnic University, **4**, 48 (2013) [in Russian]
4. B.V. Lukutin, I.O. Muravlev, I.A. Plotnikov, *Power supply systems with wind and solar power plants* (Publishing house of Tomsk Polytechnic University, Tomsk, 2015) [in Russian]
5. M.H. Cintuglu, A. Altamirano, O.A. Mohammed, 12th Latin American and Caribbean Conference for Engineering and Technology, (2014)
6. N. Abd Rahim, H. W. Pinga, J. Selvaraja, Procedia Environmental Sciences **17**, 537 (2013)
7. J.R. Rodriguez, F. Ruiz, D. Biel, F. Guinjoan, Simulation and analysis of distributed PV generation in a LV network using MATLAB-Simulink// The Spanish Ministry of Science and Innovation through the grants RUE CSD2009-00046, Consolider-Ingenio 2010 Programme, DPI2006-15627-C03-01 and DPI2007-62582.
8. B.V. Lukutin, E.B. Shandarova, A.F. Makarova, I.B. Shvartsman, IOP Conf. Series: Materials Science and Engineering, **127** (2016)
9. J. Wood, *Local Energy Distributed generation of heat and power* (Published by The Institution of Engineering and Technology, London, 2008)
10. K. Amaresh, V. Sankar. International Journal of Engineering and Innovative Technology, **3** (3), 251 (2013)

Mathematical modeling of heat transfer in production premises heated by gas infrared emitters

Vyacheslav I. Maksimov[1], *Aleksey* M. Bubenchikov[2], *Tatiana* A. Nagornova[1,*], and *Alime* A. Dudak[1]

[1] National Research Tomsk Polytechnic University, 634050 Tomsk, Russia
[2] Tomsk State University, 634050 Tomsk, Russia

Abstract. The results of numerical modeling of the process of free convective heat transfer in the regime of turbulent convection in a closed rectangular region heated by an infrared radiator are presented. The system of Navier-Stokes equations in the Boussinesq approximation is solved, the energy equation for the gas and the heat conduction equations for the enclosing vertical and horizontal walls. A comparative analysis of the heat transfer regimes in the considered region for different Grashof numbers is carried out. The features of the formation of heated air flows relative to the infrared emitter located at some distance from the upper horizontal boundary of the region are singled out.

1 Introduction

Despite the indisputable prospect of using gas infrared emitters (GIE) [1] to create a comfortable thermal regime in the local zones of large-sized production facilities, such heating systems are rarely used until now. This is due to the lack of methods for calculating the temperature fields in the areas of operation of the GIE, taking into account the main significant factors. Mathematical models [2] and the corresponding calculation methods are developed for the conditions for placing the radiators under consideration directly on the surface of the upper floors of the rooms, in which the heated GIE local zones are allocated. In real practice, the most efficient radiators can be displaced relative to the upper boundary of the heated area (Figure 1) and as close as possible to the specific workplace (or places).

* Corresponding author: tania@tpu.ru

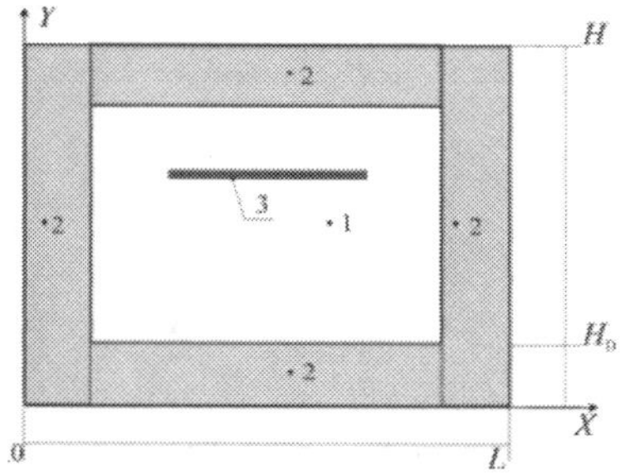

Fig. 1. Area of solution of the problem under consideration: 1 - air; 2 - enclosing structures; 3 - infrared emitter; *L*, *H* – size of the area in the direction *X* and *Y*.

When implementing this option, a significant reduction in energy costs for heating one workplace is possible. For these reasons, it is of interest to develop models [2] on the region of more complex free-convective flows corresponding to the geometry shown in Fig. 1.

The purpose of this work is to simulate the mathematical simulation of the conjugate heat transfer in a region limited by vertical and horizontal enclosing structures of finite thickness with a gas infrared radiator that is substantially displaced vertically relative to the upper boundary of the analysis area (Fig. 1).

2 Statement of the problem and method of solution

In the formulation of the problem, it was assumed that the air in the region under consideration is transparent to the radiation by the medium, and heat transfer is due to natural convection. It is also assumed that the radiation flux generated by the GIE is completely absorbed at the lower boundary of the region. Heating of the vertical walls is carried out by the flow of streams heated at the lower boundary of the air in the *Y* direction. The heat sink was taken into the enclosing structures and the accumulation of energy in them. The axisymmetric problem was solved. The gas infrared emitter was a rectangle of finite thickness. Based on the experimental data, boundary conditions of the first kind were established at all boundaries of the GIE. It is established that during operation such radiators are heated up to sufficiently high temperatures (up to 250 °C), which remain constant for long periods of time.

The results of solving a group of analogous problems of conjugate heat transfer in closed regions of small dimensions with local energy sources [3-5] have shown that the formation of temperature fields in regions with walls accumulating heat is significant. Therefore, the problem of non-stationary heat transfer was solved. The thermal conductivity and convection in the gas, as well as the thermal conductivity in the enclosing structures were taken into account.

An analysis of the ranges of changes in the main characteristics and parameters of the processes under consideration, as well as the typical sizes of the regions in which GIE is possible, showed that the Gr numbers in real conditions exceed 10^{10}. That is, the regimes of thermogravitational convection under the conditions under consideration are turbulent. Accordingly, the statement of the problem was carried out taking this factor into account. A k-ε model of turbulence was used [5, 6].

3 Mathematical model and method of solution

In the framework of the physical formulation described above, a mathematical model is formulated that includes Reynolds averaged Navier-Stokes equations, the energy equation for the gas, and the heat conduction equations for the enclosing structures with the corresponding boundary conditions below.

$$\frac{\partial\Omega}{\partial\tau}+U\frac{\partial\Omega}{\partial X}+V\frac{\partial\Omega}{\partial Y}=\frac{\partial^2}{\partial X^2}\left[\left(\frac{1}{\sqrt{Gr}}+\frac{1}{\mathrm{Re}_t}\right)\Omega\right]+\frac{\partial^2}{\partial Y^2}\left[\left(\frac{1}{\sqrt{Gr}}+\frac{1}{\mathrm{Re}_t}\right)\Omega\right]+\frac{\partial\Theta}{\partial Y},$$
$$+2\frac{\partial U}{\partial Y}\frac{\partial^2}{\partial X^2}\left(\frac{1}{\mathrm{Re}_t}\right)-2\frac{\partial V}{\partial X}\frac{\partial^2}{\partial Y^2}\left(\frac{1}{\mathrm{Re}_t}\right)+2\left(\frac{\partial V}{\partial Y}-\frac{\partial U}{\partial X}\right)\frac{\partial^2}{\partial X\partial Y}\left(\frac{1}{\mathrm{Re}_t}\right) \tag{1}$$

$$\frac{\partial\Theta}{\partial\tau}+U\frac{\partial\Theta}{\partial X}+V\frac{\partial\Theta}{\partial Y}=\frac{\partial}{\partial X}\left[\left(\frac{1}{\mathrm{Pr}\sqrt{Gr}}+\frac{1}{\mathrm{Pr}_t\mathrm{Re}_t}\right)\frac{\partial\Theta}{\partial X}\right]+\frac{\partial}{\partial Y}\left[\left(\frac{1}{\mathrm{Pr}\sqrt{Gr}}+\frac{1}{\mathrm{Pr}_t\mathrm{Re}_t}\right)\frac{\partial\Theta}{\partial Y}\right], \tag{2}$$

$$\frac{\partial^2\Psi}{\partial X^2}+\frac{\partial^2\Psi}{\partial Y^2}=-2\Omega, \tag{3}$$

$$\frac{\partial E}{\partial\tau}+U\frac{\partial E}{\partial X}+V\frac{\partial E}{\partial Y}=\frac{\partial}{\partial X}\left[\left(\frac{1}{\sqrt{Gr}}+\frac{1}{\sigma_\varepsilon\mathrm{Re}_t}\right)\frac{\partial E}{\partial X}\right]+\frac{\partial}{\partial Y}\left[\left(\frac{1}{\sqrt{Gr}}+\frac{1}{\sigma_\varepsilon\mathrm{Re}_t}\right)\frac{\partial E}{\partial Y}\right]+,$$
$$+c_{1\varepsilon}\left(\tilde{P}_k+c_{3\varepsilon}\tilde{G}_k\right)\frac{E}{K}-c_{2\varepsilon}\frac{E^2}{K}, \tag{4}$$

$$\frac{\partial K}{\partial\tau}+U\frac{\partial K}{\partial X}+V\frac{\partial K}{\partial Y}=\frac{\partial}{\partial X}\left[\left(\frac{1}{\sqrt{Gr}}+\frac{1}{\sigma_k\mathrm{Re}_t}\right)\frac{\partial K}{\partial X}\right]+\frac{\partial}{\partial Y}\left[\left(\frac{1}{\sqrt{Gr}}+\frac{1}{\sigma_k\mathrm{Re}_t}\right)\frac{\partial K}{\partial Y}\right]+.$$
$$+\tilde{P}_k+\tilde{G}_k-E, \tag{5}$$

$$\frac{1}{\mathrm{Fo}}\frac{\partial\Theta}{\partial\tau}=\frac{\partial^2\Theta}{\partial X^2}+\frac{\partial^2\Theta}{\partial Y^2}, \tag{6}$$

where $\tilde{P}_k=\frac{1}{\mathrm{Re}_t}\left[2\left(\frac{\partial U}{\partial X}\right)^2+2\left(\frac{\partial V}{\partial X}\right)^2+\left(\frac{\partial U}{\partial X}+\frac{\partial V}{\partial X}\right)^2\right]$; $\tilde{G}_k=-\frac{1}{\mathrm{Re}_t\mathrm{Pr}_t}\frac{\partial\Theta}{\partial Y}$;

The model constants are given as: $c_\mu = 0{,}09$; $c_{1\varepsilon} = 1{,}44$; $c_{2\varepsilon} = 1{,}92$; $c_{3\varepsilon} = 0{,}8$; $\sigma_\varepsilon =1{,}3$; $\sigma_\kappa =1$.

Initial conditions for the system of equations (1–4):

$$\Psi(X,Y,0)=\Omega(X,Y,0)=\Theta(X,Y,0)=K(X,Y,0)=E(X,Y,0)=0 \tag{7}$$

Boundary conditions at the three external borders of region solutions:

$$\frac{\partial\Theta(X,Y,\tau)}{\partial n}=0. \tag{8}$$

At the interfaces of solid walls and gas the following conditions are fulfilled:

$$\frac{\partial\Psi(X,Y,\tau)}{\partial n}=0,\ \frac{\partial\Theta_1(X,Y,\tau)}{\partial n}=\lambda_{1,2}\frac{\partial\Theta_2(X,Y,\tau)}{\partial n},\ \Theta_1(X,Y,\tau)=\Theta_2(X,Y,\tau). \tag{9}$$

On the surface of the radiator:

$$\Theta_1(X,Y,\tau)=1, \tag{10}$$

At the boundary H_D of the radiation heat source is taken into account:

$$\frac{\partial\Psi(X,Y,\tau)}{\partial Y}=0,\ \frac{\partial\Theta_1(X,Y,\tau)}{\partial Y}=\frac{\partial\Theta_2(X,Y,\tau)}{\partial Y}+\mathrm{Ki},\ \Theta_1(X,Y,\tau)=\Theta_2(X,Y,\tau) \tag{11}$$

For the κ-ε model near the solid surface the following is accepted:

$$\frac{\partial K}{\partial n}=0;\ E=\frac{c_{\mu}^{3/4}K^{3/2}}{k\Delta n}.$$

where X, Y – dimensionless Cartesian coordinates; τ – dimensionless time; t – time, s; t_0 – time scale, s; U, V – dimensionless velocity; V_0 – velocity scale (convection velocity), m^2/s; Θ – dimensionless temperature; T – temperature, K; T_0 – temperature at the initial time, K; T_{it} – heat source temperature, K; g – acceleration created by the mass forces, m/s^2; β – thermal coefficient of volume expansion, K^{-1}; ψ – stream function, m^2/s; ψ_0 – current function scale, m^2/s; Ψ – dimensionless analog stream function ψ; ω – swirl of vorticity, 1/c; ω_0 – vorticity scale, 1/s; Ω – dimensionless analog ω; K – dimensionless analog of turbulent kinetic energy; E – dimensionless analog of the dissipation rate of turbulent kinetic energy; $\mathrm{Ki}=qL^3/\lambda(T_{it}-T_0)$ – Kirpichev number; $\mathrm{Gr}=g\beta L^3(T_{it}-T_0)/\nu^2$ – Grashof number; $\mathrm{Re}_t=E/c_{\mu}K^2$ – turbulent Reynolds number; ν_t – coefficient of turbulent kinematic viscosity, m^2/s; $\lambda_{1,2}=\lambda_2/\lambda_1$ – relative thermal conductivity coefficient; $\mathrm{Pr}=\nu_t/a$ – turbulent Prandtl number; $\mathrm{Pr}=\nu/a$ – Prandtl number; $\mathrm{Fo}=at_0/L^2$ – число Фурье; a – thermal diffusivity, m^2/s.

Equations (1-6) with the corresponding initial and boundary conditions (7-11) are solved by the method of finite differences [7, 8] on a uniform grid. To solve the system of difference analogs of the initial nonlinear partial differential equations with nonlinear boundary conditions, the method and algorithm developed [3-5] for solving problems of thermogravitational convection in closed regions with local energy sources were used. To substantiate the reliability of the results of numerical modeling, the conservativeness of the difference scheme was estimated by a technique developed for solving nonlinear heat conduction problems [9] and developed for ignition of water-coal fuels under conditions of intense local injection of fumes into the region filled with an oxidizer [10, 11].

4 Analysis of numerical simulation results

Numerical studies were carried out for free-convective flows corresponding to the turbulent regime for two Grashof numbers: $\mathrm{Gr}=4\cdot10^{12}$ and $\mathrm{Gr}=10^{10}$. The results obtained with these values of Gr sufficiently clearly demonstrate the main regularities of the processes under consideration. Other used dimensionless and dimensional values (L = 4 m, t_0 = 5 s, T_0 = 293 K, T_{it} = 353 K), were also chosen to correspond to the actual values of the main characteristics when infrared emitters operate. Numerical studies were carried out at the values of the heat flux arriving at the H_D surface (Table 1).

Table. 1. The values of the heat flux over the surface of the concrete base in the section $Y = H_D$ for $L/2 < X < 0{,}875L$ (fig. 1).

Section	X=0.5L	X=0.563L	X=0.685L	X=0.688L	X=0.75L	X=0.813L	X=0.875L
q, W/m^2	190.5	180.2	182.1	172.2	134.2	122.2	102.1

The analysis of thermal regimes in the quite typical region under consideration can be carried out both with respect to the distributed characteristics of the process (inhomogeneous in real practice and nonstationary temperature fields) and by the averaged ones (the temperature averages over the region at fixed instants of time). In the first case, the temperature gradients can be estimated and the maximum or minimum air temperature zones identified. In the second case (which is most often used in practice), it is possible to easily and quickly compare, by average temperature values, several heating regimes, differing by the intensity of radiation coming from the GIE or by the position of the radiators relative to the upper ceiling of the room.

Below are the results of numerical modeling of temperature fields in the analysis area in the form of distributions Θ and mean temperature values.

In Fig. 2 and 3 show typical results of numerical simulation illustrating the main regularities of heat transfer in the heating system under consideration using gas infrared emitters.

Comparing the results of numerical studies for various Grashof numbers, it can be noted that at $Gr = 10^{12}$, a much more intensive circulation of air currents occurs (Fig. 2). In turn, this leads to an increase in heat transfer in the considered region, as evidenced by the mean temperatures (Table 2).

An analysis of the results of numerical modeling of the processes under consideration gives grounds also for conclusions about the possible optimization of heating schemes for workplaces using infrared emitters. So, for example, it is obvious that there is a possibility of significant energy savings in the heat insulation of the air zone between the upper surface of the GIE (Figure 1) and the overlap. It is clearly seen that as a result of thermogravitational convection, the air masses heated up by the heat sink from the lower enclosing structure rise upward, flow around the GIE, move parallel to the upper boundary of the region, giving energy to the overlaps (then an intense heat dissipation into the atmosphere is possible), and descend along the vertical enclosing structures. Obviously, a significant part of the energy is absorbed by the slabs because the air reaches them with the highest possible temperature. In addition to the energy transmitted from the heated GIE of the lower ceiling of the room, the air masses when heated up are additionally heated when flowing around the radiator housing.

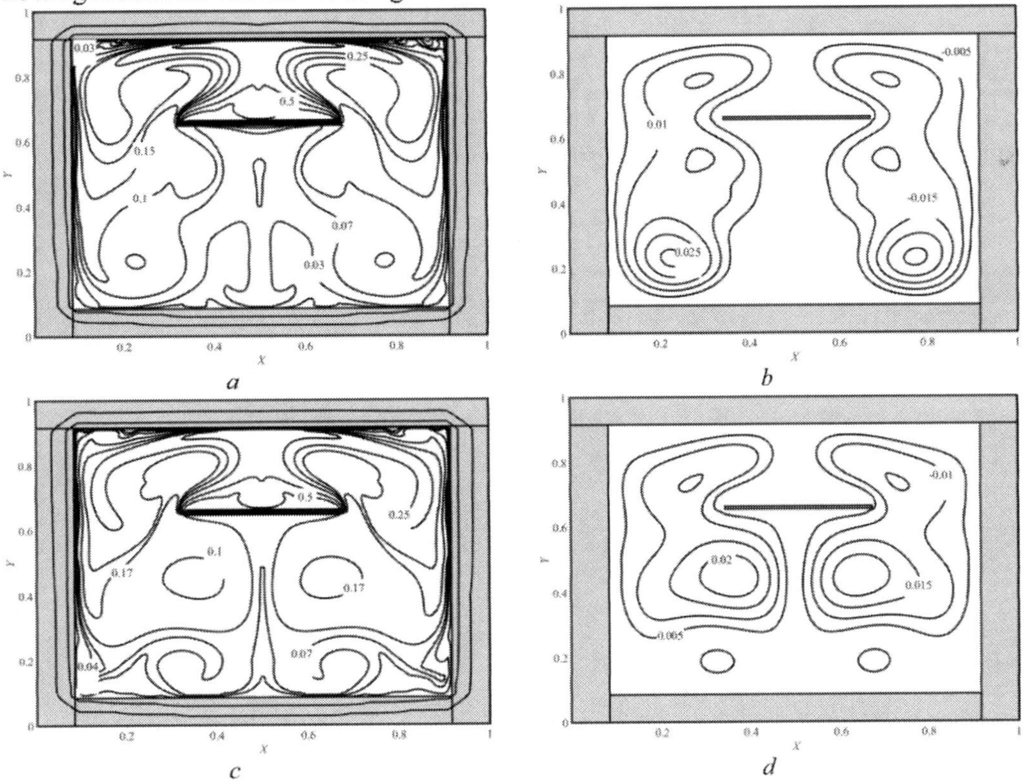

Fig. 2. Fields of temperatures (*a*, *c*) and isolines of the current function (*b*, *d*) for natural convection conditions in a closed region for $Gr = 4\cdot10^{12}$ for different values of τ: $\tau = 200$ (*a*, *b*), $\tau = 300$ (*c*, *d*).

Table. 2. Mean temperature values Θ for different time instants.

τ	Gr=10^{12}	Gr=$4{\cdot}10^{10}$
200	0.147	0.148
300	0.171	0.188

Isotherms (Fig. 2, 3) illustrate well that the region of the highest temperatures in a room heated by GIE is formed far beyond the working area (located near the lower overlap). In essence, to a large extent, GIE works on heating the air in the upper part of the room and the plates of the upper ceiling. At the same time, the losses of thermal energy are great.

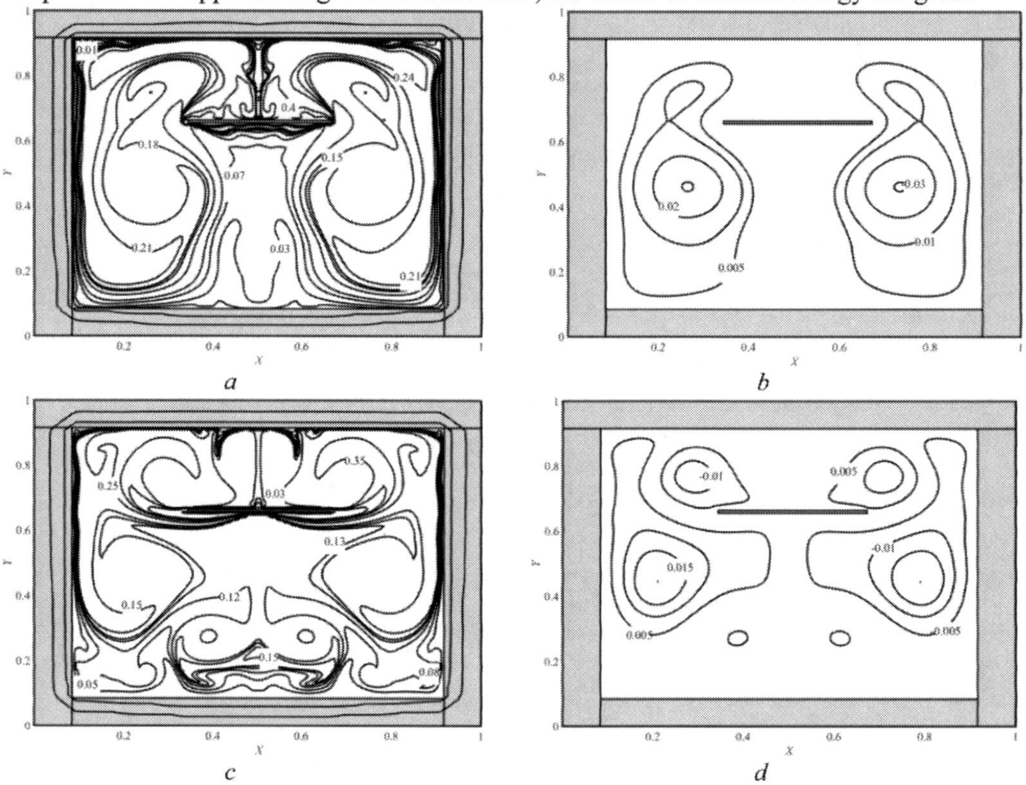

Fig. 3. Fields of temperatures (*a*, *c*) and isolines of the current function (*b*, *d*) for natural convection conditions in a closed region for Gr = $4{\cdot}10^{10}$ for different values of τ: τ = 200 (*a*, *b*), τ = 300 (*c*, *d*).

It seems reasonable to at least maximally effective thermal insulation of the upper surface of the GIE corps. A significantly greater energy saving effect can be achieved by eliminating the circulating region of hot air above the GIE casing or (at least) the thermal insulation of the plate on their lower surface. Eliminate the circulation of heated air in the upper part of the room by placing obstacles in the path of the air masses moving upwards parallel to the body of the GIE.

Conclusion

The results obtained during the solution of the problem of thermogravitational convection of air under conditions of intensive energy supply due to the operation of gas infrared radiators and heat removal into the enclosing structures are the basis for the development of methods for calculating the thermal regimes of local working zones in large-sized production rooms with heating systems based on GIE. The possible high efficiency of using such radiators for local heating of single and group work stations is shown.

The regularities of heat transfer established under the results of numerical simulation in the conditions of infrared emitters are also a basis for the development of models of the conjugate heat transfer in the part of the transition from traditional productions of computational fluid mechanics problems [3-5] to the tasks of industrial heat and power engineering and energy saving.

The work was supported by the Russian President's grant (Scientific School project - 7538.2016.8).

References

1. K.-N. Rhee, B.W. Olesen, K.W. Kim, Building and Environment, **112**, 367 (2017)
2. G.V. Kuznetsov, N.I. Kurilenko, V.I. Maksimov, G.Ya. Mamontov, T.A. Nagornova, J. Eng. Phys. Thermophys. **86** (3), 519 (2013)
3. G.V. Kuznetsov, M.A. Sheremet, Comput. Therm. Sci. **1** (3), 341 (2009)
4. V.I. Maksimov, T.A. Nagornova, I.A. Shestakov, EPJ Web Conf. **82**, 01048 (2015)
5. G.V. Kuznetsov, M.A. Sheremet, Int. J. Heat Mass Transfer, **53**, 163(2010)
6. V.I. Maksimov, T.A. Nagornova, V.P. Glazyrin, EPJ Web Conf. **110**, 01038 (2016)
7. D.S-K. Ting, *Basics of engineering turbulence* (Academic Press, USA, 2016)
8. T. Cebeci, *Analysis of Turbulent Flows with Computer Programs* (Netherlands, Elsevier Ltd., 2013)
9. V.L. Strakhov, A.N. Garashchenko, G.V. Kuznetsov, V.P. Rudzinskij, Combustion, Explosion, and Shock Waves, **37** (2), 178 (2001)
10. G.V. Kuznetsov, V.V. Salomatov, S.V. Syrodoy, Combustion, Explosion and Shock Waves, **51** (4), 409 (2015)
11. G.V. Kuznetsov, V.V. Salomatov, S.V. Syrodoy, Thermal Engineering, **62** (10), 703 (2015)

Efficiency of steam reheating in heat recovery steam generator at thermal power plant with gas turbine topping

Olga Romashova[1,*], *Aleksandr* Minor[1], *Vladimir* Martyshev[1], *Aleksandr* Tubolev[1],and *Aleksandra* Katalevskaya[1]

[1]National Research Tomsk Polytechnic University, 634050 Tomsk, Russia

Abstract. Possibility of using alternative option to the displacement of regenerative heating is analyzed in this paper: application of heat recovery steam generator's (HRSG) thermal power for reheating of working medium that has passed steam turbine.

Introduction

Technical re-equipment and modernization of existing Thermal Power Plants (TPPs) with implementation of combined-cycle technologies lead to significant reduction in specific consumption of reference fuel for electricity generation. At the same time, indices of TPPs thermal efficiency after modernization are mainly determined by adopted technical solutions [1-4].

Along with the most efficient, but also more costly option for constructing new modern Combined Cycle Power Plants (CCPPs) at the site of an old TPP, there is a transition to CCPP with application of operating stations equipment. In principle, there are two options: 1) installation of gas turbine (GT) and HRSG in existing building with dismantling of existing steam boiler; 2) integration of GT into the thermal scheme of steam-power unit (at coal-fired power plants where natural gas is used as the second fuel). In many cases, option 2 is more preferable, since it allows full utilization of existing plant infrastructure and reduction of investment costs.

There are three possibilities in principle for integrating GT in cycle arrangement: 1) application of GT exhaust gases as oxidizer for fuel combustion in steam boiler; 2) inclusion of GT in a parallel scheme for production of steam in HRSG with its further supply to steam line along with the working fluid of steam boilers; 3) the use of GT exhaust gases warmth for heating of feed water and condensate.

Schemes with heating of feed water and condensate in HRSG are most widespread and have been called GT topping with displacement of regenerative heating [5-7]. Disadvantage of such modernization for TPPs is a significant deterioration of steam turbine plant (STP) thermal efficiency: the share of heat losses in condenser increases, consumption of main

* Corresponding author: roma@tpu.ru

steam decreases compared to the nominal value due to limitation of steam passage to condenser with corresponding drop of STP electrical power.

Possibility of using alternative option to the displacement of regenerative heating is analyzed in this paper: application of heat recovery steam generator's (HRSG) thermal power for reheating of working medium that has passed steam turbine.

1 HRSG options for displacing steam reheating and regenerative heating

Thermal efficiency of CCPP schemes with displacement of steam reheating and regenerative heating from steam-turbine part for different options of HRSG heat application is compared: scheme 1 (Fig. 1) - displacement of steam reheating from steam boiler; scheme 2 (Fig. 2) - partial displacement of regenerative heating (high-pressure heaters (HPHs)); scheme 3 (Fig. 3) - displacement of steam reheating with partial displacement of regenerative heating; scheme 4 (Fig. 4) - complete displacement of regenerative heating.

As an example, CCPP calculations are applied to steam-turbine power unit with steam reheating, for which the initial data is presented in Table 1. For the initial scheme (scheme 0), efficiency of electricity production is 43 %, thermal load of steam reheating is Q_{sr} = 39 MW, thermal load of HPHs is assumed equal to Q_{sr}, thermal load of low-pressure heaters (LPHs) is 27 MW, steam flow to condenser is 55.7 kg/s.

Table 1. Initial data for STP scheme calculation.

parameters of the main steam		parameters of reheated steam		deaerator pressure	temperature of feed water	pressure in the extraction for low-pressure heater	condenser pressure	electrical power	heating in deaerator
P_0	t_0	P_{sr}	t_{sr}	P_d	t_{fw}	P_{ex}^{lp}	P_c	N_E	Δt_d
MPa	°C	MPa	°C	MPa	°C	MPa	MPa	MW	°C
14	550	2.1	550	0.6	260	0.46	0.005	100	10

Additional initial data for calculation is taken from the prototype - GT Siemens V 64.3A: exhaust gas temperature is 571°C, gas temperature before GT is 1315°C, compression ratio in compressor is 16.2 [8]. In all options pressures in the steam tapping are recalculated using Stodola-Flugel dependence according to the methods of STP variable-load operation. Passage of steam through the last stages is maintained equal to its value in the initial scheme.

For scheme 1, thermal power of HRSG is considered equal to the thermal load of steam reheating in the initial scheme.

For scheme 2, passage of steam through the last stages is limited; it is taken from the initial scheme. Thermal power of HRSG remains unchanged compared to the first option and corresponds to the thermal load of initial circuit high-pressure heater.

Schemes 3 and 4 allow lowering the temperature of exhaust gases due to more complete utilization of HRSG heat. For these schemes, temperature of waste gases is assumed to be 100°C.

Combination of two alternative options is applied at scheme 3. Gases flowrates are determined from steam reheater heat balance in HRSG. Temperature of feed water remains the same. Heating of water in deaerator does not change compared to the initial scheme.

For scheme 4, HRSG completely replaces regenerative heaters. Its thermal power corresponds to the total heat load of HPHs and LPHs. Consumption and temperature of the gases in HRSG at the outlet from HPH are determined from the heat balance equations.

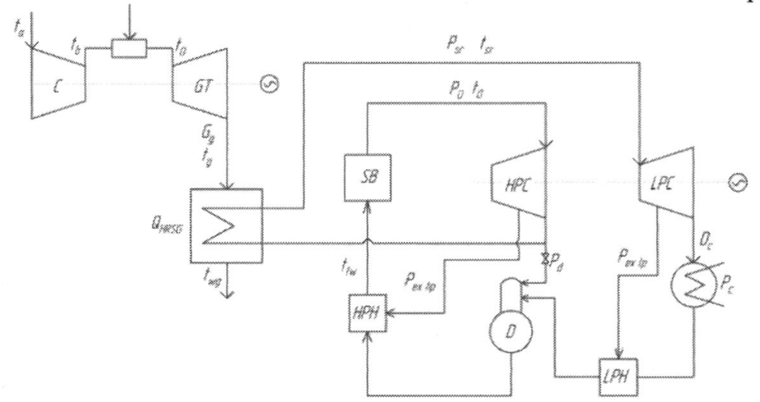

Fig. 1. Scheme 1 (steam reheating displacement from steam boiler).

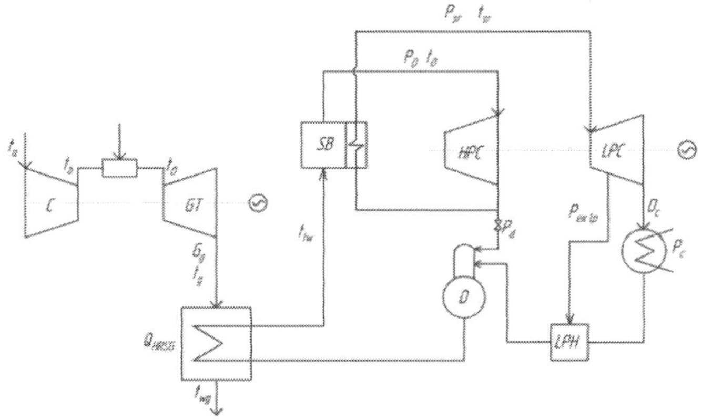

Fig. 2. Scheme 2 (partial regenerative heating (HPHs) displacement).

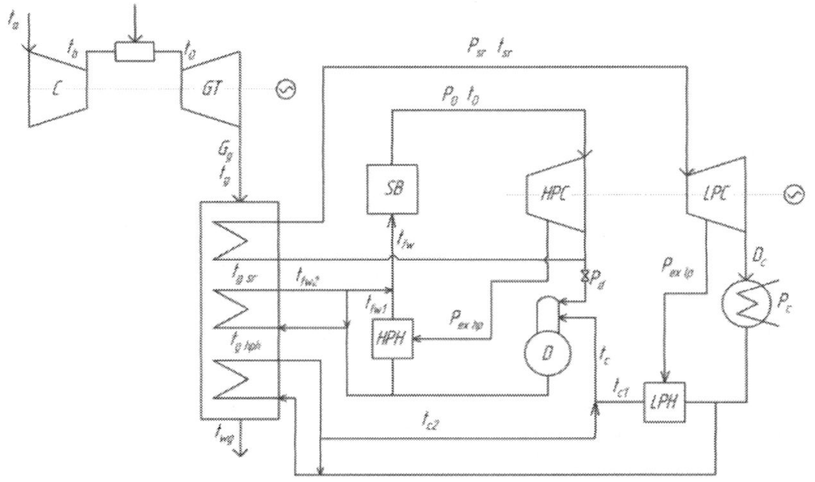

Fig. 3. Scheme 3 (steam reheating displacement from steam boiler and partial displacement of regenerative heating).

DOI: 10.1051/matecconf/201711001054

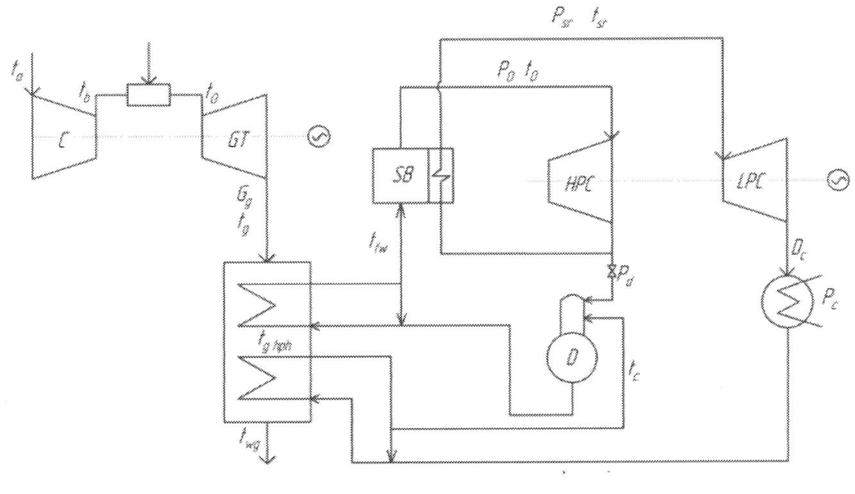

Fig.4. Scheme 4 (complete displacement of regenerative heating).

2 Results of calculations

Results of GT-toppings various options comparison are presented in Fig. 5-7.

Components of the total delivered heat of CCPP binary cycle are shown in Fig. 5 $Q_1 = Q_{SB} + Q_{HRSG}$ (steam boiler plus HRSG) for all compared options, where $\overline{N}_{E\,STP} = N_{E\,STP} / Q_1$; $\overline{N}_{E\,GT} = N_{E\,GT} / Q_1$, $\overline{Q}_{WH} = Q_{WH} / Q_1$, $\overline{Q}_C = Q_C / Q_1$ - proportions of steam turbine electric power, gas turbine, waste heat losses and losses in condenser. The largest losses of heat in condenser belong to schemes 2 and 4. Displacement of HPH leads to an increase in losses in the cold source by almost 23% in comparison with the initial scheme of STP, and complete displacement – by almost 50%.

Option of HRSG heat power application for steam reheating with complete preservation of regenerative heating (scheme 1) keeps heat loss in condenser at the nominal level. However, the losses of heat with exhaust gases (up to 15% from Q_1 of the total balance against (3–5)% for schemes 2 and 4). The reason is increased temperature of exhaust gases, which is determined by the temperature of steam in the cold thread of the steam reheating. Transition to scheme 3 with combination of HRSG application for steam reheating and regenerative heating leads to decrease in heat loss with exhaust gases to the minimal values with a certain increase in heat loss in condenser compared to STP scheme.

Fig. 6 represents graphical comparison of CCPP electrical efficiency values for the studied options. The greatest thermal efficiency has scheme with displacement of steam reheating and regenerative heating (scheme 3) - an increase in efficiency is 14.7% compared with initial scheme. The other schemes differ little from each other according to thermal efficiency. The increase in absolute electrical efficiency for scheme 2 is 5.5%, scheme 4–6.0%, scheme 1–7%.

The greatest value of CCPP electric power is characteristic of scheme 1 because of nominal electric load preservation of steam turbine in comparison with other options (Fig. 7). Electric power values of gas turbine for schemes 3 and 4 are comparable due to the close values of working medium flow rate of gas turbine cycle. The smallest electric capacity of steam-power cycle belongs to scheme 4 because of the lowest steam consumption per turbine.

 DOI: 10.1051/matecconf/201711001054

Decrease in the efficiency of steam boiler due to the change in its operating mode is estimated in all schemes approximately the same, since new values of steam boiler thermal capacity in CCPP composition lie in all options in the range (70–85)% of the nominal load.

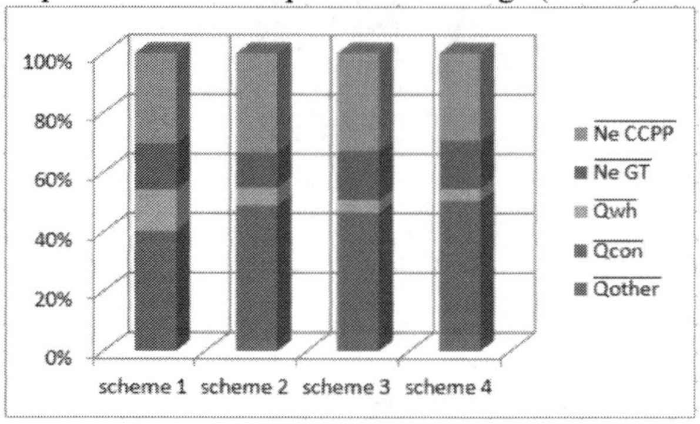

Fig. 5. Components of delivered heat in CCPP cycle.

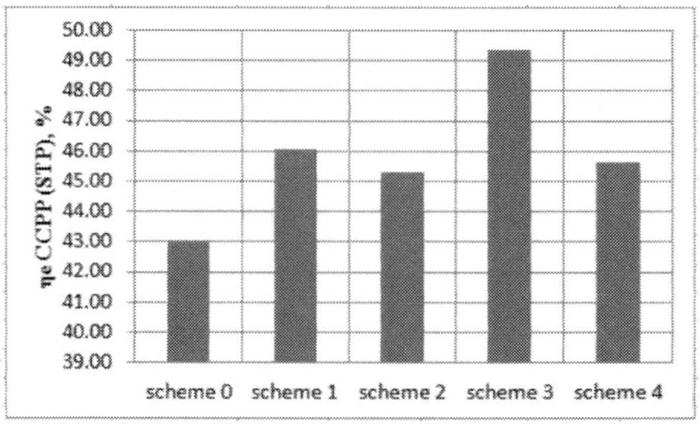

Fig. 6. Absolute electrical efficiency of CCPP (STP).

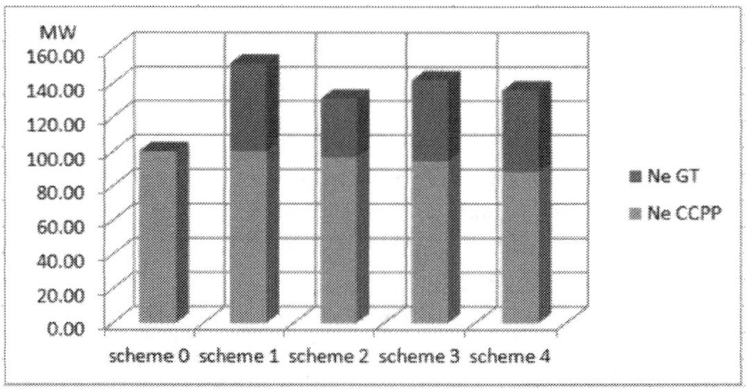

Fig. 7. Electrical capacity of CCPP (STP).

3 Choice of the optimum steam reheating pressure in the HRSG

Previous calculations were conducted at optimal design pressure of steam cycle reheating, which for STP scheme with single steam reheating is (15–20) % of the initial P_0.

This paper presents analysis of factors influencing the choice of optimal value of steam reheating pressure in CCPP schemes, as well as calculation of optimal value for the simplest scheme of STP (without regenerative heating) and for scheme 3.

DOI: 10.1051/matecconf/201711001054

For the simplest cycle of STP with steam reheating in HRSG, optimal steam reheating pressure is insignificant portion of P_0 - (1.6–2) % (Fig. 8). At the same time, as the pressure of superheating increases, temperature of exhaust gases continuously increases, because it is determined by steam temperature in cold train of steam reheating (Fig. 9) and at an optimal value greatly exceeds minimal possible value (more than 150°C against 80–100°C).

In the design of STP for scheme 3 the choice of optimal steam reheating pressure depends on assigned temperature of exhaust gases. With a decrease in exhaust gases temperature, optimal values of steam reheating pressure shift towards higher magnitudes (Fig. 11). At the same time, absolute electrical efficiency of CCPP is also increasing (Fig. 12). In the range of assigned values for exhaust gases temperature (80-120)°C, optimal steam reheating pressure is (24–27)% of P_o (Fig. 10), which exceeds maximal values of optimal parameters range of steam reheating pressure in traditional schemes of STP.

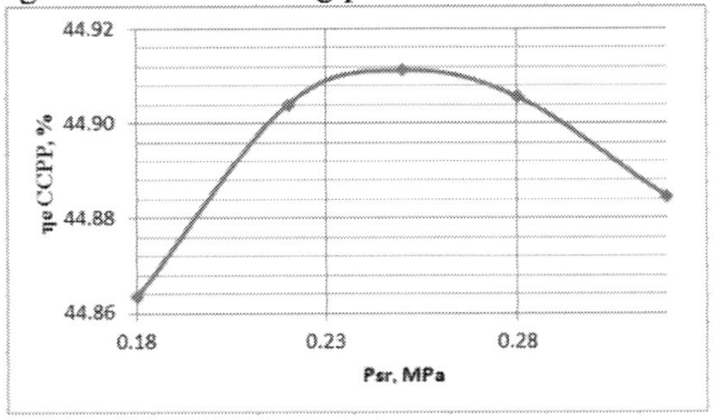

Fig. 8. Dependence of absolute electrical efficiency for the simplest cycle of STP with steam reheating in HRSG from SR pressure.

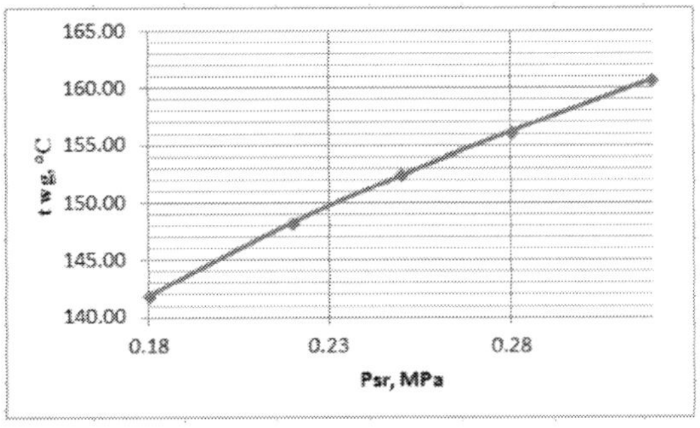

Fig. 9. Dependence of exhaust gases temperature for the simplest cycle of STP with steam reheating in HRSG from SR pressure.

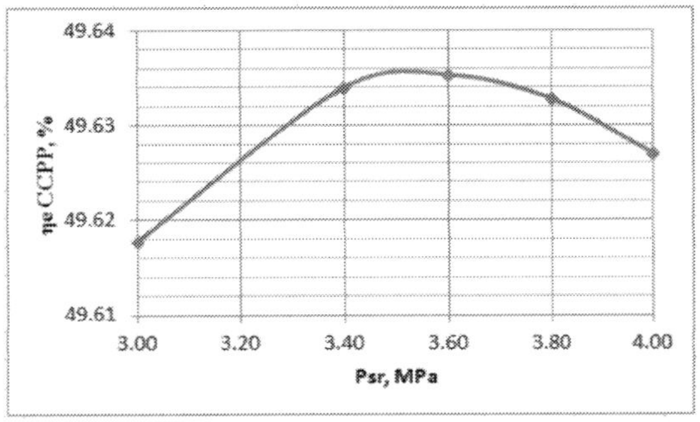

Fig.10. Dependence of absolute electrical efficiency of CCPP from SR pressure (scheme 3) (t_{wg} = 100°C).

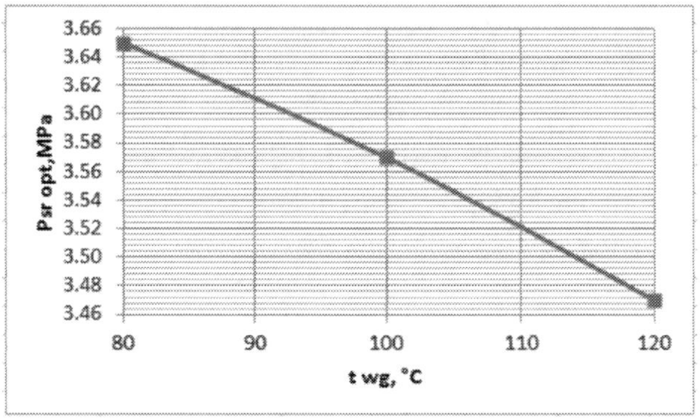

Fig.11. Dependence of optimal steam reheating pressure from exhaust gases temperature (scheme 3).

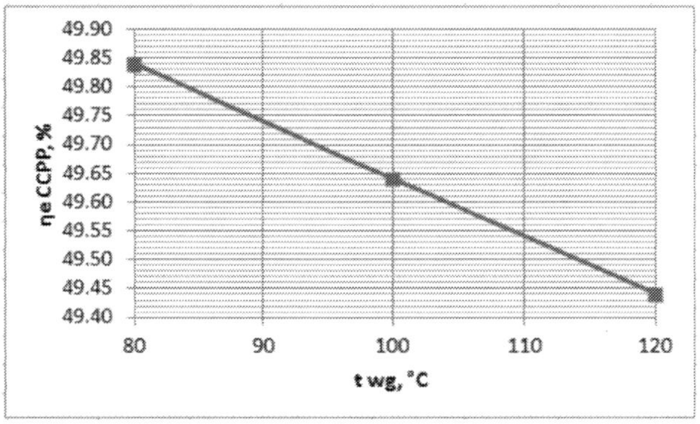

Fig.12. Dependence of absolute electrical efficiency of CCPP from exhaust gases temperature at optimal steam reheating pressure (scheme 3).

Conclusion

To sum up:

- displacement of HPH from steam turbine plant leads to an increase of heat losses in condenser by 24% compared with initial scheme of STP; complete replacement of regenerative heating increases losses by 50%. Steam flowrate to the turbine decreases, which leads to a decrease in electrical capacity.
- increase in CCPP efficiency, in comparison with initial option when HPH is displaced, is 5.5%. The use of regenerative heating complete displacement leads to insignificant increase of efficiency (by 6%) despite the decrease of exhaust gases temperature.
- displacement of steam reheating from steam boiler makes it possible to reduce heat losses in condenser in comparison with displacement of regenerative heating and to keep steam flowrate to the turbine at the level of nominal value.
- efficiency of CCPP for scheme with displacement of steam reheating is 7% higher than efficiency of the initial STP. At the same time, heat losses with exhaust gases reach maximal value in comparison with other options because of increased temperature of the gases at HRSG's cold end, which is determined by steam temperature in the cold train of steam reheating.
- the greatest increase in efficiency (up to 15%) is provided by combination of steam reheating displacement method with partial displacement of regenerative heating.

- optimal pressure of steam reheating in HRSG for the simplest cycle of STP is (1.6–2.0)% of the initial pressure, which is significantly lower than these values for single reheating scheme in STP (15–20)%.
- in the design of CCPP with full displacement of steam reheating and partial regenerative heating from steam turbine part to HRSG the choice of optimal steam reheating pressure depends on assigned temperature of exhaust gases. With decrease of exhaust gases temperature, at equal other conditions, optimal pressure and efficiency of CCPP increase. In the range of exhaust gases assigned temperatures (80–120)°C optimal steam reheating pressure is in the range of (24–27)% of P_o.

References

1. S.V. Tsanev, C.V. Tsaneva, V.D. Burov, A.N. Remezov, *Gas turbine and combined-cycle plants of thermal power plants* (Publishing House MEI, Moscow, 2006)
2. N. Galashov, S. Tsibulskii, A. Matveev, V. Masjuk, EPJ Web Conf. **110,** 01019 (2016)
3. Galashov, S. Tsibulskiy, A. Kiselev, MATEC Web Conf. **72,** 01029 (2016)
4. A.I. Gabdullina, N.N. Galashov, S.A. Tsbuskiy, D.V. Melnikov, I.A. Asanov A.S. Kiselev, MATEC Web Conf. **91,** 01004 (2017)
5. P.A. Berezinets, Electr. station, **7**, 64 (1999)
6. O.Yu. Romashova, MATEC Web Conf. **91**, 5 (2017)
7. O.Yu. Romashova, Y.A. Skrebatun, E.Ya. Sokolova, MATEC Web Conf. **19**, 5 (2014)
8. A.G. Kostyuk, V.V. Frolov, *Steam and Gas Turbines for power station* (Publishing house MEI, Moscow, 2008)

Influence of the local gas injection on the flow structure in the boundary layer

Evgeny Maslov[1, 2,*], *Nadezhda* Savkina[1], *Irina* Zharova[1], *Valery* Faraponov[1], *Evgeny* Kozlov[1], and *Vladislav* Matskevich[1]

[1]Tomsk State University, 634050 Tomsk, Russia
[2]National Research Tomsk Polytechnic University, 634050 Tomsk, Russia

Abstract. The results of an experimental and numerical study of the interaction of supersonic flow around an axisymmetric model, M = (2 ÷ 5), with a jet of gas injected locally into the boundary layer are presented.

1 Introduction

The maneuverability of the supersonic aircraft on the flight trajectory can be realized by means of local gas injection from the surface into the boundary layer. Thus, at the drawing-board stage it becomes necessary to solve the problem of determining the main parameters and flow structure in the boundary layer depending on the blowing conditions.

The aim of this paper is an experimental-theoretical study of the influence of the local gas injection into a boundary layer under the flow around an axisymmetric model in the range of Mach numbers M = (2 ÷ 5).

2 Methodology and results of experimental studies

Experiments to determine the drag coefficient when supersonic air flow around an axisymmetric body were carried out on a pulsed aerodynamic installation [1]. The range of realizable flow modes: the number M = (2 ÷ 5); the stagnation pressure at the diffuser nozzle section forming the supersonic flow, $P_0 = (0.15 \div 0.3)$ MPa at static pressure $P = (0.03 \div 0.07)$ MPa; the stagnation temperature of the incoming air is $T_0 = (17.5 \div 250)$ °C. The durability of the pulsed aerodynamic installation, depending on the test conditions, is (1.0 ÷ 3.0) s.

The models in the form of a cone with a half angle of 15 ° were subjected to tests on the aerodynamic installation (Fig. 1, a). The blowing of gas into the boundary layer was carried out through sampling holes. The general view of the model is shown in Fig. 1, b. In Fig. 1, c the photo of the model installed in the test section of the aerodynamic installation is presented.

* Corresponding author: maslov_eugene@mail.ru

DOI: 10.1051/matecconf/201711001055

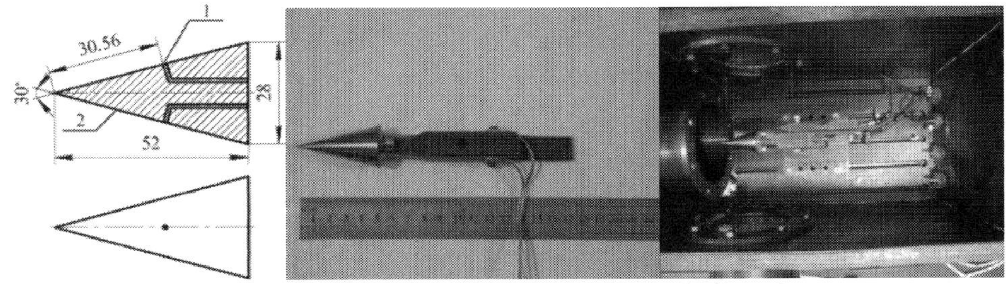

a b c

Fig. 1. Scheme (a), general view (b) and photo of the model installed in the test section of the aerodynamic installation (in): 1 – sampling hole; 2 – the surface of the model.

To determine the value of the drag coefficient in experiments using the tensometric balance, the force acting on the surface of the model when supersonic air flow around model was continuously measured in the range of Mach numbers M = (2 ÷ 5). As a control-test in experimental studies, the results of tests for flow around a cone without injection in the considered range of Mach numbers were used. Fig. 2 shows the results of measurements and the data presented in [3, 4]. Comparison of the obtained results with the data of other authors showed satisfactory agreement – the maximum difference was < 15 %.

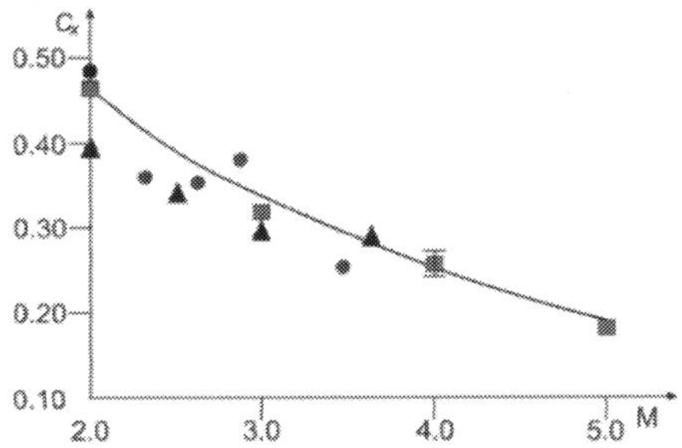

Fig. 2. Drag coefficient of a cone: ■ – results of experiments; ● – data [3]; ▲ – data [4].

In table. 1 the value of the drag coefficient of the cone averaged over the series of experiments results at M = 3 is presented.

Table 1. The drag coefficient, M = 3.

Mach number	C_x	$\overline{C}_x$	ΔC_x	δ, %
3	0.326 0.325 0.322	0.324	0.006	1.85
				δ_{mean} = 3.64 %

The structure and dynamics of the flow, the characteristic features of the boundary layer formation and the interaction of flows in the boundary layer in the presence of blowing were obtained by the method of shadow visualization. In Fig. 3 video frames of flow around the cone by an air flow (M = 3) with injection through the hole located at the bottom of the model (Fig. 3, a) and without injection (Fig. 3, b) are demonstrated.

DOI: 10.1051/matecconf/201711001055

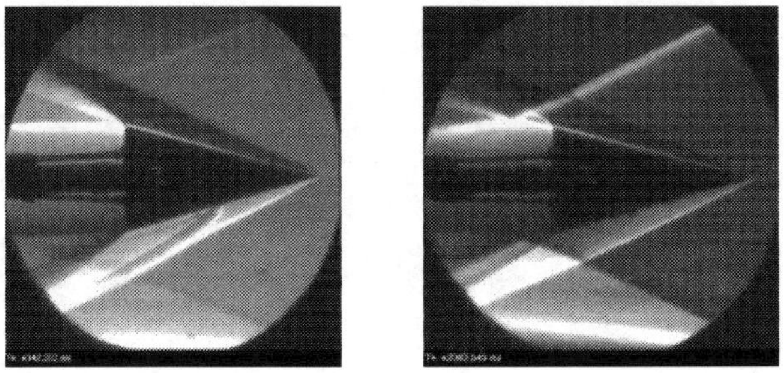

a b

Fig. 3. Flow around the cone by an air flow (M = 3) with injection (a) and without injection (b).

Comparison of the patterns of flow around the cone without injection and during local blowing through one hole showed a qualitative difference between the interacting supersonic air flow and the surface of the cone. When flowing over a cone with a local blowing through one hole, the angles between the Mach cone and the model surface are different. When a local blowing through one hole, the Mach cone is asymmetric.

3 Numerical study of supersonic flow around an axisymmetric model under local gas injection

The temperature and pressure fields for supersonic flow around a cone with allowance for local gas injection into the boundary layer were determined based on numerical simulation of the process. Thermogasdynamic flow characteristics were obtained using the Ansys Fluent software package. The flow of a continuous medium (gas) was modeled by the Navier-Stokes system for turbulent flow of a viscous compressible gas. As a model of turbulence, the SST-model (Menter's model of shear stresses) was used, based on the vortex viscosity hypothesis. Numerical implementation of the solution was carried out on an unstructured grid with different grid resolution (Fig. 4). The grid was clustered near the body surface.

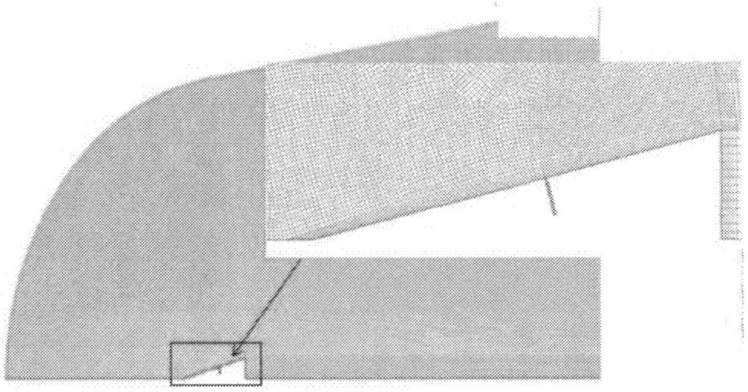

Fig. 4. Grid domain of the problem solution.

The number of grid nodes was chosen from the experimentally determined value of the drag coefficient C_x.

As an example, in Table. 2 the data of the grid convergence (the number of grid nodes) in the numerical study of the supersonic flow around a cone (M = 3) is given, $\delta C_x = \frac{\left|C_x^{calc} - \bar{C}_x\right|}{\bar{C}_x}$ – relative error of the calculated value of the drag coefficient, $\bar{C}_x = 0.324$ (Table. 2).

Table 2. Grid convergence in supersonic flow around a cone, M = 3.

Cells	C_x^{calc}	δC_x
92 532	0.3189	0.0157
138 798	0.3212	0.0086
208 197	0.3239	0.0003
320 500	0.3241	0.0003

On the basis of numerical studies, data of the temperature and pressure distributions when the supersonic flow around the cone were obtained, M = (2 ÷ 5), with considering the local gas injection into the boundary layer.

In Fig. 5 a graph of the temperature distribution near the cone surface when local blowing is presented, M = 3, and the angle of attack is α=0 °. An analysis of the obtained results showed the effect of injection on the temperature distribution in the boundary layer. The temperature distribution is essentially nonmonotonic: the highest temperature – from the top of the cone to the sampling hole, a sharp decrease of temperature at the injection segment, a monotonic increase of the temperature from the sampling hole to the base of the cone.

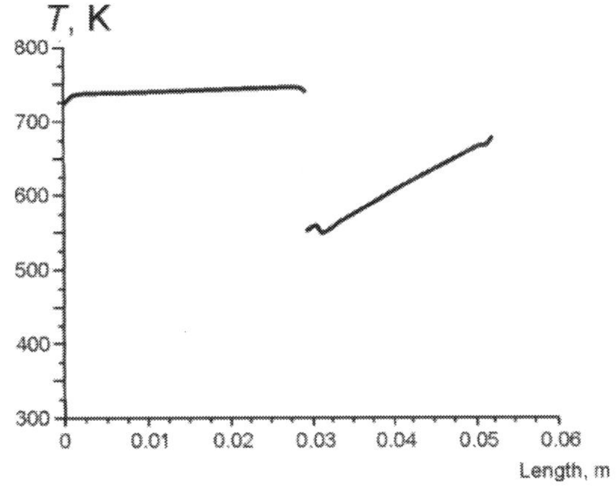

Fig. 5. Temperature distribution near the surface of the cone in the presence of local injection.

Fig. 6 shows the data of the pressure field obtained numerically (Fig. 6, a, c), and the flow structure obtained experimentally (Fig. 6, b). The data presented correspond to the flow around a cone with blowing through two symmetrically located holes at M = 3.

The analysis of the experimental and numerical results showed their conformity with the main features of the flow structure in the considered conditions: the formation of the Mach cone, the dynamic perturbation of the flow in the field of injection, the formation of detached flow from the surface at the base of the cone.

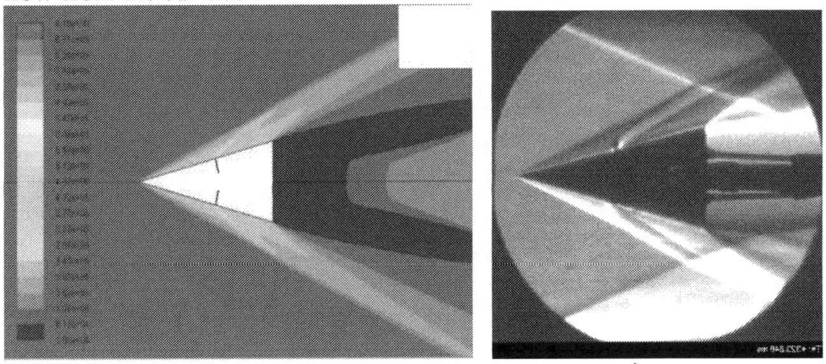

a b

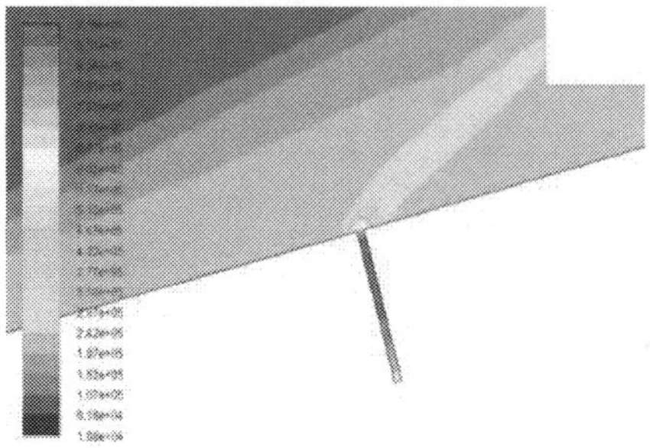

c

Fig. 6. The pressure distribution (a, c) and the flow structure (b) for flow around a cone with blowing through two symmetrically located holes, M =3.

4 Conclusions

Based on the results of an experimental-theoretical study of the effect of local gas injection into the boundary layer when the flow around an axisymmetric model in the range of Mach numbers M = (2 ÷ 5), the main thermogasdynamic flow characteristics are determined.

A qualitative analysis of the results obtained when flow around a cone with and without injection showed a significant influence of the injection on the dynamics of the boundary layer and the Mach cone formation.

This research carried out in 2017 was supported by «The Tomsk State University Academic D.I. Mendeleev Fund Program», grant N 8.2.05.2017.

References

1. E.A. Maslov, V.V. Klochikhin, I.K. Zharova, MATEC Web Conf. **23**, 01026 (2015)
2. E.A. Maslov, V.V. Faraponov, N.N. Zolotorev, A.V. Chupashev, V.V. Matskevich, S.Yu. Chizhov, MATEC Web Conf. **92**, 01056 (2017)
3. V.G. Artonkin, P.G. Leutin, K.P. Petrov, Scientific works CAI, **92**, 1413 (1972)
4. Yu.V. Sheludko, *Physical and gas-dynamic ballistic studies* (Science, Leningrad, 1980) [in Russian]

Investigation of the effect of a gap between the cylindrical substrate and curvilinear ring on the regimes of liquid film flow

Anton V. Meleshkin[1], *Valery* V. Ovchinnikov[1,*], and *Ruslan* A. Dekhtyar[1]

[1]Kutateladze Institute of Thermophysics, 630090 Novosibirsk, Russia

Abstract. Experimental studies have been carried out to reveal the peculiarities of the influence of the arrangement of the curved ring on the flow regimes of a liquid film along a cylindrical surface. Regimes under which the estimated thickness of the gap between the flowing film and the cylinder may not be sufficient. Recommendations are given to determine the minimum value of the distance between the ring and the surface along which the film flows, so that at any flow of liquid there is no contact between the liquid and the ring.

Introduction

The phenomenon of physical absorption, accompanied by the release of a large amount of heat at water vapor absorption in used in the absorption heat exchangers. The use of liquid films in absorbers, condensers, and heat exchangers makes the studies of film flow structure and effect of this structure on local and average heat and mass transfer characteristics relevant [1-9]. Intensification of heat and mass transfer processes occurs both because of an increase in the film area and due to a change in the velocity profile in the film [10, 11]. To increase the efficiency of film absorbers, it is very important to organize the uniform flow of absorbent over the heat-and-mass transfer surface without formation of dry spots and absorbent rivulets.

The aim of this work is to reveal the peculiarities of the effect of curvilinear ring arrangement on the regimes of liquid film flow along the cylindrical surface, and determine the minimal distance between the ring and the surface, where the film flows so that at any liquid flow rates there is no contact between liquid and ring.

Experimental setup and methods

Experiments were carried out at a setup, whose scheme is shown in Fig. 1a. When a circular jet from a nozzle with the diameter of 10 mm flowed onto a cylinder of 40 mm in diameter, a liquid film was formed. The distance from the nozzle cut to the cylinder surface was varied from 1 to 10 mm. Changing the height of the nozzle cutoff location, we could

* Corresponding author: mav@itp.nsc.ru

DOI: 10.1051/matecconf/201711001056

change the rate of jet flow onto the cylinder, at the same flow rate of liquid. The flow rate was set by means of a needle valve on the bypass line. The flow rate was previously calibrated by the rate of change in the mass of liquid in the control volume. The curvilinear ring was made of a copper wire with diameter D_H = 2 mm and its shape was close to the shape of the curve formed, when two cylinders of the same diameter D_B, equal to $D_B = D + 2 \cdot h + D_H$ = 46 mm, cross (Fig. 1b). The patterns of the film flow on the cylinder were recorded by the Nikon J1 camera with the frame rate from 60 to 1000 fps. The ring was mounted to make gap h one and half times higher than the maximal film thickness in the studied range of flow rates and initial velocities of water in the jet (Fig. 1c). Film thickness δ was determined for a liquid film, freely flowing along the cylinder (without a ring).

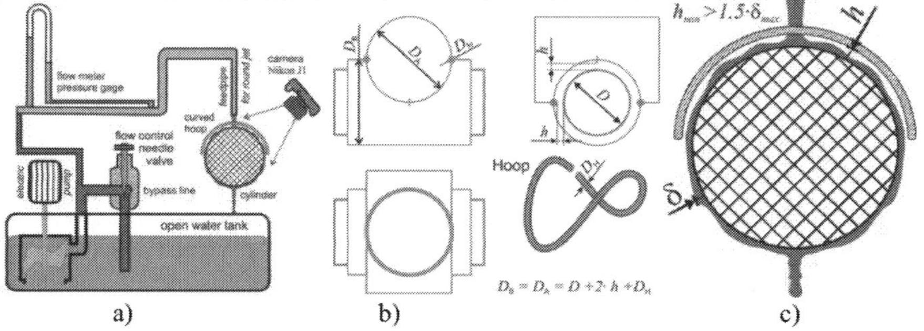

a) b) c)

Fig. 1. Scheme of setup a); ring shape b); location of ring relative to the cylinder c).

There was a series of experiments with a change in the flow rate in the jet G from 4 to 70 g/s (the velocity of jet flow onto the cylinder $V_0 = 0.05 \div 1.5$ m/s). Usually the value of a minimal gap is the sum of the film thickness and maximal wave amplitude on the film surface. With a smooth change in the flow rate from the minimum to the maximum for a set gap between the ring and cylinder, the film flow regime did not change and corresponded to the regime of a freely flowing liquid film. However, the experiments have shown that at a sharp pulsation of the liquid flow (the pulsation time is less than 2s, more than a 50% decrease in the flow rate), a qualitative change in the liquid film pattern is possible. This is because during the transition regime, a liquid bridge is formed between the ring and liquid film (liquid bridge) and this structure is stable for a long time. An example of the liquid bridge formation and resulting change in the pattern of a liquid film flow, are shown in Fig. 2. Time reading starts from the beginning of the flow rate restoration (growth) in the incoming jet. The form and conditions for the existence of a liquid bridge between the spheres are theoretically described in [12]. The experimental results and approximation formula describing the liquid bridge shape, are given in [13]. The liquid bridge height is comparable to the Laplace constant.

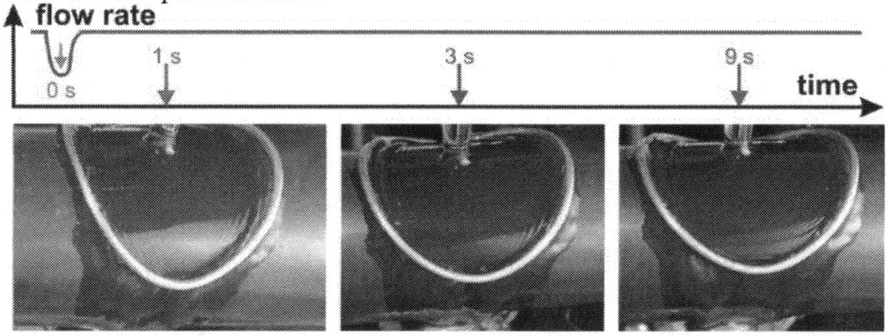

Fig. 2. Formation of a liquid bridge between the ring and liquid film after sharp flow pulsation.

After liquid bridge formation, the character of the film flow on the cylinder starts changing. In particular, the wave patterns on the film surface vary greatly within a region limited by a curved ring. Outside this region, formation of the rolls on the film surface is observed downstream (Fig. 2). As a sequence, after formation of a liquid bridge between the ring and the liquid film, dependence between the film flow regimes on the flow rate in the incident jet changes as compared with the liquid film flowing freely along the cylinder. An example of formation of the rivulet flow regime (49s) at a smooth change in the flow rate after liquid bridge formation (10s) is shown in Fig. 3.

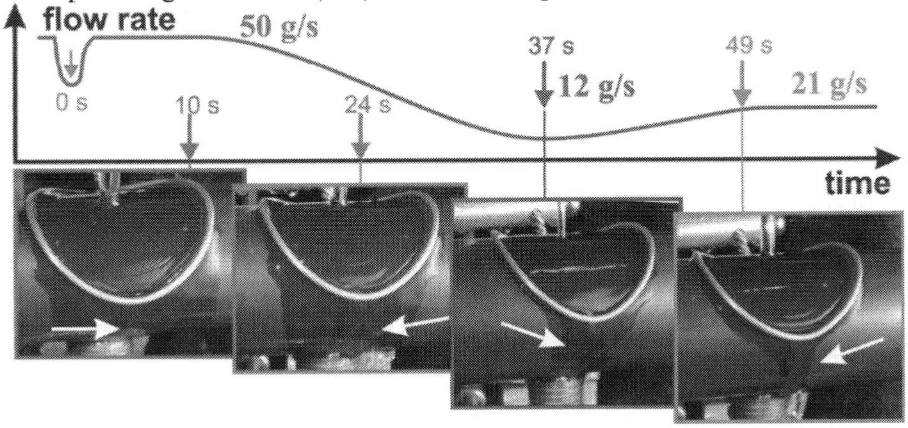

Fig. 3. Formation of a rivulet in the lower part of the cylinder after liquid bridge formation.

Conclusion

The study showed that the minimal distance at which the liquid film does not touch the outer ring is determined not only by the film thickness, but also by the height of a liquid bridge that can exist between the ring and the film. The effect of the ring on the film flow regime is determined by the fact that the ring plays the role of an indicator of nonlinear outbursts of perturbation height on the film surface. This occurs because the outbursts, short-lived on the ring surface due to the influence of wetting forces, can transform into the stable liquid elements such as liquid bridges. The existence of such structures leads to a change in the film flow patterns both inside the ring region and beyond this region downstream. As a sequence, when the liquid flow rates in the flowing jet in the lower part of the cylinder change, the film flow regime can transform into the rivulet flow.

The study was supported by the grant of the Russian Science Foundation (project No. 15-19-10025).

References

1. V.E. Nakoryakov, N.S. Bufetov, R.A. Dekhtyar', J. Appl. Mech. Tech. Phys. **45** (2), 276 (2004)
2. N.S. Bufetov, R.A. Dekhtyar, N.I. Grigoryeva, V.E. Nakoryakov, J. Appl. Mech. Tech. Phys. **44** (2), 111 (2003)
3. N.S. Bufetov, R.A. Dekhtyar, N.I. Grigoryeva, V.E. Nakoryakov, I.V. Marchuk, Int. J. Heat Mass Transfer, **47** (6-7), 1525 (2004)
4. D.V. Zaitsev, O.A. Kabov, V.V. Cheverda, N.S. Bufetov, High Temp. **42** (3), 450 (2004)
5. V.E. Dontsov, A.A. Chernov, Int. J. Heat Mass Transfer, **52** (21-22), 4919 (2009)

6. A.A. Chernov, V.E. Dontsov, Int. J. Heat Mass Transfer, **54** (19-20), 4307 (2011)
7. A.A. Chernov, A.A. Pil'nik, D.S. Elistratov, I.V. Mezentsev, A.V. Meleshkin, M.V. Bartashevich, M.G. Vlasenko, Scientific Reports, **7**, 40809 (2017)
8. A.A. Chernov, D.S. Elistratov, I.V. Mezentsev, A.V. Meleshkin, A.A. Pil'nik, Int. J. Heat Mass Transfer, **108**, 1320 (2017)
9. N.S. Bufetov, R.A. Dekhtyar', Polzunovsky vestnik, **1**, 53 (2004)
10. S.V. Alekseenko, V.E. Nakoryakov, B.G. Pocusaev, *Wave Flow of Liquid Films* (Begell House, New York, 1994)
11. V.E. Nakoryakov, S.Yu. Misyura, S.L. Elistratov, J. Eng. Thermophys. **22** (1), 1 (2013)
12. S.V. Alekseenko, P.A. Kuibin. Proc. STS-XXV. 16 (1996)
13. B.V. Perepelitsa. Book of Abstracts ISHM-II, ISHM-III. 31 (2015)

Spheroid droplets evaporation of water solutions

S.Y. Misyura[1,*], V.S. Morozov[1]

[1]Institute of Thermophysics of SBRA, 630090 Novosibirsk, Russia

Abstract. Droplet film boiling on a horizontal heating surface was studied experimentally. The heat transfer coefficient of droplet water solution in the spheroidal state decreases with a rise of wall overheating and spheroid diameter. Evaporation of small spheroid (diameter $d < 5$ mm) is markedly different from the large one ($d > 20$ mm). At the evaporation of large spheroids a spheroid shape changes in time that significantly affect coefficients of generalizing curves that use dimensionless numbers.

Introduction

The high temperature boiling is widely used in engineering apparatus. There is boiling crisis at high heat fluxes. During nucleate boiling the formed vapor layer reduces the heat transfer by several orders, resulting the wall overheating. The lowest temperature boundary of this regime with stable vapor layer is referred to as the Leidenfrost point (LFP). Evaporation of pure liquids differs significantly from solutions [1-9]. Acceleration of the flow affects the transport mechanisms in the gas mixture [10]. Influence of the wettability and physicochemical wall properties on the droplet evaporation is considered in the works [11]. The boiling crisis in mini channels lead to flow locking [12]. Phase transitions rates strongly depend on both the external and internal local heat fluxs [13-19] and on impurities [20, 21]. Thermal imager measurement procedure is presented in [22].

Experimental data

Heat transfer between a spheroidal droplet and a heated wall is generalized by the dependence of Nusselt number ($Nu = \alpha l/\lambda$) on Prandtl (Pr), Reynolds (Re), Grashof (Gr) numbers in the form of $Nu = cGr^{n_1}\,\mathrm{Pr}^{n_2}\,\mathrm{Re}^{n_3}$ (α-heat transfer coefficient, λ-liquid thermal conductivity).Neglecting liquid circulation in a spheroid, we can simplify the expression for $Nu = f\left(\frac{c_{pv}\Delta T_w}{r}\right)$. If we take the capillary constant as characteristic size l, then we can

* Corresponding author: misura@itp.nsc.ru

obtain $A = c_1(B)^n$, where parameters B and A will be $A = \frac{\alpha}{\lambda_v}\sqrt{\frac{\sigma}{g(\rho_l - \rho_v)}}$, $B = \frac{c_{pv}\Delta T_w}{r}$.

In Fig. 1(a) experimental data on heat transfer for different liquids and mixtures are shown in dependence on wall overheating ΔT_w (the difference between the wall temperature and the saturation temperature). Significant changes in the thermal-physical properties of liquids and ΔT_w lead to variation of α by the factor of 4.

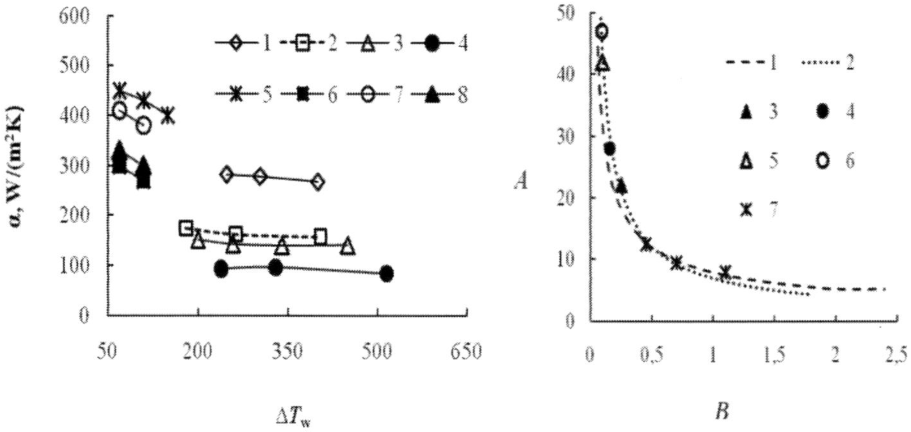

(a) (b)

Fig. 1. a) A change in α for the droplet spheroidal state vs. ΔT_w of different liquids: 1–water; 2–ethanol; 3–CCl_4; 4–C_6H_6; 5–water; 6-8: water solution of ethanol 6-(C_0=92%); 7–(C_0 = 27%); 8–(C_0=64%),
b) A change in α for the droplet spheroidal state in generalizing coordinates A and B: Curve 1 for large spheroids – $A=7.85(B)^{-0.58}$; Curve 2 for small spheroids – $A = 6{,}9(B)^{-0.81}$; 3–water solution of ethanol C_0=92%, ΔT_w=110°C; 4–water solution of ethanol (70%, 110°C); 5–water, 110°C; 6–water, 70 °C; 7–CCl_4, C_6H_6.

In Fig. 1(b) experimental data are generalized by the empirical criteria dependence in the form of

$$A = c_1(B)^n \tag{1}$$

For curve1 in Fig. 1(b) data correspond to large spheroids (diameter d is greater than 20 mm). For curve2 all spheroids were small (diameter d is less than 5 mm). Moreover, α was determined as the average value for the whole period of evaporation, when the diameter decreased from the maximal to zero values. It is characteristic, points 7 were obtained for high values of ΔT_w=200-500°C. Points 3-6 for low values of ΔT_w=60-70 °C, i.e., the data were carried out near the transitional crisis area. Generalizing curve kind varies when approaching the transitional crisis area.

This work was supported by the Russian Science Foundation (grant N 15-19-10025).

References:

1. S.Ya. Misyura, Int. J. Therm. Sci., **92** (2015)
2. G.V. Kuznetsov, D.V. Feoktistov, E.G. Orlova, Thermophys. Aeromech. **23** (2016)
3. G.V. Kuznetsov, D.V. Feoktistov, E.G. Orlova, J. Eng. Phys. Thermophys. **89** (2016)

4. V.E. Nakoryakov, S.Y. Misyura, S.L. Elistratov, J. Eng. Thermophys. **22** (1), (2013)
5. V.E. Nakoryakov, S.Y. Misyura, S.L. Elistratov, J. Eng. Thermophys. **20** (4), (2011)
6. V.E. Nakoryakov, S.Y. Misyura, S.L. Elistratov, Therm. Sci. **16** (4), (2012)
7. V.E. Nakoryakov, S.Y. Misyura, Doklady Physics, **59** (10) (2014)
8. S.Y. Misyura, Int. J. Heat Mass Transfer, **71** (2014)
9. S.Y.Misyura, Appl. Therm. Eng. **113** (25), (2017)
10. V.P. Lebedev, V.V. Lemanov, S.Y. Misyura, V.I. Terekhov, Fluid dynamics, 28 (1993)
11. J. Wang, Scientific Reports, **5**, 8474 (2015)
12. V.E.Nakoryakov, S.Y.Misyura, S.L. Elistratov, R.A. Dekhtyar, J. Eng. Thermophys. **23** (4), (2014).
13. A.K. Rozentsvaig, Ch.S. Strashinskii, Int. J. Heat Mass Transfer, **102**, 555 (2016)
14. A. Pavlenko, V. Zhukov, N. Pecherkin, V. Chekhovich, O. Volodin, A. Shilkin, C. Grossmann, AIChE J. **60** (2), (2014)
15. S.Y.Misyura, Scientific Reports, **6**, 30324 (2016)
16. A.A. Chernov, Int. J. Heat Mass Transfer, **52** (2009)
17. V.E.Nakoryakov, S.Y.Misyura, S.L. Elistratov, J. Eng. Thermophys. **22** (3), (2013)
18. S.Y. Misyura, Energy, **103** (2016)
19. V.E. Nakoryakov, S.Y. Misyura, Energy Fuels, **27** (2013)
20. R.S. Volkov, G.V. Kuznetsov, P.A. Strizhak, Int. J. Heat Mass Transfer, **96** (2016)
21. R. Iegorov, T. Valliulin, P. Strizhak, Appl. Therm. Eng. 106 (2016)
22. S.L. Elistratov, O.V. Vitovskii, E.Yu. Slesareva, J. Eng. Thermophys. **24** (1), (2015)

DOI: 10.1051/matecconf/201711001058

Study on thermal oxidation and combustion of aluminum ultrafine powder in high-energy material

Alexander G. Korotkikh[1,2,*], *Konstantin* V. Slyusarskiy[1], *Konstantin* A. Monogarov[1], and *Ekaterina* A. Selikhova[1]

[1]National Research Tomsk Polytechnic University, 634050 Tomsk, Russia
[2]Tomsk State University, 634050 Tomsk, Russia

Abstract. Aluminum is a promising metal fuel for solid propellants due to the high heat of combustion and propulsion specific impulse values. In current study the results of aluminum-containing high-energy material oxidation and combustion processes were received by means of the thermogravimetry and a constant pressure bomb. Ammonium perchlorate-based high-energy materials with aluminium powder with different particles size and genesis (produced by methods of electrical explosion of wire is ultrafine Alex and spraying is micron ASD) were studied. The oxidation process mechanism and activation energy were obtained using Freidman and Kissinger methods. It was found that the average activation energy values for tested HEM samples were in the range of 120–160 kJ/mole and the burning rate for HEM sample with Alex was 2.5–3.0 times higher than HEM sample with ASD-1.

1 Introduction

Last 15 years the intense research on new generation of solid and gel-like high energetic systems with aluminium powders of different dispersity [1, 2]. These research is coordinated array of fundamental and applied studies in sphere of development of perspective solid propellants (SP) based on aluminum ultrafine powder (UFP) [3]. They are expected to allow to create and implement SP with energetic characteristics on principally new level. There are several methods of aluminum UFP production, two of them are the most widespread: method of pneumatic molten metal spraying and electrical explosion [3]. Obtained by these methods powders have different dispersity and particle form. This has significant influence on ignition and combustion process characteristics.

Simultaneous thermal analysis allows to obtain reliable data on oxidation process kinetics with high preciseness. Many presented in literature articles are devoted to analysis of oxidation kinetics in air medium of aluminium powder with different dispersity and genesis [4, 5]. In majority of high energy materials (HEM) the metal powders are used together with solid oxidizer and binder. Combustion kinetics is significantly affected by oxidizer type too. There are few studies of this type in scientific literature and all of them

* Corresponding author: korotkikh@tpu.ru

are limited by quite narrow range of HEMs like in [6], where kinetic analysis of HEM oxidation based on HTPB and aluminium nanopowder in air medium. For propulsion applications the HEM linear combustion rate is also important. To define linear combustion rate the bomb with constant pressure and method of burning wires were used as one of the most precise.

The combustion characteristics of HEM based on ammonium perchlorate and aluminium powders produced by electrical explosion and pneumatic spraying in argon medium via simultaneous thermal analysis and method of burning wires.

2 Materials and methods

2.1 HEM samples

The HEM samples content was two fractions of ammonium perchlorate (AP) (sieve fractions less than 50 μm and 160–315 μm with mass ratio 40/60), butadiene rubber plasticized by transformer oil (19.7 wt. %), and aluminum powder (15.7 wt. %). Metal fuel consists of 85–99 wt. % of active aluminum depending on the powder dispersity. In tested HEM samples we used aluminum different grade: micron powders ASD-4 (HEM with this powder further in text would be referred as ASD-4), ASD-6 (ASD-6), ASD-10 (ASD-10) and obtained by electrical explosion of wire UFP Alex (Alex).

Characteristic sizes of aluminum powder particles were defined using experimental histograms from manufacturer. Physical-chemical properties defining revealed that poured density of produced by pneumatic method aluminum powder with dispersity 0.3–0.5 μm is 0.54 g/sm^3. It is 3.6 times higher compared to UFP Alex and 2.0 times lower than ASD-4 powder. This difference may be connected to particle structure and oxide layer thickness. According to images of electron X-ray microscope and X-ray diffraction analysis of Alex UFP with particle size 0.1 μm its oxide layer has poured structure and with gas content up to 3 wt. %. Dispersed composition of aluminum powders is given in Table 1.

Table 1. The characteristic diameters of aluminum powder particles.

Aluminum powder	d_{10}, μm	d_{20}, μm	d_{30}, μm	d_{32}, μm	d_{43}, μm
ASD-4	1.23	1.66	2.28	4.34	7.34
ASD-6	0.85	1.17	1.60	3.01	4.72
Alex	0.12	0.13	0.14	0.16	0.18

2.2 Thermal analysis

Thermal analysis of samples was carried out via simultaneous TG/DSC/DTA analyzer Netzsch STA 449 F3 Jupiter in the temperature range of 30–1000 °C with three heating rates: 2, 4 and 10 °C/min. Measurements were made with hinge mass 2.5 mg in argon medium with the volumetric flow rate of 10 ml/min. Before each measurement the chamber of analyzer was blown with argon to eliminate remaining air.

2.3 Kinetic analysis

Kinetic analysis of HEM oxidation process was realized according to presented in [7] methodic using isoconversional algorithms of Friedman and Kissinger-Akahira-Sunose.

Differential method of Friedman is based on following equation:

DOI: 10.1051/matecconf/201711001058

$$\ln\left(i\frac{d\alpha}{dT}\right)_{\alpha,i} = \ln\left[f(\alpha)A_\alpha\right] - \frac{E_\alpha}{RT_{\alpha,i}},$$

where α is the extent of conversion; $f(\alpha)$ is a kinetic function; A_α is the frequency factor, wt. %/s; E is the activation energy; J/mole; R is the universal gas constant, J/(mole K); T is the temperature, K; i is the heating rate, K/min.

The activation energy was defined using approximation curve slope at fixed conversion for each heating rate.

Integral method of Kissinger-Akahira-Sunose is based on following approach to solution of temperature integral:

$$\ln\left(\frac{i}{T_{\alpha,i}^2}\right) = Const - \frac{E_\alpha}{RT_{\alpha,i}}.$$

Simultaneous application of both integral and differential methods allows to decrease negative influence of each separate method.

2.4 Constant pressure bomb

The burning rate experiments were carried out using bomb of constant pressure via method of burning wires in argon in the nitrogen pressure range of 1–8 MPa. Cylindrical samples with diameter 10 mm and height 30 mm with fortified side surface. The used aluminum powders content was changed from 0 to 20 wt. %.

3 Results and discussion

3.1 Thermal analysis

Thermal analysis data for tested HEM samples were obtained. Form of TG and DTG curves is given in Fig.1 on example of HEM sample with Alex.

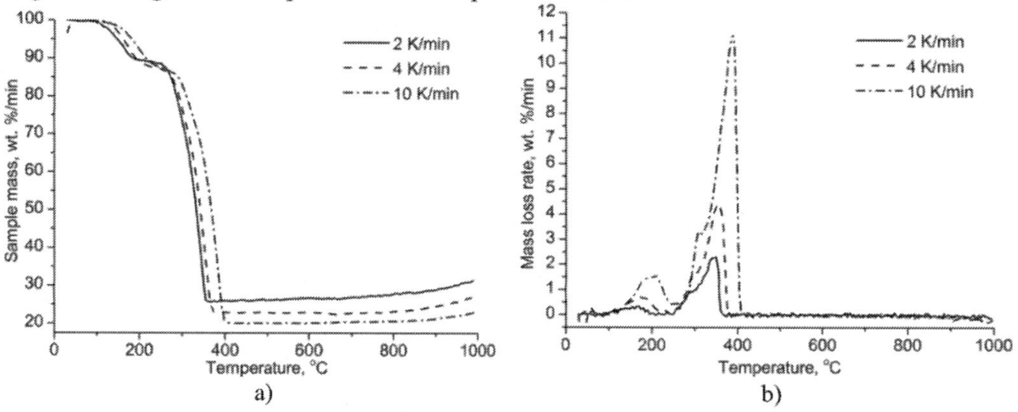

Fig. 1. TG (a) and DTG (b) curves for tested HEM with Alex.

Form of DTG-curves shows that HEM thermal decomposition and aluminum oxidation has two stages. The first stage takes place in the temperature range of 100–250 °C and connected to primary decomposition of oxidizer and binder. Second stage takes place in the temperature range of 250–400 °C and caused by aluminum oxidation with high-temperature

DOI: 10.1051/matecconf/201711001058

decomposition of oxidizer and binder. Additional peak during the second stage at temperatures close to 300 °C may be connected to further decomposition of ammonium perchlorate.

Experimental dependences of extent of conversion for tested HEM samples vs. the temperature at two heating rates values are given in Fig. 2.

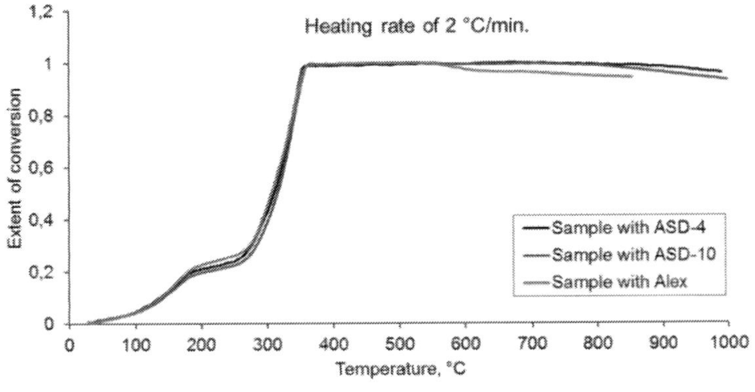

a)

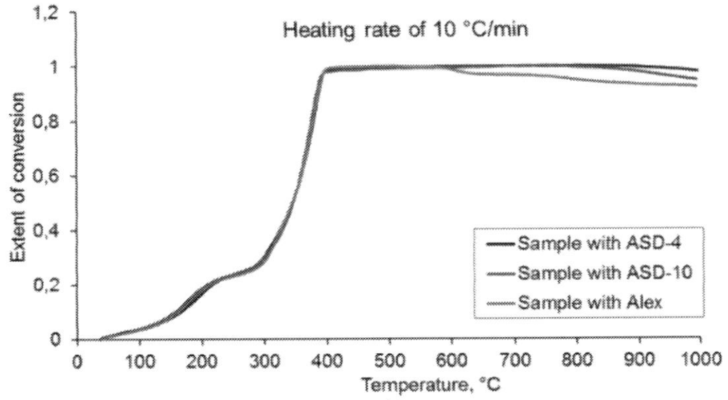

b)

Fig. 2. The HEM samples extent of conversion vs. temperature at heating rate of 2 K/min (a) and 10 K/min (b).

Fig. 2 illustrates two-staged character of HEM oxidation process and it does not changes with heating rate. At the same time the curve position shifts into area of higher temperatures at 30 °C. It results into moving of the first peak from 170 to 200 °C and the second peak from 360 to 390 °C.

Comparison of extent of conversion vs temperature graphs for all samples reveals that at 10 K/min the differences between HEM with different powders are insignificant in the temperature range of 50–400 °C. At 2 °C/min the oxidation rate of Alex HEM sample is higher than for ASD-4 and ASD-10 HEM samples. At the same time the mass gain rate for Alex sample is much higher compared to other samples. It could be concluded that at close to equilibrium conditions of TG-chamber the kinetic of HEM oxidation does not depend on dispersed composition of metal powder.

3.2 Activation energy

The activation energy dependences of oxidation process for tested HEM samples were obtained using two methods (Friedman and Kissinger-Akahira-Sunose) and are presented in Fig. 3.

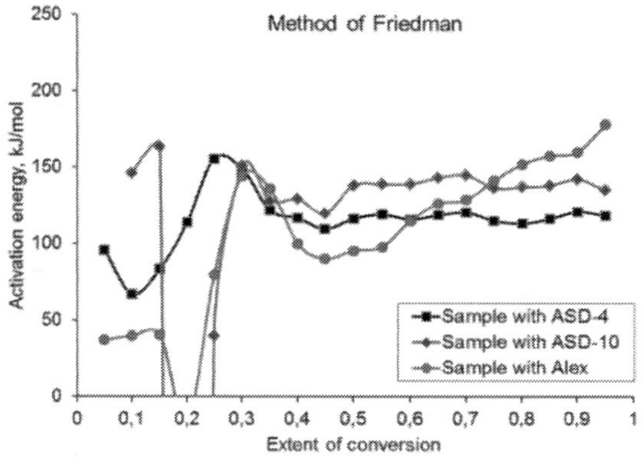

a)

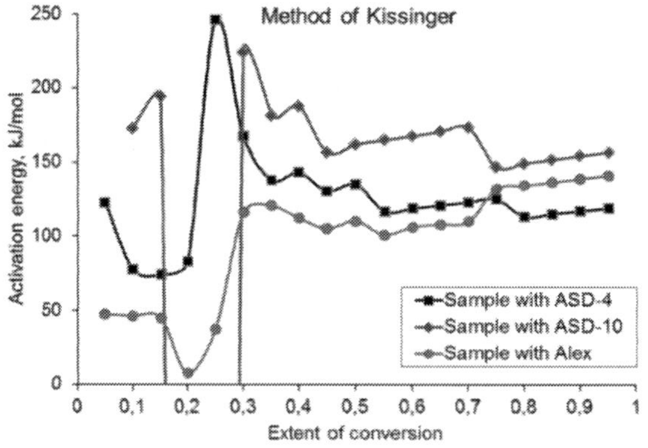

b)

Fig.3. The HEM activation energy vs. extent of conversion by Friedman (a) and KAS (b) methods.

While the activation energy for processes with several parallel reactions described by Friedman and KAS methods with significant error of the activation energy in the first stage varies in the wide range of values (from 0 to 250 kJ/mole) in the conversion range of 0.05–0.35 where organic binding decomposition takes place. In the conversion range of 0.4–0.95 the aluminum oxidation takes place and the activation energy values varies in more narrow range of 120–160 kJ/mole. These values are in good correspondence with literature data [8]. The average activation energy values during HEM oxidation are presented in Table 2.

Table 2. The average activation energy for HEM oxidation process.

Calculation method	HEM sample		
	Alex	ASD-4	ASD-10
Friedman	129	117	137
KAS	120	123	162

In the conversion range of 0.05–0.60 the activation energy of HEM with Alex is lower compared to other samples. At the higher conversion values the activation energy of HEM with Alex increases. It may be caused by interference of secondary aluminum oxidation that is less intense compared to primary oxidation.

3.3 Burning rate

The burning rate values and combustion lows for HEM samples are presented in Fig. 4.

 DOI: 10.1051/matecconf/201711001058

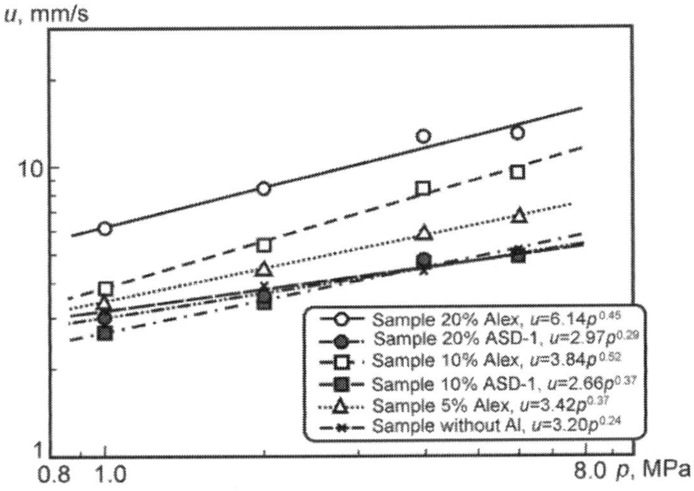

Fig. 4. The burning rate for tested HEM samples with aluminum powder.

The burning rate for tested HEM samples increases with growth of pressure in bomb. The burning rate for HEM sample with ASD-1 powder does not depend on metal content while for HEM with Alex this dependence is linear – lowering of metal content in HEM causes decreasing of the burning rate. It could be seen, that the burning rate for HEM samples with 20 wt. % of Alex UFP is 2.5–3.0 times higher than for HEM with ASD-1 micron powder. With decrease in the metal content in HEM sample this difference decreases as well.

Influence of aluminium powder particle size on HEM combustion rate may be described by thermal theory of combustion and particle oxidation mechanism. If aluminium powder particles quickly ignite and burn near HEM sample surface than overall combustion rate of HEM will increase. To reach it the particle size should be rather small while temperature and composition of combustion products should be advantageous for aluminium particles ignition and combustion. Increasing of combustion rate with addition of aluminium UFP into HEM is connected to increasing of chemical reaction rate due to higher temperature in gas phase.

4 Conclusion

The study on thermal decomposition, oxidation and combustion for high-energy materials based on ammonium perchlorate, butadiene rubber and aluminium ultrafine or micron powders is presented. Thermal analysis data for tested HEM samples were obtained via Netzsch STA 449 F3 Jupiter analyser at the heating rates of 2, 4 and 10 °C/min in argon. The combustion experiment was carried out using the bomb of constant pressure via method of burning wires.

The TG and DTG-curves for HEM oxidation process were obtained as well as the activation energy dependences on extent of conversion. The HEM activation energy value varies in the range of 0–250 kJ/mole at the 0.05–0.35 extent of conversion and in the range of 120–160 kJ/mole at the 0.40-0.95 extent of conversion. The mass loss rate and the activation energy values are close to each other for HEM samples with aluminium powder of different dispersity.

The burning rate tends to increase at growth of ambient pressure and Alex UFP content in HEM sample while the burning rate for sample with ASD-1 micron powder does not change with variation of metal content and close to values of HEM without metal. At equal conditions the burning rate of HEM sample with 20 wt. % Alex are 2.5–3.0 times higher than it with 20 wt. % ASD-1. With decrease of aluminum content the burning rate difference between HEM samples with Alex and ASD powders decreases. It reveals that in unsteady conditions the dispesity and content of aluminum fuel in HEM play important role in the combustion process.

The reported study was supported by Russian Scientific Foundation, Grant 16-19-10316.

References

1. V.A. Arkhipov, A.G. Korotkikh, V.T. Kuznetsov, A.A. Razdobreev, I.A. Evseenko, Russ. J. Phys. Chem. B, **5**, 616 (2011)
2. F. Maggi, A. Bandera, L. Galfetti, L.T. De Luca, T.L. Jackson, Acta Astranaut, **66**, 1563 (2010)
3. Y. Li, C. Hu, Z. Deng, C. Li, H. Sun, Y. Cai, Acta Astranaut, **133**, 455 (2017)
4. A.A. Il'in, R.N. Rumyantsev, V.V. Veisgaim, A.P. Il'in, Russ. J. Phys. Chem. A, **90**, 764 (2016)
5. A.B. Vorozhtsov, M. Lerner, N. Rodkevich, H. Nie, A. Abraham, M. Schoenitz, E.L. Dreizin, Thermochim. Acta, **636**, 48 (2016)
6. X. Lu, M. Zha, Z. Ma, F. Zhao, S. Xu, H. Xu, Combust. Sci. Technol. **189**, 312 (2016)
7. S. Vyazovkin, K. Chrissafis, M.L. Di Lorenzo, N. Koga, M. Pijolat, B. Roduit, N. Sbirrazzuoli, J.J. Suñol, Thermochim. Acta, **590**, 1 (2014)
8. Q.S.M. Kwok, R.C. Fouchard, A-M. Turcotte, P.D. Lightfoot, R. Bowesand, D.E.G. Jones, Propellants Explos. Pyrotech. **27**, 229 (2002)

Evaporation and boiling crisis of droplets alcohol solution

S.Y. Misyura[1,*], *V*.S. Morozov[1]

[1]Institute of Thermophysics of SBRA, 630090 Novosibirsk, Russia

Abstract. Evaporation and boiling crisis of droplets ethanol aqueous solution were studied experimentally. The evaporation intensity depends on the nucleate boiling, solution diffusion, a change in physical characteristics with time and droplet interfacial surface area. At nucleate boiling in a droplet, most evaporation relates to a growth in the droplet surface area and only 20 % relates to the diffusion effect and a variation in the thermophysical coefficients. At boiling crisis, experimental dependence for vapor layer height on overheating was observed. At Leidenfrost temperature, the height of the vapor layer was many times higher than the surface microroughness value of the wall. There are oscillates of liquid-vapor interface, and this increases the transitional temperature range associated with a boiling crisis of droplets.

Introduction

The evaporation of solutions is widely applied in microelectronics, energy apparatuses and chemical technologies. There is boiling crisis at high heat fluxes. The formed vapor layer during the crisis reduces the heat transfer coefficient by several orders, resulting the wall overheating and wall destruction. The lowest temperature boundary of this regime with stable vapor layer (without touching the wall and the liquid) is referred to as the Leidenfrost point (LFP). Predictions for LFP can be obtained on the basis of understanding of boiling key mechanisms. Condensation and evaporation of pure liquids differ significantly from absorption and desorption of salt solutions or phase transition of alcohol solutions [1-9]. At desorption of droplet solution there are salt spot and crystals, which depend on the wettability (droplet contact angle). Influence of the contact angle, wettability and surface wall properties on the droplet evaporation is presented in the work [10]. Vapor film formation and boiling crisis in mini channels lead to flow locking and apparatus breakages [11]. Acceleration of the flow affects the transport mechanisms in the gas mixture [12].

Phase transitions rates (evaporation, crystallization, dissociation) strongly depend on both the external heat flux and internal local heat exchanges [13-19] and on impurities [20-21]. The measurement procedure in non-isothermal flow is presented in [22].

* Corresponding author: misura@itp.nsc.ru

Experimental data

Evaporation curves is shown in Fig. 1. The presentation of the experimental data in Fig. 1(a) is not suitable because there are several factors affecting evaporation: the interface area; diffusivity, evaporation heat, heat transfer coefficient. The droplet wetting surface for alcohol and pure water differ by several times of magnitude. The curves in Fig. 1(b) are given taking into account the interface area. Thus at nucleate boiling in a droplet, the largest part of the evaporation relates to a growth in the droplet surface area and only 20 % relates to the diffusion effect (formation of a high transverse concentration gradient) and a variation in the thermophysical coefficients.

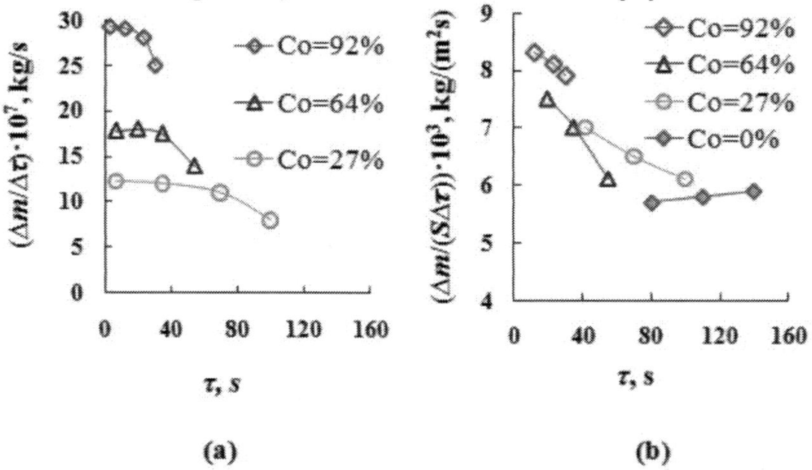

Fig. 1. a) Changing in the droplet mass without taking into account the evaporation area; b) Evaporation droplet rate considering the interface area (C_0 - mass alcohol concentration).

The vapor layer height (at the boiling crisis) δ_v increases with wall overheating ΔT_w and repeatedly exceeds the height of the wall surface microroughness (Fig. 2). Repeatedly increased thermal images and high speed shooting indicate the presence of long-wave oscillations on the droplet interphase surface.

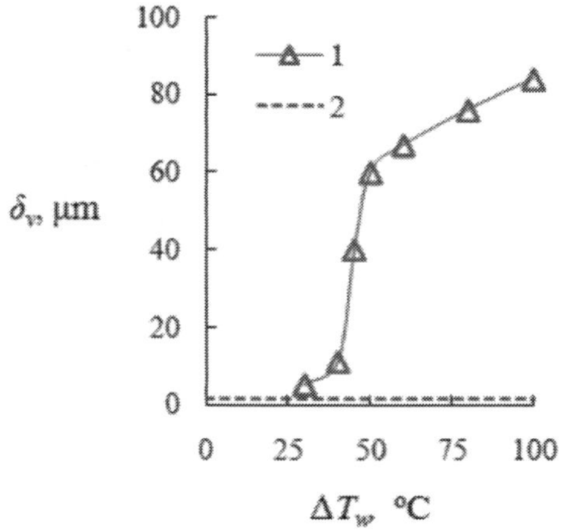

Fig. 2. 1 - Vapor layer thickness δ_v vs. overheating ΔT_w, 2 – maximal microroughness height of the wall surface.

This work was supported by the Russian Science Foundation (grant N 15-19-10025).

References:

1. S.Ya. Misyura, Int. J. Therm. Sci. **92**, (2015).
2. G.V. Kuznetsov, D.V. Feoktistov, E.G. Orlova, Thermophys. Aeromech. **23** (2016)
3. G.V. Kuznetsov, D.V. Feoktistov, E.G. Orlova, J. Eng. Phys. Thermophys. **89** (2016)
4. V.E. Nakoryakov, S.Y. Misyura, S.L. Elistratov, J. Eng. Thermophys. **22** (1), (2013)
5. V.E. Nakoryakov, S.Y. Misyura, S.L. Elistratov, J. Eng. Thermophys.**20** (4), (2011).
6. V.E. Nakoryakov, S.Y. Misyura, S.L. Elistratov, Thermal Science, **16** (4) (2012)
7. V.E. Nakoryakov, S.Y. Misyura, Doklady Physics, **59** (10) (2014)
8. S.Ya. Misyura, Int. J. Heat Mass Transfer, **71** (2014)
9. S.Y.Misyura, Appl. Therm. Eng. **113** (25), (2017)
10. J. Wang, Scientific Reports, **5**, 8474 (2015)
11. V.E. Nakoryakov, S.Y. Misyura, S.L. Elistratov, R.A. Dekhtyar, J. Eng. Thermophys. **23** (4), (2014)
12. V.P. Lebedev, V.V. Lemanov, S.Y. Misyura, V.I. Terekhov, Fluid dynamics, 28 (1993)
13. A.K. Rozentsvaig, Ch.S. Strashinskii, Int. J. Heat Mass Transfer, **102**, 555 (2016)
14. A. Pavlenko, V. Zhukov, N. Pecherkin, V. Chekhovich, O. Volodin, A. Shilkin, C. Grossmann, AIChE J. **60** (2), (2014)
15. S.Y.Misyura, Scientific Reports, **6**, 30324 (2016)
16. A.A. Chernov, Int. J. Heat Mass Transfer, **52** (2009)
17. V.E.Nakoryakov, S.Y.Misyura, S.L. Elistratov, J. Eng. Thermophys. **22** (3), (2013)
18. S.Y. Misyura, Energy, **103** (2016)
19. V.E. Nakoryakov, S.Y. Misyura, Energy Fuels, **27**, (2013)
20. R.S. Volkov, G.V. Kuznetsov, P.A. Strizhak, Int. J. Heat Mass Transfer, **96** (2016)
21. R. Iegorov, T. Valliulin, P. Strizhak, Appl. Therm. Eng. **106** (2016).
22. S.L. Elistratov, O.V. Vitovskii, E.Yu. Slesareva, J. Eng. Thermophys. **24** (1), (2015)

Heat transfer in heated industrial premises with using radiant heating system

Tatiana A. Nagornova[1,*], *Dmitriy* A. Lipchinskiy[1]

[1]National Research Tomsk Polytechnic University, 634050 Tomsk, Russia

Abstract. The results of mathematical modeling of heat transfer processes in a closed air volume surrounded by enclosing constructions, heated by supplying energy to the upper contour of gas infrared radiators are represented. Regimes of turbulent natural conjugate convection in the region bounded by solid walls are investigated. Two-dimensional nonstationary problem is solved in the framework of the Navier -Stokes equations for gas and thermal conductivity for solid walls. Nonstationary processes of heat propagation in course of time and essential heterogeneity of temperature fields and heat fluxes are established.

1 Introduction

In modern industrial production in many cases only a small part of the area used for workshops equipment maintained by the staff. Accordingly, the use of traditional systems of convective heating for the heating of industrial premises is not in most cases economically justified. Expedient use of infrared gas radiators for optimal thermal regime in certain areas large space for production purposes.

Introduction in practice of heat radiation heating systems until recently, severely limited by including in connection with the lack of science-based technologies for supplying heat energy from radioactive sources to heat objects. Thus, in particular, are not carried out any theoretical or experimental investigation of the laws of heat transfer in an area surrounded by special walling. No information about the temperature fields bounded by the outer contour of the air volume with the radiation flux at the upper boundary of the heating region.

The basis of the known methods for calculating thermal regimes objects heated by infrared gas emitters are balanced mathematical models [1]. Thus, as a rule, is not considered as heterogeneity and non-stationary temperature fields of the object. At the same time, these two factors in most real options radiant heating systems can play a decisive role. Informed choice of technology management heating industrial premises is impossible without analysis of temperature fields of the heating region. The purpose of this research is mathematical modeling of heat transfer in a closed air volume heated by supplying energy to the upper contour of gas infrared radiators.

* Corresponding author: tania@tpu.ru

2 Physical and mathematical statement of the problem

The region of a rectangular cross-section is considered. At the upper boundary region is located the infrared source. The intensity of the radiation is uniformly distributed along the transverse coordinate X (Fig. 1).

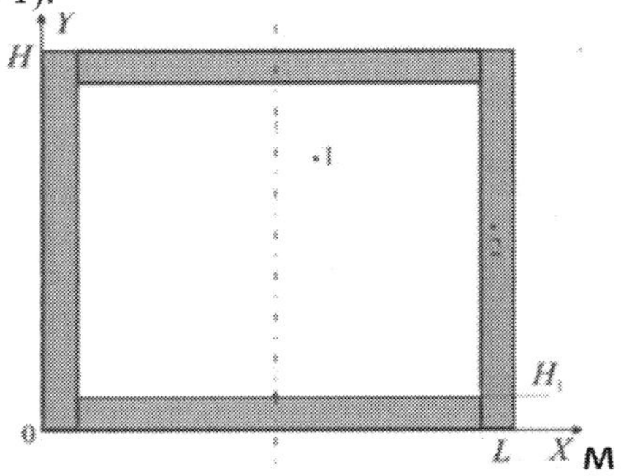

Fig. 1. Area of solution of the problem under consideration: 1 - air; 2 - enclosing structures.

The space inside the considered area is filled with air (1), which is surrounded by heat-conducting walls of finite thickness (2). Source of radiant heating in the top of the decision does not stand - it was assumed that the thickness is small compared with the characteristic dimensions of the field. Problem was considered in the plane formulation for a region with a heating zone, located at the upper boundary. The air was assumed optically transparent medium. The main mechanisms of heat transfer were considered: in the gas turbulent natural convection and thermal conductivity, in the enclosing structures - thermal conductivity. When the problem was assumed that the energy of infrared emitter proceeds to boundary H_1, which divides lower concrete base (2) and the heating region (1). Heat from the gas in the vertical enclosing structure (2) is performed due to turbulent convection and conduction.

As a basic system of differential equations, is accepted mathematical model [2-4] has been developed to describe the processes of convection and heat conduction in a closed rectangular region with heat-conducting walls and a local heating source at the lower boundary and developed later for solving heat transfer problems under radiant heating [5]. The process of heat transfer in the analyzed solution region (Fig. 1) is described by a system of non-stationary two-dimensional partial differential equations [5].

Numerical analysis is performed to change the numbers in the range from 10^6 to Gr 10^{11}. To solve the problem algorithm was used [3-5], developed for the numerical solution of natural convection in a closed rectangular regions. Turbulence in the flow of heated air taken into account in accordance with the basic provisions of the [6-8] change viscosity with increasing air velocity. Calculation of viscosity is carried out within the Prandtl model [6-8].

3 Analysis of numerical simulation results

Numerical studies were carried out at the values of the heat flux arriving at the H_1 surface (Table 1).

Fig. 2–3 shows the results of numerical simulations illustrating the basic laws of heat transfer in the system heating using gas infrared heaters.

Isotherms Fig. 2 *a*, *c*, *e* illustrate the process of cooling the heated air near the floor of the concrete mass at relatively cold reduction along the vertical walls. This results in a symmetrical circulation of heated air for $X = 0.5$ (Fig. 2 *b*, *d*, *f*).

Table. 1. The values of the heat flux over the surface of the concrete base in the section $Y = H_1$ for $L/2 < X < 0.875L$ (Fig. 1).

Сечение	X=0.5L	X=0.563L	X=0.685L	X=0.688L	X=0.75L	X=0.813 L	X=0.875 L
q, W/m^2	190.5	180.2	182.1	172.2	134.2	122.2	102.1

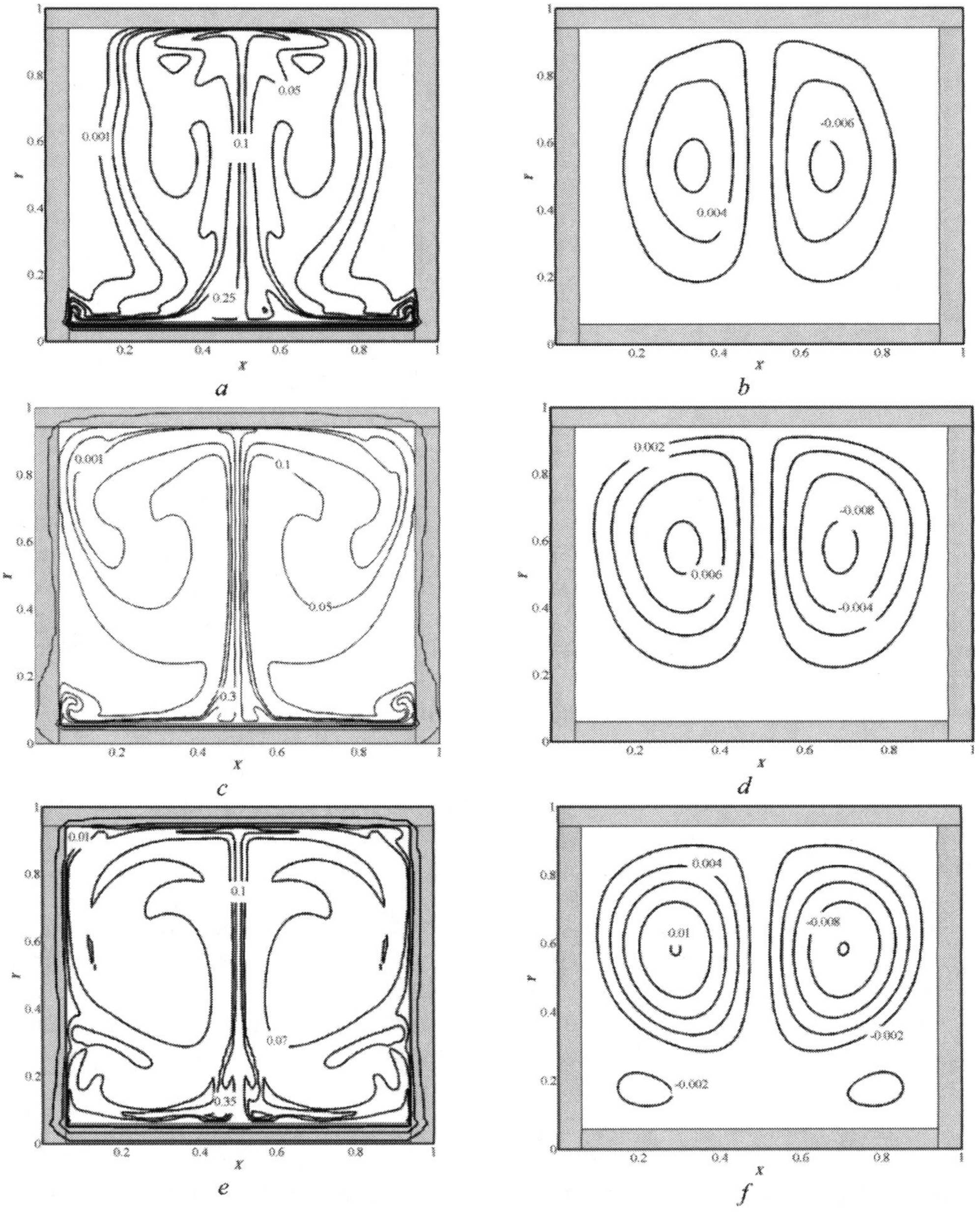

Fig. 2. Fields of temperatures (*a*, *c*, *e*) and the contours of the stream function (*b*, *d*, *f*) for the model of natural convection in a closed area with radiative heating of the lower boundary "gas - solid wall" for different values of τ: $\tau = 70$ (*a*, *b*), $\tau = 120$ (*c*, *d*), □$\tau = 150$ (*e*, *f*)

Figures 2 and 3 well illustrate the nonstationary character of the heat transfer process in the considered rather typical solution region. Well as a comparison of the provisions specific isotherms in Fig. 2 *a*, *c*, *e* and temperature distributions (Fig. 3) allows you to make

 DOI: 10.1051/matecconf/201711001060

a conclusion about the high values of the temperature gradients, both in height and in the transverse direction. Based on the above we can conclude essentially stationary in time and inhomogeneous fields of temperature and heat flow in the heating gas infrared emitters.

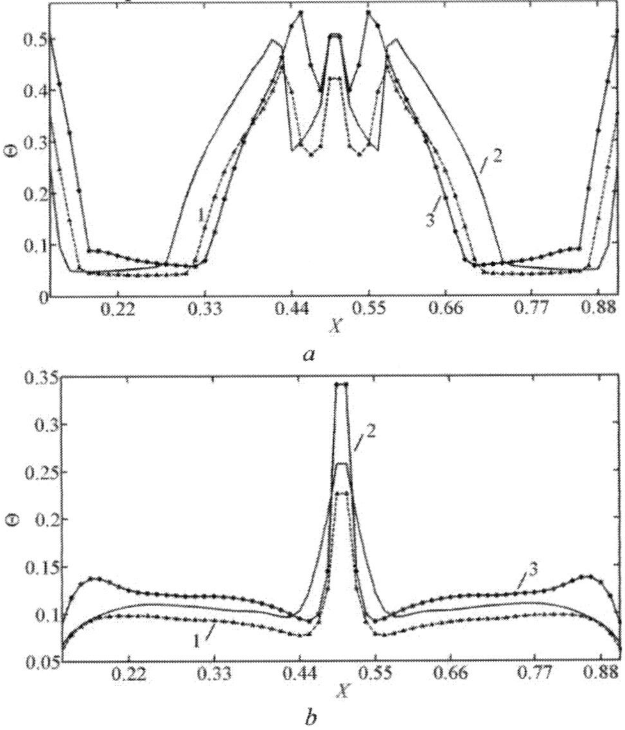

Fig. 3. Temperature profiles in the cross section $Y = 0.1$ (*a*) and $Y = 0.5$ (*b*) at different time points: 1 – $\tau = 70$; 2 -$\tau = 120$; 3 – $\tau = 150$.

The regularities are due to the specifics of heat transfer processes in the production premises, the height of which is usually 4 – 5 m or more. With such large volumes of heated air and cooling surface area (building envelope), coming from the surface of a heated concrete floor intensively cooled air as it rises to the top of the heated space. Heatsink on vertical surfaces intensifies circulation flow, increasing the rate of removal of accumulated in a layer of concrete floor heat, on the one hand. But, from other side in this case the substantial part of the energy is expended on heating of the vertical enclosing constructions.

It may be noted that the distributions (Fig. 3) generally correspond fairly well established experimentally [11] in a similar temperature dependence on the transverse coordinate in the typical cross-section.

The results obtained are objective grounds for withdrawal of the need to meet the challenges of turbulent natural convection and heat conduction within at least flat models in the evaluation of the effectiveness of local radiation heating the working areas of industrial premises.

The solution of problems of free convection is advantageously carried out in the combined formulation based on the heat sink in the building envelope and accumulation in them of energy. Balanced type model [1] thus can serve as a tool for the lower estimates of the energy required for the effective use of radiant heating local jobs.

It may be noted that in spite of the low thermal conductivity of most traditional building materials used for the production of walling (brick, concrete, reinforced concrete), the characteristic times of the processes involved are so large (up to 10 hours), high heat capacity materials that the walls and floor leads to the accumulation of too much fraction of the energy dissipated in the heating. These losses are likely, can be significantly reduced by using heat-insulating coatings on vertical surfaces walling industrial premises.

Conclusion

According to the results of solving the above problem of heat transfer in the local object of heat supply, heated using gas infrared radiators can conclude substantial heterogeneity and non-stationary temperature fields in the region heated by turbulent convection and conduction. The theoretical investigation also constitute grounds for withdrawal of the feasibility solutions conjugate heat transfer problems in the analysis of thermal conditions of local heating objects using radiant heating systems.

The work was supported by the Russian President's grant (Scientific School project - 7538.2016.8).

References

1. G. Gong, J. Liu, X. Mei, Energy Buildings, **138**, 193 (2017)
2. G.V. Kuznetsov, M.A. Sheremet, Fluid Dynamics, **41** (6), 881 (2006)
3. G.V. Kuznetsov, M.A. Sheremet, Comput. Therm. Sci. **1** (3), 341 (2009)
4. V.I. Maksimov, T.A. Nagornova, I.A. Shestakov, EPJ Web Conf. **82**, 01048 (2015)
5. G.V. Kuznetsov, N.I. Kurilenko, V.I. Maksimov, G.Ya. Mamontov, T.A. Nagornova, J. Eng. Phys. Thermophys. **86** (3), 519 (2013)
6. D.S-K. Ting, *Basics of engineering turbulence* (Academic Press, 2016)
7. T. Cebeci, *Analysis of Turbulent Flows with Computer Programs* (Elsevier Ltd., Butterworth-Heinemann, 2013)
8. J. Blazek, *Computational fluid dynamics. Principles and Applications* (Elsevier Ltd., Butterworth-Heinemann, 2015)

MATEC Web of Conferences **110**, 01061 (2017) DOI: 10.1051/matecconf/201711001061
HMTTSC-2017

Modeling of agglomerates formation and evolution at combustion of aluminized propellants in intra-chamber environments

Valery A. Babuk[1,*], *Alexander* A. Nizyaev[1]

[1]Baltic State Technical University "VOENMEH", 190000 Saint Petersburg, Russia

Abstract. In the present paper, the solution of the problem of agglomerates parameters calculation in combustion chamber environments is described. For this, a method of two-dimensional axisymmetric multiphase quasistationary flow calculation has been developed. The method uses previously developed models of agglomerates formation and their evolution within multiphase flow with considering of interactions between agglomerates and gaseous combustion products. Parametric study of the developed models with respect to intra-chamber environments has been performed.

1 Introduction

Agglomeration (enlargement of condensed combustion products within surface layer of burning propellant) is one of phenomena at aluminized propellant burning. The resulting agglomerates can significantly exceed in size initial Al particles. Agglomerates consist of Al, Al_2O_3 and have a complex structure including metal and oxide drops as well as gaseous cavities. Agglomerates represent coarse fraction of condensed combustion products. Another, fine fraction of condensed combustion products is represented by *smoke oxide particles* (SOP).

Within a multiphase flow, the *evolution of agglomerates* is carried out resulting in formation of final composition of combustion products. The evolution is understood as a set of physical and chemical phenomena that lead to changing of sizes, chemical composition, structure and mass fraction of agglomerates. As a result of the evolution, the coarse fraction of final condensed combustion products is formed.

The main features of agglomerates evolution in combustion chamber are multidimensional nature of multiphase flow, distributed inflow of agglomerates and gaseous products, mutual influence of agglomerates evolution and properties of gas flow.

Presence of agglomerates has significant influence on processes associated with slag accumulation, thermal and chemical erosion, and pressure oscillations. In that regard, the problem of numerical simulation of agglomerates formation and evolution is relevant. This led to appearance of a large number of works dedicated to modeling of agglomerates

* Corresponding author: babuk@peterlink.ru

DOI: 10.1051/matecconf/201711001061

formation [1]. A significant number of works is devoted to numerical simulations of multiphase flows containing agglomerates, for example works [2-4] and others.

In the present work, previously developed models of agglomerates formation [5, 6] and their evolution [7] are used for calculation of agglomerates parameters. The main feature of these models is the ability to calculate not only sizes and mass fraction of agglomerates but also chemical composition and structure parameters. Below the developed method for usage of the models in intra-chamber conditions is presented with attention focused on multidimensional nature of flow and influence of agglomerates evolution on properties of gaseous phase.

2 Calculation of agglomerates parameters

2.1 Agglomerates parameters on surface of burning propellant

The agglomeration model [5, 6] can be used to determine the properties of emerging agglomerates at burning propellant surface. The model is based on considering of agglomeration regularities depending on propellant type, modeling of initial propellant structure, and modeling of evolution of enlarging agglomerating particles on propellant surface. The model allows determining the following main agglomerates characteristics:

- $f_m(D)$ – mass function of size distribution density of agglomerates;
- D_{43} – mass-medium size of agglomerates, μm;
- Z_m^a – fraction of initial metal used to form agglomerates;
- η – mass fraction of Al_2O_3 in agglomerates;
- Parameters of agglomerates structure.

Input data of the model are the type and mass fraction of propellant components, sizes of oxidizer particles in propellant, operating conditions (pressure, acceleration).

2.2 Agglomerates parameters within multiphase flow of combustion products

The model of agglomerates evolution [7] can be used for calculating of agglomerates parameters within multiphase flow. The model is based on the description of following phenomena:

- Burning of Al in gas-phase mode;
- Chemical interaction between condensed Al and Al_2O_3 resulting in formation of gaseous products;
- Agglomerates structure changes;
- Coagulation of agglomerates and SOP;
- Agglomerates movement within the flow.

Agglomerates evolution simulation consists in integrating of the ordinary differential equations describing movement and changes in parameters of single agglomerates at time. The model allows determining the changes in agglomerates parameters from the moment of agglomerates detachment from propellant surface until they hit the chamber walls or enter into a nozzle.

2.3 Agglomerates evolution modeling in intra-chamber environments

The gaseous phase and SOP cloud are considered as a continuum with parameters determined using Eulerian approach while agglomerates parameters are described using discrete trajectories according to Lagrangian approach. Such method was used in works

[3, 4]. As the input data the geometric parameters of combustion chamber and grain, propellant composition and burning rate, and agglomerates initial parameters are used.

2.3.1 Modeling of gaseous phase flow

Numerical simulation of gaseous flow is carried out by solving the Navier-Stokes and continuity equations of uncompressible viscous fluid in stationary conditions. Gas properties are assumed to be a constant. The flow is assumed to be laminar and axisymmetric. Boundary conditions are determined by geometric parameters of calculation domain, mass flux of gas supply, and pressure in outlet region. Mass flux (kg/m^2/s) of gas supply at propellant surface is determined by gasification of propellant components:

$$m_{\mathrm{g}} = u_{\mathrm{p}}\left\{1 - z_{\mathrm{m}}^{\mathrm{p}}\left[\frac{2\mu_{\mathrm{Al}} + 3\mu_{\mathrm{O}}}{2\mu_{\mathrm{Al}}}\left(1 - Z_{\mathrm{m}}\right) + Z_{\mathrm{m}}\right]\right\} \Big/ \sum_{n} \frac{z_n}{\rho_n}, \quad (1)$$

where u_{p} – linear burning rate of propellant grain, m/s; $z_{\mathrm{m}}^{\mathrm{p}}$ – mass fraction of Al in propellant; μ_{Al}, μ_{O} – molar masses of elements Al and O kg/mole; Z_{m} – ratio between unburned Al in agglomerates and initial Al in propellant; z_n, ρ_n – mass fraction and density of n-th propellant component. Pressure in outlet section is determined by operating conditions.

2.3.2 Determination of agglomerates parameters in computational domain

The agglomerates parameters are determined using the discrete trajectory approach. According to this, the agglomerates phase is represented by set of discrete particles. The problem of agglomerates evolution modeling is solved by calculation of agglomerates trajectories and determining of agglomerates parameters along the trajectories. Herewith the continuous function $f_{\mathrm{m}}(D)$ is sampled for a several fractions with agglomerates sizes D_i and corresponding mass fractions z_i, where i – fraction number. Propellant surface is sampled for a several fragments from which agglomerates of all fractions are injected. Thus the mass flow (kg/s) of agglomerates of i-th size fraction from j-th propellant fragment is determined by equation:

$$M_{\mathrm{a}i,j} = u_{\mathrm{p}} \cdot \rho_{\mathrm{p}} \cdot z_{\mathrm{m}}^{\mathrm{p}} \cdot Z_{\mathrm{m}}^{\mathrm{a}} \cdot z_i \cdot F_j\left[1 + \frac{3\mu_{\mathrm{O}} \cdot \eta}{2\mu_{\mathrm{Al}} + 3\mu_{\mathrm{O}} \cdot (1 - \eta)}\right], \quad (2)$$

where ρ_{p} – propellant density, kg/m^3; F_j – area of j-th propellant fragment, m^2. As results of evolution modeling are the dependencies of agglomerates parameters including coordinates on time for each i,j-th agglomerate (of i-th size fraction injected from j-th propellant surface fragment).

2.3.3 Computational domain sampling

The sampling of computational domain is performed by creation of curved mesh connected with streamlines of gaseous phase flow. A cell of such mesh is represented by a stream tube. Within the each cell, a curvilinear coordinate system is constructed. Axis of such coordinate system is coincident with a streamline. Origin of the coordinate system corresponds to the surface of propellant grain. Analysis of agglomerates trajectories allows obtaining dependencies of agglomerates parameters on curvilinear coordinates, figure 1.

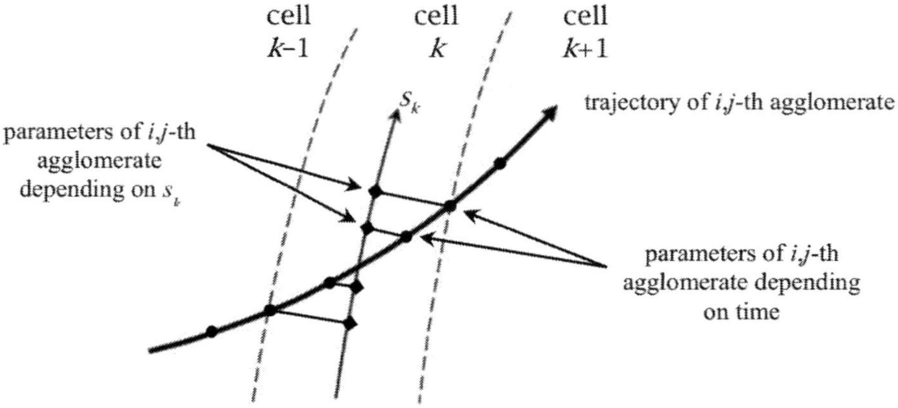

Fig. 1. Cell and curvilinear axis of computational domain.

While agglomerate appears within k-th cell, the projection of its parameters on corresponding curvilinear axis s_k is performed. Thus, parameters of each i,j-th agglomerate are determined as functions of curvilinear coordinates. Interpolation of parameters between adjacent axes allows determining the fields of agglomerates parameters in rectangular coordinate system. In other words, this procedure allows determining the parameters of agglomerates cloud using Eulerian approach.

2.3.4 Calculation of parameters of combustion products mixture

As an individual mixture components, the 22 species including compounds of elements C, H, O, N, Cl, Al are considered. The mixture parameters are calculated assuming thermodynamic equilibrium state of gaseous and condensed products excluding unburned Al in agglomerates. The evolution of agglomerates leads to decrease of fraction of unburned Al in agglomerates and to increase of Al combustion completeness. The combustion completeness of Al is characterized by parameter X_m – ratio between mass of burnt Al and mass of initial Al within the considering computational cell. It allows representing the mixture parameters (temperature, composition, mass and heat transfer properties) as functions of parameter X_m.

Within the k-th computational cell the function $X_\mathrm{m}(s_k)$ is determined by equations:

$$X_\mathrm{m}(s_k) = 1 - Z_\mathrm{m} + \sum_i \sum_j \left(M^0_{\mathrm{m}i,j} - M_{\mathrm{m}i,j}(s_k) \right) \Big/ M_{\mathrm{m}k}\,, \tag{3}$$

$$M_{\mathrm{m}i,j} = (1 - \eta_{i,j}) M_{\mathrm{a}i,j}\,, \tag{4}$$

$$M_{\mathrm{m}k} = u_\mathrm{p} \cdot \rho_\mathrm{p} \cdot z^\mathrm{p}_\mathrm{m} \cdot F_k\,, \tag{5}$$

where $M^0_{\mathrm{m}i,j}$ – mass flow of unburned Al of i,j-th agglomerate at its entry into k-th cell, kg/s; F_k – propellant surface area in the base of k-th cell, m^2. Interpolation of values of X_m between cell axes allows determining the mixture parameters using Eulerian approach.

The calculation continues through global iterations on which sequential recalculations of the fields of gaseous phase and agglomerates parameters are performed. The calculation ends at achievement of required accuracy of X_m at several control points within computational domain.

3 Results

Parametric study was performed applying to a combustion chamber with axisymmetric shape of propellant grain. Geometric parameters of the grain corresponded to the working time 15 s. Total burning surface area at this time was 6.71 square meters. Operating pressure was 6.0 MPa. Propellant parameters were: $z_m^p = 0.24$; $u_p = 8.1$ mm/s, ρ_p=1830 kg/m^3; agglomerates parameters were: $D_{43} = 250$ µm, $\eta = 0.4$, $Z_m^a = 0.344$ [8].

On the figure 2, the stream lines of gaseous phase and trajectories of agglomerates with initial sizes 50µm and 250 µm are presented. On the figure the sections corresponding to propellant surface (A-B), chamber wall (B-C) and outlet section (C-D) are presented.

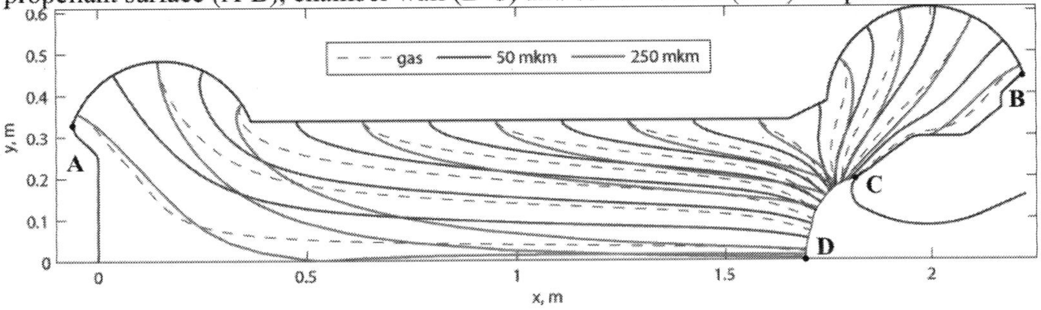

Fig. 2. Gas streamlines and agglomerates trajectories.

On the figure 3, the functions $f_m(D)$ at sections A-B, B-C, and C-D are presented. Also agglomerates mass flows (M_a) and values of η at these sections are indicated. It should be noted that values of M_a and η at section B-C can be used for estimation of slag accumulation rate and its composition.

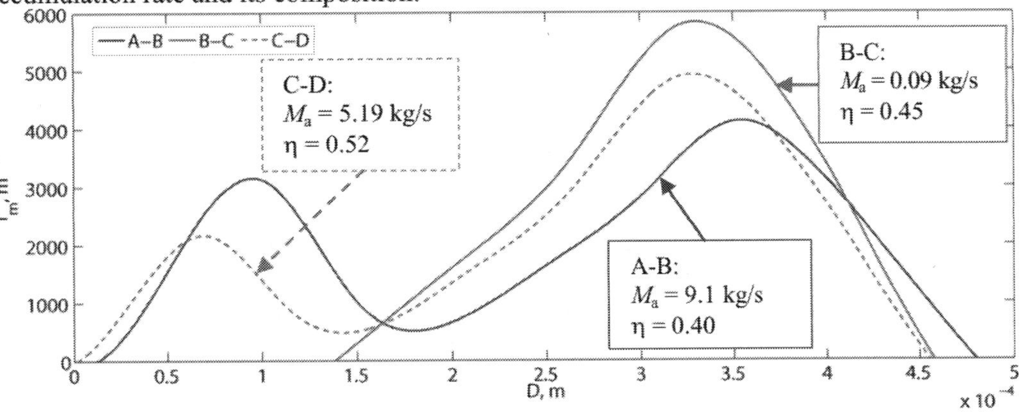

Fig. 3. Size mass density distribution function of agglomerates at sections A-B, B-C, and C-D.

The figure 4 represents the temperature field of combustion products mixture excluding agglomerates.

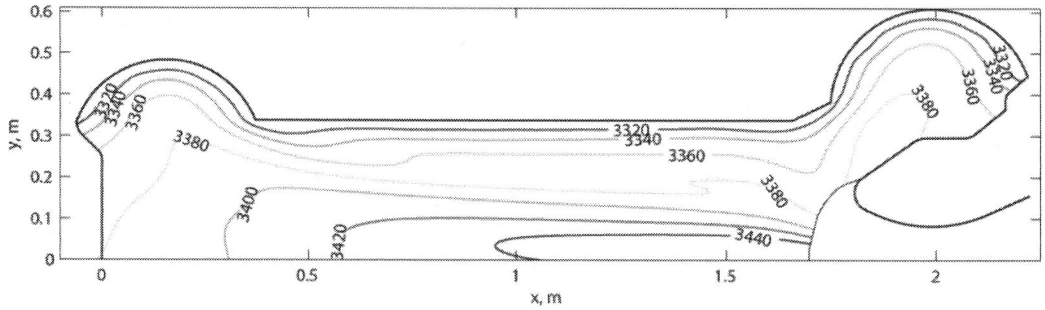

Fig. 4. Temperature field of gaseous phase and SOP.

Analysis of modeling results allows making conclusions about influence of agglomerates evolution on parameters of agglomerates and gaseous combustion products. Herewith determining of Al combustion completeness, as well as estimating of slag accumulation rate is possible. Correctness of calculated parameters of agglomerates is provided by usage of the agglomerates evolution model which was verified by experimental study results [7].

4 Conclusions

The usage of the models of agglomerates formation and evolution was considered. The method for calculation of agglomerates and gaseous phase parameters was described. Interactions between evolution process of agglomerates and gaseous phase properties as well as two-dimensional multiphase flow aspects were taken into account. Results of performed parametric study allowed to make conclusion about possibility of usage of developed models in intra-chamber conditions.

References

1. M.W. Beckstead, *Proc. of 50th Annual Israeli Aerospace Sciences Conference* (Tel Aviv, 2010)
2. J. Dupays, Y. Fabignon, P. Villedieu, G. Lavergne, J.L. Estivalezes, Prog. Astronautics Aeronautics, **185**, 859 (2000)
3. T. Shimada, Y. Daimon, N. Sekino, Proc. of the 8th International Symposium on Experimental and Computational Aerothermodynamics of Internal Flows, July 2007, Lyon, France, paper ISAIF8-00109 (2007)
4. A. Attili, B. Favini, M.Di Giacinto, 45th AIAA/ASME/SAE/ASEE Joint Propulsion Conference & Exhibit 2 - 5 August 2009, Denver, Colorado, 5507 (2009)
5. V.A. Babuk, I.N. Dolotkazin, A.A. Nizyaev, Progress in Propulsion Physics, **4**, 33 (EDP Sciences, Torus Press, 2013)
6. V.A. Babuk, A.N. Ivonenko, A.A. Nizyaev, Combust., Expl., Shock Waves, **51** (5), 549(2015)
7. V.A. Babuk, V.A. Vasilyev, J. Propul. Power, **18** (4), 814 (2002)
8. V.A. Babuk, V.A. Vasilyev, M.S. Malakhov, J. Propul. Power, **15** (6), 783 (1999)

An experimental study of the influence of a thermosyphon filling ratio on a temperature distribution in characteristic points along the vapor channel height

Atlant Nurpeiis[1,*], *Evgeniya* Orlova[1], and *Konstantin* Ponomarev[1]

[1]National Research Tomsk Polytechnic University, 634050 Tomsk, Russia

Abstract. Results of experimental studies of heat transfer in a thermosyphon illustrating the influence of the filling ratio and the heat load on the temperature distribution in the vapor channel, evaporation and condensation zones are presented. The thermosyphon was made of copper and was 161 mm high with side walls 1.5 mm thick, bottom cover 2 mm thick, an internal dimmer of the evaporation part of 54 mm and an internal diameter of the vapor channel of 39.2 mm. Based on the results of experimental studies, temperature dependences were established in the characteristic cross sections of the thermosyphon on the heat flux value supplied to the bottom cover. In addition, a well-appearing thermosyphon self-regulation property has been found – the growth of the heat load in the evaporation zone in the range from 1940 to 7685 W/m^2 does not lead to a decrease in the heat removal intensity from the heat-release region.

Introduction

Closed diphasic thermosyphons (CDTs) are considered by many authors [1, 2] as sufficiently promising, reliable heat transfer heat exchangers due to their autonomy, structural flexibility, ease of manufacture, lack of moving parts, the need for electricity, pumps, etc. In this connection, it is crucial to study the possibility of using thermosyphons as the main element of the cooling systems of devices, apparatus and equipment [3, 4]. In many studies the CDT generally is divided into three zones: an evaporator, an adiabatic region, and a condenser. But such division is rather conditional, because the transport of liquid and vapor, as well as phase transformations occur in all zones [4–7]. It should be noted that the results of studying the heat transfer processes in the CDTs and their heat transfer capacity are usually conclusions about the advantages (or disadvantages) of specific thermosyphons, coolants, structural layout schemes, technical or technological solutions [8, 9]. For the physical analysis of the processes taking place in the CDT, data on the temperature fields of the characteristic zones of the thermosyphon is necessary. However due to the objective difficulties of such measurements, in most cases experimental studies [10, 11] are focused on the analysis of temperature changes only on certain sections

* Corresponding author: nurpeiis_atlant@mail.ru

of the external surface of the heat exchanger. Due to the intense heat transfer across the case of the CDT both along the transverse and longitudinal coordinates, the measurements of the temperatures of the external surfaces are insufficient to analyze the processes taking place in the evaporation, condensation zones and vapor channel. For an objective analysis of the of heat transfer mechanisms, it is necessary to obtain the data on the temperature distributions (T) or at least the gradients of T in individual sections corresponding to the evaporation, vapor transport, and condensation zones. Carrying out the analysis of the current state of the problem under consideration, active investigations in recent years of hydromechanics and heat transfer processes taking place in thermosyphons within the framework of rather complex mathematical models should be noted [12-14]. But the results of mathematical modeling obtained even in recent years using widespread ANSIS FLUENT packages [15, 16] in many cases are not confirmed by the results of an experimental study of the main characteristics of thermo physical processes, primarily temperature fields. In this connection, an experimental study of the temperature fields of the working zone of thermosyphons is a crucial problem.

The aim of the work is an experimental study of the behaviors of temperature changes in the characteristic sections of the thermosyphon working zone under conditions of its cooling along the outer contour.

Method of experimental research

To carry out studies of the thermosyphon operating modes under various heat supply conditions, an experimental set-up was developed (Fig. 1) [17].

Heat transfer studies were carried out in a thermosyphon with a constant cross-sectional area made of copper (height is 161 mm, thickness of side walls and bottom cover are 1.5 mm and 2 mm, respectively, internal diameter of evaporation part is 54 mm, internal diameter of vapor channel is 39.2 mm). The top cover of the heat exchanger is made at an angle of $\gamma = 3°$ to the plane of the base to ensure the condensate movement along one wall. A valve is installed on the thermosyphon cover designed to regulate the pressure in the vapor channel. The heat was supplied with a heating element with a voltage (up to 100 V) and alternating current (from 0.08 to 0.2 A) from a single-phase autotransformer. The measuring circuit of the set-up (Fig. 1) made it possible to simultaneously record the temperatures: the liquid layer at the center of symmetry (thermocouple No 1), the lower (thermocouple No 2) and the upper (thermocouple No 3) boundaries of the vapor channel, in the lower and upper parts of the lateral vertical wall (Thermocouple No 4, 5). The input power and pressure in the internal cavity of the thermosyphon were also recorded. According to the results of preliminary experiments, it is established that external factors (air movement, operation of ventilation systems, room temperature, etc.) significantly influence the thermosyphon operation. For this reason, to reduce such effects on the results of the studies the device was placed in a glass box (Fig. 1).

Measurements (controlling and recording) of the temperature of the liquid, vapor, and the vertical walls of the CDT were conducted using Omega HYP0 thermocouples (temperature ranges 0-673 K, maximum permissible error ± 0.015). The thermocouple readings were recorded using analog-to-digital converter NI 9214. The pressure measurement was recorded using NI USB-6001 analog-to-digital converter and specially developed NI LabView SignalExpress software.

 DOI: 10.1051/matecconf/201711001062

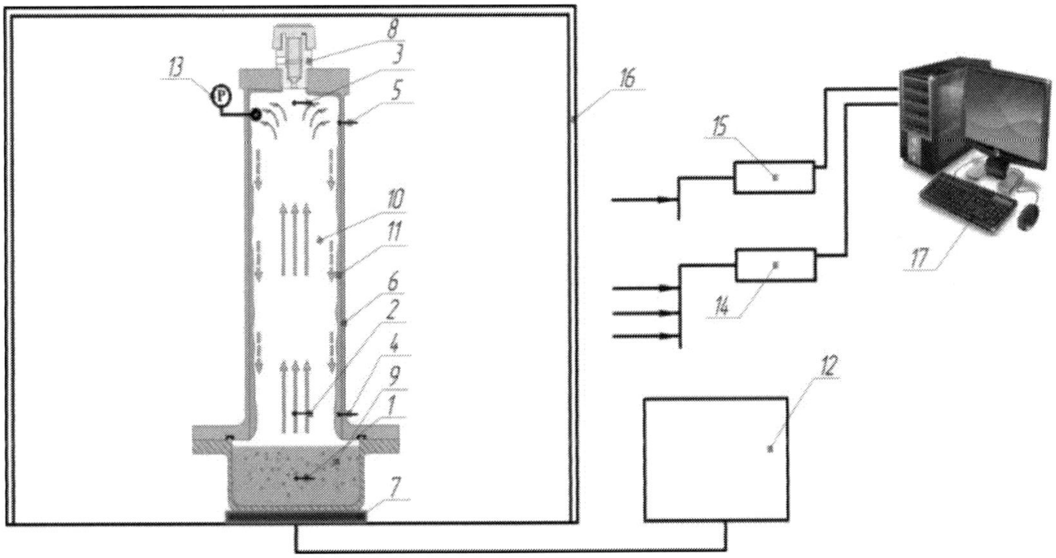

Fig. 1. Scheme of the experimental set-up: 1, 2, 3, 4, 5 – thermocouples; 6 – thermosyphon case; 7 – heating element; 8 – regulation valve; 9 – working liquid; 10 – vapor channel; 11 – condensate film; 12 – power source; 13 – pressure sensor; 14 – analog-to-digital converter NI 9214; 15 – analog-to-digital converter NI USB-6001; 16 – glass box; 17 – personal computer.

Filling ratio of thermosyphon was defined as:

$$\varepsilon = \frac{V_L}{V_E + V_C}$$

where V_L is the volume of coolant, m^3;

V_E is the volume of evaporation section, m^3;

V_C is the volume of condensation section, m^3;

Table 1. Values of filling ratios for different volumes of coolant.

V_L, cm^3	ε, %
40	19
50	24

The classical methodology of conducting experimental studies was applied for conducting research. According to this methodology the influence of each parameter was evaluated. Despite time-consuming, this approach was the most justified in the study of heat exchange inside the thermosyphon.

Results and discussion

The most important characteristics of the CDT functioning are specific heat fluxes and characteristic temperatures [1-5] for different filling ratios (ε,%).

Fig. 2 shows typical thermograms obtained during experiments in the heat loads range from 1940 to 7685 W/m2 at filling volumes of the evaporating section of ε = 19-24%.

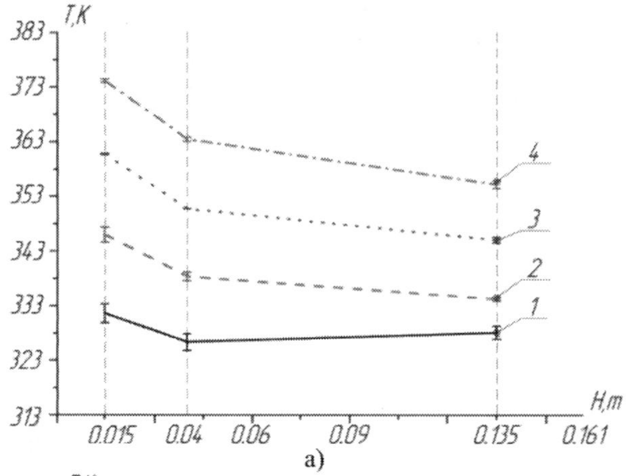

a)

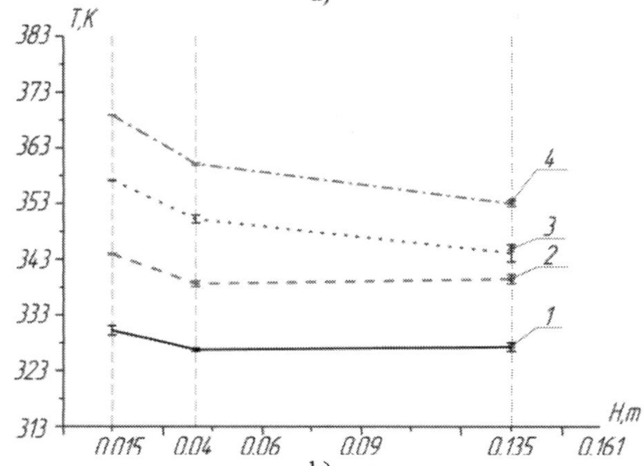

b)

Fig. 2. Temperature distributions along the height (H) of thermosyphon at filling ratios: a) 19 %; b) 24 % and supplied heat fluxes: 1 – 1940 W/m^2; 2 – 3293 W/m^2; 3 – 5332 W/m^2; 4 – 7685 W/m^2.

The temperature distributions in Fig. 2 show that the increase in the heat flux supplied to the bottom cover of the thermosyphon nearly by four times leads to the increase in the temperature in the vapor channel by 35 K. The heat transfer in the thermosyphon is carried out as follows. Due to supplying energy the lower cover and liquid warm up. An increase in temperature leads to an intensive increase in the kinetic energy of the individual molecules of the liquid, which is necessary to overcome the attractive forces of the rest molecules, and thus to carry away these molecules (pairs) from the coolant surface. Steam moves upward due to the pressure difference in the channel and condenses on the top cover and on the side walls of the CDT releasing the phase-transition heat. A liquid film forms.

The condensate flows down the side wall of the thermosyphon under the gravitational forces into the evaporation zone. At a height of H=0.135 m with the filling ratios of 19% and 24% and the heat load of 1920-3293 W/m^2 (Fig. 2 a) a slight increase in temperature was registered which is due to the release of latent heat of condensation at the upper boundary of the vapor channel. But this temperature deviation from monotonic dependence does not exceed 1 K. The same effect is noticeable in Fig. 2 b on curves 1 and 2 (the temperature deviations do not exceed 0.5 K and 0.9 K).

Conclusions

Experimental data on temperatures of liquid and vapor phases obtained under conditions of reliable metrological support are the basis for the development of heat transfer models in thermosyphons describing a complex of interrelated processes in the evaporation, vapor transport and condensation zones with intensive heat removal not only from the surface of the top cover of the thermosyphon but also from its lateral surface.

The reported study was supported by the grant of President of Russian Federation for the government support of young Russian scientists (MK-6810.2016.8).

References

1. P. Byrne, J. Miriel, Y. Lénat, Appl. Therm. Eng. 3072 (2011)
2. M.A. Hakeem, M. Kamil, I. Arman, Appl. Therm. Eng. 1572 (2008)
3. M.K. Bezrodnyi, S.S. Volkov, V.F. Moklyak, *Biphasic thermosyphons in industrial thermal engineering* (Vishcha-shkola, Kiev, 1991) [in Russian]
4. H. Jouhara, A.J. Robinson, Appl. Therm. Eng. 201 (2010)
5. M.K. Bezrodnyi, I.L. Pioro, N.O. Kostyuk, *Transfer processes in the biphasic thermal siphon systems* (Fact, Kiev, 2005) [in Russian]
6. K.O. Ponomarev, E.G. Orlova, A.E. Nurpeiis, MATEC Web Conf. **92**, 01006 (2017)
7. D.V. Feoktistov, E.G. Orlova, V.D. Velicanov, MATEC Web Conf. **23**, 01056 (2015)
8. S.H. Noie, Appl. Therm. Eng. 495 (2005)
9. C.B. Sobhan, R.L. Rag, G.P. Peterson, Int. J. Energy Res. 664 (2007)
10. A.A. Chehade, H. Louahlia–Gualous, S. Le Masson, I. Victor, N. Abouzahab–Damaj, Energy Convers. Manage. 671 (2014)
11. H. Jouhara, A.J. Robinson, Heat Transf. Eng. 1041 (2009)
12. B. Jiao, L.M. Qiu, Z.H. Gan, X.B. Zhang, Heat Mass Transfer, 1043 (2012)
13. G. Huminic, A. Huminic, Energy Convers. Manage. 393 (2013)
14. A.E. Nurpeiis, MATEC Web Conf. **72**, 01077 (2016).
15. K.Y. Leong, R. Saidur, T.M.I. Mahlia, Y.H. Yau, Int. Comm. Heat Mass Transfer, 523 (2012)
16. B. Fadhl, L.C. Wrobel, H. Jouhara, App. Therm. Eng. 122 (2013)
17. D.V. Feoktistov, E.A. Vympin, A.E. Nurpeiis, MATEC Web Conf. **72**, 01081 (2016)

Modeling of filling gas centrifuge cascade for nickel isotope separation by feed flow input to different stages

Alexey A. Orlov[1], *Anton* A. Ushakov[1,2,*], and *Victor* P. Sovach[2]

[1]National Research Tomsk Polytechnic University, 634050 Tomsk, Russia
[2]JSC "PA Electrochemical Plant", 663690 Zelenogorsk, Russia

Abstract. The article presents results of research filling gas centrifuge cascade by process gas fed into different stages. The modeling of filling cascade was done for nickel isotope separation. Analysis of the research results shows that nickel isotope concentrations of light and heavy fraction flows after filling cascade depend on feed stage number.

1 Introduction

During the operation of gas centrifuge (GC) cascade for the multicomponent isotope mixture (MCIM) separation there are nonstationary hydraulic and separation processes. It is necessary to ensure safety of the equipment and to minimize losses of cascade productivity during nonstationary processes. In this regard, actual task is full-scale research of nonstationary processes. It is advisable to study the nonstationary processes by mathematical modeling. Known mathematical models [1-4] describe nonstationary hydraulic processes for only long cascade and nonstationary separation processes in the case of stationary hydraulic parameters of cascade. For elimination of these disadvantages we had developed the mathematical model of nonstationary hydraulic and separation processes occurring in GC cascade for the MCIM separation [5-7]. Earlier we had done verification of developed mathematical model as an example silicon and germanium isotope separation. Isotope separation mode in GC cascade is preceded by its filling with process gas. So far a modeling of filling cascade was not carried out.

The results of research filling GC cascade for nickel isotope separation by feed flow input to different stages is shown in this article. Nickel isotopes are used for nuclear physics experiments and production radioactive isotopes (for example, ^{62}Ni stable isotope is used as source material to produce ^{63}Ni radioactive isotope).

2 Description of the mathematical model

The basic equations of nonstationary processes during MCIM separation is written below.

* Corresponding author: ushakovaa2015@sibmail.com

MCIM separation takes place into a cascade (Fig. 1), consisting of S separation stages numbered by i index ($i=\overline{1,S}$). The cascade has three flows: input feed flow F, output light fraction flow P and heavy fraction flow W.

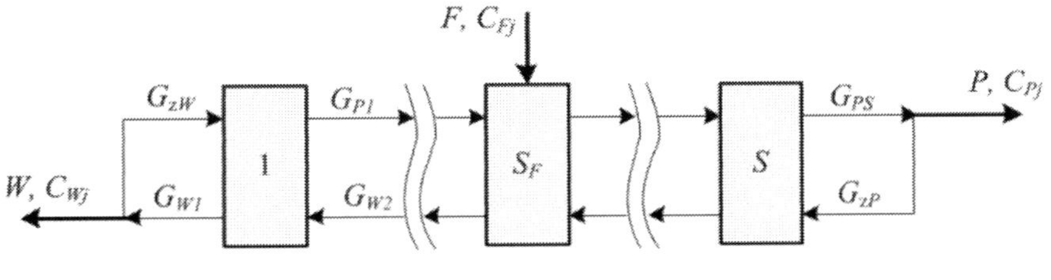

Fig. 1. Separation cascade.

During stationary hydraulic process the flows of cascade are interrelated by material balance equation:

$$F = P + W\,, \tag{1}$$

$$FC_{Fj} = PC_{Pj} + WC_{Wj}\,. \tag{2}$$

The stage has three flows: input feed flow G_{Fi}, output light fraction flow G_{Pi} and heavy fraction flow G_{Wi}:

$$G_{Fi} = G_{Pi} + G_{Wi}\,, \tag{3}$$

$$G_{Fi}C_{Fij} = G_{Pi}C_{Pij} + G_{Wi}C_{Wij}\,. \tag{4}$$

The MCIM separation in stage is determined by equations:

$$\chi_{ijl} = \frac{C_{Pij}C_{Wil}}{C_{Pil}C_{Wij}}\,, \tag{5}$$

$$\chi_{ijl} = \chi_{0i}^{M_j - M_l}\,, \tag{6}$$

where χ_{ijl} is separation factor of jth и lth components; χ_{0i} is overall separation factor per unit of mass numbers difference; M_j, M_l – is mass numbers of jth и lth components.

Nonstationary processes are described by differential equation systems. The solution algorithm is written on publications [5, 6].

3 The results of calculation cascade for nickel isotope separation

We have researched filling square cascade consisted of 60 stages (S=60) with equal flow rate G_F for all the stages during stationary mode. The overall separation factor χ_0 is equal 1.1. The ratio of cascade feed flow and stage flow F/G_F is equal 0.13. The trifluorophosphinenickel ($Ni(PF_3)_4$, molar mass is 411 kg/kmol) was used as process gas for nickel isotope separation in GC cascade.

At the initial time the cascade holdup (quantity of process gas into GC and pipelines) is equal 0. Constant feed flow is input to cascade, values of light and heavy fraction flows of cascade are equal 0 during filling cascade. At the same time the values of pressures, holdups and flows of process gas into GC and pipelines is being increased to steady values.

 DOI: 10.1051/matecconf/201711001063

We have done modeling of filling cascade with process gas by feed flow input to different stages. The nickel isotope concentrations in light and heavy fraction flows of cascade are shown on fig. 2 and 3.

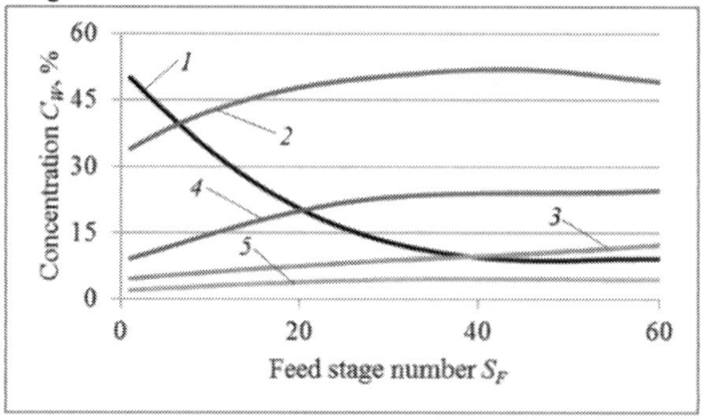

Fig. 2. Nickel isotope concentrations in heavy fraction flow of cascade after its filling depending on feed stage number: *1* – ^{58}Ni, *2* – ^{60}Ni, *3* – ^{61}Ni, *4* – ^{62}Ni, *5* – ^{64}Ni.

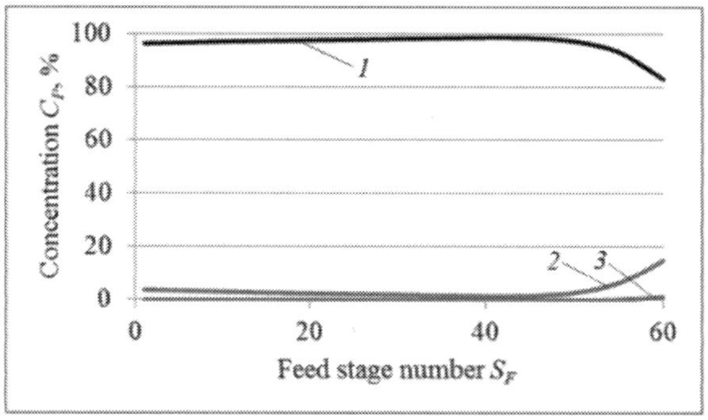

Fig. 3. Nickel isotope concentrations in light fraction flow of cascade after its filling depending on feed stage number: *1* – ^{58}Ni, *2* – ^{60}Ni, *3* – ^{61}Ni.

As it's seen on Fig. 2 and 3, there is optimal feed stage number for each nickel isotope when isotope concentration in light or heavy fraction flows of cascade takes on maximal value after filling cascade (see Table 1).

Table 1. Optimal feed stage numbers.

Nickel isotope	Output cascade flow	Optimal feed stage number S_{Fopt}	Maximal concentration value, %
^{58}Ni	light fraction	43	98.5
^{60}Ni	heavy fraction	43	51.8
^{61}Ni	heavy fraction	41	4.7
^{62}Ni	heavy fraction	60	24.6
^{64}Ni	heavy fraction	60	12.4

For example, ^{58}Ni isotope concentration in light fraction flow takes on maximal value (98.5 %) when feed flow is entered to stage with number S_{Fopt}=43. Optimization of feed stage number might make it possible to increase concentration of target isotope in light or heavy fraction flow after filling cascade. It will decrease duration of further nonstationary process of cascade start-up (establishment of steady-state isotope concentrations).

4 Conclusion

We have done research filling GC cascade for nickel isotope separation by feed flow input to different stages.

According to analysis of research results nickel isotope concentrations on light and heavy fraction flows after filling cascade depend on feed stage number.

The research have indicated that there is optimal feed stage number for each nickel isotope when isotope concentration in light or heavy fraction flows of cascade takes on maximal value after filling cascade with process gas.

Feed stage number optimization lets to decrease duration of further nonstationary process of cascade start-up (establishment of stationary isotope concentrations).

References

1. D.M. Levin, Yu.P. Zabelin, Physical and chemical processes during separation of atoms, molecules and in laser, plasma and nanotechnologies conference report collection (2008)
2. E.V. Levin, N.I. Laguntcov, G.A. Sulaberidze, J. Eng. Phys. **43**, 456 (1982)
3. V.M. Vetcko, O.A. Devdariani, E.V. Levin, G.A. Sulaberidze, G.L. Partcachishvili, Isotopenpraxis, **18**, 288 (1982)
4. A.Yu. Smirnov, A.K. Bonarev, G.A. Sulaberidze, V.D. Borisevish, S. Zeng, D. Jiang, Y.N. Zhang, IOP: Conf. Series, **751**, 012006 (2016)
5. A.A. Orlov, A.A. Ushakov, V.P. Sovach, J. Eng. Phys. Thermophys. **90** (2), 258 (2017)
6. A.A. Orlov, A.A. Ushakov, V.P. Sovach, MATEC Web Conf. **92**, 01033 (2017)
7. A.A. Orlov, A.A. Ushakov, V.P. Sovach, MATEC Web Conf.**72**, 01106 (2016)

Critical heat flux density in diphasic thermosyphons

Konstantin O. Ponomarev[1,*], *Evgeniya* G. Orlova[1], and *Atlant* E. Nurpeiis[1]

[1]National Research Tomsk Polytechnic University, 634050 Tomsk, Russia

Abstract. The paper presents an analysis of known dependencies for determining the critical heat flux density in diphasic thermosyphons. The critical heat flux density for the created experimental model of thermosyphon were calculated on the basis of the theoretical contributions of 1) the occurrence of a "flooding" regime in a thermosyphon characterized by a disturbance of the hydrodynamic stability of the phase interface and the entrainment of the liquid phase by the gas flow; 2) the mutual influence of gravitational forces and surface tension; 3) S.S. Kutateladze hydrodynamic theory of the heat transfer crisis during boiling. It is found that the existing theoretical contributions which can be used to calculate the critical heat flux density and subsequently determine the minimum filling ratio of a thermosyphon are conditionally applicable.

1 Introduction

Accident-free operation of electronic, energy-generating devices is ensured by monitoring their temperature within the required limits designated by the manufacturer. Cooling of heat-loaded elements is possible with traditional systems using relatively large volumes of coolant, as well as systems based on diphasic thermosyphons – compact, resource-saving devices, capable of removing high local heat dissipations. It is known [1–5] that the efficiency of a thermosyphon is affected by the chemical composition of the coolant, filling ratio, geometric dimensions of a thermosyphon (height, internal cross-sectional area), inclination angle, material of construction, and the cooling conditions of the condensation zone.

When operating a thermosyphon, emergency operation modes may occur due to the heat and mass transfer crisis which is characterized by the absence of contact between the coolant and the surface of a thermosyphon evaporative part. The consequences of such an operation of a thermosyphon cooling system are overheating and ignition of a thermostabilized device or apparatus.

The known causes of the heat and mass transfer crisis [6–7] are breakdown of condensate from the thermosyphon walls, limiting steam content in the wall layer, formation of dry spots on the internal surface of the evaporator. Determination of the critical heat flux density $q_{cr}, kW / m^2$ at which the heat and mass transfer crisis occurs is necessary to calculate the minimum filling ratio of a thermosyphon with the coolant.

* Corresponding author: kop.tpu@gmail.com

The aim of this work is to determine the critical heat flux density for the created experimental thermosyphon model.

2 Determination of the critical heat flux density

Three parts are relatively distinguished in the thermosyphons [8–9]: evaporative, adiabatic (transport), and condensate (Fig. 1).

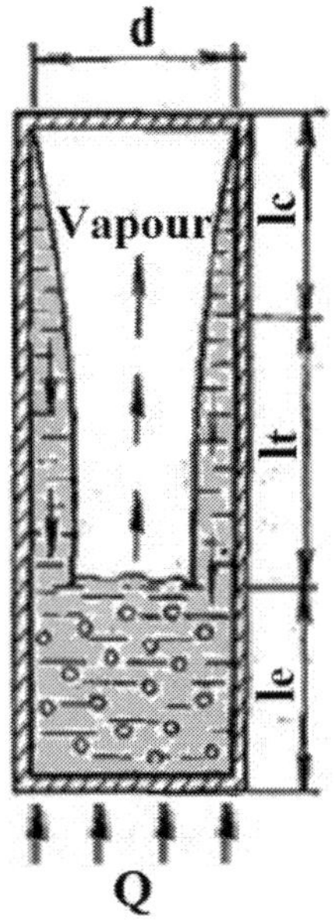

Fig. 1. Diphasic thermosyphon. l_e, l_t, l_c are lengths of evaporative, adiabatic (transport), and condensate parts, respectively; d diameter of a thermosyphon; Q heat flow.

The heat flow is removed from the heat-loaded element to the evaporative part of a thermosyphon filled with the coolant which heats up. The resulting vapor from the evaporative part moves into the adiabatic and condensation parts. Here the vapor condenses emitting the latent heat of the phase transition to the cooling medium. Condensate under the gravitational forces is transported to the evaporative part along the walls. The processes in a thermosyphon proceed continuously-cyclically which ensures heat transfer from the heat-loaded element to a thermosyphon.

A thermosyphon is a device with relatively small geometric dimensions. Therefore, at the ratio of the inner diameter d to the length of the evaporative part l_e $d/l_e < 0.2$ the boiling mechanism in it refers to the boiling in a small volume, if $d/l_e > 2$ the volume is considered as large [10].

In order to avoid a heat and mass transfer crisis in a thermosyphon, it is necessary to ensure its filling with a coolant in an amount greater than the minimum determined by the dependences [8, 11]:

$$\varepsilon_e = C_1 + \frac{0.8 \cdot l_c + l_t}{l_e} \cdot \frac{4}{d} \cdot \left(\frac{3 \cdot \mu_l \cdot l_e \cdot q_{cr}}{\rho_l^2 \cdot g \cdot r} \right)^{\frac{1}{3}} + \frac{\rho_v}{\rho_l} \cdot \left(\frac{l_c + l_t}{l_e} - \frac{0.8 \cdot l_c + l_t}{l_e} \cdot \frac{4}{d} \cdot \left(\frac{3 \cdot \mu_l \cdot l_e \cdot q_{cr}}{\rho_l^2 \cdot g \cdot r} \right)^{\frac{1}{3}} \right). \quad (1)$$

where l_e, l_t, l_c are lengths of evaporative, adiabatic (transport), and condensate parts, respectively, m; d internal diameter, m; μ_l dynamic viscosity of liquid, Pa·sec, q_{cr} heat flux density referred to the surface area of the heat supply, W/m^2; ρ_l liquid density, kg/m^3; ρ_v vapor density, kg/m^3; g acceleration of gravity, m/sec^2; r latent heat of vaporization, J/kg; C_1 coefficient which depends on the pressure of the coolant, the diameter of a thermosyphon, the thermo-physical properties of a liquid, and is taken from 0.2 to 0.33 [8].

$$\varepsilon_e = 1 - \frac{1}{1 + C_2 \cdot \left[\frac{q_{cr}^2}{2 \cdot \sigma \cdot \rho_v \cdot r^2} \cdot \left(\frac{3 \cdot q_{cr} \cdot \mu_l \cdot l_e}{\rho_l \cdot (\rho_l - \rho_v) \cdot g \cdot r} \right)^{\frac{1}{3}} \right]^{\frac{3}{4}}} \quad (2)$$

where σ is surface tension, N/m; C_2 coefficient dependent on the conditions of the surface heating and the thermo-physical properties of a liquid; for water is equal to 447 [11].

In formulas (1–2) the critical heat flux density q_{cr} is used. It depends on the thermo-physical properties of the liquid and the geometric parameters of a thermosyphon.

According to the analysis results of the known dependences, three groups of equations can be conventionally identified for determining q_{cr} [12–19] based on: 1) the theoretical contributions [12–15] of the occurrence of a "flooding" regime in a thermosyphon characterized by a disturbance of the hydrodynamic stability of the phase interface and the entrainment of the liquid phase by the gas flow. Regime takes place an intermediate position between the regions of a stable descending and stable ascending flow of a liquid film; 2) the mutual influence of gravitational forces and surface tension [16–18]; 3) S.S. Kutateladze hydrodynamic theory of the heat transfer crisis during boiling [19].

The first group includes the dependencies [12–15]:

$$q_I^{cr} = 0.25 \cdot \frac{r \cdot d^{1.5} \cdot \sqrt{g \rho_v \cdot (\rho_l - \rho_v)}}{l_e \cdot \left(1 + \sqrt[4]{\frac{\rho_v}{\rho_l}} \right)^2}, kW/m^2; \quad (3)$$

$$q_{II}^{cr} = 0.8 \cdot \left(\frac{d}{l_e} \right) \cdot r \cdot \sqrt{\rho_v} \cdot \sqrt[4]{\sigma \cdot g \cdot (\rho_l - \rho_v)}, kW/m^2; \quad (4)$$

$$q_{III}^{cr} = 0.526 \cdot \frac{r \cdot \sqrt{d} \cdot \sqrt{g \rho_v \cdot (\rho_l - \rho_v)}}{\left(1 + \sqrt[4]{\frac{\rho_p}{\rho_l}} \right)^2}, kW/m^2; \quad (5)$$

$$q_{IV}^{cr} = 3.2 \cdot r \cdot \left(th \cdot (0.5 \cdot \sqrt[4]{d} \cdot \sqrt[8]{\frac{g \cdot (\rho_l - \rho_v)}{\sigma}} \right)^2 \cdot \frac{\sqrt[4]{\sigma \cdot g \cdot (\rho_l - \rho_v)}}{\left(\frac{1}{\sqrt[4]{\rho_l}} - \frac{1}{\sqrt[4]{\rho_v}} \right)^2}, kW/m^2. \quad (6)$$

The second group includes [16-18]:

$$q_V^{cr} = 0.09 \cdot \left(\frac{d}{l_e} \right)^{0.9} \cdot \frac{r \cdot \sqrt{\rho_v} \cdot \sqrt[4]{\sigma \cdot g \cdot (\rho_l - \rho_v)} \cdot \sqrt{Bo}}{\left(1 + \sqrt[4]{\frac{\rho_v}{\rho_l}} \right)^2}, kW/m^2; \quad (7)$$

$$q_{VI}^{cr} = 0.1 \cdot \frac{r \cdot \sqrt{\rho_v} \cdot \sqrt[4]{\sigma \cdot g \cdot (\rho_l - \rho_v)}}{1 + 0.491 \cdot \left(\frac{l_e}{d}\right) \cdot \frac{1}{Bo^{0.3}}}, kW/m^2; \quad (8)$$

$$q_{VI}^{cr} = 0.8 \cdot \left(\frac{d}{l_e}\right) \cdot r \cdot \sqrt{\rho_v} \cdot \sqrt[4]{\sigma \cdot g \cdot (\rho_l - \rho_v)} \cdot \left(\frac{th\left(\frac{\sqrt[4]{Bo}}{2}\right)}{1 + \sqrt[4]{\frac{\rho_l}{\rho_v}}} \right)^2, kW/m^2; \quad (9)$$

where $Bo = d / \sqrt{\sigma / (g \cdot \rho_l - \rho_v)}$ is Bond number.

The third group includes [19]:

$$q_{cr} = K \cdot r \cdot \sqrt{\rho_v} \cdot \sqrt[4]{g \cdot \sigma \cdot (\rho_l - \rho_v)}, kW/m^2 \quad (10)$$

where K is stability criterion of a two-phase wall layer.

A stability criterion K in Eq. (10) depends on the thermo-physical properties of the heat-removing surface, the coolant and is in the range of $K = 0.09 \div 0.2$. The recommended values of K according to experimental data are $K = 0.16$ [20], $K = 0.12$ [21], $K = 0.13$ [22], $K = 0.14$ [23]. In [24–29], dependencies were obtained by definition of the stability criterion K.

In the present work calculations of the critical heat flux density for the created experimental thermosyphon model are performed. Table 1 shows the results obtained under conditions of filling thermosyphon with water at the following parameters: $d = 0.039\, m$; $l_e = 0.021\, m$; $\rho_l = 958.1\, kg/m^3$; $\rho_v = 0.597\, kg/m^3$; $g = 9.81\, m/\sec^2$; $r = 2260000\, J/kg$; $\sigma = 0.05904\, N/m$; $Bo = 15.56$; $\mu_l = 0.000279\, Pa \cdot \sec$.

Table 1. Critical heat flux density.

Dependences from references used on the basis of:	q_{cr}, kW/m^2
theoretical contributions of the occurrence of a "flooding" regime in a thermosyphon	
[12]	11571.9
[13]	12589.6
[14]	13110.1
[15]	11638.1
mutual influence of gravitational forces and surface tension	
[16]	3915.8
[17]	605.2
[18]	134.9
S.S. Kutateladze hydrodynamic theory of the heat transfer crisis during boiling	
[20]	1355.8
[21]	1016.9
[22]	1101.6
[23]	1186.3
[24]	1109.0
[25]	1264.0
[26]	1440.3
[27]	1207.8
[28]	1229.1
[29]	1105.3

Based on the analysis of the data presented in Table 2, it was found that the divergence between the values of the critical heat flux density obtained on the basis of: 1) the theoretical contributions of the occurrence of a "flooding" regime in the thermosyphon is 23.15%; 2) the mutual influence of gravitational forces and surface tension is 96.55%; 3) S.S. Kutateladze hydrodynamic theory of the heat transfer crisis during boiling is 29.4%.

In addition, using the dependencies on the basis of: 1) the theoretical contributions of the occurrence of a "flooding" regime in a thermosyphon the values of the critical heat flux density are an order of magnitude greater than the values obtained from the dependences on the basis of the mutual influence of the gravitational forces and S.S. Kutateladze hydrodynamic theory of the heat transfer crisis during boiling. It can be concluded that the existing theoretical contributions which can be used to calculate the critical heat flux density and subsequently determine the minimum filling ratio of the thermosyphon are conditionally applicable.

According to the results of the literature analysis [1–29] Russian scientific groups are found to apply Eq. (10) the most often, the foreign groups use the dependence of q_{cr} derived by Zuber N. [24] based on S.S. Kutateladze hydrodynamic theory of the heat transfer crisis during boiling.

3 Conclusions

There are a lot of researchers considering various aspects of the critical heat flux density and the list of authors presented in the paper is not exhaustive. It was found that many formulas based on the "flooding" and similarity theory are not suitable for determining the critical heat flux density during boiling of water in a diphasic thermosyphon with certain geometric parameters ($d = 0.039\ m$; $l_e = 0.021\ m$).

The divergence between the values of the critical heat flux density obtained on the basis of: 1) the theoretical contributions of the occurrence of a "flooding" regime in the thermosyphon is 23.15%; 2) the mutual influence of gravitational forces and surface tension is 96.55%; 3) S.S. Kutateladze hydrodynamic theory of the heat transfer crisis during boiling is 29.4%.

It can be concluded that at present the scientific basis for the design of energy-efficient, resource-saving cooling systems for heat-loaded elements based on thermosyphons is not developed at the level of prognostic modeling. To solve this scientific problem, it is necessary to conduct complex experimental studies of the heat and mass transfer, convection, boiling in thermosyphon using modern low-inertia, high-precision equipment for recording temperature, photo and video recording systems.

The reported research was supported by Russian Federation President Grant for state support of the Russian Federation leading scientific schools SS-7538.2016.8 (No 14.Y31.16.7538-SS).

References

1. A. Amiri, R. Sadri, M. Shanbedi, G. Ahmadi, B.T. Chew, S.N. Kazi, M. Dahari, Energy Convers. Manage. **92**, 322 (2015)
2. C.C. Chien, C.K. Kung, C.C. Chang, W.S. Lee, C.S. Jwo, S.L. Chen, Energy **36**, 415 (2015)
3. N.C. Chami, A. Zoughaib, Energy Build. **42**, 1267 (2010)
4. S. Jaisankar, T.K. Radhakrishnan, K.N. Sheeba, S. Suresh, Energy Convers. Manage. **50**, 2638 (2009)
5. H. Mirshahi, M. Rahimi, Iran J. Chem. Eng. **6**, 15 (2009)

6. M.K. Bezrodnyi, D.V. Alekseenko, Teplofizika Vysokikh Temperatur **15**, 370 (1977)
7. H. Imura, K. Sasaguchi, H. Kosai, S. Namata, Int. J. Heat Mass Transfer. **26**, 1181 (1983)
8. D.V. Feoktistov, E.A. Vympin, A.E. Nurpeiis, MATEC Web Conf. **72**, 01081 (2016)
9. K.O. Ponomarev, E.G. Orlova, A.E. Nurpeyis, MATEC Web Conf. **92**, 01006 (2017)
10. I.L. Pioro, Industrial Heat Engineering **7**, 24 (1985)
11. S. Rösler, M. Takuma, M. Groll, S. Maezawa, Heat Recov. Syst. CHP **7**, 319 (1987)
12. G. Wallis, *One-dimensional two-phase flows*, ed. by I.T. Alad'ev (Mir, Moscow, 1972) [in Russian]
13. O.L. Pushkina, Y.L. Sorokin, Heat Transfer Soviet Research **1**, 56 (1969)
14. R.S. Sakhuja, Mechanical Engineering **96**, 80 (1974)
15. C.L. Tien, K.S. Chung, Proc. 3rd Int. Heat Pipe Conf. 36 (1978)
16. C.L. Tien, K.S. Chung, AIAA J. **17**, 643 (1978)
17. Y. Katto, Trans. Japan. Soc. Mech. Engs. **44**, 3908 (1978)
18. M. Shiraishi, Prep. 6lh Int. Heat Pipe Conf. 609 (1987)
19. S.S. Kutateladze, *Fundamentals of the heat transfer theory* (Atomizdat, Moscow, 1979) [in Russian]
20. S.S. Kutateladze, Kotloturbostroenie **3**, 10 (1948)
21. E.A. Kazakova, Izvestiya AN SSSR. OTN **1**, 64 (1949)
22. Y.P. Chang, *An analysis of the critical conditions and burnout in boiling heat transfer* (USAEC Rep. TID-14004, Washington, DC., 1961)
23. O.N. Man'kovskij, A.R. Tolchinskij, M.V. Aleksandrov, *Heat exchange equipment for chemical production* (St. Petersburg, Himiya, 1976)
24. N. Zuber, Trans. Of ASME. J. Heat Transfer **80**, 711 (1958)
25. J.H. Lienhard, V.K. Dhir, J. Heat Transfer **95**, 152 (1973)
26. W. Bailey, E. Young, C. Beduz, Y. Yang, Thermal and Thermomechanical Phenomena in Electronics Systems, 2006. ITHERM'06. The Tenth Intersociety Conference on. – IEEE, 599 (2006)
27. V.M. Borishanskiy, Technical Physics. The Russian Journal of Applied Physics **26**, 452 (1956)
28. Y.P. Chang, N.W. Snyder, Chem. Eng. Prog. Symp. Ser. **56**, 25 (1960)
29. Y. Haramura, Y. Katto, Int. J. Heat Mass. Transfer. **26**, 389 (1983)

Structural Characteristic of the German-English Hybrids of the Information Security Sphere

Darya S. Poendaeva [1,*], *Juri V.* Kobenko[1], and *Elena S.* Riabova[2]

[1]National Research Tomsk Polytechnic University, 634050 Tomsk, Russia
[2]Samara University, 443086 Samara, Russia

Abstract. The article describes the results of structural analysis of the German-English hybrid terms of the information security sphere. The obtained data testify to the serious impact of the English influence on German primarily in the peripheral sphere of the German language stock. The hybridity is stated as a result of aggregation of the literary German lexical system and language mixing owning to an invasive borrowing process.

1 Introduction

In the course of its evolution German to some extent has been subjected to the influence of more prestigious languages. Being the language of international communication English has had the greatest influence over German in last decades, which has led to invasive borrowing and mixing of the elements of the contacting languages in the system of the recipient-language. This phenomenon is known as hybridization or language mixing, which is the process of combining the elements of the recipient language with donor language citing [1]. Hybridization results from heterogeneity of a person's consciousness [2]. Every field of knowledge has its own definition of hybridity, what is common for all of them is that hybrids are created as a result of mixing objects with different properties and therefore acquires unique features and new functions. Particularly mixing of the languages of various arts can enrich the texts of a specific semiotic nature [3]. This also can be applied to different languages of all speech communities, which evolution and further existence depends upon this process. Hybridization is inseparable from the language nature as one of the word formation mechanisms [4]. Hybridization is mainly caused by extralinguistic factors, including interrelation between speech communities and local peculiarities of technological development. Hybridization is a complicated process as it can be applied to all levels of the language system. A great illustration of this statement is the Yiddish language, which initially was a mix of one of the German dialects, Slavic languages and Hebrew [5]. This example illustrates the leading role of extralinguistic factors as it clearly shows connection between structural changes in the systems of the languages and their relations. These kinds of processes are natural because borrowing is a condition for any language development [6]. Furthermore, the periods of active borrowing from a language

* Corresponding author: poendaevads@ya.ru

testify to the leading position of this language. However, it is the character of borrowing which defines an extent of the impact of the more prestigious language on the less prestigious and not the borrowing itself. Invasive borrowing occurs when a number of items, which were borrowed or are comprised of borrowed elements, prevails on a number of items formed by means of a recipient language [2]. To define a speech community status, a character of borrowing should be taken into consideration as borrowing is a part of any language development.

A language status is defined not only by the character of borrowing, but also by social, historical and cultural processes. A contact between two languages is always characterized by a non-equivalent mixing due to the degree of the donor language prestige among the recipient language speakers [4]. Diachronic description of the functional paradigms of literary languages at the different stages of their development is always based on the social processes and specific historical and cultural situation [7]. More explicitly hybridization can be seen at the lexicological level, where it is represented by terminological hybrids, which are new words, built by means of the elements, belonging to different languages [8]. Peripheral layers of the vocabulary, which contain the elements with weak or non-assimilated borrowed items, are most exposed to this process. A number of borrowed language elements in peripheral spheres may prevail on a number of autochthonous elements [6]. Information security is a relatively new field of knowledge, which as well as legal terminology uses terms of Internet and information technology spheres and thus can be considered peripheral.

Currently German-English terminological hybrids of the information security sphere are not fully examined. From this perspective studying functional characteristics of the terminology of the peripheral sphere appears to be required. The results would allow assessing more exactly, how considerable is the influence of the English language, being a donor, over the functional characteristics of the German language, being a recipient.

2 Hybridity research in the German terminology of the information security sphere

The database of terms was compiled from 12 documents and standards, published by Federal Office for Information Security of the Federal Republic of Germany [9] and official documents of the Organization for Economic Cooperation and Development (OECD) [10]. The terms were selected by means of continuous sampling method.

Survey of the selected text corpus revealed 104 terminological hybrids, including 9 comprised of one constituent (Cyber-Raum, Schadsoftware, Botnetz), 79 comprised of two constituents (Fault-Attacke, Bit-Folge, hardware-Plattformen) and 16 comprised of more than two constituents (Public-Key-Infrastruktur, Challenge-Response-Verfahren, related-Key-Angriffe).

Structural analysis of the hybrids revealed the following.

Firstly, there are 74% of determinate items comprised of two constituents (Flash-Speicher, Push-Nachricht, OpenPGP-Zertifikat) (fig. 1) among all terminological hybrids of the German information security sphere. Items comprised of more than two constituents (IT-Grundschutz-Kataloge, MAC-Management-Nachricht) amount to 18% of the total number of the hybrids. Those comprised of one constituent (Botnetz, Trojaner) come to 8% of the total number of terms, including 4 items formed by means of prefixation (Cyber-Raum), 2 items formed by means of clipping a final part of the word (Schadsoftware), 2 items formed by means of composing stems of the words (Botnetz) and 1 item formed by adding an English suffix to a German stem (Trojaner).

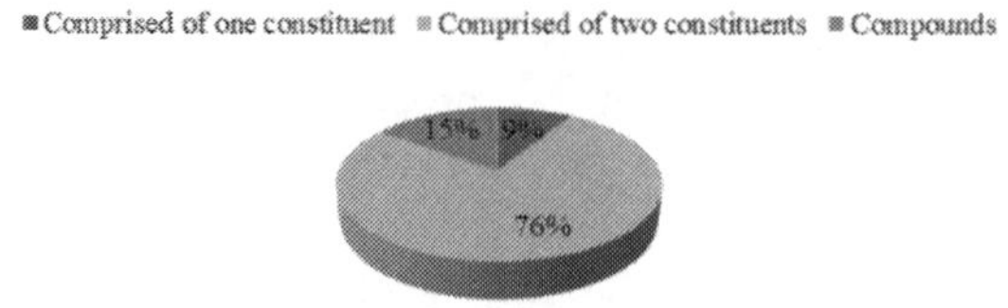

Fig. 1. Hybrids.

Secondly, 48% of the hybrids comprised of two or more constituents are terms which consist of simple constituents (Client-Server-Architektur, Unternehmens-Service), items comprised of only one compound constituent (SDN-Architektur, Full-Motion-Inhaltsfreigabe) represent 40% of the total (fig. 2); items comprised of two or more compound constituents (IP-Anlagenanschluss, IT-Grundschutz-Vorgehensweise) represent 12% of the total. Thirdly, one of the constituents of the compound terminological hybrids can be a numerical acronym (B2B-Bereich) or a clipped final part (E-Government-Dienst), initial constituent of which is an acronym and two final constituents are compounds (VoIP-Kommunikationslösung).

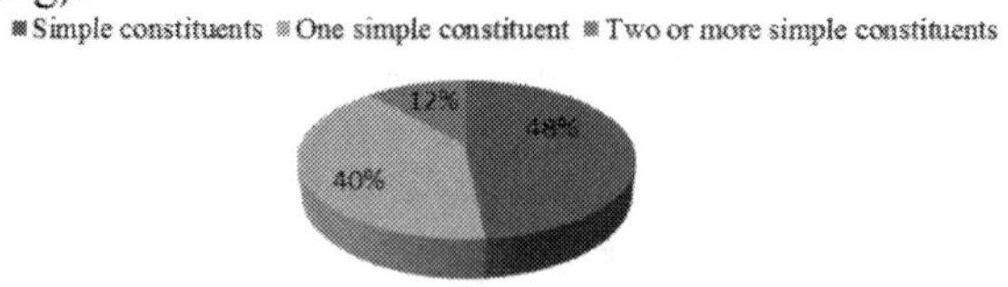

Fig. 2. Hybrids comprised of two and more constituents.

Compositional analysis of the hybrids showed, that all selected terms are determinates, 84% of which have a German determiner (DNS-Angriff) and 16% of which have an English determiner (AES-verschlüsselte E-Mail). Furthermore, 2% of the com-pound terms have one hybrid determiner (IT-Sicherheitsmanager) (fig. 3).

In total, 52% of the English elements were borrowed to the German terminology of the information security sphere. Overall, 7 following hybrid patterns were discovered: E+D (Mobile Dienste), D+E (Schadprogramm), E+E+D (IT-Governance-Beauftragten), E+D+D (IP-basierte Dienst), E+D+E (Bluetooth-Protokoll-Stack), D+E+D (Empfänger-E-Mail-Adresse) and E+E+E+D (Internet Control Message Protokoll). The largest share is comprised of E+D hybrids, which represent 70% of all hybrids. Almost equal shares of 11% and 10% are formed by E+E+D and D+E hybrids. E+D+D и E+D+E hybrids make up 4% and 3%. Finally, both D+E+D and E+E+E+D hybrids amount to 1% of the total.

German-English Hybrid Patterns of the Information Technology Sphere:

1) E+D (Mobile Dienste);
2) D+E (Schadprogramm);
3) E+E+D (IT-Governance-Beauftragten);
4) E+D+D (IP-basierte Dienst);
5) E+D+E (Bluetooth-Protokoll-Stack);
6) D+E+D (Empfänger-E-Mail-Adresse);
7) E+E+E+D (Internet Control Message Protokoll).

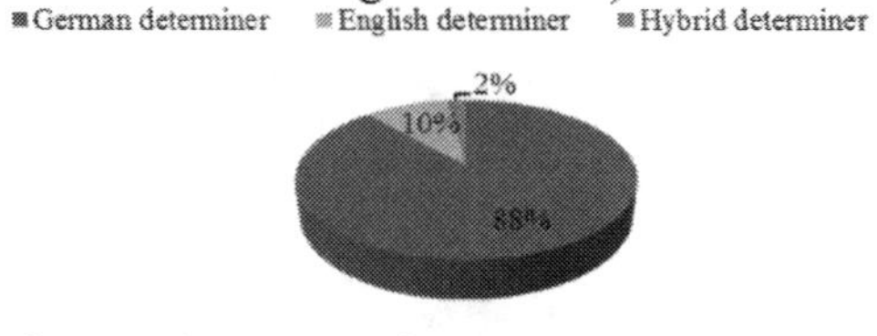

Fig. 3. Hybrids comprised of two and more constituents.

3 Conclusion

Language mixing can result in new sociolinguistic situations, causing a big number of research problems to appear. There is a real need for empirical research into these hybrid language situations. But it is plain that the emergence of hybrid trends and varie-ties raises all kinds of theoretical and pedagogical questions: They blur the long-standing distinctions between "first", "second" and "foreign" language. They make us reconsider the notion of "standard", especially when we find such hybrids being used confidently and fluently by groups of people who have education and influence in their own regional setting [11].

However hybridization is not just a process going beyond existing paradigms, but mainly a source of new standarts. This feature of hybrids was described by Stuart Hall and Homi K. Bhabha, who were the first to develop the theory of cultural hybridization: "The process of cultural hybridization gives rise to something different, something new and unrecongnisable, a new area of negotiation of meaning and representation" [12]. The literary works of V. Vertlib, who immigrated from the Soviet Union to Austria, can serve as an illustration of this statement. Vertlib created hybrid characters and features that combine elements of two different cultures and enrich the stylistic properties of his books [3].

The given statistics testifies to the special case of language mixing, which occurs in all levels of the system of a language. However, even examination of terminological hybrids is important in conveying an idea of the state of the language in the Ger-man speech community. Obtained results will help define a role played by English borrowings in the German language develop-ment. Nevertheless, structural change of the language system is only a superficial manifestation of the process, which has deeper roots. The obtained data can be used to solve strictly linguistic issues as well as implemented in the adjacent fields. Particularly, the results can be helpful to explore the rules of the speech communities functioning and development. The main conclusion which every researcher can draw when considering the data is that hybridization is inevitable and an indispensable condition of the further existence of any speech community.

References

1. Y. V. Kobenko, *Linguistic situation in the Federal Republic of Germany: Americanization and Exoglossal tendencies* (Tomsk Polytechnic University publishing house, Tomsk, 2014) [in Russian]
2. Y. M. Lotman, *Rhetoric is the sense creating mechanism* (History, Moscow, 1996). [in Russian]
3. V. G. Zusman, Russian Germanic Studies: Yearbook of the Russian Germanic Scholars Union. **10**, 400 (2013)
4. Y. V. Kobenko, I. V. Sharapova Meditteranean Journal of Social Sciences MCSER Publishing, **6**, 1 (2015)
5. K.A. Shishigin, Vestnik KemGUKI, 29, 111 (2014) [in Russian]
6. S. V. Grinyov-Grinevich, Terminology studies, manual for graduate students, **304**, (2008) [in Russian]
7. Y.V. Kobenko, Language and the environment. Practice of classifying the data of the interdisciplinary research, **214**, (2017). [in Russian]
8. A. V. Kravchenko, Cultural Interaction and Language Hybridization, Homo Communicans: a person in the intercultural communications, **80** (2012) [in Russian]
9. Documents published by Bundesamt für Sicherheit in der Informationstechnik (BSI) [Online version] URL: https://www.bsi.bund.de/
10. OECD Richtlinien für die Sicherheit von Informationssystemen und -Netzen: Organisation for Economic Co-operation and Development (OECD), Paris, (2003)
11. D. Crystal, English Today **15**, 2 (1999)

12. I. A. Ushanova, Russian Germanic Studies: Yearbook of the Russian Germanic Scholars Union **10**, 400 (2013) [in Russian]

Experimental study of loss in stability of the drop shape in the approach air flow

Nikolay Zolotorev[1], *Ksenia* Perfilieva[1,*], and *Sergey* Polenchuk[1]

[1]Tomsk State University, 634050 Tomsk, Russia

Abstract. A new experimental scheme of installation to study the deformation of drops by aerodynamic forces was described. Results of experimental studies of the deformation of drops in the air flow were shown. Video shoots of the deformation of the drops for different values of Weber numbers were obtained. Experimental dependence of the degree of deformation of the drop from Weber number was obtained and an approximating formula for determining the degree of deformation of the drop was obtained.

1 Introduction

The processes of loss in stability of drops in approach gas flow leading to their deformation and fragmentation play an important role in the hydrodynamics of two-phase flows [1]. These processes are of practical importance in meteorology (the formation of a spectrum of droplet sizes of atmospheric precipitation [2]), in engine building (the dispersion of fuel droplets in internal combustion engines and liquid rocket engines [3, 4]), in ecology problems (evolution of a cloud of droplets of toxic components Liquid rocket fuels, formed during depressurization in the atmosphere of fuel tanks of carrier rockets [5]) and in a number of other branches of engineering and technology. In this paper, we consider an experimental facility for studying the deformation of a stationary droplet in an air flow. The main criterion for determining the deformation of the droplets in the gas flow is the Weber number that characterizes the correlation of forces of the dynamic pressure of the gas (aerodynamic force) to the force of surface tension. With increasing Weber number the drops deformation increases, and with some critical value it is splitting due to the instability of Kelvin-Helmholtz [1].

2 The scheme of the test installation for studying the deformation of drops by aerodynamic forces

To improve the accuracy of measurement of the degree of drop deformation and of the parameters that is included in the Weber number a new test installation for studying the deformation of drops by aerodynamic forces was proposed [6]. The scheme of installation is shown in Fig. 1. The installation includes a vertically arranged dropper with the capillary

* Corresponding author: k.g.perfiljeva@yandex.ru

1, the system of flow which directs drops vertically upwards and visualization system. The air supply system contains a battery of cylinders 2 with compressed air connected by tube through a reduction gear 3 with a control manometer 4 regulating valve 5 and a flowmeter 6 with the lower access of the cylindrical pipe 7, is located coaxially with dropper 1. A flow shaper 8 is in pipe 7, which is made in the form of at least six symmetrically arranged nozzle plates. The imaging system includes a video camera 9, which can check the original drop 10 in the cut of the capillary dropper 1, and two high-speed video camera 11, located with the possibility of registration of deformed drops 12 in perpendicular surfce in the output section of the nozzle 7.

For rgistration of original spherical drop 10 the digital video camera "the Panasonic HDC – SD60" was used, and for the registration of deformed drops 12 two high-speed cameras "Citius C 100" were also used. Video recording was performed with a spatial resolution of 384×790 pixels with a rate of 300 frames per second and exposure time (0.5÷2.0)MS. To control the distance made by the droplet, a scale with divisions of 1 mm was used. To measure the volumetric air flow a turbine flowmeter, calibrated with drum gasostatic GSB – 400 class was used, 1 [7] (measurement error rate of no more than 1%).

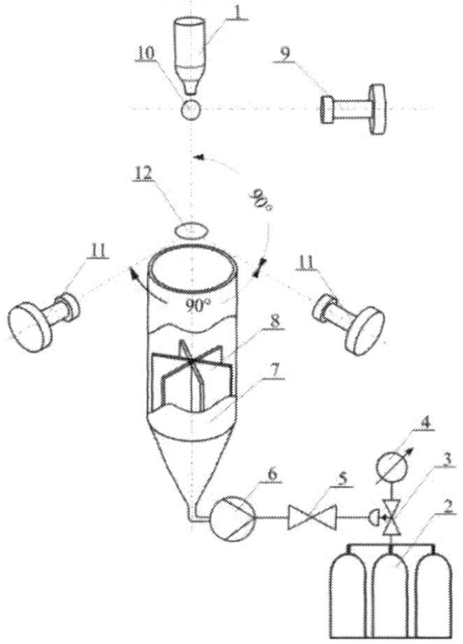

Fig. 1. The scheme of the test installation for studying the deformation of drops by aerodynamic forces.

The installation works like this. By means of gear 3 a predetermined volumetric flow rate of air Q is set, measured by the flowmeter 6. Using dropper with capillary 1 the original spherical droplet of the working fluid is formed (water-glycerol solutions, silicone oil, castor oil, etc.). At the time when a drop 10 separates its shape and diameter is determined by the processing of the video shoots, obtained by camera 9.

After separation of the original drop 10 from the capillary its gravitational sedimentation under the action of gravity happens. The counter equal flow of air from the nozzle 7 affects the drop, under which the drop is deformed. The size and the speed of the deformed drop 12 is recorded by two high speed video cameras 11, situated near the outlet section of the nozzle 7. The speed of movement of the drop is determined by frame-by-frame video processing.

Thus, this installation provides improved accuracy measurement of the degree of deformation of the drop and parameters that is included in the Weber number due to providing the well-controlled conditions of blowing the original spherical drop with a significant simplification of the experimental equipment of the installation and the alignment of the measuring apparatus.

3 Experimental study of the drop deformation in an air flow

Let`s consider the deformation of drops of glycerin in an air flow in conditions of indoor temperature. Let`s carry out the calculations of the parameters of the installation.

The diameter of the capillary is determined from the relation [8]:

$$d_C \le \frac{1}{6}\sqrt{\frac{\sigma \cdot \mathrm{Bo}_{cr}^3}{\rho_p \cdot g}}, \tag{1}$$

where σ is a surface tension coefficient of the liquid; Bo_{cr} the critical value of Bond number; ρ_p is the fluid density; g is the gravitational acceleration.

The maximum value of the diameter of the initial drop is determined from the relation:

$$D_0 \le \sqrt{\frac{\sigma \cdot \mathrm{Bo}_{cr}}{\rho_p \cdot g}}. \tag{2}$$

The diameter of the nozzle is calculated from the relation:

$$d_n \ge 5D_0. \tag{3}$$

The length of the pipe is determined from the relation:

$$l_n \ge 10d_n. \tag{4}$$

The number of Weber is the main criterion for determining the deformation of the drop in the gas flow [1]:

$$\mathrm{We} = \frac{\rho_g u^2 D_0}{\sigma}, \tag{5}$$

where ρ_g is the air density; u is the speed of the blowing of drops.

The rate of the blowing of drops is calculated by the formula:

$$u = u_p + u_g, \tag{6}$$

where u_p the speed of the drop in the outlet section of the nozzle, determined by a specific frame-by-frame processing of results of high-speed photography; u_g is speed of air flow in the outlet section of the nozzle.

The speed of air flow in the outlet section of the nozzle is calculated by the formula

$$u_g = \frac{4Q}{\pi d_n^2} = \frac{Q}{S_n}, \tag{7}$$

where Q is the volumetric consumption of air; S_n is the square of the sectional area of the pipe.

The Reynolds number is calculated by the formula

$$\mathrm{Re} = \frac{\rho_g u D_0}{\mu_g}, \quad (8)$$

where μ_g - is the dynamic viscosity of the air.

The Bond number is calculated by the formula:

$$\mathrm{Bo} = \frac{\rho_p D_0^2 g}{\sigma}. \quad (9)$$

The value of the degree of deformation is determined by the formula:

$$\varepsilon = \frac{D_m}{D_0}, \quad (10)$$

where Dm - is the diameter Miteleva cross-section of the deformed drop, which defined by the formula:

$$D_m = \frac{D_{m1} + D_{m2}}{2}, \quad (11)$$

where, D_{m1}, D_{m2} - are the diameters of Miteleva section of the drop, measured by the results of high-speed video shooting in two perpendicular surfaces.

Calculations, required for physical characteristics of air and glycerol at a temperature of 20 ° C, required for calculations, are shown in table 1.

Table 1. The physical properties of air and glycerol.

Parameter	**Air**	**Glycerol**
ρ, kg/m^3	1.205	1260
μ, Pa·s	$1.808\cdot10^{-5}$	–
σ, N/m	–	$63\cdot10^{-3}$

Parameters of a installation are shown in table 2.

Table 2. Parameters of a installation.

D_0, **mm**	d_c, **mm**	d_n, **mm**	l_n, **mm**
4.24	2.5	24	240

A series of experiments for different values of air flow consumption in the range Q=(1.11 to 2.78) dm^3/s is carried out and shown on a installation. The results of measurements and calculations are shown in table 3.

DOI: 10.1051/matecconf/201711001066

Table 3. Conditions and results of the experiment (Bo = 3).

Q, dm^3/s	u_p, m/s	u_g, m/s	u, m/s	We	ε	Re
0	2.98	0	2.98	0.66	1.01	777
1.11	2.76	2.46	5.22	2.03	1.07	1360
1.67	2.68	3.69	6.37	3.02	1.08	1657
2.22	2.44	4.91	7.35	4.02	1.15	1915
2.78	1.87	6.15	8.02	4.79	1.43	2087

Video shoots of deformed drops are shown in Fig. 2 for different values of Weber numbers.

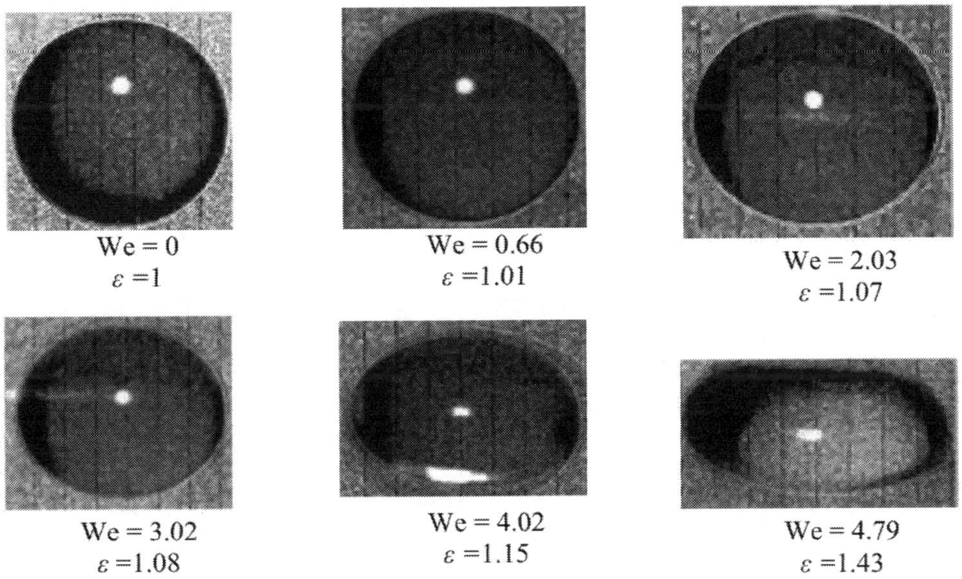

Fig. 2 Video shoots of deformed drops for different values of Weber numbers.

Experimentally obtained dependence of deformation degree of drops on Weber number (We) is shown in Fig. 3. From the above graph it follows that with increasing speed a blowing flow (or Weber number) the degree of deformation of drops monotonically increases.

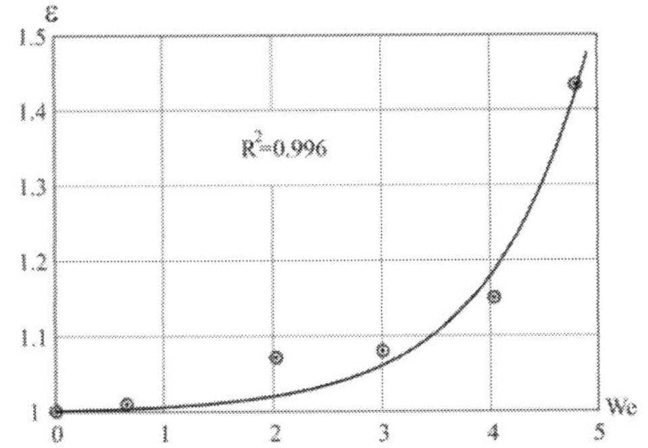

Fig. 3. The dependence of the degree of drop deformation of the number of Weber.

Approximation of the experimental data allowed to obtain the analytical formula (the coefficient of determination R2 =0.996):

$$\varepsilon = 1 + 2.5 \cdot 10^{-3} \cdot \exp(1.07 \cdot \mathrm{We})\,. \qquad (12)$$

4 Conclusion

- A modified scheme of installation for studying the deformation of drops by aerodynamic forces was presented.
- Video shoots of deformed drops for different values of Weber numbers were obtained.
- It was determined that by increasing the speed of a blowing flow (or Weber number) the degree of drop deformation monotonically increases.
- On the approximation of the experimental data, an analytical formula for calculating of the degree of drop deformation was obtained.

The study was supported by the Russian Science Foundation (No 15-19-10014).

References

1. R.I. Nigmatulin, *Dynamics of multiphase media* (Science, Moscow, 1987) [in Russian]
2. A.T. Matveev, *Fundamentals of General Meteorology* (Gidrometeoizdat, Leningrad, 1965) [in Russian]
3. A.P. Vasiliev, V.M. Kudryavtsev, V.A. Kuznetsov, *Fundamentals of the theory and calculation of liquid rocket engines* (Higher School, Moscow, 1983)
4. V. Arkhipov, A. Antonnikova, S. Basalayev, I. Zharova, S. Orlov, EPJ Web Conf. **110**, 01002 (2015)
5. V.A. Arkhipov, E.A. Kozlov, S.S. Titov, A.S. Tkachenko, A.S. Usanina, I.K. Zharova, Arabian J. Geosciences, **9**, 114 (2016)
6. The application on patent RU N 2016149916, IPC B01L 99/00. Test installation for studying the deformation of drops by aerodynamic forces was suggested / V. A. Arkhipov G. R. Schrager, A. S. Usanina, S. A. Basalayev, S. N. Polenchuk, K. G. Perfilyeva, (Filing date: 19.12.2016) [in Russian]
7. V.M. Plotnikov, V. A. Podreshetnikov, L. N. Teterevyatnikov, *Devices and means of the accounting of natural gas and condensate* (Nedra, Leningrad, 1989) [in Russian]
8. R. Clift, J.R. Grase, M.E. Weber, *Bubbles, drops and particles* (London, Academic Press, 1978)

Numerical analysis of the heat loss of stop valves of heat networks

Vyacheslav Polovnikov[1,*], *Anatolij* Vergun[1], and *Vasilij* Sergeenko[1]

[1]National Research Tomsk Polytechnic University, 634050 Tomsk, Russia

Abstract. The results of mathematical modeling of heat regimes of stop valves of heat networks were given. It was revealed that for the stop valves with a relatively small nominal bore the equivalent pipe length can be reduced more than twice.

1 Introduction

Stop valves are used in various industries [1-5]. It is known [5] that heat loss of a thermal network are made up of linear (heat loss of pipeline sections that do not have fittings and fittings) and local heat loss (heat loss of shaped parts, stop valves, supporting structures, flanges).

2 Problem statement

Heat loss of bends, bent compensators and other parts are calculated by the formulas for straight pipes [6]. Heat loss of flanges, fittings and fittings are determined in equivalent pipe lengths of the same diameter by the formula [5]:

$$Q_l = q \cdot l_e \qquad (1)$$

Heat loss from a non-insulated valve or gate are assumed to be equal to heat loss of an isolated pipeline of length l_e = 12-24 m of the same diameter. The equivalent length of a valve or gate isolated on 3/4 of a surface, depending on the diameter of the pipeline and the temperature of the heat carrier, can be taken as 4-8 m of an insulated pipeline. The equivalent length of the uninsulated flange can be taken as 4-5 m of an insulated pipeline.

This approach to the assessment of local heat loss is empirical, based on years of operation of heat networks, but has no theoretical basis. The main reason for the absence of theoretical studies of the thermal regimes of fittings and shaped parts (flanges, latches, etc.) of heat pipes of thermal networks is their rather complicated geometry.

This problem can be solved with using modern approaches to modeling thermal conditions and heat loss of heat transfer systems [5], based on numerical modeling of physical processes.

* Corresponding author: polov@tpu.ru

The purpose of this paper is to analyze of heat loss of ball valves, as one of the types of typical stop valves of heat networks.

3 Mathematical model and method of solution

As an example, we consider not an isolated ball valve of the company Broen Ballomax of the lineup 62.102.000. The heat transfer on the external surfaces of the ball valve is carried out under conditions of radiative heat transfer and natural convection, the ball valve is in the open state, and the internal surfaces are maintained at a constant temperature equal to the temperature of the heat carrier.

For the case under consideration, a three-dimensional stationary heat conduction problem is solved in the "ball valve – environment" system. The mathematical model for the problem under consideration includes the heat conduction equation for the valve body with the corresponding boundary conditions setting the problem. The solution of the problem was obtained using tools and functions of the COMSOL Multiphisics modeling software package. The studies were carried out on a non-uniform finite element mesh. The number of elements was chosen from the conditions for the convergence of the solution.

4 Initial data

Investigations were carried out for ball valves with nominal conduit diameters of 125, 250 and 500 mm. Ball valves are made of steel ST 37, the degree of blackness of the surface was assumed equal to 0.8. The temperatures of the heat carrier and the environment were 100 °C and 25 °C, respectively The average value of the heat transfer coefficient on the outer surface of the valve body was assumed equal to 5 W/(m2·°C) [5, 6].

5 Results of numerical simulation

The main results of the numerical study of thermal conditions and heat loss of the ball valves are given in the table in the figure.

Validity and reliability of the obtained results follow from tests of the methods for convergence and stability of solutions on multiple meshes, fulfillment of the energy balance conditions at boundaries of the calculation domain. The relative calculation error in all versions of the numerical analysis did not exceed 0.1%, which is acceptable for investigations of thermal conditions of the system under consideration.

The table presents the results of a numerical calculation of the heat loss from non-isolated ball valves Q, the local heat losses Q_l calculated from formula (1), the values of the standard linear heat losses q [6], the equivalent tube lengths l_e, chosen on the basis of recommendations, equivalent lengths pipes l, calculated as a function of Q, as well as a comparison of Q and Q_l with each other.

The results of numerical simulation of heat loss from non-isolated ball valves indicate an increase in heat loss with an increase in the diameter of the conditional passage of a ball valve.

Comparison of the results of numerical modeling Q and local heat losses Q_l (table) allows us to conclude that the discrepancy between them is from 6.9 % to 123.3 %. Therefore, it is possible to reduce the estimated equivalent length of the pipeline l_e for relatively small conventional diameters.

Recalculation of the equivalent lengths of pipelines l in accordance with the results of numerical simulation (table) allows us to say that for ball valves with relatively small conditional passages, le can be reduced by more than two times.

Table. Results of numerical analysis.

Nominal conduit diameters, mm	**125**	**250**	**500**
Q, W	204.2	633.9	1894.6
Q_l, W	456	855	1764
q, W/m [5]	38	57	98
l_e, m	12	15	18
$l = \dfrac{Q}{q}$, m	5.37	11.12	19.33
$\dfrac{\lvert Q - Q_l \rvert}{Q} \cdot 100$ %	123.3	34.9	6.9

In the figure, as an illustration of the simulation results, a typical temperature field of a ball valve with a nominal conduit diameter of 500 mm is given for the conditions considered in this paper.

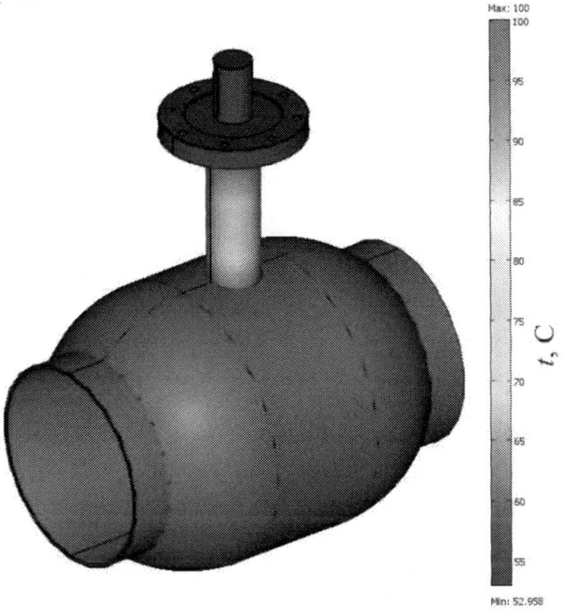

Fig. 1. The temperature field of a ball valve with a nominal conduit diameter of 500 mm.

6 Conclusion

Based on the numerical simulation of thermal regimes of typical stop valves of thermal networks, the scales of heat loss of ball valves were theoretically established for the first time.

It is shown that for ball valves with relatively small conditional passages the equivalent length of the pipeline can be reduced by more than two times.

7 Notations

Q_l – local heat loss, W; q – linear heat loss, W/m; l_e – equivalent pipe length, m.

The reported study was funded by RFBR according to the research project No 16-08-00229.

References

1. A.S. Drogalev, V.S. Balashkov, A.P Vergun, Proceedings - 2012 7th International Forum on Strategic Technology, 6357761 (2012)
2. A. Razva, M. Vasilevsky, A. Pleshco, A. Asan, MATEC Web Conf. **72**, 01122 (2016)
3. M. Vasilevsky, A. Razva, B. Roman, A. Alibek, EPJ Web Conf. **110**, 01054 (2016)
4. V.Y. Polovnikov, F.T. Makhsutbek, Z.F. Ozhikenova, EPJ Web Conf. **110**, 01049 (2016)
5. V.Y. Polovnikov, N.V. Razumov, EPJ Web Conf. **110**, 01064 (2016)

MATEC Web of Conferences **110**, 01068 (2017) DOI: 10.1051/matecconf/201711001068
HMTTSC-2017

Impact of the ash deposits from coal combustion on thermal conditions of the furnace pipes

Alexander Subbotin[1,*], *Igor* Razov[2], *Elena* Popova[1], *Alexander* Dolgih[1], and *Anna* Mazanik[1]

[1]National Research Tomsk Polytechnic University, 634050 Tomsk, Russia
[2]Tyumen industrial University, 625000 Tyumen, Russia

Abstract. One of the problems of thermal power plants is the contamination of heating surfaces that arises during the boiler unit operation while combusting the solid fuel. The main part of the energy is consumed by pipes forming the furnace. Non-combustible mineral part of the fuel in the form of the ash and slag accumulates on the heating surfaces and thereby prevents heating of the working liquid. Therefore, boiler requires higher fuel consumption to maintain the operating conditions. Violation of the thermal conditions of heating surfaces in consequence of contamination causes the negative effects on metal pipes right up to failure. This problem is urgent due to the fact that information about the process of slagging and its effect on the thermal conditions of the heating surface operation enables the possibility to minimize the negative consequences. This possibility is achieved by adjusting the thermal conditions of the combustion process in the furnace with the periodic cleaning of the heating surfaces.

1 Formulation of the problem

One of the problems of thermal power plants is the contamination of heating surfaces that arises during the boiler unit operation [1]. Ash and slag accumulates on the heating surfaces and thereby prevents heat output. Therefore, boiler requires higher fuel consumption to maintain the operating conditions [2, 3]. We consider a furnace pipe to have a layer of deposits on its surface. The diameter of the furnace pipe is much larger than its thickness (60 × 6 mm). Thus, the influence of the curvature of the wall can be neglected [2]. In this case, the model of a flat homogeneous wall is implemented. The combustion process is adopted to be unsteady. For a steady-state solution we must specify a calculation time that will not affect the result [4, 5]. We use the Fourier-Kirchhoff differential equation to describe the unsteady heat transfer by thermal conductivity:

$$\rho \cdot c \cdot \frac{\partial T}{\partial t} = \frac{\partial}{\partial x}\left(\lambda \cdot \frac{\partial T}{\partial x}\right) + \frac{\partial}{\partial y}\left(\lambda \cdot \frac{\partial T}{\partial y}\right) + \frac{\partial}{\partial z}\left(\lambda \cdot \frac{\partial T}{\partial z}\right), \tag{1}$$

* Corresponding author: subbot@tpu.ru

where ρ – density, c – specific heat, λ – coefficient of thermal conductivity.

Equation (1) describes temporal and spatial temperature variation at any point of the body. Since the considered model is a flat wall, and the heat flux occurs in one direction, we can move to the one-dimensional equation of non-stationary heat transfer:

$$\rho \cdot c \cdot \frac{\partial T}{\partial t} = \frac{\partial}{\partial x}\left(\lambda \cdot \frac{\partial T}{\partial x}\right), \tag{2}$$

To solve this problem, we apply the numerical method of finite differences. The solid body is represented as a plurality of nodes that are described by the algebraic equation of temperature. For nodes located on the boundary of a solid body we use separate equations for boundary conditions. As a result, algebraic equations are obtained for n nodes to be solved by computer-aided technologies [4, 5]. When a single-layer wall is considered, one boundary condition is written for a contaminated wall, depending on the number of layers, two or more. By the iterative approximations, the values of the temperature gradient in the layers are found. At the boundary of contamination and metal there is a contact resistance, because of which two boundary conditions of the third kind at known ambient temperatures T_{B1}, T_{B2}, the contact resistance r_k and the heat transfer coefficient to the steam-water mixture α_2 are recorded.

The first layer:

$$x = 0, \quad -\lambda \frac{\partial T}{\partial x} = q_{C1}, \quad t > 0; \tag{3}$$

$$x = \delta, \quad -\lambda \frac{\partial T}{\partial x} = \frac{1}{r_k} \cdot \left(T_{B2}^{k} - T\right), \quad t > 0, \tag{4}$$

where q_{c1} – heat flux density of thermal radiation:

$$q_{C1} = \varepsilon\sigma\left(\left(\frac{T_1}{100}\right)^4 - \left(\frac{T_{B1}}{100}\right)^4\right) x = 0, \tag{5}$$

The second layer:

$$x = 0, \quad -\lambda \frac{\partial T}{\partial x} = \frac{1}{r_k} \cdot \left(T_{B1}^{k} - T\right), \quad t > 0; \tag{6}$$

$$x = 0, \quad -\lambda \frac{\partial T}{\partial x} = \alpha_2 \cdot \left(T_{B2} - T\right), \quad t > 0. \tag{7}$$

The temperatures T_{B1}^{k} , T_{B2}^{k} at the boundary are :

The first layer:

$$T_1^{j+1} = T_2^{j+1} + q_{C1}\frac{h}{\lambda_1}, \quad j = 0, 1, 2, \ldots, \tag{8}$$

$$T_N^{j+1} = \frac{T_{N-1}^{j+1} + Bi_1 \cdot T_{B2}^{k}}{1 + Bi_1}, \quad j = 0, 1, 2, \ldots; \tag{9}$$

DOI: 10.1051/matecconf/201711001068

The second layer:

$$T_1^{j+1} = \frac{T_2^{j+1} + Bi_1 \cdot T_{B1}^k}{1 + Bi_1}, \quad j = 0, 1, 2, \ldots, \tag{10}$$

$$T_N^{j+1} = \frac{T_{N-1}^{j+1} + Bi_2 \cdot T_{B2}}{1 + Bi_2}, \quad j = 0, 1, 2, \ldots, \tag{11}$$

Where $Bi_1 = \frac{h}{r_k \cdot \lambda_1}$, $Bi_2 = \alpha_2 \cdot \frac{h}{\lambda_2}$ is the grid analog of Bio, h is the step of the difference grid.

Inside the layer, to determine the temperature at each point, it is calculated both under the boundary conditions of the first kind:

$$T_i^{j+1} = T_i^j + \frac{\lambda \cdot \tau}{\rho \cdot c \cdot h^2}\left(T_{i+1}^j - 2T_i^j + T_{i-1}^j\right). \tag{12}$$

Boundary conditions of the third kind require the application of stringent conditions for the stability of the algorithm:

$$\frac{\lambda \cdot \tau}{h^2}\left(1 + Bi\right) < 0.5. \tag{13}$$

2 Analysis of numerical results

The algorithm for calculating the three layers is similar to the calculation of the two layers with addition of the new data. The initial data (Table 1) [6] were the thermal conductivity of annulus contamination. This contamination is caused by burning of lignite of the Kansko-Achinsk basin at Krasnoyarsk district (Russian Federation). The coefficient of thermal conductivity for all samples depends on the temperature and the composition of the contamination. For sulfate-bound deposits the coefficient of thermal conductivity is slightly less than 0.1 W/mK, for ferriferous it lies in the range of 1.4 to 1.5 W/mK.

Table 1. – Initial data.

Data / Layer	Coefficient of thermal diffusivity, m^2/s	Coefficient of thermal conductivity, W/(mK) [4]	Thickness, m	Amount of points	Calculation time, s	Flare temperature, K	Temperature of the working fluid, K	Temperature of liquid ash state [5]
Slag	$8.33 \cdot 10^{-7}$ - $3.13 \cdot 10^{-6}$	1-1.6	0.001	20	600	2000	473	1503
Contamination	$4.74 \cdot 10^{-7}$ - $1.76 \cdot 10^{-6}$	0.8-1.5	0.001-0.003	20-80	600			
Pipe	$1.38 \cdot 10^{-5}$	50	0.006	120	600			

The contact resistance according to the experimental data [6] is a function in terms of temperature:

$$r_k = -7.247 \cdot 10^{-7} \cdot T + 0.000647 \tag{14}$$

The heat transfer coefficient on the outside surface is $\alpha_2 = 4000 W/(m^2K)$. The estimated time is 600 seconds and enables to come to a steady state.

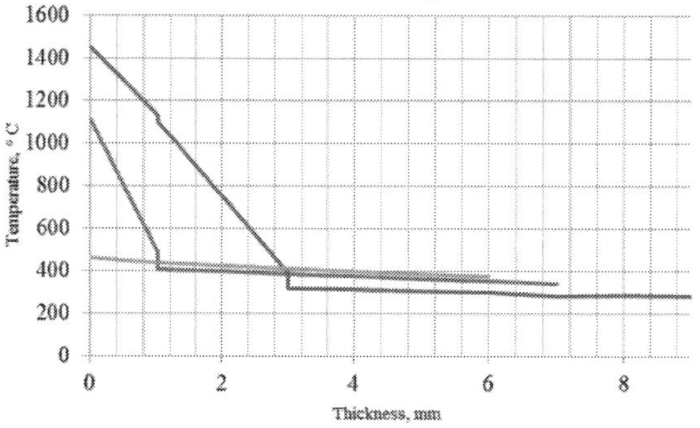

Fig. 1. Temperature distribution in the sediment layer (A).

This model does not take into account the heat sink by the steam-water mixture and is limited by the melting temperature of the slag. Different heat transfer acts if the slag turns into a fluid state. Therefore, a thin liquid non-flowing barely 1 mm layer of the slag was investigated over the impurities. Using the computer-aided technology, the developed algorithm is rewritten in the Pascal programming language. The results of the calculations are shown in Figures 1-3.

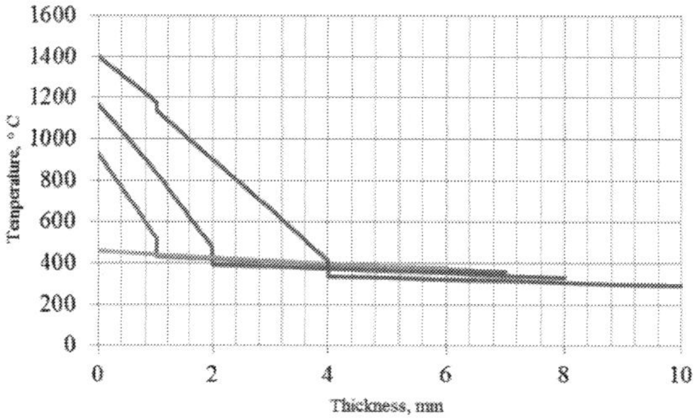

Fig. 2. Temperature distribution in the sediment layer (B).

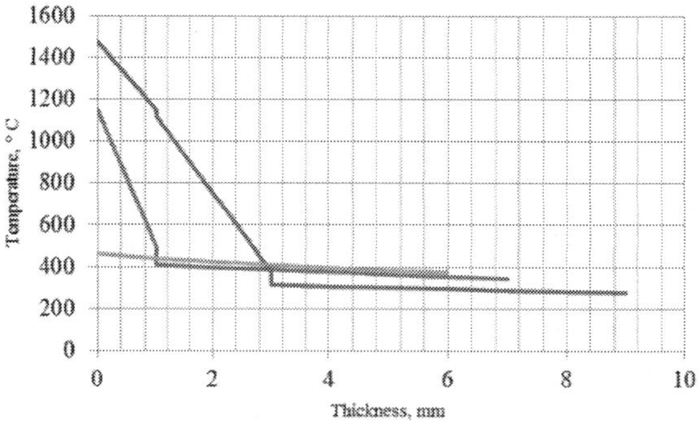

Fig. 3. Temperature distribution in the sediment layer (C).

Fig. 1-3 show that the temperature on the external surface rises sharply, but the temperature in the metal wall decreases significantly due to the contact resistance. Higher thickness of the contaminants causes the higher heat losses. The maximum thickness of contaminants reaches up to 0.5 m. At this thickness value the heat exchange is minimum or absent depending on the composition of the deposits. Iron deposits (B) can have a large thickness with equal temperature distribution with sulfate-bound (A) and silicate (B). But this fact is useless, since they are hard to remove. Periodic cleaning will minimize the negative impact of pollution.

References

1. S.A. Khaustov, Y.A. Belousova, K.V. Buvakov, A.Y. Dolgih, R.N. Kulesh, 11th International Forum on Strategic Technology (IFOST 2016) Proceedings **2**, 548 (2016)
2. K.Y. Shcherbakova, P.S. Gergelizhiu, B.V. Lebedev, MATEC Web Conf. **23**, (2015)
3. K.Y. Orlova, B.V. Lebedev, MATEC Web Conf. **92**, 01044 (2016)
4. G.V. Kuznetsov, V.Y. Polovnikov, J. Eng. Thermophys. **20**, 2 (2011)
5. G.V. Kuznetsov, A.E. Sitnikov, Inzhenerno-Fizicheskii Zhurnal, **75**, 4 (2002) [in Russian]
6. A.S. Zavorin, Y.Y. Rakov, Therm. Eng. **47**, 12 (2000)

Heat Transfer in Human Skin Exposed to Radiation from Forest Fire Taking Into Account Moisture Evaporation

Nikolay Poptsov[1], *Nikolay* Baranovskiy[1,*]

[1]National Research Tomsk Polytechnic University, 634050 Tomsk, Russia

Abstract. Thermal radiation is one of the damaging factors of forest fire. As a result of exposure to radiant heat flow, thermal damage of the human skin is possible. A layer of skin is considered with effective thermophysical characteristics. Thin layer of sweat excretion situated on top of the skin. A mathematical model of heat transfer in the system of "skin-excretion" is developed taking into account evaporation of moisture. Typical simulation results are presented.

1 Introduction

The problem of the damaging factor impact of the forest fire on living organisms has recently become relevant. One of these factors is radiation from a forest fire. Thermal damage can occurs as a result of exposure to radiant heat flux on the skin [1].

Over the past decade, the scope of intensive research and application of heat exchange phenomena has greatly expanded. It includes both the leading directions of technology (chemical technology, metallurgy, construction, oil production, engineering, agricultural technology) and the basic natural sciences (biology, geology, atmospheric and ocean physics). At present time theoretical research of heat exchange processes is largely based on their numerical simulation using computers. This became possible due to significant progress in the development of computational methods to solve partial differential equations and increasing the performance of up-to-date computers.

The goal of the present work is to numerically study the thermal effect on the skin of radiation from forest fires taking into account moisture evaporation. To achieve this goal, the effect of heat radiation on the human skin near the fire is studied.

2 Mathematical Setting-Up

The skin is a complex structure with its own peculiarities in various parts of the body. It consists of three layers (Fig. 1) - epidermis, dermis and subcutaneous tissue (hypodermis) [1].

* Corresponding author: firedanger@tpu.ru

The epidermis includes mostly dead cells that act as a protective barrier against moisture, ultraviolet radiation and high temperatures [2]. Sweat glands pass through the epidermis. The epidermis and dermis have a thickness of about 1-2 mm. These two layers are subcutaneous tissue (hypoderma), which contains sweat glands and large blood vessels, allowing the skin to adapt to local heating or cooling, but only within certain limits [3].

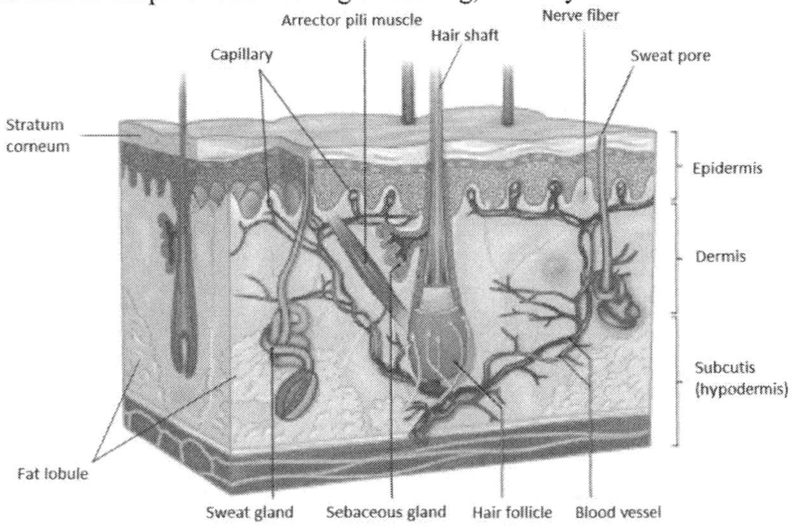

Fig. 1. Structure of the skin.

It is assumed that the examined area of the skin is affected by convective and radiant heat flux from the forest fire front. Thermophysical characteristics do not depend on temperature. The monolithic structure of the skin is considered. The heat transfer is leaded within the skin by conduction. Skin is considered as one layer with effective thermophysical characteristics.

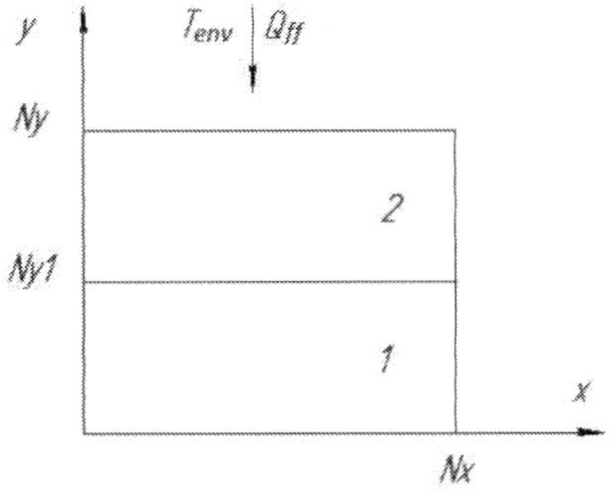

Fig.2. Geometry of the solution area: 1 – skin; 2 – sweat layer.

The geometry of the solution area is shown in Fig. 2. Heat transfer is described by non-stationary differential heat equations with the corresponding initial and boundary conditions:

$$c_1\rho_1 \frac{\partial T_1}{\partial t} = \lambda_1 \left(\frac{\partial^2 T_1}{\partial x^2} + \frac{\partial^2 T_1}{\partial y^2} \right), \quad (1)$$

$$c_2\rho_2 \frac{\partial T_2}{\partial t} = \lambda_2 \left(\frac{\partial^2 T_2}{\partial x^2} + \frac{\partial^2 T_2}{\partial y^2} \right), \quad (2)$$

Initial conditions :

$$t = 0, T_i = T_{i0}, \quad (3)$$

Boundary conditions :

$$x = 0, \lambda_i \frac{\partial T}{\partial x} = 0 , \quad (4)$$

$$x = Nx, \lambda_i \frac{\partial T}{\partial x} = 0 , \quad (5)$$

$$y = Ny, -\lambda_2 \frac{\partial T}{\partial y} = \alpha \left(T_{env} - T \right) + \varepsilon\sigma \left(T_{env}^4 - T^4 \right) - Q_{ev} \cdot W_{ev} + Q_{ff} \quad (6)$$

$$y = Ny_1, \lambda_1 \frac{\partial T_1}{\partial y} = \lambda_2 \frac{\partial T_2}{\partial y}, T_1 = T_2 , \quad (7)$$

$$y = 0, \lambda_1 \frac{\partial T_1}{\partial y} = 0 , \quad (8)$$

$$W_{ev} = \frac{A \left(P_0 \cdot \exp\left(-\frac{E}{RT} \right) - P_{atm} \right)}{\sqrt{\frac{2\pi RT}{M}}}, \quad (9)$$

Where T_i, ρ_i, c_i, λ_i – temperature, density, heat capacity, heat conduction coefficient (i=1 – sweat layer, 2 – skin layer). x, y – spatial coordinates. t – time. α – heat transfer coefficient. Q_{ev} – heat effect of evaporation. Q_{ff} – heat flux from forest fire. ε – emissivity coefficient. σ – Stefan-Boltzmann constant. W_{ev} – evaporation mass rate. A – accommodation coefficient. P_0 – pre-exponent factor. E – activation energy. P_{atm} – atmospheric pressure. R – universal gas constant. M – molecular weight.

3 Results and Discussion

Partial differential equations of the parabolic type are used for the mathematical description of the heat transfer process. The numerical implementation is carried out using the finite-difference method [4,5]. As a result, temperature distributions were obtained in the "skin-excretion" structure. Numerical simulation was carried out for the following scenario. A person deprived of personal protective equipment situated at some distance from the front of a forest fire. Heat releases as a result of combustion of forest combustible materials. Heat transfer occurs to the human body through radiation [6].

For definiteness, the effect of radiant heat flux on a specific part of the body is considered (namely, the frontal part of the human hand palm [7]). Table 1 shows the initial data.

Table 1. Initial data.

Layer	Skin and sweat
Temperature of the environmnet, T_{env}, °C	900
Initial temperature, T_0, °C	36,6
Layers thickness, *mm*	1

According to the obtained results it was found that when the sweat layer evaporates at the time, the surface temperature of the skin reaches 54 ° C. Thickness of the heating is small in comparison with the dimensions of the layer under study. Comparative analysis allows to draw a conclusion about the correct operation of the presented software implementations for the mathematical model which describes heat transfer in the skin structure under the influence of radiant heat flux from the forest fire front [6].

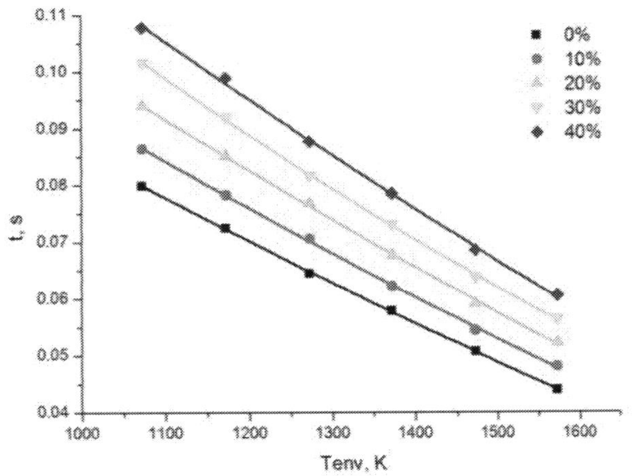

Fig. 3. Dependence of the pain threshold when the sweat evaporates from the skin surface from the ambient temperature at different salt content.

Analysis of the dependences in Fig. 3, 4 allows us to conclude that increasing the mass content of salt in the sweat leads to increase of evaporation time. The onset of the pain threshold increases the more caused by the lower ambient temperature. The pain threshold time ranges from 0.04 s (ambient temperature of about 1573 K and mass concentration of the salt in the sweat of 0%) to 0.1 s (ambient temperature of about 1073 K and mass concentration of the salt in the sweat of about 40 %). The time for complete moisture evaporation varies from 0.05 s (ambient temperature of about 1573 K and mass concentration of the salt in the sweat of 0 %) to 0.15 s (ambient temperature of about 473 K and mass concentration of salt in the sweat of 40%). The pain threshold times and complete evaporation grow with increasing concentration of salt in sweat.

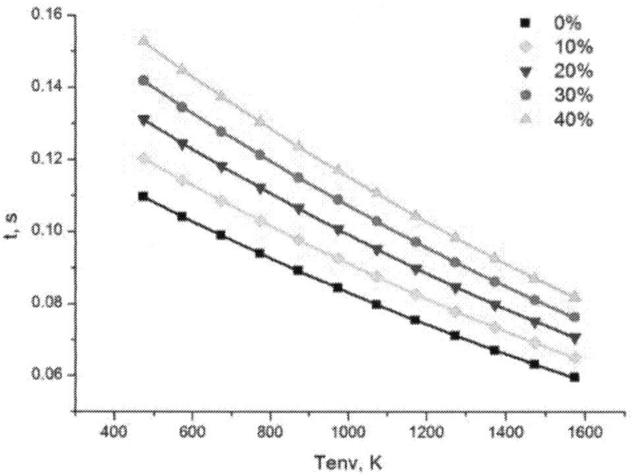

Fig. 4. Dependences of time on salt content for evaporation of whole sweat layer.

4 Conclusion

Mathematical model of heat transfer in the skin structure has been developed. Algorithm and program implementation of the skin warming has been carried out taking into account radiation and convection with evaporation of the sweat layer in a high temperature environment. The software components have been created to simulate the heat transfer in the skin structure. These might be applied to the development of a new generation medical information systems [8,9] and also for integration with GIS of forest fire hazard prediction [10,11] based on deterministic-probabilistic approach [12].

Authors are glad to express thanks to Dr. Semen V. Syrodoy for useful comments and discussion of present work.

Work is impelemnted within the project N 7538.2016.8 (Granted by President of Russian Federation within Leading Scientific Schools Program)

References

1. J.T. Whitton, J.D. Everall, Br. J. Dermatol. **89** (1973)
2. F. Xu, T.J. Lu, K.A. Seffen, Acta Mechanika Sinica, **24** (2008)
3. A.E. Pushkareva Laser selective heating of skin tissue elements: PhD thesis. Saint petersburg: ITMO University. (2006) [In Russian]
4. E.N. Valendik, I.V. Kosov, Siberian Ecological Journal, **4** (2008) [In Russian]
5. D.V. Korobkina, N.V. Baranovskiy, MATEC Web Conf. **19** (2014)
6. A.A. Samarskii, P.N. Vabishchevich, *Computational Heat Transfer. Vol. 1. Mathematical Modelling* (Chichester, Wiley, 1995)
7. A.A. Samarskii, P.N. Vabishchevich, *Computational Heat Transfer. Vol. 2. The Finite Difference Method* (Chichester, Wiley, 1995)
8. S. Petter, A. Fruhling, Int. J. Medical Inf., **80** (2011)
9. N. Bricon-Souf, C. Verdier, A. Flory, M.C. Jaulent, IRBM, **34** (2013)
10. E.P. Yankovich, N.V. Baranovskiy, SGEM, **1**, 2 (2014)
11. E.P. Yankovich, N.V. Baranovskiy, K.S. Yankovich, 2014 9th International Forum on Strategic Technology, IFOST (2014)
12. N.V. Baranovskiy, J. Autom. Inf. Sci. **47** (2015)

MATEC Web of Conferences **110**, 01070 (2017) DOI: 10.1051/matecconf/201711001070
HMTTSC-2017

Mathematical modeling of melting during laser heating of metal plate

Mikhail Purin[1,*], *Arkadii* Zakharevich[1], and *Nadegda* Gutareva[1]

[1]National Research Tomsk Polytechnic University, 634050 Tomsk, Russia

Abstract. The mathematical modeling of heat transfer processes proceeding together under the conditions of intense phase transformations (melting and metal evaporation) under laser action on a metal plate has been carried out. The analysis of the results has showed that increasing of the laser radiation (q_l) power leads to acceleration of the melting process. However, it has been found that the change in the melting dynamics with variation of q_l is characteristic only for the plates made of iron and copper. At the same time, it has been found that an increase in q_l does not affect significantly the melting rate of the aluminum plates. It has been established that when the laser is exposed to a plate, zones with large temperature gradients are formed which can cause thermal stresses.

1 Introduction

The modern industry is linked inextricably with the metal processing operations, such as cutting, welding, micro-remelting and others. Traditionally, they have been performed mechanically on the metalworking machines, however, due to constantly stricter requirements for precision and reliability of the machined parts, a new type of metal processing — laser treatment is becoming very popular and in demand today. The latter has a number of the undoubted advantages [1], such as high accuracy of processing, the ability to work with the complex geometric shapes, the absence of the mechanical damage to the workpiece, etc. The high density of power of the laser in the area of the radiation scatter spot ensures the performance of the process in the combination with the high quality of surfaces treatment.

Laser metals processing operations are linked inextricably with melting and crystallization processes, and the effective studies of the phase transformations are impossible without the computational experiments that allow modeling of the various conditions of action prior to the onset of the costly in-situ experiments.

The mathematical modeling of the processes of metals laser processing is of great interest. The russian scientists V.I. Mazhukin [2-4] and A.A. Samarskii [5,6] have made a significant contribution to the solution of the Stefan problem in the multidimensional formulation, but the methods for solving the Stefan problem proposed earlier are based on the explicit separation of the fusion front, are very complex, require the use of a complex

* Corresponding author: mvp21@tpu.ru

mathematical apparatus. It should also be noted that, for example, the models [2-4] require the use of the special functions that determine the dynamics of diffusion of the nodes of the difference grid. Their values can only be found from an experiment. For this reason, we can say that the development of a mathematical apparatus for the melting of metals under the conditions of local action of laser radiation is an actual and not solved problem to date.

The aim of the work is mathematical modeling of the processes of heat and mass transfer proceeding together under the conditions of intense phase transformations (melting, evaporation of metal) under the action of a laser.

2 Physical statement of the problem

Laser radiation begins to act on the bounded region of the surface of the metal plate at the initial moment of time. As a result, the metal begins to heat up. When the phase transition temperature is reached in the heating region, the melting process is initiated. The fusion front moves deeper into the material. As a result, a heterogeneous structure is formed, consisting of a solid and a liquid phase with the different thermophysical characteristics. As a result of continuous heating, the evaporation process is initiated.

When solving the problem, it has been assumed that the thermal conductivity, heat capacity and density of the metal depend on the position of the melting front. The thermophysical characteristics of the metals under study are given in Table 1. The heat transfer in the wall (to the metal plate) gas region has been taken into account in the simulation. The thermophysical characteristics of the latter are given in Table 2.

It has been assumed that the laser beam is directed perpendicular to the surface under consideration, the radiation flux density in the heating zone is distributed regularly.

Table 1. The thermophysical characteristics of metals [7].

	Solid phase			Fluid phase		
	c, J/(kg·K)	λ, W/(m·K)	ρ, kg/m^3	c, J/(kg·K)	λ, W/(m·K)	ρ, kg/m^3
Aluminum	951.3	240	2712	1090	90,7	2368
Iron	450	80	7870	835	39	7069
Copper	385	380	8933	545	165,6	8000

Table 2. The thermophysical characteristics of gas [8].

	c, J/(kg·K)	λ, W/(m·K)	ρ, kg/m^3
Argon	519	0.0394	1.661

Table 3. Heat of fusion, heat of vaporization and the melting point of metals.

	Q, kJ/kg	Q_{melt}, kJ/kg	T_{melt}, °K
Aluminum	393	10900	873
Iron	277	6300	1810
Copper	213	4800	1356

3 Mathematical formulation of the problem

The mathematical model of the process under study includes the energy equation for a metal, with taking into account the dependence of the thermophysical characteristics on temperature:

$$c(T)\rho(T)\frac{\partial T}{\partial t} = \frac{\partial}{\partial x}\lambda(T)\frac{\partial T}{\partial x} + \frac{\partial}{\partial y}\lambda(T)\frac{\partial T}{\partial y} - \frac{QW}{h_d}\delta(x_{melt}, y_{melt}); \quad (1)$$

- energy equation for gas:

$$c\rho\frac{\partial T}{\partial t}=\frac{\partial}{\partial x}\lambda\frac{\partial T}{\partial x}+\frac{\partial}{\partial y}\lambda\frac{\partial T}{\partial y}, \qquad (2)$$

here c is the heat capacity, J/(kg·K); λ is the coefficient of thermal conductivity, W/(m·K); ρ — density, kg/m^3; Q — heat of fusion, J/kg; W is the mass melting rate of the metal, kg/(m^2·s); h_d is the melting front smearing parameter, m; δ (x_{melt}, y_{melt}) is the Dirac delta function. The system (1) - (2) has been solved under the following boundary conditions (Figure 2):

$$t=0;\ 0<x<l;\ 0<y<H;\ T(x,y)=T_s;$$

$$0<x<l;\ 0<y<H;\ T(x,y)=T_g$$

$$\left.\frac{\partial T}{\partial x}\right|_{x=0}^{0<y<H_g}=\left.\frac{\partial T}{\partial x}\right|_{x=l}^{0<y<H_g}=\left.\frac{\partial T}{\partial y}\right|_{y=0}^{0<x<l}=\left.\frac{\partial T}{\partial y}\right|_{y=H_g}^{0<x<l}=0;$$

$$\lambda(T)\left.\frac{\partial T}{\partial y}\right|_{y=H}^{0<x<x_1}=\lambda_g\left.\frac{\partial T}{\partial y}\right|_{y=H}^{0<x<x_1};$$

$$\lambda(T)\left.\frac{\partial T}{\partial y}\right|_{y=H}^{x_1<x<x_2}=\lambda_g\left.\frac{\partial T}{\partial y}\right|_{y=H}^{x_1<x<x_2}+q-W_{isp}\cdot Q_{isp};$$

$$\lambda(T)\left.\frac{\partial T}{\partial y}\right|_{y=H}^{x_2<x<l}=\lambda_g\left.\frac{\partial T}{\partial y}\right|_{y=H}^{x_2<x<l};$$

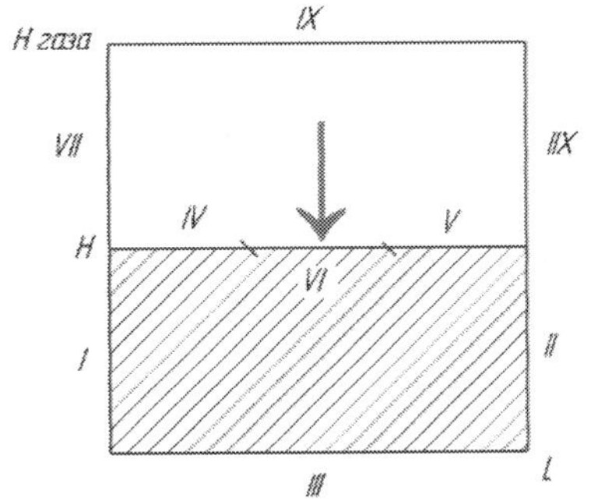

Fig. 2. Scope of the problem.

The problem has been solved by the method of finite differences with using an implicit four-point difference template with the use of the discontinuous coefficients [9, 10]. The nonstationary multidimensional Stefan problem has been solved by the method of implicit separation of the fusion front [11].

4 Results

The numerical studies have been carried out for aluminum, copper and iron plates. The choice of these metals is due to their wide application in the industries requiring high accuracy and reliability of material processing.

Figure 3 shows the melting times of the plates of iron, copper and aluminum. The analysis of the dependences of Fig. 3 shows that an increase in the power of the heat flux leads to a decrease in the melting time of the iron and copper plates, that is, to an increase in their melting rate. Thus, with an increase in the heat flux from 2 to 4.5 MW / m^2, the melting time of the iron and copper plates decreases from 210 to 150 and from 150 to 75 seconds, respectively. Other regularities are manifested when the aluminum plate melts. The increase in the power of the heat flux does not change practically the melting time of the sample. The latter is explained by the significant differences in their heat capacities: the heat capacity of aluminum is more than 2 times higher than the heat capacity of iron and almost 3 times higher than the heat capacity of copper (Table 1).

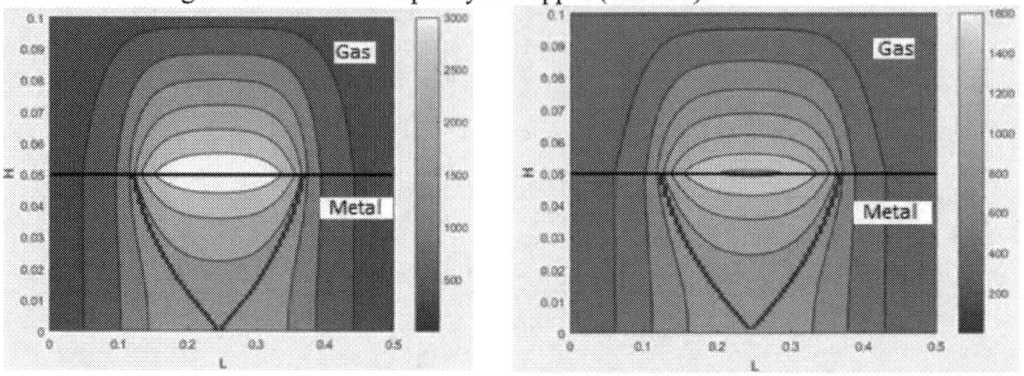

a b

Fig. 3. The typical temperature field in the "gas-metal" system: an iron plate (Fig. 4, a); an aluminum plate (Figure 4, b).

Figure 4 shows the typical temperature fields in the "gas-metal" system. One can note the formation of a characteristic bath of the molten metal in the laser action zone. The high temperature gradients are clearly visible, illustrating the possibility of forming zones of local thermal stresses in high-energy laser processing of metals. The latter can lead to the destruction of the metal blank. Also, the analysis of the temperature fields shows that in the aluminum blank, the temperature gradients are substantially less than in the iron blank. The latter is due to the higher thermal conductivity (Table 1) of Al compared to Fe. This can lead to high energy costs for laser processing of the aluminum surfaces, due to intensive heat removal through aluminum.

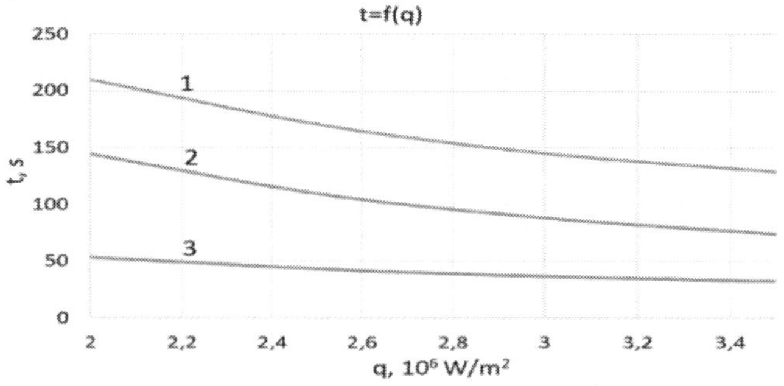

Fig. 4. Dependence of the time t of melting of plates from different metals on the magnitude of the incident heat flux q.

5 Conclusion

According to the results of the mathematical modeling of laser heating of a metal plate, the integral characteristics of the melting process of a metal have been established. It has been shown that a change in the radiation flux density can affect the melting conditions of the metal plates. However, the degree of influence is different and depends essentially on the material. So increasing the laser radiation power from 2 to 4.5 MW/m^2 (approximately 2 times) reduces the melting time of iron from 210 to 150 seconds (by 25%) and a copper plate from 150 to 75 seconds (by 50%). At the same time, the results of the numerical analysis have showed that the same increase in the radiation flux density does not have a significant effect on the time characteristics of aluminum melting (a decrease of $t_{пл}$ is not more than 10%). The latter is most likely due to the high heat capacity and thermal conductivity of aluminum. As a result of numerical simulation, it has been established formation in the plate of the high local temperature gradients in the zone of laser heating. Accordingly, it can reasonably be assumed that laser heating may cause the local temperature stresses leading to the destruction of the metal.

The work was supported by the scientific schools grant NSH-7538-2016.8.

References

1. A.G. Grigoriants, A.A. Sokolov, A.A. Grigor'yants, *Laser cutting of metals.* (Higher education. Shk., 1988)
2. V.I. Mazhukin, A.V. Shapranov, *Mathematical modeling of the processes of heating and melting metals* (Part I. Model and computational algorithm, 2012)
3. V.I. Mazhukin, A.V. Shapranov, *Molecular-dynamic modeling of heating and melting of metals* (Part II. Computational Experiment, Preprints of IPM, 2012)
4. O.N. Koroleva, V.I. Mazhukin, Zh. Vychisl, Math. And Math. Fiz. **46**, (2006)
5. A.A. Samarsky, P.N. Vabishchevich, *Computational heat transfer* (Editorial URSS, 2003)
6. A.A. Samarskii, B.D. Moiseenko, Zh. Vychisl, Math. and Math. Fiz. **5,** (1965)
7. V.S. Chirkin, *Thermophysical properties of materials* (FIZMATGIZ, 1959).
8. N.B. Vargaftik, *Reference book on thermophysical properties of gases and liquid.* (Moscow, 1972).
9. S.V. Syrodoy, N.Y. Gutareva, R.I. Taburchinov, MATEC Web Conf. **72,** (2016)
10. S.V. Syrodoy, N.Y. Gutareva, R.I. Taburchinov, MATEC Web Conf. **72**, (2016)
11. S.V. Syrodoy, N.Y. Gutareva, K. A. Bugaeva, MATEC Web Conf. **72**, (2016)

MATEC Web of Conferences **110**, 01071 (2017) DOI: 10.1051/matecconf/201711001071
HMTTSC-2017

The characteristic and separation effects in a cylindrical cyclone dust collector

Aleksandr Naumkin[1], *Aleksandr* Razva[1,*], *Michail* Vasilevsky[1], and *Vladimir* Arkhipov[2].

[1]National research Tomsk Polytechnic University, 634050 Tomsk, Russia
[2]National research Tomsk State University, 634050 Tomsk, Russia

Abstract. A comparison of the level of circular velocities and velocities of turbulent transfer in a coaxial channel and a countercurrent cylindrical cyclone is performed. Comparison of dust separation efficiency in these apparatuses is made.

1 Introduction

Cylindrical cyclone chambers are used as technological devices for burning, melting, drying, high-temperature reactions [1-7], they are used as dust concentrators in gas cleaning systems [8, 9], in installations of dust removal of air of systems of aspirations [10, 11]. Technical characteristics on devices are efficiency of dust removal of gases, coefficient of hydraulic resistance of the device which causes the capacity of the purified gas. Efficiency is defined as the ratio of the amount of dust entering the cyclone to the amount of dust trapped by the cyclone over a fixed period of time.

There is a wide variety of devices with built-in elements forming coaxial channels with constant or variable cross-sections for separating particles from a swirling gas stream: cyclones of LIOP (Leningrad Institute of Labor Protection), Institute of Siindmashcoalenrichment [9], cyclone with intermediate dust extraction [12]. In the elements of battery cyclones, there is a stabilization section of the swirling flow, which is a coaxial channel, in which the particles are "pressed" to the periphery [13]. Calculations of the effects of dust separation in a coaxial channel along particle trajectories have shown that efficiency decreases with decreasing distance between cylinders. The existing results of the operation of cyclones are inconsistent when assessing their effectiveness in cleaning gases and, therefore, it is required to clarify the effect of hydrodynamic factors on the separation of particles. In [14], the turbulent motion of an aerosol in a coaxial channel is considered (Fig. 1).

* Corresponding author: razva@mail.ru

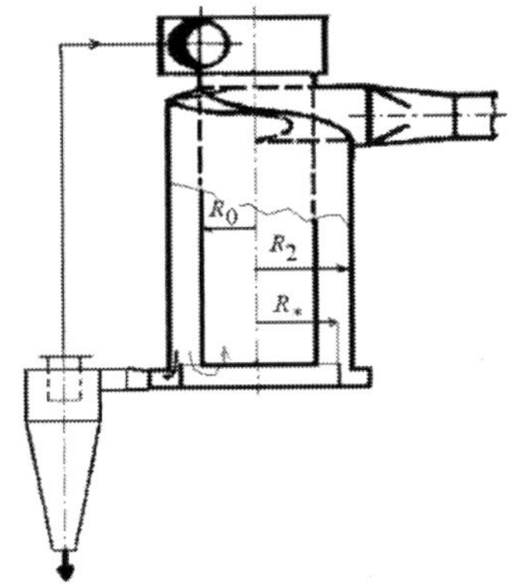

Fig. 1. Coaxial channel–1. Separation scheme with concentrator.

In the cylindrical part of the cyclone with an extended dip tube, the gas makes several revolutions in the coaxial channel, the variation in the flow parameters in the circumferential direction is insignificant and, on the whole, it is believed that the aerosol motion is axisymmetric. In [8], there are data on the optimum penetrations of the central branch pipe, from the point of view of separation efficiency, however, no physical justification is presented. The aim of the study is to distribute the circumferential velocities and to determine the separation efficiency for cylindrical cyclones with different coaxial channel lengths.

2 Experimental part

Measurement of speed and pressure in vortex chambers, associated with a significant disturbance of the flow of measuring equipment. The introduction of particles into the rotating flow leads to the deceleration of the tangential velocities, and the particle speed differs from the flow speed.

To investigate the causes of the disturbance and to determine their magnitude, studies were made of the errors of a radially injected pneumatic probe [14] in a chamber with an end-to-end injection. When the probe is inserted, the pressure on the chamber covers changes, which makes it possible to observe the influence of the probe on the vortex flow in the chamber.

The tangential speed in the vortex chamber is determined by the pressure gradient:

$$W_{\varphi}^{2} = \frac{R}{\rho}\frac{dp}{dR}. \tag{1}$$

Methods based on the sounding of speed fields are laborious and require the introduction of corrective coefficients to eliminate the effect of the probe on the velocity field of the flow. In order for the influence to be negligible, it is necessary to have a probe surface less than 2 percent of the cross-sectional area of the chamber and no more than 10 percent of the cross-section of the gas outlet.

For example, in the experiments of Smulsky I.I. [14] a capillary with an outer diameter of 1.4 mm was installed, installed in the diametrical direction of the chamber. For a chamber having a diameter of 160 mm and a diameter of the air outlet opening 29 mm in diameter, the ratio of the surface of the capillary to the exit area is $100 \cdot 1.4 \cdot 160/(29^2/4)$ = 106%.

Other methods based on use of thermo anemometers don't allow defining fields of pressure. Besides, the existing methods aren't suitable for sounding of the stream containing the weighed particles.

In the curvilinear channel distribution of district speeds in the radial direction differs from uniform at internal and external surfaces [15]. In this regard, quality standards of value of district speeds in the coaxial channel on measurement of pressure difference on the internal and external cylindrical surfaces of the coaxial channel have been carried out.

In Fig. 2 shows the measurement scheme

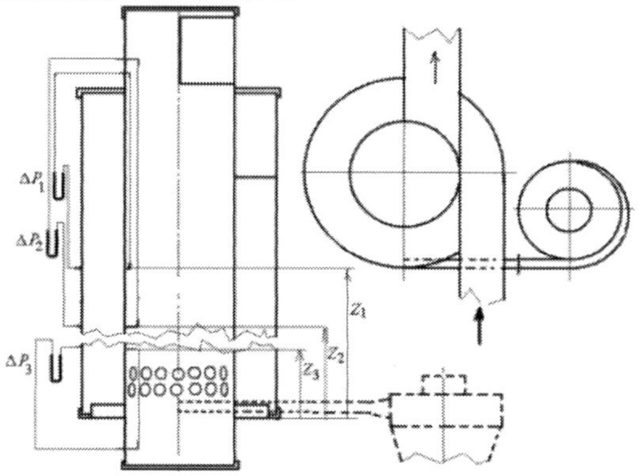

Fig. 2. Scheme of measurement of pressure differences.

The case of the cyclonic concentrator had a joint, allowing carrying out turn of section with an entrance branch pipe concerning unions of selection of pressure (fig. 3).

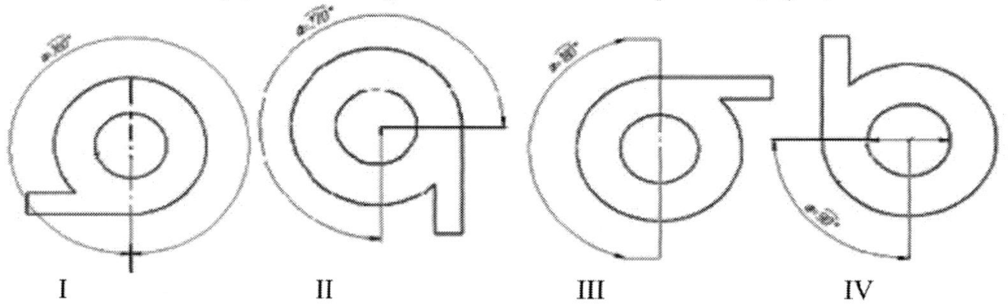

Fig. 3. Orientation of an entrance branch pipe concerning unions of selection of pressure (a vertical axis).

Water manometers were in inclined situation with a corner β to the horizon with coefficient $K=\sin\beta\approx0.1$. $\Delta P=(\Delta L\cdot K)\rho_f g\approx1.18\Delta L$ Pa, where ΔL – a difference of counting of position of meniscuses on scales of manometers in mm.

In fig. 4 the scheme of connection of devices is provided to an exhaust collector.

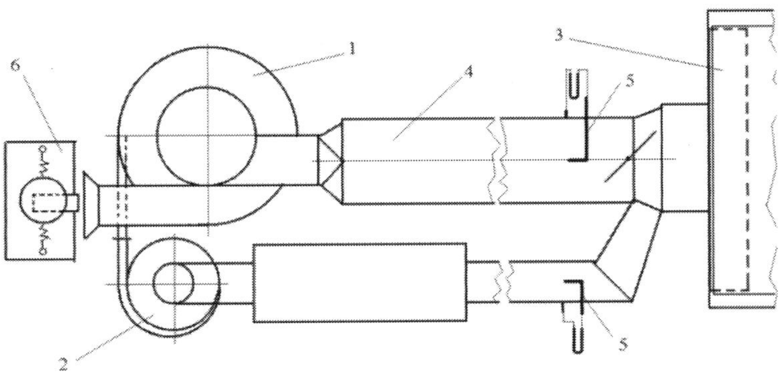

Fig. 4. The variant of the layout stand: 1 - investigated cyclone concentrator, 2 - remote dust collector, 3 - exhaust manifold with filter, 4 - connecting pipeline, 5 - flow meters, 6 - dust collector.

The air consumption through the cyclone-concentrator is calculated on a formula:

$$Q = 0.95F_Q\sqrt{\frac{\Delta P_Q}{\rho}} \approx 1.0F_Q\sqrt{\Delta L}\ (m^3/s), \quad (2)$$

where F_Q - the area of the branch pipe, where the flow is measured Q, ρ - air density (ρ=1.2 kg/m^3). The input speed was calculated from the relation $W_{in}=1.22Q/F_{in}$.

The flow through the cyclone dust collector is 22% of the Q.

From formula (1), assuming that in the radial direction the circumferential component has a constant value, we obtain $W_\phi = \sqrt{\frac{\Delta P_i}{\rho \ln\frac{R_2}{R_0}}} \approx 1.09\sqrt{\Delta L}$, where ΔP_i (Pa) - The pressure difference between the outer and inner cylinders in the *i* section in Fig. 2. The effect of the solid phase was investigated by introducing a well-flowing, classified, alumina M100 powder (96% of 100 μm particles). The concentration of particles in the input stream was calculated using the formula $C=G_{in}/1{,}22Q$. Separation tests were carried out on brand 400 cement with a median size of particles δ_m =23 micron and the variance of the distribution lg σ= lg3.0=0.477. In such dust particles less than 7 μm contains 17%, particles less than 3 μm contain 4%.

The efficiency of separation of particles in the concentrator was determined by the formula $\eta_1 = \frac{G_{sed}}{G_{in}\eta_2}$, where G_{sed}, G_{in} - the amount of dust entering the cyclone concentrator and trapped in the cyclone dust collector 2. In this case, in Fig. 1 R_0= 50 mm, R_2=100 mm. Attitude R_0/R_2=0.5. Data of measurements and calculations are provided in table 1. Results of measurement of speeds of air when using of the classified large M100 alumina are given in the left columns. In the right column of table 1 are given efficiency in polydisperse cement dust.

In [16], a method is given for studying a counterflow concentrator with a small penetration of the inner tube, a spiral inlet, a bottom cylinder. In this case, experimental efficiency is also given for cement dust of grade 400. Calculations carried out using the technique with allowance for turbulent particle transport [17] showed their correlation with the experimental results. The efficiency of air dedusting proved to be more than 95%. From Table 1 it follows that the swirling flow is asymmetric, which manifests itself in the differences in speed when the inlet is turned relative to the measuring nipples. However, the efficiency of air dedusting depends little on the rotation of the inlet relative to the nozzles. And the efficiency calculated on trajectories of particles, has appeared much above, than measured. In [18], a hydrodynamic calculation of annular channels with continuous and local spiral swirls is presented. An estimate is made of the turbulent effective viscosity, which is orders of magnitude greater than the molecular viscosity. The effective viscosity is calculated from the ratio $\mu^{ef}_{r\phi} = \rho V^*(r)l(r)$ where ρ - density of gas, $V^*(r)$ - local dynamic speed, $l(r)$ - scale length of the transfer of turbulent moles, depending on the radial coordinate. At idealization of a real current on condition of prevalence of forces of inertia over viscosity forces in [18] the uniform screw stream for which solutions of distribution of speeds of gas are known is considered. The solutions include a parameter $K = \frac{2W_\phi R_2}{RW_z}$.

Table 1.

No scheme	No experience	C g/m³	Win m/s	W1 m/s	W2 m/s	W3 m/s	Efficiency
I	1	0	17.8	9.0	9.9	14.0	Cement, C=0.6 g/m³
	2	15	13.7	6.05	4.9	8.1	W_{in}=18 m/s, η_2=93%
	3	90	17.4	11.1	9.0	14.0	η_1=88%
II	1	0	18.1	19.3	14.9	8.1	Cement, C=0.6 g/m³
	2	15	13.2	11.4	6. 7	5.5	W_{in}=18 m/s, η_2=93%
	3	90	14.6	13.1	6.6	5.6	η_1=89%
III	1	0	17.3	15.9	11.064	8.5	Cement, C=0.6 g/m³
	2	15	13.0	9.0	6.02	5.41	W_{in}=18 m/s, η_2=93%
	3	90	16.3	8.05	6.05	5.3	η_1=84%
IV	1	0	16.6	14.8	11.064	13.0	Cement, C=0.6 g/m³
	2	15	13.4	8.1	12.6	4.9	W_{in}=18 m/s, η_2=93%
	3	90	13.4	4.4	11.6	5.6	η_1=90%

In [16] the technique of a research of the counterflow concentrator with small deepening of an internal pipe, spiral input, the ground cylinder is given. At the same time experimental efficiency is also given for cement dust of brand 400. Calculations carried out using the technique with allowance for turbulent particle transport [17] showed their correlation with the experimental results. The efficiency of air dedusting proved to be more than 95%. Which determines the nature of the distribution of gas flow rates depending on the ratio R_0/R_2. When R_0/R_2=0.5 and K>6.4 gas flow in the axial direction on internal and external area of a stream are opposite. In [18] detailed distributions of components of speeds of gas for different values are given R_0/R_2 and K. Thus, in a coaxial channel turbulent mixing of gas moles in the radial direction and convective mixing in the axial direction occur.

When the flows in the lower part of the channel are divided into a flow into the central tube and the flow into the remote cyclone dust collector, the generation of turbulent vortices increases, which leads to an increase in the turbulent transfer speed. According to [19], the value of the turbulent transfer rate $V^* = \sqrt{W_{\phi 0} W_{r0}}$, где $W_{\varphi 0}$, W_{r0} - circumferential and radial speed of gas at a radius R_0. Estimates of the speed distribution over the channel height and in the lower part show that V^* is in the range 5–20 m/s, While in devices with a small depth of the central tube in the range 2–5 m/s. In [20] calculations of the efficiency are given for turbulent transport of particles.

3 Conclusion

In separators with tangential input and an elongated coaxial channel, the swirling flow is asymmetric, the tangential velocities are less important than the input speed and are not symmetrical, the separation of particles is effected by centrifugal forces and turbulent transfer, which plays a decisive role. A stabilizing role in the swirling flow of the cyclone is played by the spiral inlet and the device of the flow divider in the bottom part, in the area of concentrating the concentrate into a remote dust collector.

References

1. A. Gupta, D. Lilly, N. Sayred, *Swirling flows* (Mir, Moscow, 1987)
2. A.N. Shtym, *Aerodynamics of cyclone-vortex chambers* (Far Eastern University, Vladivostok, 1985)
3. S.S. Kutateladze, E.P. Volchkov, V.I. Terekhov, *Exchange in Bounded Vortex Flows Institute of Aerodynamics and Heat and Mass* (Novosibirsk, 1987)
4. E.N. Saburov, S.V. Karpov, S.I .Ostashev, *Heat exchange and aerodynamics of a swirling flow in cyclonic devices* (Leningrad University, Leningrad, 1989)
5. V.Y. Polovnikov, A.M. Habibulin, V.A. Arkhipov, I.K. Zharova, MATEC Web Conf. **23**, 01033 (2015)
6. E.E. Bulba, A.A. Malinovsky, MATEC Web Conf. **23**, 01053 (2015)
7. M.A. Ulchiekov, E.E. Bulba, MATEC Web Conf. **23**, 01032 (2016)
8. B. Straus, *Industrial cleaning of gases* (Chemistry, Moscow, 1981)
9. A.A. Rusanova, *Handbook on dust and ash collecting* (Energy, Moscow, 1983)
10. M.V. Vasilevsky, A.S. Razva, V.I. Romandin, E.G. Zykov, Occupational Safety in Industry, **31**, 4 (2016)
11. V.A. Reznik, N. N. Prokofiev, E.L. Alexandrov, Turning the boiler dust pipeline. (patent Russian № 2059932 10.05.1996)
12. V.S. Aslamova, (Dissertation of Doctor of Technical Sciences Tomsk, 2009)
13. T.T. Aliev, S.N. Belyaev, A.S. Galakhar, *Simulation in the ANSYS environment of the gas purification phase of a multicomponent gas mixture through the cyclone cyclone of the dust collector GP-628* (News of higher educational institutions, Moscow, 1981)
14. Smulsky, *Aerodynamics and processes in swirl chambers* (IN “Science”, Novosibirsk, 1992)
15. B.P. Ustimenko, *Processes of turbulent transport in rotating flows* (Science, Alma-Ata, 1977)
16. M.V. Vasilevsky, A.S. Razva, *Workshop on methods of protecting the atmosphere from harmful emissions: a training manual* (publishing house of Tomsk Polytechnic University, Tomsk, 2012)
17. Karpov S.V., Saburov E.N. TOKCT, **24**, (5), 651 (1990)
18. O.V. Mitrofanova, *Hydrodynamics and heat transfer of swirling flows in the channels of nuclear power plants* (PHYSMATHLIT, Moscow, 2010)
19. S.G. Ushakov, N.I. Zverev. *Inertial dust separation* (Energy, Moscow 1974)
20. V.E. Mizonov, S.G. Ushakov. *Aerodynamic separation powders* (Chemistry, Moscow, 1989)

MATEC Web of Conferences **110**, 01072 (2017) DOI: 10.1051/matecconf/201711001072
HMTTSC-2017

Comparison of space antennas mirror reflectors parameters made of composite materials

S.V. Reznik[1,*], *P.V.* Prosuntsov[1], and *A.D.* Novikov[1]

[1]Bauman Moscow State Technical University, 105005 Moscow, Russia

Abstract. Communication satellites today are fitted with one or more parabolic antenna reflectors. The unique mechanical and thermal characteristics mean that space antenna reflectors (AR) are most frequently fabricated from carbon-fibre reinforced plastics (CFRP). The reflector design must meet the stringent requirements for low linear density and high dimensional stability.

1 Introduction

At present, the linear density values for AR approach 3 kg/m^2, while dimensional tolerance should not exceed 1/16 wavelength. The development of communication satellite systems resulted in a number of approaches to satisfy these requirements: honeycomb-filled three-layered structures [1], dual gridded structures [2] and rib structures [3-8].

As was already demonstrated in [5-8], the rib-stiffened structures have the lowest specific density. The following rib reinforcement patterns on the convex surface were studied: isogrid pattern [5, 6], five-point star and six-point star [7, 8]. There are reasons to believe that the linear density for such AR can be lower than 2 kg/m^3. However, the issues of the ribs configuration, their height and thickness remain to be resolved.

The aim of this research was to determine the AR design variant with the lowest specific density, when ribs of various thickness and height are used in the five-point star reinforcement pattern.

2 The simulation problem statement

The reflector geometric model was a gently sloping parabolic shell, 1000 mm in diameter and 125 mm construction depth with ribs on the convex surface (Fig. 1). The ribbing comprised a perimeter rib and a system of ribs laid as a five-pointed star over the parabolic shell surface.

The following design parameters were selected for analysis: two variants of ribs height – 30 and 50 mm, two variants of the shell and rib thickness – 0.6 and 1.0 mm. The perimeter rib is the same as the other ribs in height and thickness.

[*]Corresponding author: sreznik@bmstu.ru

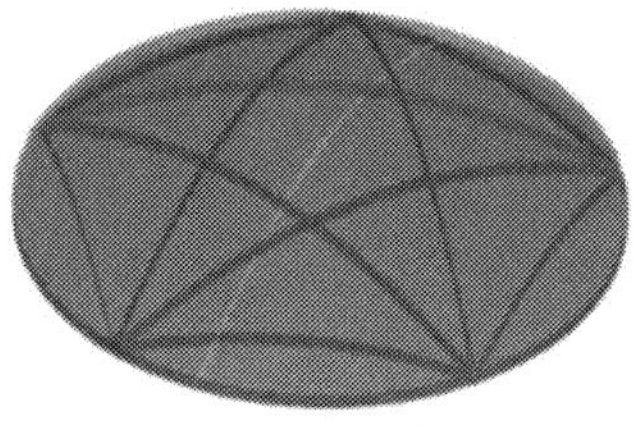

Fig. 1. Geometrical model: five-star pattern reinforcement.

The material for simulation was CFRP with the following mechanical and thermal physical characteristics [9]: heat transfer coefficient 31 W/(m·K); specific heat capacity 1000 J/(kg·K); density 1550 kg/m^3; emissivity 0.85; absorptivity 0.735; CLTE $5.27 \cdot 10^{-7}$ 1/K; Young modulus 140 GPa; Poisson ratio 0.3.

Taking into account the correlation between the reflector linear dimensions and the shell and ribs thickness, as well as the fact that the reinforcement pattern (0°/45°/0°/0°/45°/0°) enabled virtually uniform values of heat transfer coefficients, Young modules, Poisson ratios and shear modules, a two-dimensional finite elements model was used, where the thermal resistance perpendicular to the reinforcement plane and the inter-layer strength effect on the stress-strain behavior can be neglected.

The stress-strain and thermal behavior simulation was conducted in the Femap software package (Siemens PLM Software). The geostationary orbit motion was modelled. The modelling results analysis demonstrated that the greatest temperature difference appears at the moment following the spacecraft leaving the Earth shadow, when the spacecraft rotation axis is at 150 degrees to the Sun-Earth axis. This is the reason for this case to be selected for the comparative analysis of the design variants.

3 Mathematical modelling results and analysis

Fig. 2 presents the results of the thermal behavior calculation for reflectors with 30 mm and 50 mm ribs height and 0.6 and 1.0 mm shell and ribs thickness.

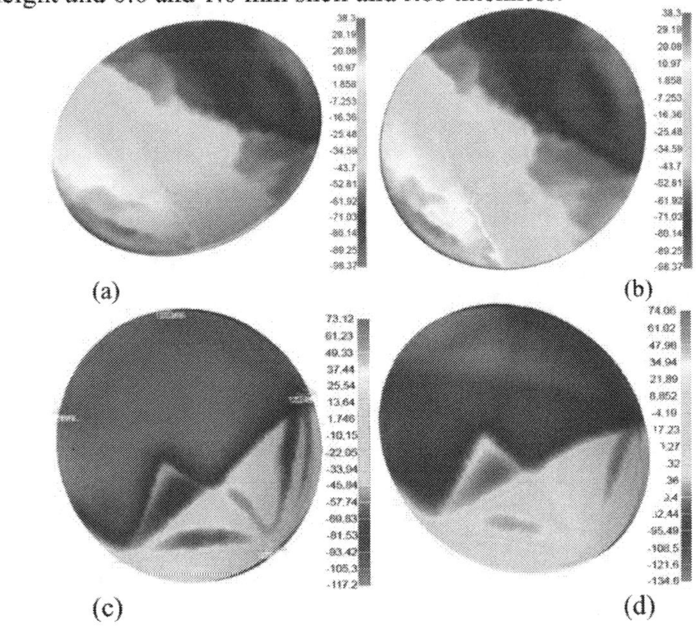

Fig. 2 Temperature distribution in the reflector shell: ribs height 30 mm, shell and ribs thickness a) 0.6 mm, b) 1.0 mm; ribs height 50 mm, shell and ribs thickness c) 0.6 mm, d) 1.0 mm.

As seen from Fig. 2, the temperature level and difference vary greatly for the reflectors with different ribs height.
The shell with the relatively high ribs have a much larger low-temperature area compared to the shell stiffened with the 30 mm high ribs. Moreover, for the shell surface with the 30 mm high ribs, the temperature difference constituted 136.7 degrees, practically not affected by the ribs thickness in comparison to the 190.3 degrees for the 50 mm high, 0.6 mm thick ribs and 208.7 for the 50 mm high, 1.0 mm thick ribs.

The temperature field was used to determine the stress-strain field (Fig. 3). Fig. 3 demonstrates that the maximum displacement for 30 mm height was about 0.07 mm for the 0.6 thickness and 5 point-fixing variant. For the 0.6 mm thickness and 1.0 mm thickness variants, the difference in displacements constituted 0.0001, which is practically negligible. The displacement for the five-star pattern with 50 mm height was 0.26 mm.

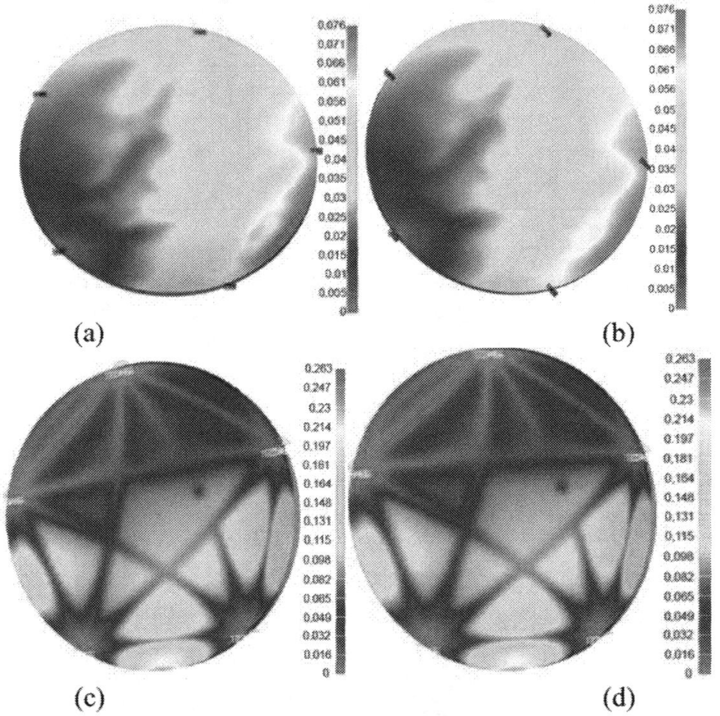

Fig. 3. Reflector shell displacement: ribs height 30 mm, shell and ribs thickness a) 0.6 mm, b) 1.0 mm; ribs height 50 mm, shell and ribs thickness c) 0.6 mm, d) 1.0 mm.

4 Conclusions

Calculations demonstrated that the shell and ribs thicknesses in the values range under investigation do not influence the structure displacement. However, the displacement is significantly affected by the ribs height: at 30 mm height the displacement was 0.076 mm, at 50 mm – 0.263 mm. In the variants under consideration the structure mass was 1.502 kg for 30 mm high, 0.6 mm thick ribs; and 2.503 kg for 1.0 mm thick ribs, after accounting for the reflector dimensions, it gives the linear density as great as 1.328 kg/m^2 and 2.221 kg/m^2 correspondingly. Similar is true for the 50 mm high ribs. For 0.6 mm thick ribs, the mass constitutes 1.715 kg; for 1.0 mm thick ribs – 2.836 kg, which in terms of linear density is correspondingly 1.559 kg/m^2 and 2.554 kg/m^2.

Thus, a conclusion may be made to the effect that the design with the 30 mm height and 0.6 thickness ribs is preferable to the other variants.

References

1. Satellite antennas for perfect performance [Electronic resource] / Invent Gmgh. – 2014. – Access mode: http://www.invent-gmbh.de/en/company/business-activities/satellite-antennas-for-perfect-performence.
2. New concepts for dual gridded reflectors [Electronic resource] / HPS GmbH. – 2014. – Access mode: http://www.hps-gmbh.com/projects-descriptions-mainmenu-79/lightweight-structures-mainmenu-51/23-kadgr.html.
3. V.A. Vechtomov, V.N. Zimin, A.N. Kuzenkov, D.V. Dronov, A.A. Kozlov. Vest. MGTU Im. Baum. Ser. "Priborostr." **7,** 70-81 (2012)
4. E.V. Bitkina, A.V. Denisov, V.E. Bitkin. Izv. Samar. Nauch. Tsent. Ros. Akadem. N. **14,** (4-2), 555 (2012)
5. S.V. Reznik, P.V. Prosuntsov, A.V. Azarov. J Eng Phys Thermophy. **88** (3), 699 (2015)
6. S.V. Reznik, P.V. Prosuntsov, A.V. Azarov. J Eng Phys Thermophy. **88** (4), 978 (2015)
7. P.V. Prosuntsov, S.V. Reznik, K.V. Mikhailovsky, A.D. Novikov. IOP Conf. Ser.: Mat. Sci. & Eng. **153** (1), 7 (2016)
8. S.V. Reznik, A.D. Novikov, MATEC Web Conf. **01012**, 5 (2017)
9. V.M. Petrov, E.N. Beletskiy, S.N. Bezpalchuk. Vest. Gos. Univer. Morsk. I Rech. Flot Im. Adm. S.O. Makarova **2**, 66 (2014)

MATEC Web of Conferences **110**, 01073 (2017) DOI: 10.1051/matecconf/201711001073
HMTTSC-2017

Inclusion of absorption heat pumps into heat power plant scheme for stage heating efficiency increase of delivery water

Olga Romashova[1,*], *Dmitrij* Luzhkovoj[1], *Aleksandr* Tubolev[1], and *Marina* Vagner[1]

[1]National Research Tomsk Polytechnic University, 634050 Tomsk, Russia

Abstract. Design analysis of absorption heat pumps use for delivery water stage heating levelling-off at ambient air temperature decrease has been reported in the present paper.

1 Introduction

Two stage heating of delivery water at heat power plant in comparison with one stage heating allows to increase electric energy generation at heat consumption. In modern heating turbines there is no opportunity to maintain equal distribution of delivery water heating within the whole heating season, which leads to decrease of efficiency of heating generation of electric energy [1].

Design research of work modes in two stage network installation in turbines of type T-110-130 and T-250-240 in respect to climate conditions in Tomsk for temperature chart of heat supply system 150/70 and extraction ratio=0.6 [2] shows that dispatch of heat enormously depends on ambient air temperature. During turbine plants work on heat graph share of water heating in low heating of water in lower line heater at decrease of ambient air temperature increases from (30–40) % to (65–75) % with corresponding decrease of heating in upper heater (fig. 1). All other conditions being equal inequality of network heating increases with decrease of delivery water discharge rate and increase of its temperature in return line [3].

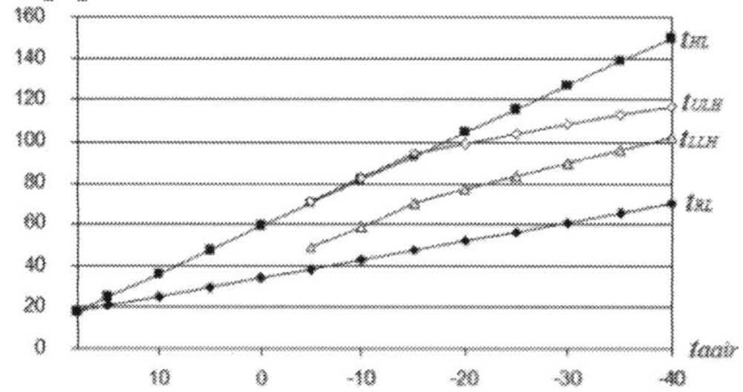

Fig. 1. Temperature chart of heat supply system (HL – heating line, RL – return line, ULH – upper line heater, LLH – lower line heater, aair – ambient air).

* Corresponding author: roma@tpu.ru

2 Statement of problem and limitations

Variant of absorption heat pumps (AHP) use for delivery water stage heating levelling-off at ambient air temperature decrease has been considered in the present paper [4]. Inclusion of absorption heat pumps into delivery water two stage heating scheme at heat power plant has been presented in fig.2. In evaporator of absorption heat pumps return wire temperature decreases due to use of heat of high-potential vapor extraction from turbine. Transmission of heat to delivery water in condenser of absorption heat pumps leads to decrease of delivery water temperature for upper line heater with corresponding decrease of vapor pressure in the upper controlled selection.

Simultaneous decrease of return wire temperature and vapor pressure in upper line heater causes redistribution of heating between upper and lower heaters and allows to approximate work mode of network installation to design mode with equal heating of delivery water. Pressure in work selection and its corresponding delivery water temperature for lower line heater are set on the base of coordinated solution of the Stodola-Flugel formula for interstage section of turbine and heat-balance equation of lower line heater.

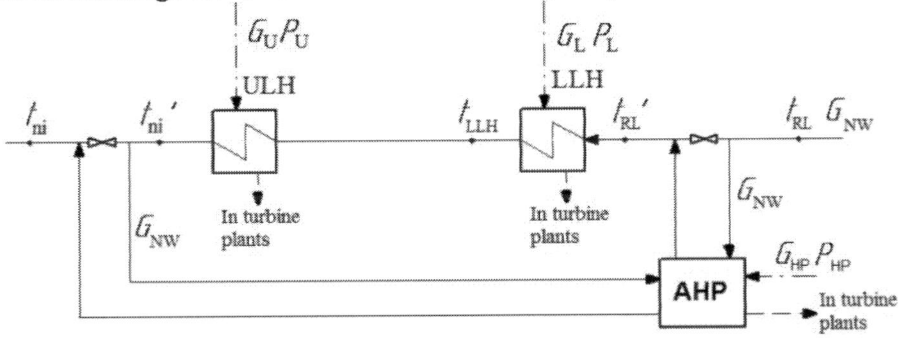

Fig. 2. Inclusion of absorption heat pumps into two-stage network installation of heating turbine scheme.

Initial data for design are presented in table 1. Design mode of heating turbine corresponds to equal heating of delivery water. It is close to the temperature of inclusion of Peak Water Boiler and it was accepted as equal to t_{aair} = -15 for Tomsk. The design shows that at minimal ambient air temperature a share of heating in the lower stage of network installation increases to 70 % of overall heating of delivery water in lower line heater and upper line heater.

Table 1. Initial data for the task.

Specifications	Value
Common specifications for all modes	
Turbine plant prototype	T-110-130
Heat supply system temperature chart t_{h}/t_{rl},°C	150/70
Extraction factor $\alpha_{heat\ power\ plant}$	0.6
Minimum design temperature of ambient air t_{aair}, °C	-40
Thermal capacity of system Q_t, MW	200
Working vapor pressure P_0, MPa	13
Working vapor temperature t_0, °C	540
Efficiency ratio of turbine η_{oi}	0.87
Product of mechanical efficiency and generation efficiency $\eta_m * \eta_g$	0.995
Design mode (t_{aair} = -15 °C)	
Temperature of delivery water behind network installation t_{ni}, °C	95
Return wire temperature t_{rw} °C	47

Share of heating in the lower stage of network installation (NI)	0.5
Minimal design ambient air temperature mode without absorption heat pumps (t_{aair} = -40 °C)	
Temperature of delivery water behind network installation t_{ni}, °C	118
Return wire temperature t_{rw} °C	70
Minimal design ambient air temperature mode with absorption heat pumps (t_{aair} = -40 °C)	
Delivery water temperature behind condenser of absorption heat pumps t_{ni}, °C	118
Return wire temperature t_{rw} °C	70
Coefficient of performance of absorption heat pumps φ	1.7
Vapor pressure of selection on absorption heat pumps P_{hp}, MPa	0.5

3 Analysis results

Analysis results are presented in Table 2.

Table 2. Analysis results.

Specification	Value		
	95/47 without absorption heat pumps	118/70 without absorption heat pumps	118/70 with absorption heat pumps
Delivery water discharge G_{dw}, kg/s	992.1		
Vapor pressure of selection on upper line heater P_u, MPa	0.1065	0.2292	0.1065
Vapor pressure of selection on lower line heater P_l, MPa	0.0422	0.1436	0.0635
Vapor feed rate of selection on upper line heater G_u, kg/s	47.07	28.15	27.19
Vapor feed rate of selection on lower line heater G_l, kg/s	47.38	65.8	41.45
Vapor feed rate of selection on absorption heat pumps G_{hp}, kg/s	-	-	26.55
Delivery water heating in condenser of absorption heat pumps Δt_{heat}, °C	-	-	23
Delivery water cooling in evaporator of absorption heat pumps Δt_{cool}, °C	-	-	9.47
Delivery water temperature behind network installation t'_{ni} ,°C	-	-	95
Delivery water temperature after evaporator of absorption heat pumps t'_{rw},°C	-	-	60.53
Delivery water temperature of lower line heater t_{llh}, °C	71	103.6	81.14
Vapor feed rate on turbine G_0, kg/s	94.45	93.95	95.19
Generated electric power N_e, MW	92.6	81.24	84.77
Specific vapor consumption on turbine d_0, kg/MJ	1.02	1.156	1.123
Specific electric energy production on heat consumption e	0.463	0.406	0.424

Design results of turbine plant efficiency at different modes of network installation are presented in fig. 3,4.

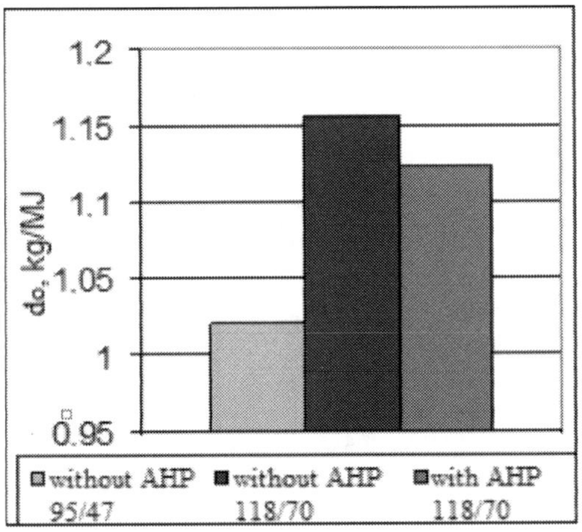

Fig. 3. Inclusion of specific vapor consumption on turbine for different modes of network installation efficiency.

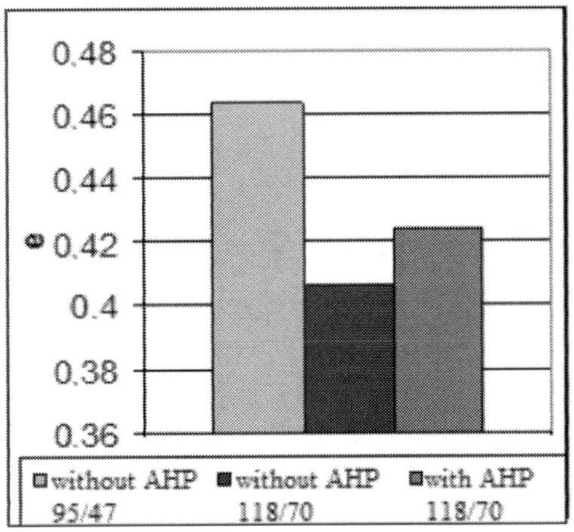

Fig. 4. Specific electric energy production at heat consumption for different modes of network installation efficiency.

4 Conclusion

To sum up:

- Delivery water heating in lower and upper heaters of two-stage network installation during heating season greatly differs from equal one. During ambient air temperature decrease thermal load of lower heat extraction and share of water heating in lower line heater grows. During delivery water discharge decrease and increase of temperature in return line at other equal conditions inequality of heating increases.
- Use of absorption heat pumps in stage heating increase of delivery water scheme of stage heating delivery water at low temperatures of ambient air allows to decrease return wire temperature at the input into lower line heater and at the output of upper line heater and approximate off-design conditions of network installation to equal one.
- Efficiency of inclusion of absorption heat pumps into two-stage network installation scheme of heating turbine without interstage interval with initial characteristics P_0=13 MPa, t_0=550 °C for mode with minimum temperature of ambient air is evaluated by increase of electric energy generation at heat consumption 2–3 % in reference to initial model.

References

1. D.P. Elizarov, Je.I. Tazhiev, Jelektricheskie stancii **2**, (1994) [in Russian]
2. O.Ju. Romashova, O.R. Gabidullin, Jenergetika: Jekologija, nadezhnost', bezopasnost'. Materialy dokladov devjatoj Vserossijskoj nauchno-tehnicheskoj konferencii - Tomsk, TPU, 3-5.12.2003 239 (2003) [in Russian]
3. O.Ju. Romashova, O.R. Gabidullin, Jenergetika: Jekologija, nadezhnost', bezopasnost'. Materialy dokladov devjatoj Vserossijskoj nauchno-tehnicheskoj konferencii - Tomsk, TPU, 8-10.12.2004 222 (2003) [in Russian]
4. D. Luzhkovoy, O. Romashova, A. Tubolev, Y. Kobenko, E. Riabova, MATEC Web Conf. **92**, 01054 (2017)

MATEC Web of Conferences **110**, 01074 (2017) DOI: 10.1051/matecconf/201711001074
HMTTSC-2017

Efficiency evaluation for the heat pump system in sharp continental climate

Vyacheslav I. Maksimov[1,*], *Amer* Saloum[1]

[1]National Research Tomsk Polytechnic University, 634050 Tomsk, Russia

Abstract. With the aim to calculate the energy efficient of water source heat pump system (HPSW) under sharp climate conditions , the ice formation processe on the surface of evaporator pipes under conditions of "autumn-winter-spring" was studied.

1 Introduction

The use of water bodies as energy source for heat pump system attracted the attention of researchers and engineers for many years [1-6]. A lot of heat pump systems were built in different places , but the use of these systems in Russian Federation was limited [7]. The lack of information about the water bodies temperatures and behavior in conditions of low ambient air temperatures ("autumn-winter-spring" period) is one of the main reasons of the limited using for this system.

The heating period ,where the heat pump system is needed is, the cold period of the year, but when the air temperature drops under 9 °C in heating season ,the water surfaces temperatures start to drop down ,and the ice covers these water bodies in the central part of Russian federation in November.

When the water starts to freeze in standing reservoirs, the temperature of the surface is about 0 °C ,and under the surface, there is gradient in temperature to reach the value of 4 °C in a defined depth [8] , where in moving water (like in rivers the average water temperature is 1 °C , and energy efficiency assessment for heat pump systems in these low temperature ranges has not been carried out up to date.

According to heat transfer principles, and the working range for the heat pump evaporator, the water will freeze on the evaporator pipe wall [9]. The efficiency analysis for heat pump system in the conditions of ice formation is recommended.

The object of this work is an experimental study for the main regularities of ice formation process on the working surfaces of evaporator pipe of HPSW, surrounded by water with average temperature 7 °C ,and the analysis of the main characteristics of the heat pump system with an ice-covered evaporator.

2 The experimental research methods

* Corresponding author: elf@tpu.ru

The experimental study was done with the use of vapor compression heat pump system , the system diagram is shown in fig.1. The Freon 137a was used as working refrigerant in the cycle. Refrigerant pressures in the evaporator and condenser are 0.09, and 0.1 MPa consequently.

The refrigerant evaporates at temperature -30 °C , and condensates at temperature +50 °C .

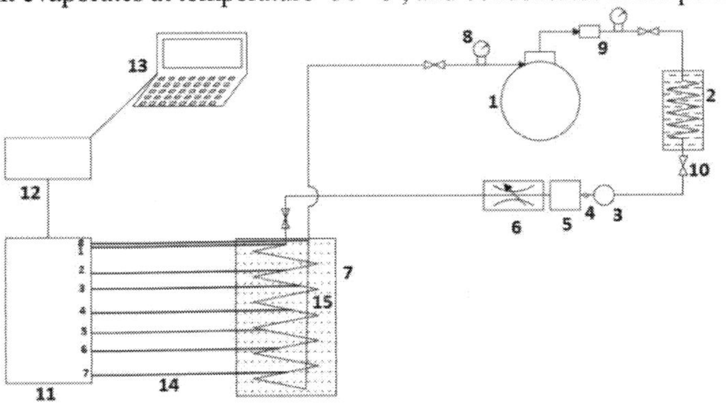

Fig1. Schematic diagram for the experiment setup: (1) - compressor; (2) - condenser. (3) - filter; (4) - Sight glass; (5) - Desiccant; (6)- throttle; (7)- evaporator; (8) – Manometer; (9)- Flowmeter; (10) – valve; (11) – Analog-to-digital converter; (12) – Network adapter; (13) – a computer; (14) - thermocouples; (15) - water.

The dimensions of the block which contain the evaporator pipes are 0.22×0.18×0.25 m, the block walls made of polypropylene with a thickness of 0.003 m (fig.2).

The evaporator pipe is made of copper with inner diameter 0.008m and length 3.11m. The block was isolated with 70 mm thickness of foam polystyrene. The water initial temperature varies between 4.7 °C.

The average room temperature was 22 °C.

Along the experiment the fixed parameters were: the voltage, current of the compressor motor, and the refrigerant flow.

To measure the temperature along the tubes of the evaporator, 8 Chlomel-Alumel thermocouples (VIII thermocouple) with 0.1 mm junction size were used .The thermocouples were Connected to an analog -digital converter with a resolution of 16 bits, and an internal cold junction temperature sensor. Thermocouples were fixed with a certain pitch along the copper pipe length (Table 1), and also at different heights in the evaporator block . A network adapter was used to connect the analog -digital converter to a personal computer, and a complete galvanic isolation between the devices was provided . A software package based on the National Instruments LabVIEW was developed to register the results and measurements, special for this type of measurements [10]. The temperature values were recorded in real time with the preservation of the obtained data.

Table 1. Location of thermocouples on the surface of the pipe.

No Thermocouple	1	2	3	4	5	6	7	8
Pipe length L, m	0	0.44	0.89	1.33	1.78	2.22	2.67	3.11

The total relative systematic error in determining the temperature values did not exceed 4.8% over the entire range of variation of the parameters.

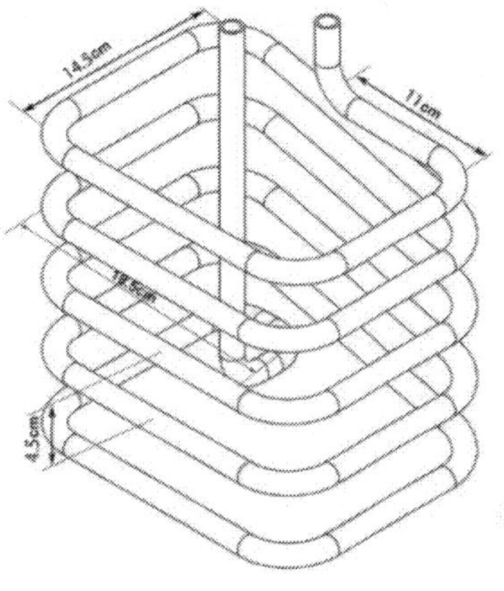

Fig2. Evaporator pipe.

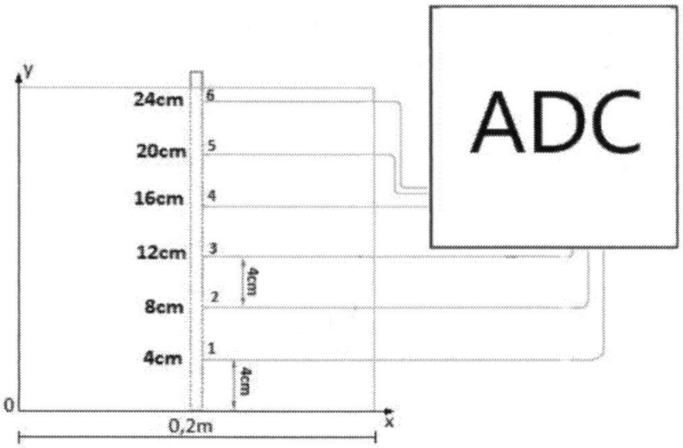

Fig3. Thermocouples location in evaporator block.

3 The results of experimental studies

The main experiments were carried out at a water temperature of +7 ° C or lower. In practical practice, higher water temperatures in water bodies in the main part of the territory of Russian Federation during the heating season (October-April) are not expected .

Fig4.5 shows the values of local temperatures on the surface of the evaporator heat exchanger tube and at local points in the experimental block, measured at initial temperature of +7 ° C.

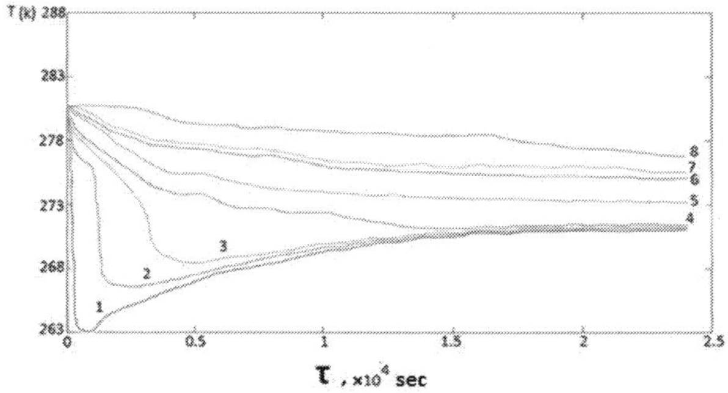

Fig 4. Temperature variation on the surface of the evaporator pipe as a function of time. (1-8) numbers of thermocouples located on the surface of the evaporator tube in accordance with Table 1.

It is noticed that after starting-up the system, a sharp drop in temperature t -10 ° C occurs in the first part of the evaporator pipe (thermocouple 1-3).This is due to the fact that the main process of evaporation for the refrigerant occurs precisely in this segment of the pipe ,in this section the temperatures gradually increase to -2 ° C with the time , and on the next length of the evaporator surface the temperature smoothly decreases, the intensity of the local phase transition process decreases, and the evaporation zone expands along the length of the pipe.Temperature values were measured at different heights , in the cross section X = 0.1 m and 0 <Y <0.2 m.(Fig.3.)

In Fig. 5. It can be seen that the temperature profiles correspond to the conditions of free-convective currents in the regime under consideration.

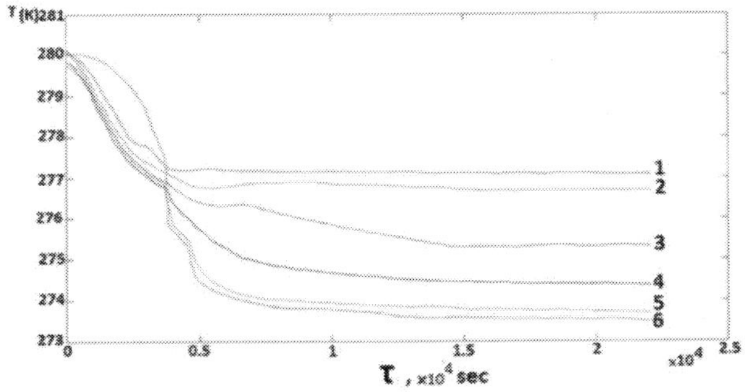

Fig 5. Dependence of water temperature on time at different points along the height of the evaporator block. 1 – 4 cm, 2 – 8 cm, 3 – 12 cm, 4 – 16 cm, 5 – 20 cm, 6 – 24 cm.

Intensive convection currents for the studied mass of water are caused by the temperature decrease in the upper part of the heat exchanger where the intensive evaporation for the refrigerant occurs . The cooled water moves down and the warm moves up.

After 4000 sec , a sharp drop in temperature is noticed near the surface of the water , this drop is related to the fact that the when the water temperature reaches the value 4 °C the density of water becomes maximum and the warm water moves down where cold water moves up.

A smooth decrease in water temperature occurs in the studied volume, because the heat exchange becomes weaker and the temperature keeps its constant value in the first part of evaporator because of ice formation on the evaporator pipe (crystallization).

This change in the flow structure explains the change in the intensity of ice formation on the evaporator pipe (Figure 6).

The decrease in the ice formation rate on surface of the lower part of the tube after 60 minutes of operating time is due to the presence of warm water in this region. On the upper part of the pipe, the formation of ice continues with a decrease in the its rate with time, in connection with the decrease in heat transfer intensity, due to the growth of thermal resistance with the increasing of ice thickness.

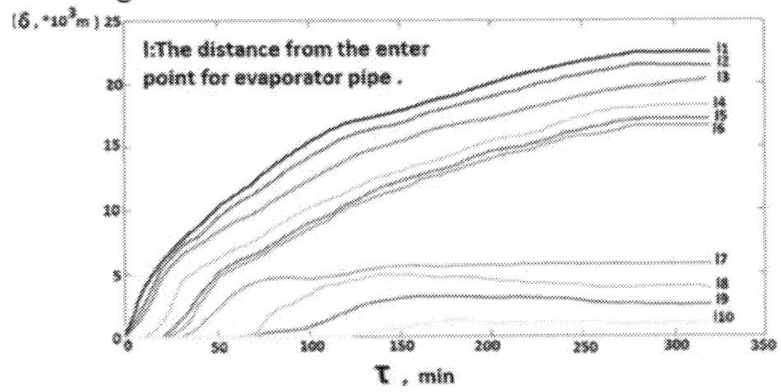

Fig 6. Dependence the ice thickness of ice along the evaporator pipe on time ,(l1-l10) - distance along the length of the evaporator tube from the enter (Table 2).

Table 2. (l1-l10) are the distances along the length of the evaporator tube from the enter.

l1	l2	l3	l4	l5	l6	l7	L8	L9	l10
0cm	4cm	10.5cm	41cm	58.5cm	63cm	96cm	119.5cm	139cm	162cm

4 Conclusion

The temperatures were measured experimentally in different points in the evaporator of heat pump system, both along the surface of evaporator pipe and at different heights in the evaporator block. Also the thickness of ice on the surface of evaporator pipe were measured.

The main regularities of ice formation process on the surface of the evaporator pipe were registered depending on the working time of the HPSW.

It is noticed that the formation of ice begins on the upper part of the heat exchanger pipe, and eventually spreads down along the length of it, the depending of ice thickness along the evaporator pipe on time was registered. The influence of convective heat transfer in the evaporator chamber (heat exchanger - evaporator - water) on ice formation rate on the evaporator pipe is revealed.it can be revealed the effect of the drop in temperature under 4 ^{o}C on the nature of heat exchange process and consequently on the process of ice formation on heat exchanger surface.

The results of the research can be used to analysis the working of the heat pump system in the conditions of ice formation on evaporator wall during "autumn-spring" period in the natural water bodies, which will improve the calculation methods for the system efficiency in such conditions.

The work was supported by the Russian President's grant (Scientific School project - 7538.2016.8).

References

1. Z. Shenghua, X. Xiaokai, Appl. Therm. Eng. **112,** 200 (2016)
2. Y. Wang., Chongqing University, **30** (2011)
3. A. Morton, A thesis submitted in partial fulfilment for the requirement of the degree Master of Science (2013)
4. L. Nan, Q. Zhang, D. Wu, C. Zhenqian, Procedia Engineering, **121,** 1880 (2016)
5. L Schibuola, M Scarpa, Energy and Buildings **113,** 182 (2015)
6. N.V. Demyanovich, V.I. Maksimov, T.A. Nagornova, MATEC Web Conf. **5** (2017)
7. G.P. Vasilyeva, N.V. Peskova, V.F. Gornova, M.V. Kolesova, Geothermics **62,** 93 (2016)
8. T.A.Kawalecki, A.Cybulski., Polish Academy of science, **10** (1995)
9. V. I. Maksimov, T. A. Nagornova, Vladimir. S. Chernyshev, MATEC Web Conf. **8** (2015)
10. G.V. Kuznetsov, V.I. Maksimov, J. Eng. Phys. Thermophys. **89,** 1241 (2016)

MATEC Web of Conferences **110**, 01075 (2017) DOI: 10.1051/matecconf/201711001075
HMTTSC-2017

Assessment of the effect of borides on the thermodynamic characteristics of heterogeneous systems

Vladimir Arkhipov, *Liliya* Savelieva, and *Pavel* Ponomarev*

National Research Tomsk State University, Tomsk, Russia

Abstract. The possibility of creating high-energy materials with increased ecological purity of combustion products based on a mixed oxidizer (ammonium nitrate, ammonium perchlorate), and inorganic fuel (boron, diboride or aluminum polyboride) is shown.

Introduction

Increasing of the requirements for the environmental cleanliness of combustion products of high-energy materials required the development of fuel systems with a minimum content of chlorine compounds in gaseous combustion products. [7, 6].

One of the directions of possible hydrogen chloride (HCl) reduction in combustion products of high-energy materials (HEM) is the replacement of the regular oxidizer of (ammonium perchlorate) with ammonium nitrate. However, nitrate formulations are characterized by a smaller unit impulse, a lower burning rate and longer fluid system ignition delay times. At present, mixed oxidizers (perchlorate and ammonium nitrate) and new metal combustibles, boron (B) and borides, mainly aluminum borides (AlB_2 and AlB_{12}) are used to increase the energy characteristics of the HEM. [4, 5, 8].

In the literature, data on the optimal ratio of ammonium nitrate (NA) and ammonium perchlorate (PA) components in a mixed oxidant are not found, allowing a single impulse of nitrate HEM to be increased in the combustion products, at a minimum content of HCl, making it closer to the characteristics of perchlorate compounds. The literature also does not show the advantage of aluminum (Al) compared to boron and aluminum borides. [1, 2].

The thermodynamic characteristics of HEM containing a mixed oxidant (NA + PA), where the maximum mass fraction of PA is 0.5 is estimated. Al and B, as well as AlB_2 and AlB_{12} borides based on them, were considered as metallic combustibles.

The aim of the work is to find the optimal composition of a HEM with a single pulse of at least 2500 m/s and a minimum content of hydrogen chloride in gaseous combustion products based on the results of calculations of the thermodynamic characteristics.

* Corresponding author: pahans-94@mail.ru

Calculations were carried out for systems with an excess of an oxidizer coefficient equal to 0.5, containing 15% Al, B and borides AlB_2 and AlB_{12} as metal fuel. The result of the work is the selection of the optimal compositions of the HEM for experimental research.

Results of thermodynamic calculations

In the work, the characteristics of 35 different HEM compositions were calculated using the "Astra-4" software complex [3], differing in the composition of the mixed oxidant and the nature of the metallic combustible. Calculations of the effectiveness of the influence of the metallic fuel nature on the specific impulse are shown in Fig. 1, as well as on the HCl content in the combustion products (Fig. 2).

Here 1 is the efficiency of the influence of Al, 2 is the efficiency of the influence of B, 3 is the efficiency of the effect of AlB_2, 4 is the efficiency of the influence of AlB_{12} on the thermodynamic characteristics of the HEM.

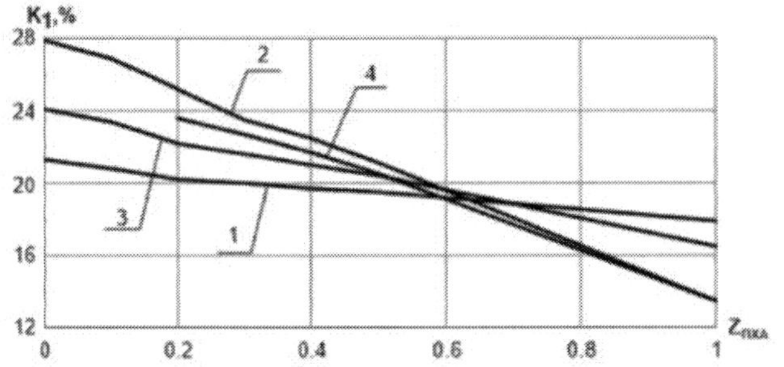

Fig.1. Effectiveness of the influence of the metallic fuel nature on the specific impulse value.

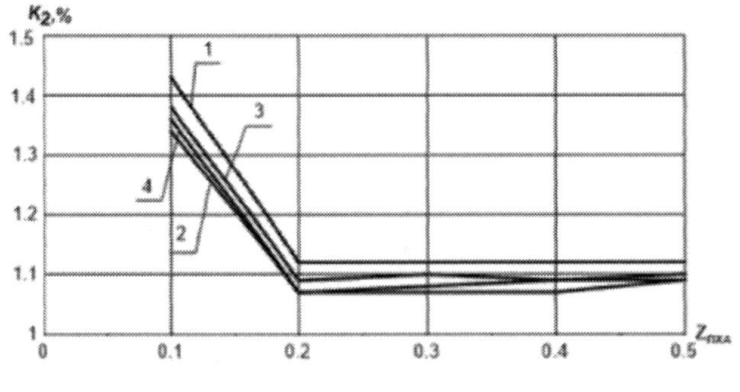

Fig.2. Effectiveness of the influence of the metallic fuel nature on the content of HCl in combustion products.

The efficiency coefficients K_1 and K_2 were calculated using the following formulas:

$$K_1 = \frac{I}{I_0}, \quad K_2 = \frac{C_0}{C},$$

where K_1 is the efficiency of the metallic fuel effect on the specific impulse, I is the specific impulse of the metallized HEM, I_0 is the specific impulse of the nonmetallic systems, K_2 is the efficiency of the metallic fuel effect on the HCl content, C is the HCl content in the combustion products of metallized HEM, C_0 - content of HCl in combustion products of nonmetallic HEM.

The mass fraction of PCA in the mixed oxidant is

$$z_{PA} = \frac{m_{PA}}{m_{PA} + m_{NA}}$$

It follows from Fig. 1 that the efficiency of metals is affected by the composition of the oxidizer of the fluid system, in such a manner for systems containing the initial ammonium perchlorate as an oxidizer, the efficiency of metals decreases in a number:

$$Al > AlB_2 > AlB_{12} = B,$$

And with an oxidizer-ammonium nitrate, the efficiency of metals varies in a series:

$$B > AlB_{12} > AlB_2 > Al.$$

The second regularity of the change in the efficiency of the metals effect is also valid for a BEM on a mixed oxidizer containing not more than 30% of PA. A further increase in the content of PA leads to a decrease in the effect of metals on the value of the specific impulse. One of the possible causes of changes in the efficiency of inorganic fuels is the temperature level in the combustion chamber (Table. 1).

Table 1. HEM characteristics.

Oxidizer	Characteristics	Value of characteristics HEM			
		Al	AlB_2	AlB_{12}	B
PA	*I*, sec	268.3	265.8	260.9	259
	T, K	3045	2845	2624	2582
NA	*I*, sec	256.9	260.1	264.2	265.7
	T, K	2655	2532	2427	2431

The effectiveness of the boron and borides influence increases when used in compositions with a combustion product temperature not exceeding 2500K.

It can be seen from Fig. 2 that the HCl content in the combustion products of the HEM is directly proportional to the amount of ammonium perchlorate in the mixed oxidant. Therefore, to solve the problem, the choice of a mixed oxidizer is limited by the content of ammonium perchlorate not more than 30%. It should be noted that the injection of the metal somewhat reduces the content of gaseous HCl in the combustion products in comparison with the nonmetallic compositions. A similar effect is probably due to the formation of chlorine-containing compounds with the metal $AlOHCl_2$, $AlCl_2$, and BCl_3, BCl_2, and others. The effect of lowering the HCl content is more pronounced when injected as a metallic combustible aluminum compared to the boron and borides.

Conclusion

Analysis of the calculated data showed that the greatest effect of metallic fuel injection both on the value of the specific impulse and on the content of HCl in combustion products is manifested for systems containing no more than 30% of PA in the mixed oxidant. The greatest decrease in HCl in combustion products occurs when the oxidizer contains 10% PA. It should be pointed out, that the effectiveness of the Al influence on the reduction of HCl in combustion products is just over than that of boron and borides AlB_2 and AlB_{12}. The efficiency of boron and borides depends almost not at all on their composition. The calculated data allow us to estimate the interval of PA content in a mixed oxidant and to choose the fluid systems for the experimental study.

Acknowledgements

This work was supported financially by the Ministry of Education and Science of the Russian Federation in terms of state order, project code 9.9752.2017.

References

1. Ya.M. Paushkin *Liquid and solid chemical rocket fuels* (Science, Moscow, 1978) [in Russian]
2. S. Sarner *Chemistry of rocket fuels* (World, Moscow, 1969) [in Russian]
3. B.G. Trusov, *Modeling of chemical and phase equilibria at high temperatures "Astra-4" version 1.06* (Publishing house of the Moscow State Technical University named after Bauman, Moscow, 1991) [in Russian]
4. M. K. King, Journal of Spacecraft and Rockets, **19** (4), 294 (1982)
5. F. Gorman, Missile technology issues, (1), 41 (1971)
6. Mirković, D., Gröbner, J., Schmid-Fetzer, R. , Fabrichnaya, O., Lukas, H.L. , J. Alloys Compd., **384**, 168 (2004)
7. A.B. Vorozhtsov, *Novel Micro- and Nanofuels: Production, Characterization, and Applications for High-Energy Materials* (Springer International Publishing, Switzerland, 2017)
8. Lempert D. B., Power capabilities of Solid rocket fuels with oxidizers based on stressed frames, 98 (2000)

The feasibility of using photovoltaic panels to illuminate the entryway to an apartment building

Liudmila Sumarokova[1], *Alena* Surovezhko[1,*]

[1]National Research Tomsk Polytechnic University, 634050 Tomsk, Russia

Abstract. The article considers the possibility of using an LED lighting system with a power source from solar modules in the climatic conditions of Siberia. The technical possibility of implementing an autonomous house lighting system is shown for example in the lighting of a residential five-story building located in Tomsk. The choice and justification of the necessary electrical equipment for solar panels was made. Calculations have been made for the energy consumption of the existing lighting system and a system with LED light sources from photovoltaic panels. The payback period of the project is determined. On the example of an autonomous system of interior lighting of an apartment building, conclusions were made about the feasibility and efficiency of using photovoltaic panels in the climatic conditions of Tomsk region.

The purpose of this paper is to evaluate the possibility of using photovoltaic panels as a power source for LED lighting fixtures in common areas in Siberia. As an object for the designing of the LED lighting system was chosen a residential five-story house located at the address: Tomsk city, Lenin Avenue, 255 house. 40 W filament lamps (FL) are used to illuminate the approach in this house, 11 lamps are in each entrance.

For the selection lighting system of common areas, monthly electricity consumption was calculated; the calculation results are given in Table 1. The data were obtained based on the average duration of day and night time during each month for 2015, the work of LED lighting fixtures with motion sensors and photosensors was taken into account, and also 1.5 minutes after switching on were included in working time. Three persons live in one apartment; there are 4 apartments on the floor, 5 floors, 4 entrances. On the average, it was also taken into account that every person who lives there goes out and goes in 4 times per day. The duration of cloudy and overcast weather was not taken into account in the calculations. Information of the duration of the day in each month was taken from the site pogodavtomske.ru.

* Corresponding author: aos16@tpu.ru

Table 1 - Monthly consumption of electricity by the lighting system in common areas on the base of LED lamps (5 W) of a residential five-story house at the address: Tomsk city, Lenin Avenue, 255 house.

Month	1	2	3	4	5	6	7	8	9	10	11	12
kW·h	28.3	23.2	22.5	18.6	16.3	14.2	15.4	17.5	21.3	23.8	27.5	29.1

Based on the calculated data, it can be concluded that the longest lighting of the entrances works in December and January, because the duration of the day is short, and the weather is mostly cloudy, which also shortens the duration of daytime. Changing FL to an LED lighting system with different sensors can reduce the power consumption in more than 30 times. The service life of LED lamps is in 60 times higher than FL. This means that the working hours of the LEDs is more than 60 000 hours, if interpret into real time then such LED lights will last more than 10 years [1].

To estimate the possibility of obtaining electricity from photovoltaic panels, calculations were made according to monthly and annual values of solar radiation, taking into account the different arrangement of the panels: horizontal, vertical and at an angle of 45° degrees. The results of the calculations are shown in Table 2.

Table 2 - Average monthly and total annual insolation for different orientations of photovoltaic panels, kW·h / m^2.

Orientation	Month						
	1	2	3	4	5	6	7
Horizontal	20.3	45.4	104.5	108.3	111	115.8	217.7
45°	27.5	61.4	141.4	146.6	150.2	156.7	294.6
Vertical	65.3	114.1	147.7	98.8	70.1	67.1	82.3
Orientation	Month						
	8	9	10	11	12	Year	
Horizontal	110	92.3	75.2	57.7	15.8	1058.2	
45°	148.9	124.9	101.8	78.1	21.4	1453.4	
Vertical	83.5	98.5	130.2	149.1	66.5	1173.1	

Having investigated solar insolation in Tomsk, it can be concluded that it is more rational to use solar energy by installing panels in vertical position (Figure 1). This will give the greatest energy output in the winter, because the maximum load on lighting is in the winter. Panels are recommended to install on the south wall of the residential building, because this is more simple and less expensive installation version.

Fig. 1. Place of panels installation on the apartment house.

Solar power plant and uninterrupted power supply (UPS) will be used to power the LED lamps. The solar power plant (SPP) will be the main source of power [2]. There is no device for converting electrical energy such as an inverter in this lighting system with power from SPP. This reduces the cost of such lighting system in general. The main components of SPP:

1) A solar panel;
2) Electrical battery;
3) Charge Controller EB;
4) LED lighting fixtures (load).

The structural scheme of a solar power plant is shown in Figure 2. The choice of this system is due to the fact that there are few sunny days in Tomsk especially in winter because of large cloudiness and precipitation. This affects the generation of electricity, because the photo panels will not work at full capacity in such days. Therefore, the system needs emergency power source, which have to replace the solar panels (SP) for a certain period of time in order to avoid underproduction of electricity. In the event if the photo panels do not cover the required load, and the battery is discharged, the load will be transfered to the 2nd power source (backup) - a switching power supply that will be connected to the supply [3]. The PSU will always be operational.

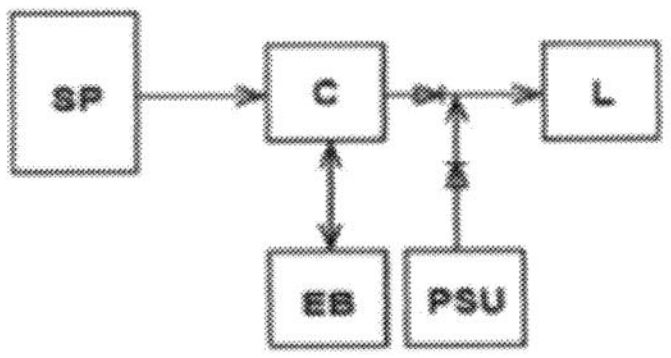

Fig. 2. Block diagram of solar power plant: SP - solar panels, EB - electrical batteries, C - charge controller EB, L - load, PSU - static power supply unit (emergency power source).

Electric power generation as applied to the conditions of the investigated object, as well as electricity consumption by LED lights are shown in the graph of Figure 3.

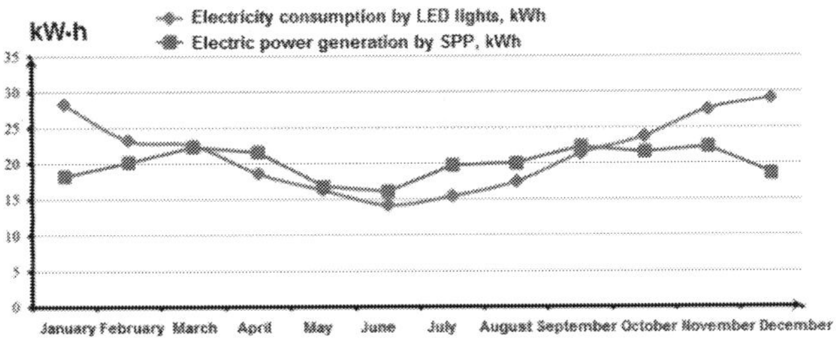

Fig. 3. Comparison of power consumption and power generation.

It can be concluded according to Figure 3, that the equipment will be able to generate enough energy to cover the load from March to September, and the rest of the time: from October to February the remaining underproduction of electricity will be covered with the PSU.

Such development may be well realized, because the payback of the project will be almost 8 years, which proves real and profitable usage of SPP for lighting common areas in Tomsk.

References

1. S.N. Udalov, *Renewable energy sources: Textbook* (Publishing House of the National Technical University, Novosibirsk, 2007)
2. L.P. Sumarokova, Naukovedenie **8**(6), 11 (2016)
3. M.A. Surkov, Naukovedenie **8**(4), 13 (2016)

MATEC Web of Conferences **110**, 01077 (2017) DOI: 10.1051/matecconf/201711001077
HMTTSC-2017

Analysis and prevention of accident-caused faults in power cable lines

Tatyana V. Susakova [1], *Arina M.* Evseeva[1,*], and *Irina A.* Chesnokova[1]

[1]National Research Tomsk Polytechnic University, 634050 Tomsk, Russia

Abstract. To ensure fault-free operation of power cable lines it is essential to make a thorough analysis of the type of faults and their location. The paper considers the most common type of faults and main reasons for cable damage. The case study of cable faults at an urban network (Zabaikalsky Krai) shows that the root cause is insulation deterioration, especially with the cables reaching the end of their 35 year lifetime. Therefore, to prevent faults and mitigate the threats it is necessary to increase surveillance of cable systems and use cables with higher reliability and functionality, e.g., XLPE cables.

1 Introduction

The draft of the new Energy Strategy up to 2035 clearly indicates that energy efficiency will be among the priorities in the energy sector [1]. According to the National Audit Chamber of Russia, the deterioration of Russian utility networks has reached 60%. This is due to ineffective management and a lack of transparency from utility companies in relation to the costs and expenditure on capital construction and renovation. Obviously, it is becoming vitally important for the Russian energy sector to focus on the need to limit installation and service costs and obtain higher performance, reliability, and asset life.

One of the factors ensuring uninterrupted energy production is the cable system reliability. Cable systems are degrading over time and subsequently more failures are recorded. The unscheduled interruption of energy production, resulting from cable failures, entails overhead and labor costs associated with downtime. Downtime can be minimized through proper handling and maintenance of cables, continuous monitoring and analysis of their condition.

Today's industrial environments need superior mechanical protection for the power cables. The higher demands by the electricity industry force manufacturers to make innovative cable designs (e.g., cross-linked polyethylene insulation, PVC insulation) to address important technical and reliability problems. In Russia, however, 95% of power cables are paper insulated and cannot ensure proper protection from oil based products as well as acids that can cause electrical cables to fail over time.

To effectively meet the challenges awareness of state of the art at Russian utilities can be of great value. The article attempts to give an overview of the type of cable faults and their typical location. The presented case study of cable faults at an urban network

* Corresponding author: averseeva@yandex.ru

(Zabaikalsky Krai) shows that the major reason for their damage is cable deterioration. The undertaken comparative study of paper-insulated cables and cables with cross-linked polyethylene insulation aims to show the ways to improve energy efficiency and to minimize the need for fault-recovery at Russian utilities.

2 Types and reasons for cable damages

The most common types of cable line damages:

- insulation damage, resulting in one-, two- or three-phase short circuit to the ground;
- insulation damage, resulting in phase-to-phase faults;
- one-, two- or three-phase fault without grounding, with grounding of broken cores, with grounding of unbroken cores.

Single-phase damages is the most common type of faults in 1-10 kV power cable lines in which one of the cable cores is short circuited to the shielding. Single-phase faults can be divided into three groups by the value of transient resistance at the short circuit location:

1) with transient resistance equal to tens and hundreds of mega ohms;
2) with transient resistance equal to up to hundreds of Ohms;
3) with resistance close to zero.

Interphase damages account for about 20 percent of all types of faults. They fall into two groups of damages: damages with transient resistance at the fault location close to zero and with resistance from some kilo ohms to hundreds of mega ohms.

The main reasons for cable faults can be traced back to:

- errors in design (deterioration of the insulation properties caused by unacceptable overheating of load currents due to low cross-section of cable cores; damages in emergency modes due to improper choice of protective equipment, etc.);
- manufacturing defects (cracks or through holes in the cable sheath; matching of several paper tapes; burrs on wires of current-carrying cores, etc.);
- cable laying defects (sharp bends of cables at turning angles of a line routing; mechanical damage (pits, cuts, twists of cables); ignoring the acceptable distance to the facilities which can affect cables performance (heat transport system, electric railways, etc.)
- cable joints defects resulting from uneven distribution of electromagnetic field strength in cable fittings, as well as non-observance of the cable joints manufacturing technology when installing cable fittings in the field;
- damages during cable maintenance (accidental mechanical cable damages; insulation deterioration; metal corrosion of cable armature or lead coating caused by ground current or chemical composition of ground, etc.) [2].

3 Statistical analysis of accident-caused damages

Scarcity of relevant data on cable performance can be a factor impeding its proper maintenance and improving its reliability. Therefore, a case study of cable faults was done at an urban network (Zabaikalsky Krai).

Table 1. Distribution of accidents according to the features of damage causes.

No	Causes of damages	Number of accidents, 2014	Number of accidents, 2015 (the first half)
1	Due to insulation deterioration:	36	13
1.1	Damage in the unbroken place	27	7
1.2	Cable joints damages	9	6

2	As a result of inappropriate installation of cable lines by the contracting organization	1	3
3	As a result of mechanical impact during excavation by third-party organizations	2	1
	Total	39	17

As is seen from the table the major reason for cable damage is insulation deterioration, i.e. damages associated with cable maintenance, mostly in the unbroken place. A positive trend is the decrease of the number of emergency outages of 6/10 kV cable lines equipment from 39 accidents in 2014 to 17 accidents in 2015 (reduction by 56%). It entailed lost energy decrease from 25,23 to 6,64 megawatt hour (reduction by 74%) and decrease of the total duration of power interruption from 34,95 to 14,82 (reduction by 58%).

To find out causes of the significant decrease in cable damages, dependence of the number of accidents on cable lifecycle was determined.

Table 2. Distribution of accidents according to the cable life time.

No	Life time / years	Over 2014 / number	Over the first half of 2015 / number
1	Up to 5	6	2
2	From 5 to 15	7	2
3	From 15 to 25	8	2
4	From 25 to 35	12	5
5	From 35 to 45	3	4
6	Over 45 years	3	2
	Total	39	17

Table 2 shows that in 2014 a surge of faults was recorded with the cables reaching the end of their 35 year lifetime. The obtained data are supported by the evidence of Federal Grid, suggesting that ageing of power cables accounts for 31% of damages on Russian utilities. The raising concern is that a considerable number of cable lines exceeding the licensed 35 year lifetime are still in service.

The first half of 2015 exhibits a slight decrease in damages, which can be attributed to more accurate monitoring and maintenance and well-timed replacement of old cables. Prevention of cable damages is an urgent task for utilities. To increase the rate of cable faults detection a variety of fault detection techniques is used. These methods fall into two groups: conventional and modern. Conventional methods are visual inspection, insulation resistance test, voltage proof test. Modern methods are partial-discharge test with oscillating wave test system, time domain reflectometry, heat monitoring, etc.

The key issues to be considered in a workable cable maintenance program are 1) choice of a proper cable design that is consistent with voltage, safety and expected performance; 2) well-managed cable installation; 3) record of causes of cable failures; and 4) prompt remedial actions based on records. Ideally, cable asset management should start at an early stage and continue through the cable lifecycle.

Currently, the globally preferred cable for both transmission and distribution is XLPE (cross-linked polyethylene insulation) cables (Japan, Finland, France, Sweden – 100% of the cable market; Germany and Denmark – 95%; Canada and the USA – 85%). Russian utilities, however, mostly rely on paper-insulated cables. Let us consider pros and cons of both types of cables.

Paper-insulated cables have shown fairly good electrical properties and have, in general, given very good service. Their main failure mode is insulation deterioration (as is shown in Table 1) due to partial discharge. Other failures include:

1) cable manufacture is rather complicated;

2) thermal runaway due to mutual heating from neighboring cables or overloading;

DOI: 10.1051/matecconf/201711001077

3) restrictions on a cable vertical run resulting from adhesive leaking;
4) inconvenience with fixing oil leakage from cable sheath and cable sealing;
5) heavy construction due to a lead sheath.

The first cables with cross-linked polyethylene insulation had a poor service record. However, arrival of developments in manufacturing techniques has improved their performance. Today, despite some drawbacks, such as instability to high temperatures and faster ageing under UV exposure, XLPE cables exhibit far better performance. Their major highlights are:

1) scarce records on cable laying defects;
2) higher cable performance owing to the increase of a core temperature limit (permissible loading current is generally 15-25% higher than with paper-insulated cables);
3) continuous load (90°C versus 70°C);
4) enhanced thermal and moisture resistance;
5) in case of short circuits conventional thermal current is up to 250°C;
6) small insulation losses (0,001 versus 0,008);
7) relative eco-friendliness (no lead, oil, bitum);
8) smaller dimensions and weight;
9) cheaper cable maintenance [3,4].

Opponents of XLPE cables, however, claim that high production costs outweigh the above-mentioned advantages. Nevertheless, to ensure reliable and uninterrupted energy supply, Russian utilities will have to reconsider their strategies for increasing energy efficiency and gradually introduce XLPE cables.

4 Conclusion

Thus, to prevent faults and mitigate the threats it is necessary to increase surveillance of cable systems with modern methods and use cables with higher reliability and functionality. A number of leading Russian power systems are targeted at the use of medium and high voltage XLPE cables when laying the new cable lines and replacing or maintaining the old ones. To establish competitive advantage of Russian power systems on the international level it is necessary to improve reliability of cable lines that will eventually contribute to enhancing efficiency of electrical networks.

References

1. A. Gusev, K.Westphal, SWP Research Paper, **5** (2015)
2. H. Sharma, M. Deshpande, R. Pandey, IJETAE, **3**, 229 (2013)
3. P. Soepboer, E. Dorison, P. Leemans, 22nd International Conference on Electricity Distribution, 0584 (2013)
4. Y. Wang, N. Davies, T&D, **48** (2015)
5. Yu. Khrushchev, N. Batseva, N. Fix, I. Chesnokova, V. Khar'kovskaya, Procedia - Social and Behavioral Sciences, Linguistic and Cultural Studies: Traditions and Innovations, **206**, 267 (2015)

Influence of the near-surface water film thickness on the characteristics and conditions of water-coal fuel particles inflammation

Samen Syrodoy[1,*], *Roman* Taburchinov[1], and *Irina* Mikhaylova[2]

[1]National Research Tomsk Polytechnic University, 634050 Tomsk, Russia
[2]Tomsk State University of Architecture and Building, 634003 Tomsk, Russia

Abstract. The results of experimental studies of thermal pretreatment processes complex and inflammation of water coal fuel's particles (WCF), which surface is covered with water film, have been presented there. The research has been carried out with using of three fuel types (cannel coal, coal B and filter cake): they have essentially different thermo physical characteristics. The results of our studies have been shown that the dynamics of the near-surface water film evaporation and fuel inflammation depends significantly on coal type. It has also been analyzed the influence of water layer thickness on the conditions of WCF particles inflammation. It has been established the time of water film evaporation is about 70% due to the whole induction period.

1 Introduction

Turbulent dynamics of pricing for main utilities (oil and natural gas) [1-3] forms the certain conditions to apply the alternative types of fuel (for example, coal) in the world and native power economy. However, the main problems, connected with coal combustion in mass-fired refuse power boilers, haven't been solved yet. To latter we include, first of all, emissions of nitrogen oxides (NO_x) and sulfur (SO_x), formed during high-temperature oxidation of coal fuel in the combustion chamber. Nowadays the developed systems of clean coal combustion (for example, with the application of the circulating fluidized bed [4], the vortex principle [5], the plasma systems [6], etc.) haven't been used widely due to the high price of the auxiliary boiler equipment. At the present time, the technology of coal combustion in the form of especially made fuel compositions on the foundation of coal-water slurry is the main perspective technical solution for using of coal fuel.

To date, there are a number of objectives and subjective reasons that significantly impede the introduction of WCF in the energy sector. As a rule, the main part of them is connected with the technology process, but has no connections with the fundamental problems of energy science.

The most acute problem is the development of physical theory and its mathematics base of WCF drops inflammation and combustion processes. The latter is connected with the difficulties of the heat transmission modeling in the essential heterogeneous structures of

*Corresponding author: syrodoy@tpu.ru

water-coal fuel. This is due to the fact that evaporation of water is the most influential for the thermal preparation process. As it is known [7], the phase transformation processes usually occur with a large endothermic effect (up to 2.5 MJ / kg) in a very narrow and constantly moving zone (the front of the phase transition). The mathematical modeling is significantly complicated by the fact that during the spraying of fuel with a fairly high probability it is possible to form both partially dried and excessively watered particles of water-coal fuel [8]. The latter particles are a drop, the surface of which is covered with a water film. As it has been established earlier [9-11], such a liquid layer can have a significant effect on the dynamics of ignition of the WCF particle. However, the estimates of the influence of the thickness of this layer on the characteristics and conditions of ignition have not been carried out yet.

The aim of our studies is to investigate the inflammation processes of water-coal fuel particles in the conditions of the intensive radioactive convection heating with the variation of water layer thickness.

2 Experiment

The experimental studies have been carried out on the stand which is the same to the installation according to the field recording procedure [12]. The fuel particle (covered with the water film) has been fastened on the suck board and has been put in into the hollow ceramic chamber, heated to the temperature T_e=1270K with the help of coordinate remote controlled device. The video fixation of the heat and mass transfer processes' dynamics has been realized with high-speed video camera Photron FASTCAM CA4 5. The period time of the beginning of heating effect (the moment of putting the fuel particle into the piston chamber) to the beginning of the firing (appearance of flame) has been considered as the time of combustion delay (t_{ign}). The sizes of the particles are varied within the limits of $1.5<\delta<3.5\cdot10^{-3}$м. The systematic error of the main measurable parameters dimensioning has been no more than 5 %. The confidence interval of this definition t_{ign} with the level of confidence 0.95 has not exceeded 18 %.

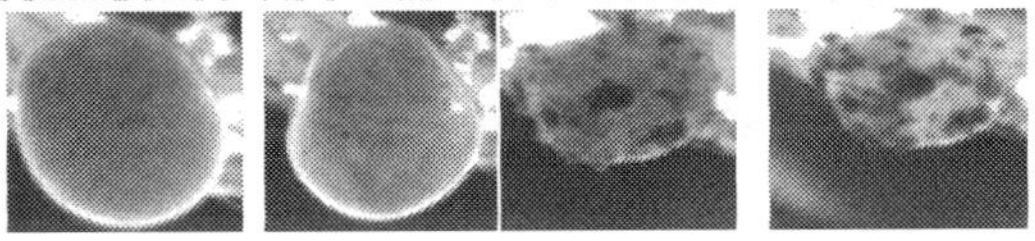

a *b* *c* *d*
t=0 s *t=1.01 s* *t=2.51 s* *t=4.08 s*

Fig. 1a. The typical video frames of the particles inflammation of water-coal fuel on the foundation of coal D covered with the water film.

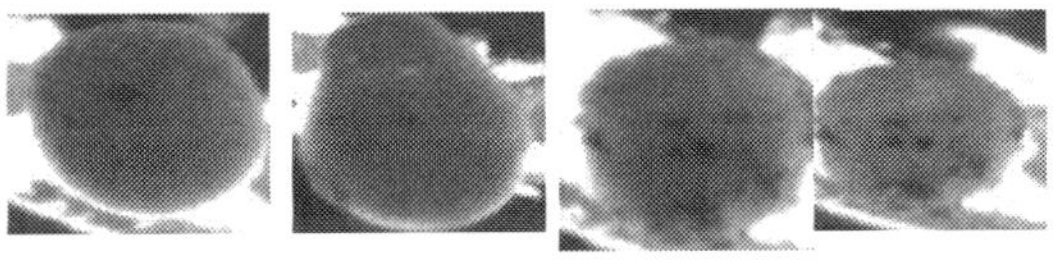

a *b* *c* *d*
t=0 s *t=0.78 s* *t=2.42 s* *t=3.87 s*

Fig. 1b. The typical video frames of the particles inflammation of water-coal fuel on the foundation of coal B covered with the water film.

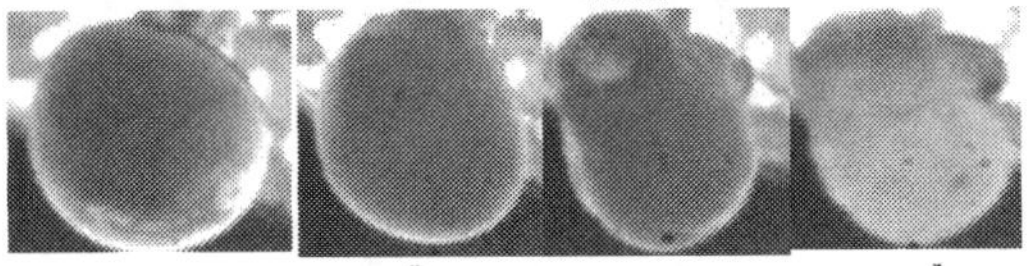

a *b* *c* *d*
t=0 s *t=1.83 s* *t=3.21 s* *t=5.01 s*

Fig.1c. The typical video frames of the particles inflammation of water-coal fuel on the foundation of the filter-KEK covered with the water film.

3 The results of the experimental studies

The typical video frames of the particles inflammation and firing processes (diameter $\delta \approx 2.5 \cdot 10^{-3}$м) of the water-coal fuel made from the variant kinds of coal with different metamorphism intensity are presented in the figures 1 (a, b, c): cannel coal (fig.1 a), brown coal (fig.1 b) and filter cake (fig.1 c), covered with water film at the temperature environment T_e=1273K. The analysis of the frames has shown the whole period of thermal treating of fuel we could relatively divide into a raw of some succeeding stages. At first (video frame a) is a water-coal aggregation covered with water film. As a result, we have evaporation process that has been instigated due to the thermal influence, when this happens we could observe the change of the size and form of the fuel drop (fig. b). The last has been caused with boiling water. There is dehydration of the main fuel layer and coal inflammation after the disposal of water film. It can be noted that the evaporation period of the water film is approximately 70% of the total ignition delay time. Most likely this is due to the fact that when the water film boils, the heating of the main fuel layer to the phase transition temperature occurs. Accordingly, the subsequent dehumidification occurs much faster than for the fuel particles without such a layer.

There are the time dependences of the water film evaporation at the fig.2 and the ignition delays of water film (t_v) of the water-coal fuel particles based on the cannel coal at the exterior temperature in the different water layers' thickness. The analysis of the dependences shows that water film influences significantly on the inflammation conditions. So the change of water film thickness from 0.2 to 0.7 leads to the double increasing of the induction period (from 4.5 to 8 seconds) at the exterior temperature (T_e). And as it is noted previously the time of water film evaporation is about 60% from t_i. The last, most likely, is determined with the heating up of the main coal-water agglomerate in the near-surface water layer evaporation process to the water boiling temperature. Consequently, the following WCF drop dehydration takes place more quickly than the particles without water film.

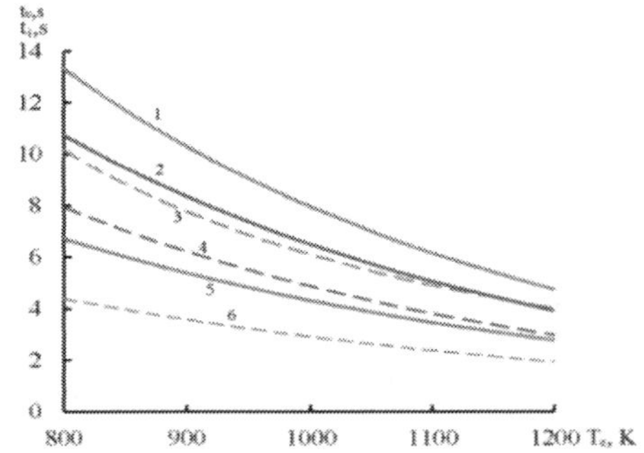

Fig. 2. Time dependence of the full water film evaporation and ignition delay of water film with diameter: $d=2 \cdot 10^{-3}$ m from the temperature, with water film: 1,3 – $\delta \approx 0.7$mm; 2,4 - $\delta \approx 0.4$mm; 5,6 – $\delta \approx 0.2$ mm. 1, 2, 5 – time of ignition delay (t_i); 3, 4, 6 – time of ignition delay of water film (t_v).

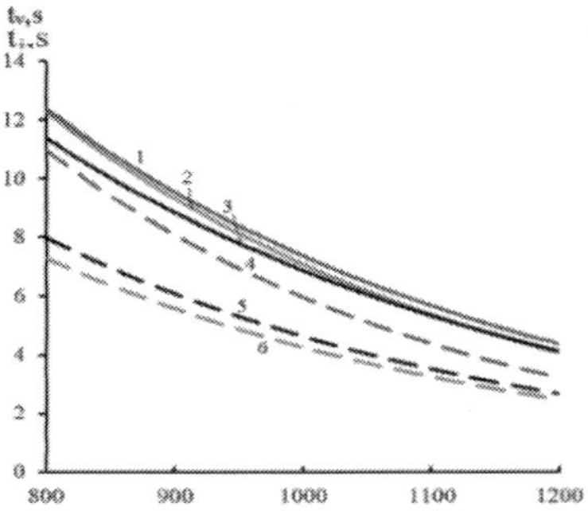

Fig. 3. Time dependence of the full water film evaporation and ignition delay of water film with diameter $d=2 \cdot 10^{-3}$ м from the temperature. WCF have been done from: 1,4 – filter cake; 2,6 – brown coal ; 3,5 – channel coal. 1, 2, 3 – dwell time of implantation (t_i); 4, 5, 6- time of ignition delay of water film (t_v).

The temporal characteristics of the evaporation processes of the near-surface water layer and the ignition of water-coal fuel particles based on coals of different grades (long-flame, brown and filter-KEK), with a film thickness of 0.5 mm are compared in the figure 3. The analysis of the dependences shows that the delay time of all three fuel types is practically identical, while the terms of t_v differ markedly. Perhaps it is connected with essentially different thermo physical characteristics (thermal conduction, heat capacitance and density) of the main fuel component (coal).

Probably the great amount of the metal and mineral existence in the structure of filter-KEK leads to the increasing its heat capacitance, because of it the heating up and water film evaporation takes place slower due to the heat efflux in the fuel structure. Also the dependences at the figure 4 confirm the noted the thesis (using at the description figure 2) about the near-surface water film time evaporation is the most part of the thermal pretreatment period. It is connected with the heating up of the main fuel layer at the water film evaporation.

4 Conclusion

The results of the experimental studies have been established the main time characteristics of the thermal pretreatment processes and the particle inflammation of water-coal fuel, which surface is covered with thin watered layer. It has been improved that the time of the water film evaporation could take about 60% from the whole induction period. The hypothesis has been put forward that when the near-surface water layer evaporates, the main fuel mass is heated to the evaporation temperature. As a result, the subsequent dehydration of the main fuel layer occurs much faster than when the WCF particles are ignited without a water film. Accordingly, it can reasonably be assumed that in order to ensure the steady ignition of drops of water-coal fuel in the furnace space, it is necessary to organize zones of the constant recycling of the WCF flame.

The work was supported by the scientific schools grant NSH-7538-2016.8.

References

1. A. Demirbas, Prog. Energy Combust. Sci. **31**, (2005)
2. S.H. Hosseini, G.H. Shakouri, Energy Policy. **91**, (2016)
3. H.B. Ghassan, H.R. AlHajhoj, Applied Energy, **169**, (2016)
4. J.K. Martin, J. A. Faaij, Prog. Energy Combust. Sci. **33**, (2007)
5. K.Y. Orlova, B.V. Lebedev, MATEC Web Conf. **92**, (2017)
6. V.E. Messerle, E.I. Karpenko, A.B. Ustimenko, Fuel, **126**, (2014)
7. V.V. Salomatov, G.V. Kuznetsov, S.V. Syrodoy, N.Y. Gutareva, Appl. Therm. Eng. **106**, (2016)
8. S.E. Nydick, F. Prochet, H.A. Steiger, J. Eng. Gas Turbines Power, **109**, (1987)
9. A.V. Zakharevich, G.V. Kuznetsov, V.V. Salomatov, P.A. Strizhak, S.V. Syrodoy, Combustion, Explosion, and Shock Waves, **52**, (2016)
10. S.V. Syrodoy N.Y. Gutareva, R.I. Taburchinov, MATEC Web Conf. **72**, (2016)
11. S.V.Syrodoy, N.Y. Gutareva, R.I. Taburchinov, MATEC Web Conf. **72**, (2016)
12. V.V. Salomatov, G.V. Kuznetsov S.V. Syrodoy, N.Y. Gutareva, Appl. Therm. Eng. **106**, (2016)

Use of the PLIF and LIP methods for definition of non-stationary temperature fields of water droplets

Roman S. Volkov*, *Maria* M. Osmolovskaya, and *Yaroslav* S. Solomatin

National Research Tomsk Polytechnic University, 634050 Tomsk, Russia

Abstract. This study illustrates the results of experiments to determine the unsteady and non-uniform temperature fields of evaporating water droplets. It is shown that during the warming up of the droplet inside it an inhomogeneous temperature field is formed. In the droplet trace, formation of a region with lower (relative to the gas flow) temperatures of the vapor-gas mixture is observed. The experiments were carried out for droplets with initial radii of 1-2 mm. The speed of the air flow was 0.5-5 m/s. The temperatures of the gaseous medium varied in the range of 20-600 °C. During the research, modern panoramic methods of non-contact flow diagnostics such as "Planar Laser Induced Fluorescence" and "Laser Induced Phosphorescence" (LIP) were used.

1. Introduction

Modern models of heat and mass transfer describing the heating and evaporation of liquid droplets in a gaseous medium are formulated on the basis of long-established hypotheses and representations by Spalding D.B. [1], Fuchs N.A. [2], Ranz W.E. and Marshall W.R. [3]. The development of these concepts over the last 30 years has allowed us to obtain a group of approximations which illustrate the corresponding dependences of Nusselt numbers on the Reynolds and Prandtl numbers. These expressions are traditionally used in the simulation of convective-conductive heat and mass transfer processes and phase transformations at the liquid-gas interface. The results of the studies [4] showed that in the region of moderate (up to 600 K) gas temperatures, the use of the expression group Nu = f (Pr, Re) makes it possible to obtain a satisfactory agreement of the theoretical values to the experimental data (the differences do not exceed 10-15%). At temperatures above 500 °C, the theoretical and experimental values of the evaporation rates of water droplets differ by several times [4]. A hypothesis [4] has been advanced that at high temperatures of a gaseous medium, the non-stationary processes of formation of a substantially inhomogeneous temperature field of evaporating water droplets can play a decisive role. With intense heating, the temperature distribution in the droplet can be substantially

* Corresponding author: romanvolkov@tpu.ru

nonuniform and nonstationary. Unfortunately, up to the present time, adequate experimental methods for determining the temperature distributions inside the droplet have not been developed.

Contactless optical methods for diagnosing thermophysical parameters (such as Particle Image Velocimetry (PIV) [2], Particle Tracking Velocimetry (PTV) [3], Planar Laser Induced Fluorescence (PLIF) [4] and others) have been actively developing over the past few years. To solve the previously identified problems and determine the temperature fields of the evaporating water droplets, it is expedient to use the methods of PLIF [4] and LIP [5, 6], since it allows the recording of the temperature of liquid and gas media with high sampling in time and space with high sampling.

The purpose of this work is to determine the features of heating and evaporation of water droplets during high-temperature heating in a gaseous medium using non-contact panoramic optical methods "Planar Laser Induced Fluorescence" (PLIF) and "Laser Induced Phosphorescence" (LIP).

2. Experimental setup and procedure

Fig. 1 show the experimental setup used in carrying out experimental researches.

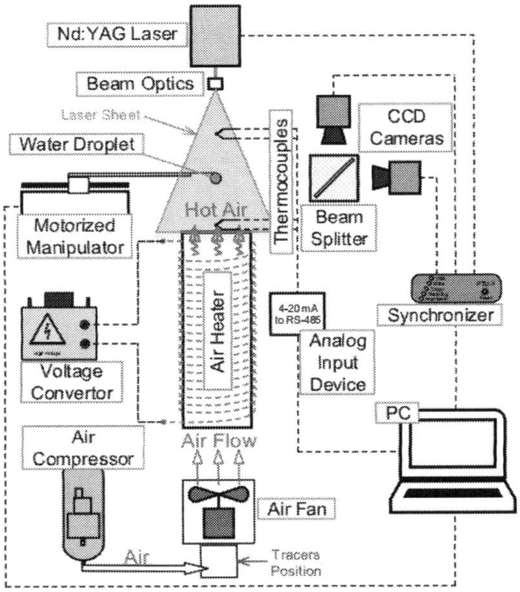

Fig.1. A scheme of experimental setup.

The main element of the setup (see Fig. 1) was the heating element. The heating element consisted of a hollow ceramic cylinder (inner and outer diameters 60 mm and 52 mm) 1 m high, the external circuit of which is provided with 6 independent windings (of nichrome wire 0.3 mm in diameter) per 1 winding for every 0.2 m of the ceramic pipe. The total maximum power consumption of nichrome spirals was 9 kW. The airflow in experiments was pumped by a fan connected to the lower inlet of the heater (ceramic tube). The speed of the air flow was 0.5-5 m/s. The air flow rate was monitored using the PIV method [2]. The temperature of the air stream (T_a) at the outlet of the heater (the upper part of the ceramic tube) was regulated in the range 20-350 ° C by changing the voltage from 50 to 250 V on the nichrome windings. Two lowering autotransformers were used. Type-K thermocouples (temperature measurement range 223–1473 K, accuracy ± 3 K, heat retention no more than 0.1 s) were used to measure temperature of hot air flow (see Fig. 1).

To generate water droplets of fixed sizes, the Finnpipette Novus electronic dispenser (0.1 μl volume variation step) was used. The initial volume (V_d) of the generated droplets ranged from 5 μl to 30 μl. This corresponds to a change in the droplet radii in the range R_d=1-2 mm.

The PLIF [4] and LIP [5, 6] method was used to determine the instantaneous temperature distributions inside the water droplet and in the trace of the evaporating droplet. The schemes of the methods are shown in Fig. 2.

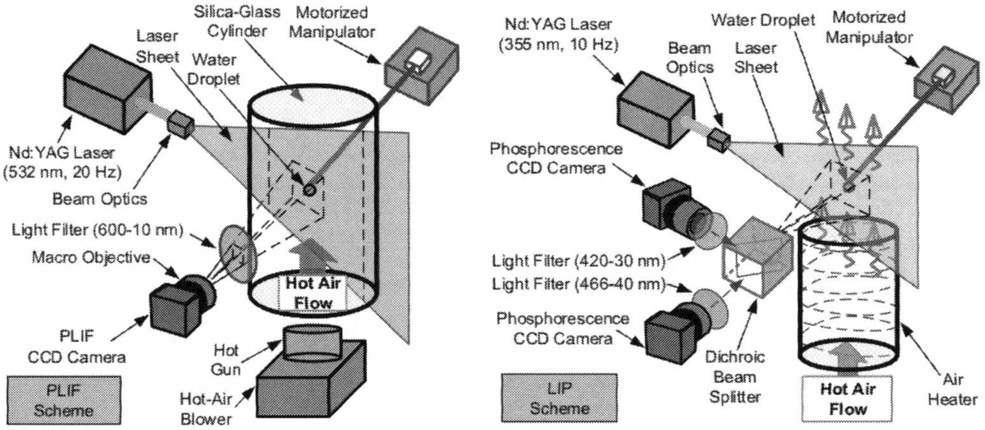

Fig. 2 The schemes of the PLIF and LIP methods.

To illuminate the water droplet in the experiments used solid-state pulsed Nd: YAG Quantel lasers (wavelength 532 nm and 455 nm). To create a laser sheet laser optics (based on cylindrical lenses) with an opening angle of 8° was used. Images of water droplets were recorded by CCD ImperX IGV-B2020M and Imager M-lite 2M cameras.

The experiments with the use of PLIF and LIP (Fig.4) were performed in two stages: the first was the temperature calibration of the system (in the absence of a drop of water in the registration area), the second - the recording of the temperature fields of the evaporating droplet [4-6].

As specialized tracers for flow in the LIP method, specialized phosphor particles of $BaMgAl_{10}O_{17}$:Eu powder (BAM: Eu) [5, 6] were used, the particle sizes were 6-8 μm. To avoid clumping of particles in the flow, they were first placed in a muffle furnace (at 100 °C) for 120 min (in accordance with the recommendations and conclusions [5, 6]). As a fluorophore in the PLIF experiments, Rhodamine B was used [4]. The selected fluorophore is sufficiently stable under the action of a laser; moreover, the intensity of the fluorescent emission of a given dye has a pronounced temperature dependence: the decrease in fluorescence intensity is about 2% per 1 °C in the range 10-70 °C. The errors in measuring the temperature using the PLIF method did not exceed 2 °C, the error of the LIP method was about 25 °C.

DOI: 10.1051/matecconf/201711001079

3. Results and discussions

Fig. 3 shows temperature field inside the droplet of water obtained using the PLIF method. PLIF images allow us to observe non-steady and non-homogeneous temperature fields of droplets throughout the long heating intervals achieving several tens of seconds. A non-steady behavior of the temperature field becomes obvious when comparing T_d values in identical droplet sections during various time intervals. These values quickly change with the increase in heating time. Thus, Fis. 3(a) illustrate the difference of 30–40 °C in the identical cross-sections, while in Fig. 6(b) this difference is not over 10 °C. From this we conclude that the increase in heating time results in practically homogeneous temperature field. Certain constant values of T_d to which the droplets heat up before their complete evaporation, are observed in each experiment. The droplet temperature field is strongly affected by the direction of the hot airflow. The higher T_a and U_a values the higher temperature rise on the droplet frontal surface unlike its back and lateral surfaces (Fig. 3).

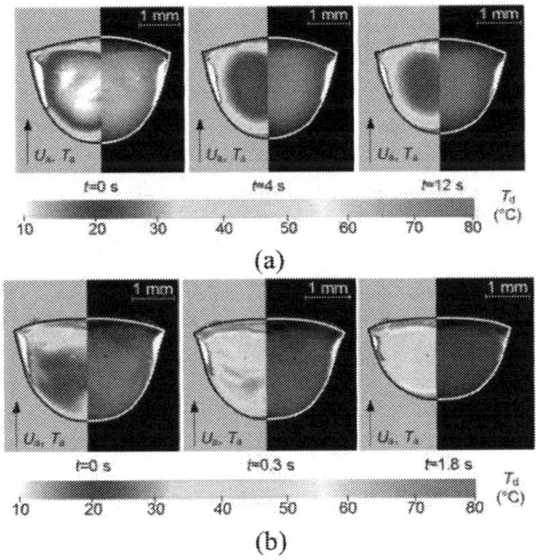

Fig. 3. Time history for droplet temperature field (PLIF method) at different heating conditions (U_a≈3 m/s, R_d≈1.53 mm): (a) – T_a≈100 °C; (b) – T_a≈500 °C.

Fig. 4 shows temperature field outside the evaporating droplet of water obtained using the LIP method. The recorded temperature trace, its longitudinal and transverse dimensions (Fig. 4) is visually shown. In contrast to the thermocouple measurements [10] in this paper, it was possible to determine the essentially non-stationary nature of the formation of a temperature field around evaporating water droplets. In particular, an intensively changing temperature trace is recorded both in the longitudinal and transverse directions relative to the incoming flow of heated air. With an increase in the heating time and a decrease in the droplet size, the latter is heated to a practically homogeneous temperature field (Fig. 4). As a consequence, heat transfer processes on the droplet surface stabilize somewhat, evaporation rates acquire maximum values. Under such conditions, the cooling of the flowing gases is due to intensive evaporation due to the high heat of vaporization. Only when a certain small value of the radius of the evaporating droplet reaches its temperature trace does it begin to decrease sharply. In Fig. 4, one can see a sharp decrease in longitudinal and transverse dimensions (when t≈35 s).

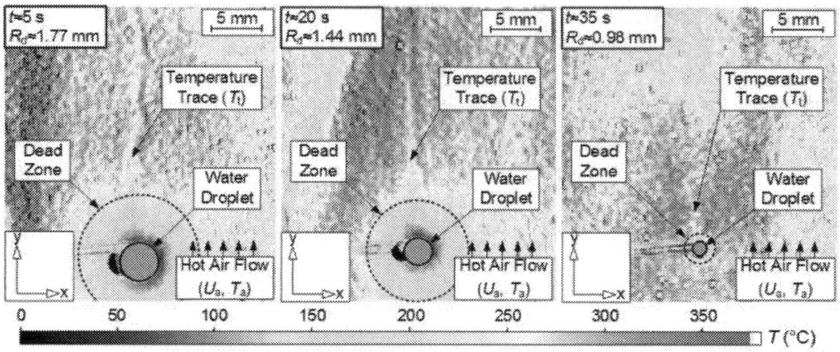

Fig. 4. Non-stationary and significantly non-uniform fields of temperature (LIP method) of a gas stream at a water drop flowing (U_a≈3 m/s, R_d≈1.81 mm).

The obtained generalized results of the conducted experiments are the fundamental basis for the development of a number of promising technologies, for example, fire extinguishing, thermal cleaning, etc. In particular, with their application, it is possible to develop perspective physical and mathematical models that allow predicting the required heat exchange conditions for a drip aerosol with a hot air flow.

4. Conclusion

In the course of the research, the possibility of applying the PLIF and LIP methods for the diagnostics of the temperature distribution of an evaporating droplet was demonstrated. The performed experiments showed the presence of a substantially non-stationary temperature field both inside the drop itself and in its trace. It is established that the temperature drop inside the water droplet can reach up to 50 °C. The decrease in temperature in the droplet trace due to its evaporation can reach values of 60-80 °C (in comparison with the gas flow temperature). In this case, the temperature trace of the droplet is retained throughout the evaporation process. The results obtained can be used to develop ideas of high-temperature evaporation of droplets, as well as the development of appropriate mathematical models

The research was supported by Foundation of President of Russian Federation (project MD–1221.2017.8).

References

1. D.B. Spalding, *Some fundamentals of combustion*, Butterworth's, London (1955)
2. N.A. Fuchs, *Evaporation and Droplet growth in Gaseous Media*, Pergamon Press, London (1959)
3. W.E. Ranz, W.R. Marshall, Chem. Eng. Prog. **48** 141, 173 (1952)
4. O.V. Vysokomornaya, G.V. Kuznetsov, P.A. Strizhak, J. Eng. Phys. Thermophys. **89** 141 (2016)
5. F. Yan, A. Rinoshika, Exp. Therm. Fluid Sci. **44** 534 (2013)
6. D. Damiani, E. Meillot, D. Tarlet, J. Therm. Spray Technol. **23** (3) 340 (2014)
7. B. Williams, P. Ewart, X. Wang, R. Stone, H. Ma, H. Walmsley, R. Cracknell, R. Stevens, D. Richardson, H. Fu, Combust. Flame **157** (10) 1866 (2010)
8. A. Charogiannis, F. Beyrau, Exp. Fluids **54** 1518 (2013)
9. C. Abram, B. Fond, A.L. Heyes, F. Beyrau, Appl. Phys. B. **111** (2) 155 (2013)
10. R.S. Volkov, G.V. Kuznetsov, P.A. Strizhak, I.S. Voytkov, Int. J. Multiph. Flow **91** 184 (2017)

Analysis of composite fuel application possibility based on coal and oats husks in industrial power engineering

Stanislav Yankovsky[1,*], *Anton* Tolokolnikov[1], and *Vladimir* Zaytsev[1]

[1]National Research Tomsk Polytechnic University, 634050 Tomsk, Russia

Abstract. Experimental studies were conducted to determine the energy and technical characteristics of composite fuels from T grade coal (Alardinsky deposit) and oats (Shegarsky district, Tomsk region). An effective concentration of composite fuel of 50% / 50% is established, at which the heat of combustion is reduced by less than 2%, the ash content is up to 44%, and the fly ash output is up to 48%.

1 Introduction

Efforts have been intensively conducted in many developed countries (China, India, the USA, Canada, Russia, etc.) over the last years to create scientific basis for so-called composite fuels application in power engineering: coal-water [1,2] and organic coal-water slurry [3,4], fuel based on products of agricultural production [5,7], etc.

This trend of recent decades is due, in the main, to fairly objective criteria: uncertainty with the prospects of using gas and oil products in the next 15-20 years; solution of the tasks of sulfur, nitrogen and carbon oxides emissions reduction into the atmosphere; objective need to burn waste of many industrial production and agriculture. The simplest and, possibly, most productive solutions in practice are technologies of composite fuels application based on coal and biomass in power engineering.

It is known [5] that coal is the main energy resource for thermal power plants and boiler houses. Large expenses for fuel supply to remote areas of the Russian Federation are the main reason of rising prices for heat consumers [5]. Energy strategy of the Russian Federation foresees an increase in the share of coal in heat and electricity production in the fuel balance of the country, as well as development of renewable energy sources [5-7]. In recent years, the share of biomass in energy resources production has been increasing [7]. The most common bioresources are wood, wood waste and waste from agricultural production of cereals, annual accumulation of which is up to one hundred million tons. In recent years, the pace of agricultural development in the Russian Federation is growing rapidly, so it is possible that utilization of waste from this sector of production will take an important place in the future [7].

A common way of agricultural waste utilization is by burning. The most developed recycling technologies are in Bangladesh, Thailand, India, Indonesia, Malaysia, USA [7].

* Corresponding author: jankovsky @tpu.ru

DOI: 10.1051/matecconf/201711001080

Oats husks have low calorific value in comparison with coal, which negatively affects its promising application in thermal and electric energy production. But on the other hand, low ash content of this biofuel is attractive for solving the problem of power engineering negative impact reduction on the environment.

The result of mixing high-ash energy coal with oats husks can allow obtaining high synergetic effect. However, the use of such composite fuels in Russia is not developed due to the lack of both general theory of their component composition formation and combustion. Technical, energy and environmental characteristics of produced composite fuels, including addition of oats husks to coal, have not been sufficiently studied experimentally.

Urgency of the problem of agro-industrial complex wastes recycling with the use of composite fuels is confirmed by numerous publications (for example, [8-14]), which give results of fuels properties study based on biomass and coals during combustion in power plants. But results of [8-14] are not the basis for consistencies analysis of composite fuels thermal decomposition processes with a change in the ratio of biomass and coal concentration. Analysis of composite fuels creation possibility with energy characteristics close to the initial coal and significantly better environmental parameters (close to biofuels) is of great interest.

2 Results of conducted experiments and discussion

Technical and energy characteristics of composite fuels have been determined. Quantity of T grade coal (Alardinskoe deposit, particle size less than 80 μm) and oats husks (Shegarsky district of Tomsk region, particle size less than 200 μm) were loaded at controlled mass concentrations into galvanized drum of Pulverisette 6 planetary mill [8]. Mixing process was conducted at drum rotation speed of 500 rpm for seven minutes. Technical analysis of composite fuels was carried out after mixing. Change in calorific value of composite fuels is shown in Figure 1.

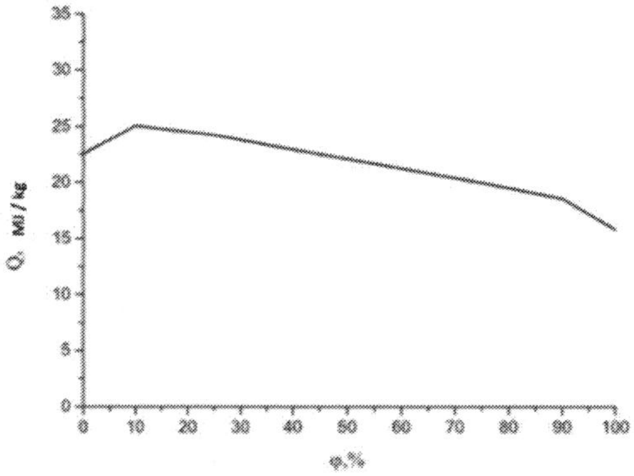

Fig 1. Effect of oats husks concentration on calorific value of composite fuel in relation to homogeneous coal (φ – oats husks share).

Analysis of the results shown in Fig. 1 describes that increase in oats husks concentration leads to synergistic effect. Calorific value of composite fuel relative to the initial coal with 10% biomass concentration increases by 11%, with 25% concentration – by 7%, with oats husks content of 50% calorific value of the fuel is reduced only by 2%.

Based on the foregoing, it can be concluded that an increase in oats husks concentration up to 25% leads to improvement in energy performance of composite fuels compared to coal. An increase of up to 50% reduces them by only 2%.

An important indicator of environmental impact during thermal decomposition of fuel is its ash content. Figure 2 shows dependence of change in fuel ash content obtained in experiments from increase of biomass concentration.

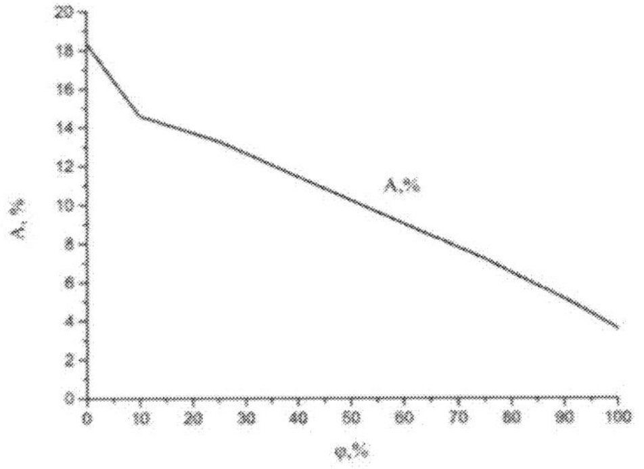

Fig 2. Effect of oats husks concentration on composite fuel ash content in relation to homogeneous coal (φ – oats husks share).

Figure 2 shows that increase in oats husks concentration in composite fuel leads to significant decrease in ash content. Thus, when concentration of biomass is increased to 50%, ash content is reduced by 44%. The obtained results allow drawing conclusion about positive influence of biomass on energy and ecological characteristics of composite fuel based on coal and oats husks.

The main and important parameter characterizing fuel is yield of volatile substances. Figure 3 shows experimentally established dependence of volatile substances release from oats husks concentration in composite fuel (relatively to homogeneous coal).

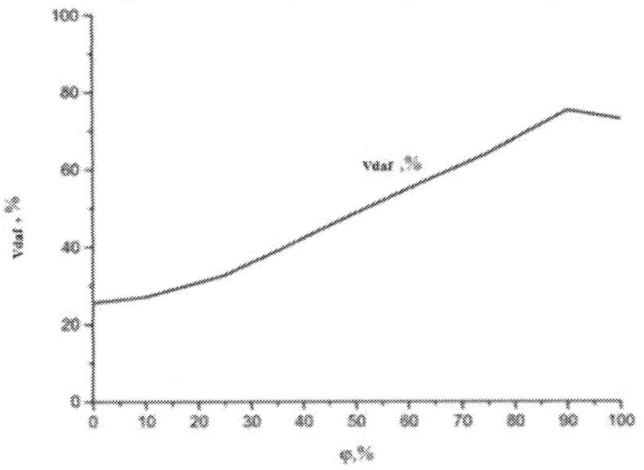

Fig 3. Effect of oats husks concentration on composite fuel volatiles yield in relation to homogeneous coal (φ – oats husks share).

Analysis of Vdaf(φ) dependence shows an increase in volatile substances yield with increase of biomass concentration in composite fuel. Thus, when husks concentration in composite fuel is 50%, volatiles yield is 90%.

3 Conclusion

Influence of oats husks concentration on technical and energy characteristics was established based on the results of conducted experiments, as well as parameters of volatiles release during thermal decomposition of composite fuel on the basis of coal and biomass.

Results of carried out experimental research allow drawing conclusion that composite fuel based on coal and oats husk can be used for combustion in furnaces of steam and water-heating boilers.

The research has been accomplished with financial support of the Russian Federation Ministry of Education and Science within implementation of the project No 13.7644.2017/BCH.

References:

1. S.V. Syrodoy, G.V. Kuznetsov, V.V. Salomatov, Therm. Eng., **62**, 10 (2015)
2. G.V. Kuznetsov, V.V. Salomatov, S.V. Syrodoy, Combustion, Explosion and Shock Waves, **51**, 4, (2015)
3. K.Y. Vershinina, D.O. Glushkov, G.V. Kuznetsov, P.A. Strizhak, Solid Fuel Chem. **50**, 88 (2016)
4. K. Y. Vershinina, J. Eng. Thermophys. **90**, 1 (2017)
5. M. Jumoke, Applied Energy **188**, 215 (2017)
6. A.V. Fedyukhin, I.L. Maikov, V.A. Sinelshchikov, Book of Abstracts of International Conference on Interaction of Intense Energy Fluxes with Matter (2011)
7. W. Yan, S. Wang, D.G. Zhou, C. Xing, Y. Zhang, G.M. Pharr, Wood and fiber science **44**, 1 (2009)
8. K.V. Slyusarskiy, K.B. Larionov, V.I. Osipov, S.A. Yankovsky, V.E. Gubin, A.A. Gromov, Fuel, **191**, 383 (2017)
9. N. David, Z. John *Cofiring Biomass and Coal for Fossil Fuel Reduction and Other Benefits—Status of North American Facilities in 2010* (USDA, Washington DC, 2010)
10. M. Van der Stelt, H Gerhauser, J. Kiel, K. Ptasinski, Biomass Bioenergy, **35**, (2011)
11. L. Nunes, J. Matias, J. Catalo, Renew Sustain Energy Rev. **40**, (2014)
12. K. Larionov, I. Korjashov, S. Jankovskij, V. Gubin, A. Zenkov, MATEC Web Conf. **37**, (2015)
13. K.B. Larionov, I.V. Mishakov, A.A. Gromov, A.V. Zenkov, MATEC Web Conf. **91**, (2016)
14. D.V. Gvozdyakov, D.V. Gvozdyakov, V.E. Gubin, Chem. Petr. Eng. **51**, (2015)

MATEC Web of Conferences **110**, 01081 (2017) DOI: 10.1051/matecconf/201711001081
HMTTSC-2017

Research of the influence of flue gas recirculation on the processes in the ltv furnace

Evgeny Tsibizov[1,*], *Boris* Lebedev[1], and *Maria* Cherckasova[1]

[1]National Research Tomsk Polytechnic University, 634050 Tomsk, Russia

Abstract. The combustion of solid low-grade fuel in LTV-boiler furnaces is a pressing research questions currently. The aim of this work is to study the influence of flue gas entering the process fully-the boiler furnace. The result of the calculation of the resulting distribution of temperatures and oxygen concentrations along the height of the furnace. The analysis of the obtained results.

1 Introduction

Low-temperature vortex combustion is a modern effective technology of using solid organic fuels. This method of burning and heating device for its implementation developed in St. Petersburg technical University [1].

The technology is based on the organizing principle of low-temperature vortex (NTV) solid fuel burning coarse grinding in the conditions of continuous circulation of the particles in a chamber furnace.

2 Object of study

In LTV-furnace with two zones of combustion. The vortex area is the volume of the lower furnace from the mouth of the furnace and to the burners. The flow combustion zone is located above the vortex area in the upper part of the furnace.

Temperature controls of gases in combustion chamber of steam boilers can be done in various ways. One of them is recirculation selected for the tail heating surfaces of flue gases.

3 Air and Recirculation modes

Aerodynamics of the vortex zone created by the interaction of two organized flows. The first stream is formed from the fuel-air mixture; the second stream consists of hot air. It is fed into the furnace through a system of bottom blowing. Flows directed towards each other and form a couple of forces. It creates a vortex motion in the lower part of the furnace [2].

* Corresponding author: TabakaevRB@tpu.ru

Table 1. Feed rates of air and flue gases.

Velocity	The air	Recirculation of flue gases			
		The upper tier of tertiary blast	The lower tier tertiary blast	The upper layer of the lower blast	The lower layer of the lower blast
	m/s	m/s	m/s	m/s	m/s
DHCp[1)]	23	23	23	23	23
Primary	30	30	30	30	30
Secondary	40	40	40	40	40
Up.Tert.Bl.	0.1	20	28	26.5	34
Low.Tert.Bl.	35	30	15	30	35
Up.Low.Bl.	30	30	30	55.40	30
Low.Low.Bl.	30	30	30	30	11.68

1) DHCp – The dust feeding system of high concentration under pressure

4 Temperature and oxygen distribution

In the furnace of the investigated boiler is the selection of flue gas behind the heater. The proportion of recirculation is 0.15, and the temperature at the point of selection is 160 C.

Evaluated the effect of recirculation gases on combustion processes. To allow comparison of the results obtained is maintained constant excess air in the furnace, equal to 1,2. Places the input gas recirculation and velocity are presented in table 1.

Graphics were constructed from the simulation results (fig. 1, 2). They represent the average values of flue gas temperatures and oxygen level at the height of the furnace depending on the introduction of recirculation gases.

In Fig. 1 shows the averaged horizontal cross-section of the temperature change of flue gases at altitude. The maximum temperature observed in the air mode in the vortex area at a height of 4.1 m, and is 1698 K. the Minimum temperature (1399) is observed in the vortex zone at the input of flue gases through the lower tier tertiary blowing. However, this regime is delaying the process of combustion, burning of fuel occurs outside the vortex zone, and the temperature in this zone of the furnace (10-20 m) than the other temperature regimes. At the exit of the furnace the maximum temperature is observed when the air mode is 1271 K. the Minimum temperature is observed at the input of flue gases through the lower layer of the lower blast is 1231 K.

In Fig. 2 shows averaged along the horizontal cross-section change in the concentration of oxygen in the flue gases at altitude. Peaks of concentrations observed in the inlet air, namely in the area of the burners and lower blast. However, when you enter the flue gas through the system of bottom blowing oxygen concentration in the lower part of the vortex is significantly reduced. The minimum value of concentration (of 0.002) is reached at a height of 6.6 m upon introduction of flue gases through the lower layer of the lower blast. The minimum concentration of oxygen (0.058) at the outlet of the furnace is observed when entering the flue gas through the upper tier tertiary blowing, and the maximum (0.072) when entering gases through the lower layer of the lower blast.

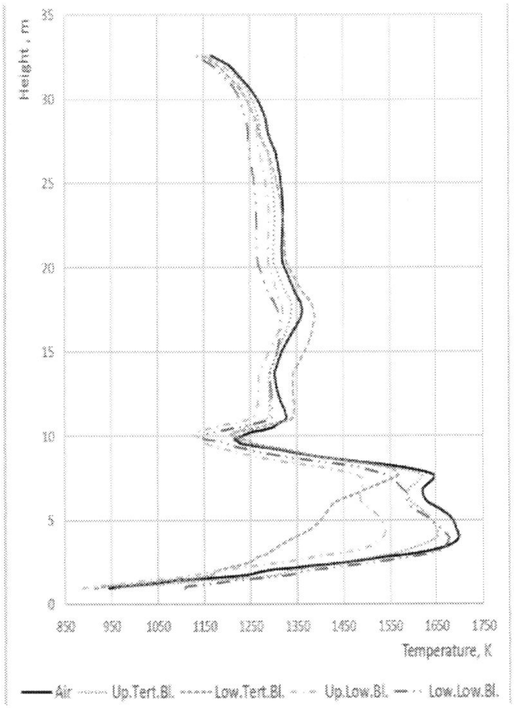

Fig. 1. The average temperature at the height of the furnace.

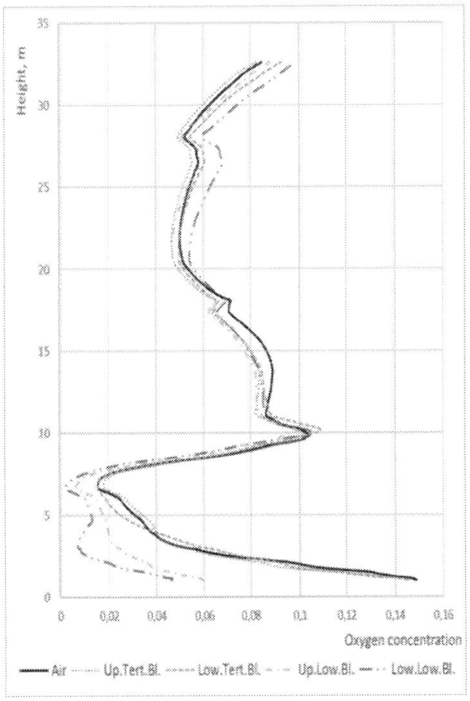

Fig. 2. Average values of oxygen concentration along the height of the furnace.

5 Conclusions

1) The temperature inside the investigated volume of the combustion chamber reaches the maximum value in the absence of recirculation gases.

2) The concentration of O2 in the lower part of the furnace is much lower when entering the flue gas through the channels of the lower blast than the other modes. The input of recirculation gases through the device of the tertiary blast has little effect on the concentration of oxygen in the vortex part of the furnace compared to the air mode.

References

1. P.S. Gergelizhiu, K.Y. Scherbakova, B.V. Lebedev, MATEC Web Conf. **23**, (2015)
2. K.Y. Orlova, B. V. Lebedev, MAREC Web Conf. **92**, (2017)
3. B.V. Lebedev, A.S. Zavorin, Bulletin of Tomsk Polytechnic University, **315**, 15 (2009) [In Russian]
4. New modernized power-generating unit was put into service on Nazarovo state district power station. (Newslab.ru, 2013) [In Russian]
5. A.M. Bubenchikov, A.V. Starchenko Numerical modeling of aerodispersive mixings' dymanic and combustion in channels, **221** (1998), ISBN 5-7511-1004-8 [In Russian]

Water drops passing through a forest fuel bed without its burning and during its thermal decomposition

Roman S. Volkov, *Alyona* O. Zhdanova*, and *Victoria* A. Kirienko

National Research Tomsk Polytechnic University, 634050 Tomsk, Russia

Abstract. This article contains the findings of experimental research into determination of the liquid volume passed through forest fuel bed (pine needles) without burning and during its thermal decomposition. The experiments were carried out using hollow cylinders with fixed geometric parameters (height h_f and diameter d_f) for creating model of fire sources. When analyzing the results we determined the dependences of the drops mass after the passage of the forest fuel bed as a function of the initial mass. All necessary conclusions about the way drop volume effects on the mass loss after passing through a forest fuel bed are included.

1. Introduction

The problem of the emergence, monitoring and prevention of forest fires has been urgent for a long time. The total area of the Earth covered with forest is about 3 866 million hectares. It occupies about one third of the land surface. Forest ecosystems are suppliers of ecological resources for all living things and sources of economically valuable products [1]. According to statistics, a forest fire is registered on the planet every 5 seconds. The average annual number of fires reaches 5.5 million [2]. Doubtless, such number of fires brings with it cost-intensive losses: human, economic, reduction of forest resources, destruction of biogeocenoses, etc. For example, according to Russian statistics, forest fires destroyed 70 million cubic meters of wood (2002), about 700 thousand hectares of forest plantations (2000), 800 thousand hectares of forests (2009). In fact, the real damage from forest fires is much higher than published in documents [3, 4]. Despite huge losses, Russian forests are still very important. Furthermore, forests have economic and global ecological significance. Therefore, the forest fires problem is under steadfast attention

The solution of forest fires problem requires different kinds of activities. It includes the prediction of fires, monitoring and control of areas under the threat, development and application of an effective extinguishing system. All aspects are developing and progressing at the moment. Many groups of scientists are working on new methods of firefighting. So, in modern realities, there are two methods of fire-extinguishing. According to the first method, firefighters should act on the fire edge (active methods). This method includes such technologies as fire beating, blowing, throwing the soil on the fire edge, quenching with water or chemicals. The second method involves creating mineralized barrier bars or ditches, as well as annealing (passive methods) [5].

* Corresponding author: zhdanovaao@tpu.ru

 DOI: 10.1051/matecconf/201711001082

The most common method of extinguishing fires in Russia is quenching using water with aviation help. However, such technology entails irrational consumption of huge amounts of water or other quenching liquids. Therefore, various studies are being conducted in order to find ways of reducing liquid losses. Some necessary characteristics have been already determined. These include droplet sizes, thickness of the liquid film layer [6], and necessary droplets position in the water shell [7].

During carrying out such studies and using its results, it is necessary to take into account the fact that part of the extinguishing liquid will not take part in the fire extinguishing process. This fact needs to be studied in details

The purpose of this research was to estimate the amount of liquid that flows into the ground and does not participate in suppressing the thermal decomposition process.

2. Experimental setup and procedure

Fig.1 shows the scheme of experimental setup used in carrying out researches. This setup provides registration of a forest fuel bed temperature in three points of standardized fire. Also, the setup enable us to fix the forest fuel (FF) sample weight as well as droplet masses after passing through the dry layer and through the layer during its destruction.

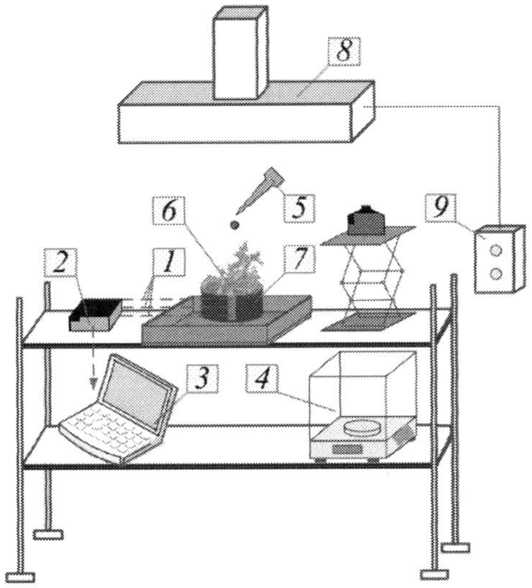

Fig. 1. A scheme of experimental setup: *1* – thermocouples; *2* – high-speed analog input module; *3* – workstation (PC); *4* – analytical microbalance; *5* – dispenser; *6* – FF; *7* – cylinder with a FF sample; *8* – ventilation system; *9* – control panel of turning on/off ventilation system.

Experimental setup consisted of a cylinder *7* with an FF sample *6* and three type K thermocouples *1* that recorded and transmitted temperature values to workstation *3* through the high-speed analog input module *2*. The droplet sizes were regulated by the dispenser (dosing device) *5*. The liquid mass that did not participate in suppressing the thermal decomposition process was recorded with laboratory microbalance *4*.

Spruce needles were chosen as a forest fuel for research. Such choice of material can be explained by structure and porosity of this FF. Using spruce needle makes it possible to register the process of water passing through the bed. For other forest fuels it is difficult to obtain such results. For example, for birch leaves, it is difficult to trace the motion of a drop across the entire layer's thickness. This phenomenon is conditioned by the structure of this forest fuel. A drop can "settle" on a birch leaf and further recording of its motion becomes impossible. The samples used for research was pre-dried for 3-5 days at a temperature of about 300 K. Immediately before the experiments we determined FF moisture using thermal drying method. The relative spruce needles moisture content γ_f was 7–10 %.

Before each experiment, the forest fuel was weighed on an analytical microbalance *4* in order to determine mass m_{f0}. After that, forest fuel was placed to the bottom of the cylinder *7*. The density of samples in each experiment remained constant.

Cylinder made of corrugated aluminum (see fig. 2) was used for create model of fire source. In our experiments cylinder had the diameter d_f=60 mm and the height h_f=40mm. The model of fire source *2* was placed on a specialized woven surface made of annealed wire of standard quality (GOST 3282-74). The size of the grid cells was1x1 mm. Below the grid *3* was a drop trap *4*. A vertical hole (3 mm in wide) was made in the cylinder to visually monitor the thermal decomposition of forest fuel (see Fig. 1).

Fig. 2. The video record of the experimental determination of the water mass during passing through FF bed: *1* – spruce needles, *2* – cylinder, *3* – stainless steel grid (tissue type, made of the metal), *4* – trap, *5* – dispenser.

Water droplets volume was varied in the range 86-900 μL during the series of experiments. The volumes and radii of drops used for the experiments are given in table 1. We conducted from 15 to 20 experiments for each droplet size. Then, we sampled the experimental values of m_{de} differing by less than 5% for identical experimental conditions. The process of the experiment was conducted in two stages. The drop was generated by the dispenser *5*. After passing through the FF bed, drop fell into the trap *4* (Fig. 2). At the first stage droplet mass after passing through the forest fuel bed to ignition was recorded. The estimation of the water mass, which lingers on the grid *3*, was determined. Before starting the experiment, water droplets of considered size were passed through the grid. We measured part of mass that passed from m_{d0} to the trap. The average mass loss through the grid surface was 22 – 24 % of the initial mass m_{f0}.

Table 1. The volumes and radii of water droplets used in the experiments.

V_d, μL	86	334	776	900
R_d, mm	2.7	4.3	5.7	5.9

During the second stage of experiment, the cylinder *7* was placed on a stainless steel grid *4* (Fig. 2). Then we set three type-K needle thermocouples on cylinder's symmetry axis (temperature measurement range 223–1473 K, accuracy ± 3 K, heat retention no more than 0.1 s). After that the model of fire source was ignited. The criterion for stable FF burning was the excess of FF temperature the thermal decomposition temperature (T_{if}) - 370 K. Further, after the readings of all thermocouples reached the value T_{if}, single droplets were fed to the standardized fire. At the last stage, the initial drop mass m_{d0} was determined using analytical microbalance *4*. We carried out series of experiments for 4 different droplet sizes.

3. Results and discussion

During experimental investigations the dependences of the droplets' mass loss on their initial masses during passage through the FF bed without combustion and in the combustion process were obtained (Fig.3). The curves were built for drops with size R_d≈4.3 mm, R_d≈5.7 mm и R_d≈5.9 mm. Weight loss value was calculated by formula

$$\Delta m = m_{d0} - m_{de} \quad (1)$$

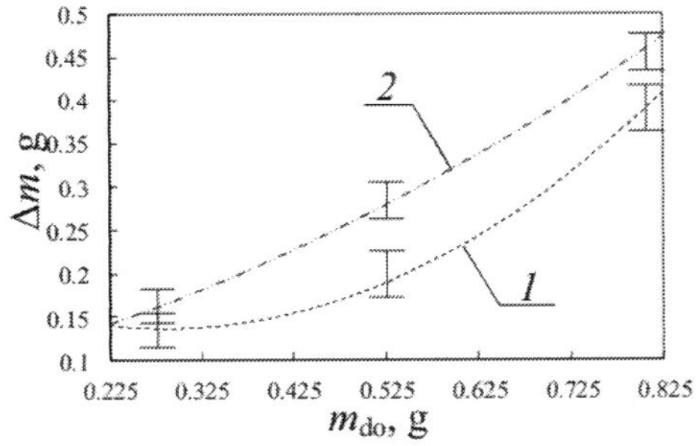

Fig. 3 Dependence of the parameter Δm on the initial mass of droplets m_{d0}, (Δm is the parameter characterizing the mass changing during the passage through forest fuel bed, taking into account the losses on the grid): *1* – water drops passing through FF bed without its burning; *2* – water drops passing through FF bed during its thermal decomposition.

The obtained results showed that droplets with a radius R_d=2.7 mm completely "settled" in FF bed (without and during its burning). Therefore, dependency curves were built for droplets with other dimensions. For them complete subsidence in the bed did not occur. Analysis of the dependencies presented in figure 3 allows us to conclude that the more droplet size the greater mass fraction it loses during the passing through the bed. This fact can be explained by large contact area with FF fragments for drops with volume V_d ≈900 μL.

Furthermore, during the analysis of the obtained experimental data, dependence of the droplets masses (R_d≈4.3 mm, R_d≈5.7mm and R_d≈5.9 mm) after passing through the FF bed on the water mass that has passed though the grid was built (fig.4).

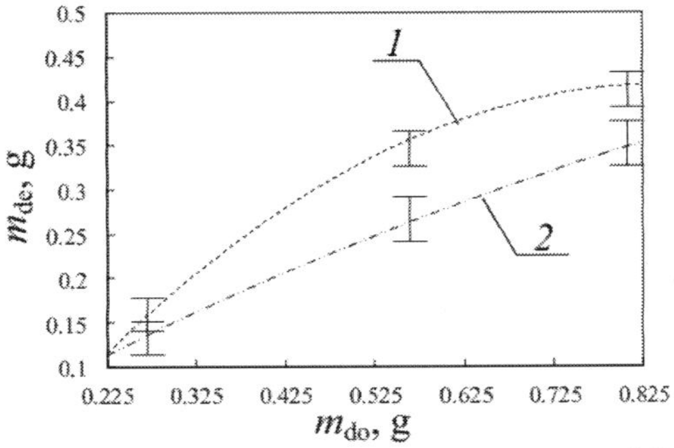

Fig.4 Dependence of the droplets masses (R_d≈4.3 mm, R_d≈5.7 mm and R_d≈5.9 mm) after passing through FF bed on the water mass that has passed though the grid: *1* – water drops passing through FF bed without its burning; *2* – water drops passing through FF bed during its thermal decomposition.

Analysis of the obtained dependences, enable us to make a conclusion about almost linear dependence between masses in case of thermal decomposition process and an increasing dependence during water drops passing through a dry forest fuel. The results of the series of experiments indicate that the most rational way of firefighting is using medium-sized drops of quenching liquid. In such case liquid can reach all FF bed and affect the process of suppressing the decomposition.

Thus, obtained during experiments dependencies can describe the process of liquid drops' mass loss during passing through a FF bed without and during its burning, taking into account the droplet size. The results of experimental studies can be used to estimate liquid volume, which is necessary to extinguish a forest fire.

4. Conclusion

As a result of the experimental studies, we established the dependencies of droplets mass loss on the initial one when passing through a forest fuel bed as well as dependencies of droplets mass loss on drops volumes. The obtained results allowed us to determine the range of droplets sizes for spraying in the stream. It will help to minimize the amount of liquid, which is not involved in suppressing the thermal decomposition process. Our conclusions and dependencies should be taken into account when creating numerical models for finding the necessary quenching liquid volume, drops size and times of extinguishing forest fires.

The investigations of heat and mass transfer processes in the suppression of forest fire were supported by Russian Science Foundation (project 14–39–00003). Experiments to determine the specific differences between the processes of thermal decomposition and combustion of various types of forest fuel were carried out within the framework of the President of the Russian Federation grant MK-1684.2017.8.

References

1. Yu. L. Vorobiev, V. A. Akimov, Yu. I. Sokolov, *Forest fires on the territory of Russia: state and problems* (Deks-Press, 2004)
2. A. I. Bobko, Fires as geoecological danger, a collection of the conference proceedings "Geography and Geoecology in the Service of Science and Innovative Education", **319** (2014)
3. L. M. Ivanova, North and the market: the formation of the economic order, **1**, 103 (2011)
4. A. V. Konstantinov, V. V. Morkovina, Actual directions of scientific research of the XXI century: theory and practice, **4**, 319 (2016)
5. D. S. Stupnikov, Actual directions of scientific research of the XXI century: theory and practice, **3**, 201 (2015)
6. A. O. Zhdanova, G. V. Kuznetsov, P. A. Strizhak, Chemical Physics and Mesoscopy, **17**, 172 (2015)
7. A. O. Zhdanova, G. V. Kuznetsov, P. A. Strizhak, Chemical Physics and Mesoscopy, **16**, 215 (2014)

MATEC Web of Conferences **110**, 01083 (2017) DOI: 10.1051/matecconf/201711001083
HMTTSC-2017

The usage of photoelectric modules in the outdoor lighting system of university building

Bator B. Tsybikov[1,*], *Lyudmila P.* Sumarokova[1]

[1]National Research Tomsk Polytechnic University, 634050 Tomsk, Russia

Abstract. The paper presents the usage of solar panels for lighting system of university building and there is a cost sheet of all necessary equipment: spotlights, controller, panels, batteries. The lighting systems are related to low-power consumers. At nights and unsunny weathers of lithium-ion batteries work after charging.

1 Introduction

The biggest part of Russian territory has favourable climatic conditions for solar energy consumption. In south regions, length of solar radiation is in the range from 2000 to 4000 hours a year and annual amount of solar energy on a horizontal surface is between 1280 and 1870 Kilowatt·hour/meter^2 [1]. The usage of solar electricity has many advantages. Nowadays, the such electricity is widely used in remote areas where is no centralized electric power supply, solar batteries are used for electricity supply of cottages, water pumps and so on [2]. The aim of this study is a operation of outdoor lighting system of 8th TPU university building based on photoelectric power sources. An autonomous solar power plant includes: photoelectric modules, storage batteries, electronic unit, comprising controller, which tracks of battery`s work, then charging unit and inverter for transformation a direct current into alternating current 220 V with frequency 50 Hz.

Rated power of photoelectric modules provides necessary energy production for uninterrupted work of street lightning during the whole day. At nights, usage of batteries is expected to. Surplus of power transfers to extra ballast resistance.

Block scheme of the autonomous electric supply`s street lighting in 8th university building can be seen in Figure 1.

Information channels are marked with stippled lines [3], Power channels with solid ones.

* Corresponding author: tytyt1993@mail.ru

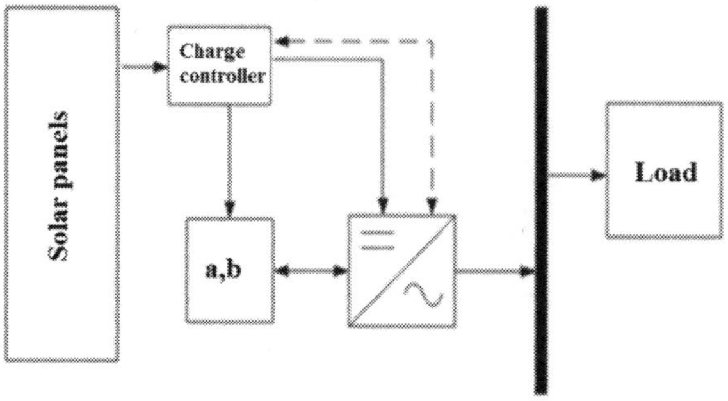

Fig. 1. Block scheme of the autonomous electric supply`s street lighting in 8th university building.

2 Calculation

For definition of the best solution of architectural highlight of university 8th building`s front, comparison study was made for three lighting units with different powers: PSD-70, NL-60, PSD-30. As a result of the comparison, it was offered to use 36 LED spotlights PSD-70, intended for indoor and outdoor lighting of industrial installation, inside courtyard space and building yard.

Rated power consumed by the LED spotlights is $P_{heat} = 0.07 \cdot 36 = 2.52$ $kWatt$.

Average consumption of electricity during seasons a day is: in spring 24, 86 kWatt·hour, in summer 18.63 kWatt·hour, in autumn 34.03 kWatt·hour, in winter 40.27 kWatt·hour.

The project of street lighting of 8th university building`s front at the side of the intersection of Usova Street and Sovetskaya street was simulated in DIALux program. It can be seen in Figure 2.

Silicone single-crystal solar units from Yingli PANDA Solar firm with modules YL300C-30b, 300 Watts were used as photoelectric modules.

Determination of necessary number of moduleswas executed in 3 stages:

1. The total square of solar modules S_{Σ}^{sp} was defined forrequired energy production and steady maintenance work of lighting system in December. This winter month is characterized by the most unfavorable conditions for generating electricity and consumption – few daylight hours and big value of energy consumption.

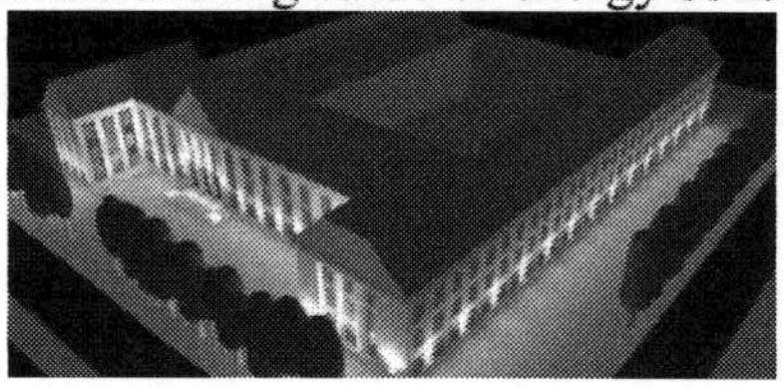

Fig. 2. Outdoor lighting of the front with spotlights PSD-70.

2. Square of one standard panel S_{1sp} is equal:

$$S_{1sp} = a \cdot b, \quad (1)$$

where a – length of a solar panel, b – width;

3. The number of solar panels is defined by division of required total square of solar panels S_{Σ}^{sp} into the value of one standard panel $S_{1cп}$.

$$n_{sp} = \frac{S_{\Sigma}^{sp}}{S_{1sp}}, \quad (2)$$

N_{sp} – number of solar panels.

As a result of the study, it was decided to set up photoelectric modules at an angle β= 80° to the horizon and south-orientation, which allowed significantly increase energy production during the period November to February. LED spotlights are used as source of light. Measures were taken to optimize energy production through disconnection half of spotlights after 22.00 hours and it was achieved to reduce energy consumption in winter season to 32.7% and 33% in autumn term.

Main technological characteristics of solar power plant: power of photoelectric modules – 21.6 kWatt, one voltage inverter, 7 charge controllers, 64 lithium-ion batteries, 36 LED spotlights each power 70 Watts and copper cables of mark VVGng 3 × 6 mm^2.

3 Conclusions

In this study, the possibility of using helio-energy resources have been analyzed in the context of street lighting of TPU university building.

The energy production of the autonomous solar power plant was determined per year with using of hourly average and daily average calculations [4]. The performed calculation shows the opportunity and appropriateness of using solar panels to generate electricity for low-power consumers in the climatic conditions of the Tomsk region.

References

1. A. Yurchenko, A. Kozlov, A. Volkov, Photovoltaic Solar Energy: Proc. of XXII European Conf. Valencia, 2989(2008)
2. L.P. Sumarokova, Naukovedenie, **8**,(6) 11 (2016) [in Russian]
3. M.A. Surkov, Naukovedenie, **8**,(4) 13 (2016) [in Russian]
4. B.V. Lukutin, E.B. Shandarova, A.F. Makarova, I.B. Shvartsman, Urgent problems of modern mechanical engineering, 12005 (2016)

High temperature x-ray diffraction of zr-2.5nb during thermal cycling in vacuum

Mikhail Tumanov[1,*], *Lyudmila* Lyubimova[1], and *Evgeniy* Puzyrev[2]

[1]National Research Tomsk Polytechnic University, 634050 Tomsk, Russia
[2]Polzunov Altai State Technical University, 656038 Barnaul, Russia

Abstract. The cyclic thermal tests in vacuum of zirconium alloy Zr-2.5Nb in the temperature range 250-350°C is established the presence of anomalies of thermal deformation of the crystal lattice, reducing the efficiency of the fuel rods.

1 Introduction

Zirconium tubes for nuclear reactors undergo irreversible deformation, which on the one hand, creates a threat of destruction of safety barriers due to the non-design stresses, on the other hand, the forming of the cladding restricts the depth of burnup of the fuel and thereby reduces the efficiency of the re-actor [1]. The relationship between the deformation due to radiation swelling that occurs with stress and performance elements of the active zone was the basis for the methodology for thermal cycling tests to determine thermal dilatations of the crystal lattice of the zirconium alloy.

As an object of research used a sample of the channel tube of zirconium-niobium alloy Zr-2.5Nb (known in Russia as E-125 alloy) with the content of niobium in the amount of 2.4-2.7 percent.

Methodology for thermal-cycle tests simulated the processes of thermal fatigue and was playing a sequence of cycles "heating – cooling – heating" with the increase in temperature in each subsequent cycle.

2 Material and methods

Experimental data on changes of the shape and size of the metals showed that the size can both increase and decrease as thermal-cycling [2].

Thermal tests were conducted in vacuum by means of high temperature x-ray diffraction (HT XRD) using x-ray diffractometer and a high temperature diffractometric installation "UVD-2000". The wavelength of the used radiation was λ_{Co} = 0.71069 Å. For analysis of the selected two x-ray diffraction line index (002) and (101), the angles of diffraction which served as the basis for determining the size and crystalline with the hexagonal lattice of the alloy of zirconium, which were determined according to the basic formula of crystallography [3].

* Corresponding author: tumanov_mihail@mail.ru

DOI: 10.1051/matecconf/201711001084

The relative error in the measurement of the lattice parameters were: $\Delta a/a \leq \pm 0.03\%$ and $\Delta c/c \leq \pm 0.3$ %.

In Fig. 1-3 shows the changes in a, c and the relationship *c/a* unit cell of an alloy of zirconium in the process of thermocycling.

For high-purity zirconium, a = 3.23118 Å, c = 5.14634 Å, the ratio *c/a* is: *c/a* = 1.59271 [4, 5].

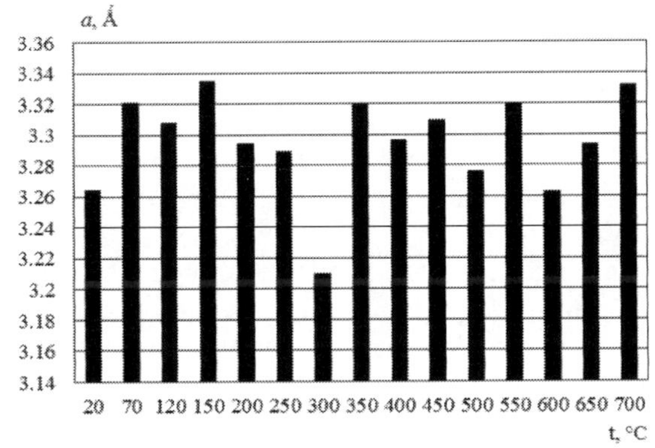

Fig. 1. The dependence of the parameter *a* zirconium alloy Zr-2.5Nb from temperature.

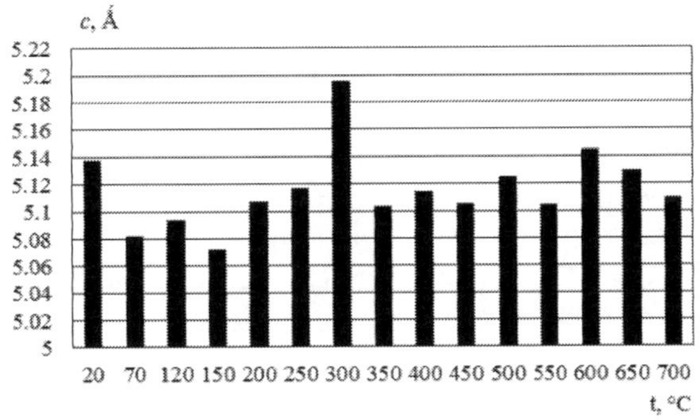

Fig. 2. The dependence of the parameter c of zirconium alloy Zr-2.5Nb from temperature.

Data on the value of *c/a* for various zirconium alloys may differ markedly from each other, because the lattice parameters are significantly influenced by alloying elements, impurities of introduction and substitution, temperature, texture, pre-strain, etc [5]. All this testifies to the need to establish patterns of behavior for samples of similar products the specific composition and shape.

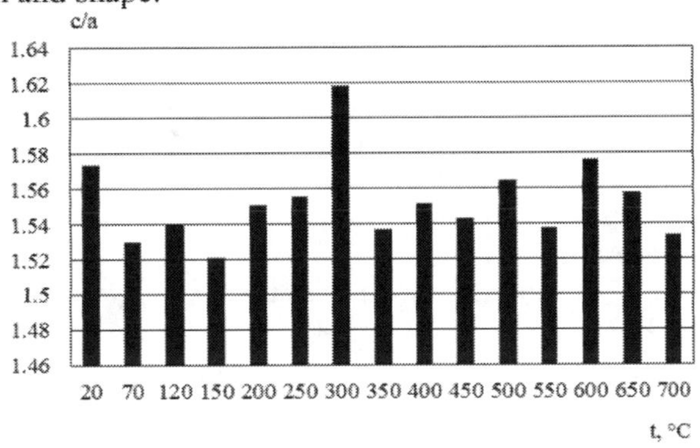

Fig. 3. The ratio *c/a* of zirconium alloy Zr-2.5Nb.

From Fig. 1-3 it is seen that the lattice parameters of the alloy reveal a very complex temperature dependence.

Thus, the lattice parameter *a* oscillates, showing some temperature negative creep (Fig. 1): about of 100°C, in the range of temperatures in the 350-400°C, 450-500°C, 550-600°C. the temperature anomaly for lattice parameter as stated in the service temperature for the alloy of zirconium 250-350°C. The temperature anomaly for lattice parameter as stated in the service temperature for the alloy of zirconium 250-350°C.

For parameter c in addition to negative effects creep there is an anomalous jump in the thermal strains in the temperature 300°C (Fig.2).

The dependence of the ratio c/a of crystal lattice is also from differs the structural instability is most pronounced for the temperature range 250-350 °C (Fig.3).

Anomalous effects of thermal deformation of the crystal lattices relates to the field of polymorphic transformations [4].

The feature of polymorphic transformations in structural materials as a kind of phase transitions is that in the process of transformations, for example, under the action of tempera-tours, a substance changes the density, while in solid-house condition. When this occurs, the change of microhardness, coefficient of hydrogen diffusion, heat capacity, there is a temperature hysteresis.

Observed anomalous changes in the properties of coefficient of Li-Nanoha the extension of the young's modulus, elongation, ductility and heat capacity in strictly certain intervals of temperature [6]. Thus the symmetry of the crystal lattice in the region of the anomalies is not changed, therefore, to attribute the anomalies to the polymorphic phase transitions cannot, accordingly, not found out the cause. It is important to note that anomalies are inherited with chemical compounds and alloys like structural polymorphism. For example, if the Zirconia has phase transitions of the first kind and the anomalous effects of thermal deformation in the studied temperature region, but the composition of the alloy is, for example, Nickel having this effect, the alloy of zirconium with the addition of Nickel inherits this feature [7]. In particular, the Nickel included in the zirconium alloy Zr-2.5Nb has an anomaly of thermal deformation of the crystal lattices in the region of 300-400 °C, i.e. approximately the same as an alloy of zirconium [4].

It is fair to assume that the presence of Nickel in alloy composition can have a negative impact on its service properties, as in the anomalous temperature points have been the following material properties and processes:

– superplasticity – a sharp decline in the tensile strength and tough-ness in zirconium alloys (iron, titanium);

– 10 to 20 – fold stepwise elongation of the metal without the necking and rupture;

– punching leads to a smaller grain;

the quenching retains the properties of the high temperature phase, which started cooling;

– annealing of the deformed metal below the abnormal point leads to small grain, and at temperatures above the anomalous pointing to large (different speed of grain growth);

– thermal cycling through the anomaly causes irreversible deformation of the products;

– increase the sorption properties of the material and azoproite the cost;

– with increasing degree of deformation increases the rate of oxidation and dissolution of metals in chemical reactions [2].

Negative creep is associated with bond breaking and formation of microcracks and porosity [8]. The occurrence of porosity and microcracks in the structure of the alloy of zirconium is of great importance, because the zirconium is practically the only metal (even titanium), dis-solving large quantities of oxygen, which forms with the zirconium as a chemical compound ZrO_2, and the phase of introduction of oxygen into the crystal lattice. Non-metallic impurities (O, H, C, N) significantly alter the properties of zirconium. The dissolution of oxygen in zirconium is an important process associated with the formation of protective oxide films. Corrosion behavior of zirconium and its alloys in a steam environment is characterized by the presence of two periods. In the first period to fracture on the surface of the zirconium oxide film is formed, providing a high protective function, and the oxidation rate of zirconium stabilizirue-most. After reaching a certain critical thickness of the oxide layer suddenly collapses – this is the second period of oxidation, which oxide layer will no longer protect the metal and it is the rapid destruction [1, 5, 9].

In this regard, one of the problems is the loss of protective proper-ties of oxide films of ZrO2. This important issue is debated until the pre-sent time. To explain the second period in the oxidation kinetics of zirconium involved various theories and mechanisms, in particular the theory of mechanical cracking of the oxide film [1, 5, 9].

On the basis of the obtained experimental results (Fig. 1-3) it is believed that the properties of the oxide films zirconium and their destruction affects the anomalous behavior of the alloy at the operating temperature and accumulation of damage in the presence of even minor temperature gradients. Accumulation of damage leads to a limit state surface, its destruction and the destruction of the protective oxide film. The trigger to the destruction of the oxide film are the processes of destruction of the zirconium alloy, which can not affect the structure of the formed protective oxide ZrO_2.

Thus, the determining process in the destruction of the oxide-film Noah, or important, may be internal structural state of the alloy, the mechanisms and rate of degradation.

Impurities, texture and nature of the structure of intra - and inter-grain boundaries will have a significant impact on the other properties, such as hydrogenation. It is known that zirconium alloys actively absorb hydrogen, and the deformation of specimens of alloy of zirconium, as shown in Fig. 1-3, will only lead to increase of hydrogen absorption.

3 Conclusions

The result of the research established the following.

1. The crystal lattice of the zirconium alloy Zr-2.5Nb he-have a complex dependence on temperature, showing some values, negative creep, which is associated with micro cracking.

2. The crystal lattice of the zirconium alloy Zr-2.5Nb exhibit the effects of abnormal thermal expansion of the crystal re-bars in the absence of phenomena of polymorphism in the region of temperatures of the experiment. The effect is most evident in the field of operating temperatures of 250-350°C and can be the cause of the destruction of protective oxide films.

Thus, speaking on the anomalous characteristics in relation to thermal stress to structural elements of nuclear power reactors can be expected that the study about anomalous phenomena provides the most complete picture of the behaviour of multicomponent metallic systems at high temperature

The reported study was supported by RFBR grant "Fundamentals of engineering Sciences" No. 15-08-99544a.

References

1. L. A. Belyanin, V. I. Lebedev, Yury V. Garusov, *Bezopasnost' AEHS s kanal'nymi reaktorami. Rekonstrukciya aktivnoj zony* [Safety of nuclear power plants with channel reactors. Reconstruction of the active zone] (MEI, Moskov, 1997) [in Russian]

2. V.F. Kovalenko, *Teplofizicheskie processy i ehlektrovakuumnye pribory* [Thermophysical processes and electrovacuum devices] (MEI, Moskov, 1975) [in Russian]
3. S. S. Gorelik, Y. A. Skakov, L. N. Rastorguev, *Rentgenograficheskij i ehlektronno-opticheskij analiz* [Radiographic and electron-optical analysis] (MEI, Moskov, 2002) [in Russian]
4. V.A. Finkel. *Vysokotemperaturnaya rentgenografiya metallov* [High-temperature radiography of metals] (MEI, Moskov, 1968) [in Russia]
5. D. Duglas, *Metallovedenie cirkoniya* [Metallography of zirconium] (MEI, Moskov, 1975) [in Russian]
6. L.L. Lyubimova, A.A. Makeev, A.A. Tashlykov, Steel in Translation, **2** (2014)
7. Y.D. Vishnyakov, D*efekty upakovki v kristallicheskoj structure* [Defects of packing in the crystal structure] (MEI, Moskov, 1970) [in Russian]
8. A.S. Zavorin, L.L. Lyubimova, A.A. Tashlykov, R.N. Fisenko, R.B. Tabakaev, Sposob rentgenometricheskoj ocenki temperaturnyh uslovij ehkspluatacii trubnyh ehlementov kotlov [Method radiometric evaluation of temperature operating conditions of tube elements of boilers]. Patent RF, no. 2012147388/28, 2006
9. B.G. Parfenov, V.V. Gerasimov, G.I. *Venediktova, Korroziya cirkoniya i ego splavov* [Corrosion of zirconium and its alloys] (MEI, Moskov, 1967) [in Russian]

MATEC Web of Conferences **110**, 01085 (2017) DOI: 10.1051/matecconf/201711001085
HMTTSC-2017

Helium passage through homogeneous ultra-fine hydrocarbon layers

Michael A. Bubenchikov[1,2,*], *Valentina A.* Poteryaeva[1], and *Anton V.* Ukolov[1]

[1]National Research Tomsk State University 634050 Tomsk, Russia
[2]LLC "Gazprom Transgaz Tomsk" 634029 Tomsk, Russia

Abstract. The present paper deals with the problem of helium atoms and methane molecules moving through a hydrocarbon layer of evenly distributed energy sources. A computational technique for integrating the Schrödinger equation based on formulation of two fundamental numerical solutions to the problem of waves passing through a barrier is suggested. A linear combination of these solutions defines the required wave function, while cross-linking with asymptotic boundary conditions allows determining the coefficients of transmission and particle reflection from the potential layer barrier.

1 Introduction

Fluid motions in nanochannels [1-3], diffusion of nanoparticles in gas [4-6], deformations of crystalline and diamond-like phases [7-12], as well as permeability of ultra-thin films [12-18], are the subject of modern nanomechanics. In this paper, we conduct theoretical research of permeability for nanoscale layers, which have a fibrous structure and are created on a polymer basis. Threads of such structures are fragments of polymer chains. The formula of a polymer molecule can be written as follows: $CH_4-(CH_2)_n, n=50\div100$. Under these values of *n,* it can be approximately assumed that a polymer thread contains twice as many hydrogen atoms as compared to carbon atoms. With no further consideration of $\mathrm{C-H}$ linkages and the character of spatial arrangements of threads, we will analyze helium atoms passing through a layer with evenly distributed carbon and hydrogen atoms.

2 Numerical model

To study the problem of particles with the mass *m* and the energy *E* passing through a potential barrier, one can solve the one-dimensional stationary Schrödinger equation.

The classical Schrödinger equation is [19-21]:

$$-\frac{\hbar^2}{2m}\Delta\psi+U\psi=i\hbar\frac{\partial\psi}{\partial t}. \quad (1)$$

*Corresponding author: michael121@mail.ru

Here ψ is the wave function, $\hbar = h(2\pi)^{-1}$, h is the Planck's constant, U is the potential energy, i is the imaginary unit. Knowing the function ψ from solving equation (1), we can find the density of particle distribution: $\rho=|\psi|^2$. If the potential energy U does not depend on time, the solution of equation (1) is found in the form of: $\psi = \Psi\exp(-iEt/\hbar)$. Then, for the function Ψ we get the stationary Schrödinger equation:

$$\frac{\partial^2\Psi}{\partial z^2} + \frac{2m}{\hbar^2}(E-U(z))\Psi = 0. \qquad (2)$$

In problems related to molecular passage through nanoporous layers, the Schrödinger equation is used, ultimately, to find the density of moving particles experiencing elastic interactions with the porous frame. Therefore, instead of the Planck's constant, it is advisable to take a value which is proportional to the size of the layer $\hbar = l\sqrt{m\varepsilon}$, where l is the transverse dimension of the layer; ε is the depth of the potential well within the interaction between molecules of the layer structure and the filtered component, m is the mass of the particle passing through the layer. Thus, equation (2) can be written as follows:

$$\frac{\partial^2\Psi}{\partial z^2} + \left(K^2 - \tilde{U}(z)\right)\Psi = 0, \text{ here } K^2 = \frac{2m}{\hbar^2}E, \ \tilde{U}(z) = \frac{2m}{\hbar^2}U(z). \qquad (3)$$

If we assume, that $\tilde{U}(z) = 0$, the particular solutions of equation (3) will be the functions e^{iKz} and e^{-iKz}, the first private function being the wave which falls on the barrier and the second one being reflected. Their combination should provide the necessary physical result, including the case of the barrier being present, however at a distance. Therefore, as mathematical boundary conditions for the function $\Psi(z)$ we can write:

$$\Psi(z) = e^{iKz} + be^{-iKz}, \ \Psi(z) = ae^{iKz}, \ |z| \approx l. \qquad (4)$$

These conditions can be called asymptotic because they must be carried out away from the barrier. The first function in (4) determines the value of the density of probability that particles are located to the left of the barrier, and the second one – to the right. The coefficients a and b are complex values that define the character of the transmission and reflection in the z-axis direction. To find them, we approximate the second derivative in equation (3) by symmetric differences:

$$\frac{\Psi_{j-1} - 2\Psi_j + \Psi_{j+1}}{\Delta z^2} + K^2\Psi_j - \tilde{U}_j\Psi_j = 0, \ j = 1,\ldots,N-1. \qquad (5)$$

By changing values of the index j, we obtain a system of algebraic equations with a tridiagonal matrix for finding the wave function at the nodes of the difference grid.

By removing any two rows of the main matrix of coefficients and applying the resulting function **null()** of the MATLAB package [22] to the matrix, we find two independent numerical solutions $\Psi_1(z)$ and $\Psi_2(z)$ to the system of equations (5). Therefore, the required wave function $\Psi(z)$ is a superposition of these solutions: $\Psi(z) = C_1\Psi_1(z) + C_2\Psi_2(z)$, $\Psi'(z) = C_1\Psi_1'(z) + C_2\Psi_2'(z)$. Equating the value of the function $\Psi(z)$ and its derivative $\Psi'(z)$ on the borders of the area under consideration and the asymptotic values (4), we obtain a system of four algebraic equations for finding the 4 un-

 DOI: 10.1051/matecconf/201711001085

known coefficients C_1, C_2, a and b. In general, all the values resulting from this system are complex. Then, the physical transmission coefficient is $D=|a|^2$.

3 Calculation results

Figure 1 shows the dependence of a layer's permeability on the value of a helium particle's energy E/k for a fixed barrier height U/k=300 K, as well as the form of the respective barrier. Here k is the Boltzmann's constant. A rectangular barrier of about 10 nm in width was selected for the study. As can be seen from Fig. 1, the resulting solution is close to analytical.

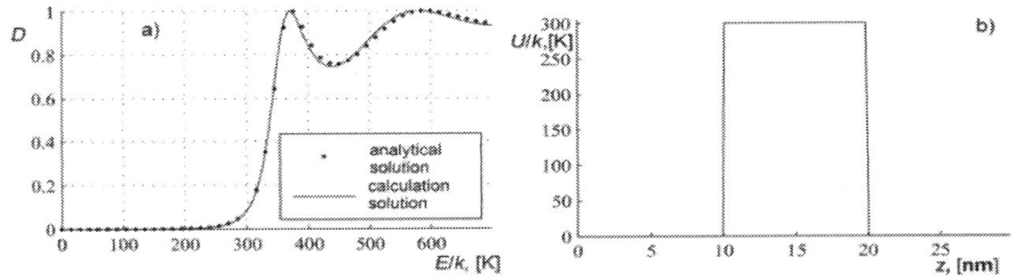

Fig. 1 Comparison of analytical [21] and calculated solutions for rectangular barrier (a) and barrier form (b).

As calculations show, the dependence of the layer's permeability on its thickness is negligible. Figure 2 presents a graph of this relation for a rectangular barrier with the height of U/k=200 K.

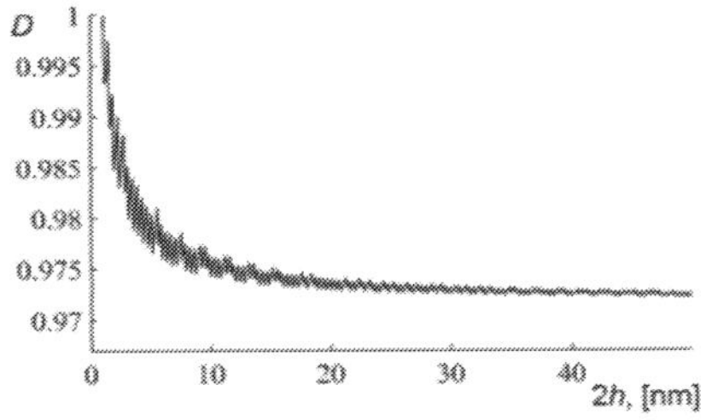

Fig. 2 Dependence of layer's permeability on its thickness at U/k=200 K for rectangular barrier.

Let us find the value of the barrier, which is defined by the resistance of carbon atoms $U_C(z)$. The potential energy, in case of uniform distribution of field sources in an infinite layer with the thickness of $2h$, is [23, 24]:

$$U_c(z) = 2\pi q \int_0^{\infty} r dr \int_{-h}^{h} \Pi_c(\sqrt{r^2 + (z-c)^2}) dc \,. \qquad (7)$$

Here q is the number of carbon atoms per cubic nanometre of the layer material, $\Pi_c(\rho) = 4\varepsilon_c(\sigma_c/\rho)\,\mathrm{th}\left((\sigma_c/\rho)^{11} - (\sigma_c/\rho)^5\right)$ – is the modified potential of pair molecular interactions built on the basis of the classical LJ-potential. σ_C and ε_C are the parameters of this potential which are defined as the average values of the relevant parameters of monomolecular interactions for carbon, helium and methane. The depth of the potential well ε in the expression for $\hbar$ is defined as follows: $\varepsilon = 0.333 \cdot \varepsilon_C + 0.666 \cdot \varepsilon_H$, where ε_C and ε_H are geometric mean values of the corresponding values for monomolecular interactions. Let us consider a hydrocarbon layer with the thickness of 10 nm consisting of evenly distributed C and H atoms. The movement of helium atoms and methane molecules occurs in the positive direction of the Oz-axis perpendicular to the layer (Fig. 3). Then, for helium atoms and methane molecules the interaction energy from the ultra-thin layer having a polymer base is equal to the sum of potentials $U(z)=U_C(z)+U_H(z)$. In Fig. 4 we show the dependence of a layer's permeability on density of packing with carbon atoms q. It can be seen that a layer becomes impermeable for methane molecules already at q=20 nm^{-3}, while helium atoms pass through it without any difficulty. The layer permeability of about 50%, as shown in Fig. 4, is achieved when q=60 nm^{-3}, which corresponds to compacted polyethylene. Layers with the thickness of 100 nm were investigated.

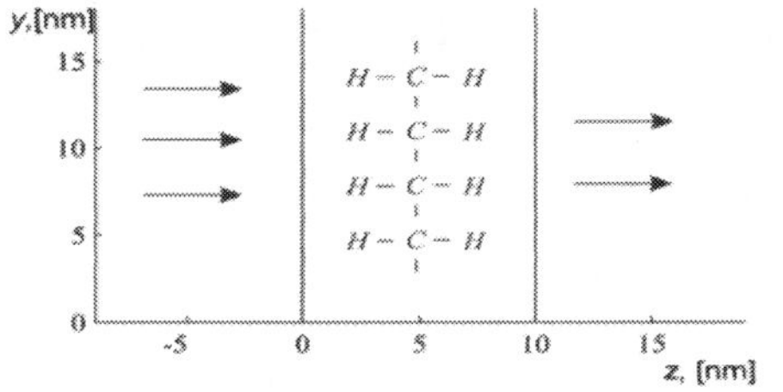

Fig. 3 Dependence of layer's permeability on its thickness at *U*/*k*=200 K.

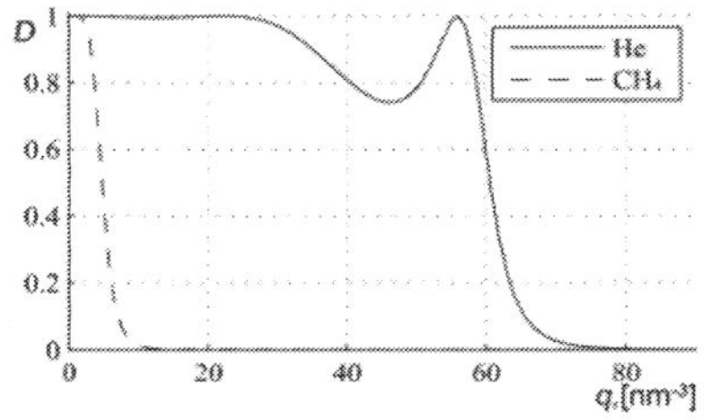

Fig. 4 Dependence of layer's permeability on density.

4 Summary

To sum up, if it is possible to create a plastic layer with the thickness of 50 ÷ 100 nm, it may well be used as a passive filter for separation of methane-helium mixtures.

This paper was written within the frame of Competitiveness Improvement Program of Tomsk State University and with financial support of the Russian Science Foundation, grant under Agreement N 16-19-00089.

References

1. V.Ya. Rudyak, S.L. Krasnolutskii, D.A. Ivanov, J. Dok. Phys. **57**, 33 (2012)
2. V.Ya. Rudyak, A.A. Belkin, V.V. Egorov, D.A. Ivanov, Int. J. of Multiphys. **5**, 141 (2011)
3. V.Ya. Rudyak, S.L. Krasnolutskii, D.A. Ivanov, J. Microfluid. Nanofluid. **11**, 501 (2011)
4. V.Ya. Rudyak, S.N. Dubtsov, A.M. Baklanov, Tech. Phys. Lett. **34**, 519 (2008)
5. V.Ya. Rudyak, J. VDI-Bericht. **1803**, 187 (2003)
6. V.Ya. Rudyak, A.A. Belkin, Tech. Phys. Lett. **29**, 560 (2003)
7. A.V. Savin, E.A. Korznikova, S.V. Dmitriev, Phys. Sol. St. **57**, 2348 (2015)
8. J.A. Baimova, B. Liu, S.V. Dmitriev, N. Srikanth, K. Zhou, Phys. Chem. Chem. Phys. **16**, 19505 (2014)
9. E.A. Korznikova, S.V. Dmitriev, Appl. Phys. **47**, 345307 (2014)
10. E.A. Korznikova, S.Y. Fomin, S.V. Dmitriev, Mat. Sc. Forum. **845**, 211 (2016)
11. J.A. Baimova, S.V. Dmitriev, Mat. Sc. Forum. 838-839 (2016)

12. A.I. Potekaev, M.A. Bubenchikov, Russ. Phys. J. **54**, 211 (2011)
13. M.A. Bubenchikov, Russ. Phys. J. **54**, 102 (2011)
14. M.A. Bubenchikov, A.I. Potekaev, A.M. Bubenchikov, Russ. Phys. J. **56**, 785 (2013)
15. M.A. Bubenchikov, A.I. Potekaev, A.M. Bubenchikov, Russ. Phys. J. **56**, 341 (2013)
16. A.I. Potekaev, A.M. Bubenchikov, M.A. Bubenchikov, Russ. Phys. J. **55**, 1434 (2013)
17. M.A. Bubenchikov, A.I. Potekaev, A.M. Bubenchikov, O.V. Usenko, A.V. Ukolov, IOP Conf. S.: Mater. Sc. and Eng. **87**, 012111 (2015)
18. M. A. Bubenchikov, A. M. Bubenchikov, O. V. Usenko, V. A. Poteryaeva, S. Jambaa, EPJ Web of Conf. **110**, 01014 (2016)
19. N. F. Mott, I. N. Sneddon, *Wave Mechanics and its applications, Clarandon Press* (Oxford, 1948)
20. H. Feshbach, Ph. M. Morse, *Methods of Theoretical Physics. Part II* (McGraw-Hill, New York, 1953)
21. L. D. Landau, E. M. Lifshitz, *Quantum Mechanics: Non-Relativistic Theory, Vol. 3* (Pergamon Press, Oxford, 1965)
22. S. R. Otto, J. R. Denier, *An Introduction to Programming and Numerical Methods in Matlab* (Springer-Verlag, London, 2005)
23. A.M. Bubenchikov, M.A. Bubenchikov, A.I. Potekaev, E.E. Libin, Yu.P. Khudobina, Russ. Phys. J. **57**, 1126 (2014)
24. A.M. Bubenchikov, M.A. Bubenchikov, A.I. Potekaev, E.E. Libin, Yu.P. Khudobina, Rus. Phys. J. **58**, 882 (2015)

MATEC Web of Conferences **110**, 01086 (2017) DOI: 10.1051/matecconf/201711001086
HMTTSC-2017

Thermal transfer in a multilayer plate with source of metabolic heat

Mikhail A. Ulchiekov[1,*], *Tatyana N.* Nemova[1]

[1]National Research Tomsk Polytechnic University, 634050 Tomsk, Russia
[2]Tomsk State University of Architecture and Building, 634000 Tomsk, Russia

Abstract. In this article heat transfer in a multi-layered plate (human skin and thermal insulation) with various boundary conditions is considered. This allows you to justify the choice of air temperature in the production premises.

1 Introduction

The air temperature (T_A) in the production room is the most significant factor during design heating and ventilation systems, the TA value is selected in such a way as to ensure the requirements of the "thermal comfort" of the worker are met [1]. But by now the choice of T_A is based on empirical data on human sensations and this is "subjective" (it is assumed that TA should be in the range *18 °C* to *24 °C* [2-4]), and the mathematical justification is "not enough" described [5]. In this connection, the mathematical modelling of heat transfer within the framework of the heat conduction model (which allows us to describe the process of heat exchange of a person with the environment quite accurately) is an actual research direction.

The purpose of this work is numerically research of the heat transfer within a multilayer plate (layers of human skin and external thermal insulation), with different heat exchange options at the boundaries of the layers.

2 Formulation of the problem

In this paper is considered two options of formulation the problem: the temperature of the main part of the human body T_1 varies in a small range of *35.5-35.8°C*; from the surface of the human body the heat flows and equal to $q=M/A$ [5] (M is the part of metabolic heat leaving the surface of the skin, and A is the area of the human body).

Human skin is considered as a natural insulation consisting of subcutaneous fat (thickness Δ_1), dermis (thickness Δ_2), epidermis (thickness Δ_3) [6]. Also, the outer insulation layer adheres to the surface of the skin, which includes an air layer (thickness Δ_4) and clothing (thickness Δ_5).

* Corresponding author: ulchieckovmihail@mail.ru

It was assumed that the temperature of the air in the production premises is constant and does not change with time ($T_A = const$), and heat exchange on the outer surface of clothing occurs due to natural convection and thermal radiation.

As a parameter that serves as a "comfort characteristic" is used the surface temperature of the T_S kin. On the basis of the theory [1] it is assumed that the worker feels "comfort" at $T_S = 306 \div 307\ K$.

Mathematical statement of the first problem:

$$\begin{cases} \rho_1 c_1 \dfrac{\partial T_1}{\partial t} = \lambda_1 \dfrac{\partial^2 T_1}{\partial x^2}, & 0 < x < x_1; \\ \rho_2 c_2 \dfrac{\partial T_2}{\partial t} = \lambda_2 \dfrac{\partial^2 T_2}{\partial x^2}, & x_1 < x < x_2; \\ \rho_3 c_3 \dfrac{\partial T_3}{\partial t} = \lambda_3 \dfrac{\partial^2 T_3}{\partial x^2}, & x_2 < x < x_3; \\ \rho_4 c_4 \dfrac{\partial T_4}{\partial t} = \lambda_4 \dfrac{\partial^2 T_4}{\partial x^2}, & x_3 < x < x_4; \\ \rho_5 c_5 \dfrac{\partial T_5}{\partial t} = \lambda_5 \dfrac{\partial^2 T_5}{\partial x^2}, & x_4 < x < L; \end{cases}$$

Here ρ_i, λ_i, c_i – the density, thermal conductivity and specific heat of the material system of *i*-layer, Ti – current temperature value ($i = 1 \dots 5$).

The boundary condition of the first kind ($T_1 = 309\ K$) is given on the left boundary $x=0$, heat transfer due to natural convection and radiation is taken into account at $x=L$, boundary conditions of the fourth kind (equality of temperatures and heat fluxes) are given on the interfaces of the layers.

The initial and boundary conditions are:

$$\begin{aligned} &x = 0:\ T = T_1,\ T_1 = const; \\ &x = x_1:\ \lambda_1 \frac{\partial T_1}{\partial x} = \lambda_2 \frac{\partial T_2}{\partial x},\ T_1 = T_2; \\ &x = x_2:\ \lambda_2 \frac{\partial T_2}{\partial x} = \lambda_3 \frac{\partial T_3}{\partial x},\ T_2 = T_3; \\ &x = x_3:\ \lambda_3 \frac{\partial T_3}{\partial x} = \lambda_4 \frac{\partial T_4}{\partial x},\ T_3 = T_4; \\ &x = x_4:\ \lambda_4 \frac{\partial T_4}{\partial x} = \lambda_5 \frac{\partial T_5}{\partial x},\ T_4 = T_5; \\ &x = L:\ \lambda_5 \frac{\partial T_5}{\partial x} = \alpha(T_L - T_A) + \varepsilon\sigma((T_L)^4 - (T_A)^4); \\ &t = 0:\ T_1 = T_1^0,\ T_2 = T_2^0, \dots\dots,\ T_{x1} = T_{x1}^0,\ 0 < x < x_3; \\ &\qquad T = T_0^A, x_3 < x < x_4; \\ &\qquad T = T_0^{cl}, x_4 < x < L; \end{aligned}$$

Here α - heat transfer coefficient, ε - reduced emissivity surface of human skin, σ - Stefan-Boltzmann constant.

The mathematical model of the second version of the formulation with the corresponding initial boundary and initial conditions has the form:

 DOI: 10.1051/matecconf/201711001086

$$
\begin{cases}
\rho_4 c_4 \dfrac{\partial T_1}{\partial t} = \lambda_4 \dfrac{\partial^2 T_1}{\partial x^2}, \ 0 < x < x_1; \\
\rho_5 c_5 \dfrac{\partial T_2}{\partial t} = \lambda_5 \dfrac{\partial^2 T_2}{\partial x^2}, \ x_1 < x < L;
\end{cases}
$$

$$x = x_2: \ \lambda_4 \frac{\partial T_4}{\partial x} = \lambda_5 \frac{\partial T_5}{\partial x}, \ T_1 = T_2;$$

$$x = 0: \ -\lambda_4 \frac{\partial T_4}{\partial x} = q_1; \ q_1 = M / A;$$

$$x = L: \ \lambda_5 \frac{\partial T_5}{\partial x} = \alpha (T_L - T_B) + \varepsilon\sigma ((T_L)^4 - (T_B)^4);$$

$$t = 0: \ T = T_0^{e}, x_3 < x < x_4;$$

$$T = T_0^{oo}, x_4 < x < L;$$

The main difference between these two statements is that in the first one only layers of external isolation are considered, and a boundary condition of the second kind (*q=M/A*) is given at the boundary *x=0*.

The heat transfer coefficient was calculated by the criterion eqution for the conditions of natural convection near a plane wall [7]:

$$Nu^{1/2} = 0.825 + \frac{0.387 \cdot Ra^{1/6}}{\left[1 + \left(\dfrac{0.492}{\Pr}\right)^{9/16}\right]^{8/27}}; \tag{1}$$

$$Ra = \Pr \cdot Gr;$$

$$Gr = \frac{\beta \cdot l^3 \cdot g \cdot (T_L - T_B)}{\nu};$$

$$Ra = \Pr \cdot Gr; \tag{2}$$

$$Gr = \frac{\beta \cdot l^3 \cdot g \cdot (T_L - T_B)}{\nu}; \tag{3}$$

Here Gr - the number of Grazgof, l - a characteristic size (height of a man), ν - kinematic viscosity coefficient (air), β - the temperature coefficient of volume expansion of the air.

From the determination of the Nusselt number [8]:

$$\alpha = \frac{Nu \cdot \lambda}{l}; \tag{4}$$

At this stage of the research it was assumed that the person is at rest (the heat given off by evaporation is small [8], because *M/A* - for a resting state of *50 W/m²*):

$$Q_u = 0.49 \cdot (M / A \cdot (1 - \eta) - 50); \tag{5}$$

The system of differential equations with appropriate initial and boundary conditions is solved by finite difference method, using an algorithm [8,9] developed for solving heat transfer problems in the areas of multi-layer discontinuous thermal conductivity.

At the solution of the problem the temperature dependence of thermal characteristics of materials and layer materials are not taken into consideration, since the latter was only a change from *1 K* to *10 K*.

3 Results of numerical modelling

When modelling TA changed in the really possible range (from 277 K to 304 K), the parameters of clothing varied.

DOI: 10.1051/matecconf/201711001086

The thermo physical characteristics of the layers of the system are accepted in accordance with the data [11, 12, 13]:

$\lambda_1 = 0.2\ W/m\cdot K,\ \lambda_2 = 0.45\ W/m\cdot K,\ \lambda_3 = 0.21\ W/m\cdot K,\ \lambda_4 = 0.024\ W/m\cdot K,$

$\lambda_5 = 0.03\ W/m\cdot K$

$\rho_1 = 850\ kg/m^3,\ \rho_2 = 1000\ kg/m^3,\ \rho_3 = 1200\ kg/m^3,\ \rho_4 = 1.2\ kg/m^3,\ \rho_5 = 120\ kg/m^3$

$c_1 = 2250\ J/kg\cdot K,\ c_2 = 3200\ J/kg\cdot K,\ c_3 = 3600\ J/kg\cdot K,$

$c_4 = 1005\ J/kg\cdot K,\ c_5 = 2260\ J/kg\cdot K,$

The thicknesses of the layers of skin accepted in accordance with [7]:

$$\Delta_1 = 0.013\ m,\ \Delta_2 = 0.0005\ m,\ \Delta_3 = 0.001\ m$$

Typical initial temperatures:

$$T_0^{cl} = 301\ K,\ T_0^{a} = 303\ K,\ T_0^{x1} = 307.$$

The duration of the modeled process chosen within the limits: $t = 7000$ s.
The degree of emissivity of human skin [13]:

$$\varepsilon = 0.7.$$

Dependences of skin surface temperature on time at different ambient temperatures for the adopted formulations of the problem are shown in Fig. 1. It is established that the T_S decreases significantly faster and depends more on T_A in a system with a boundary condition of the second kind.

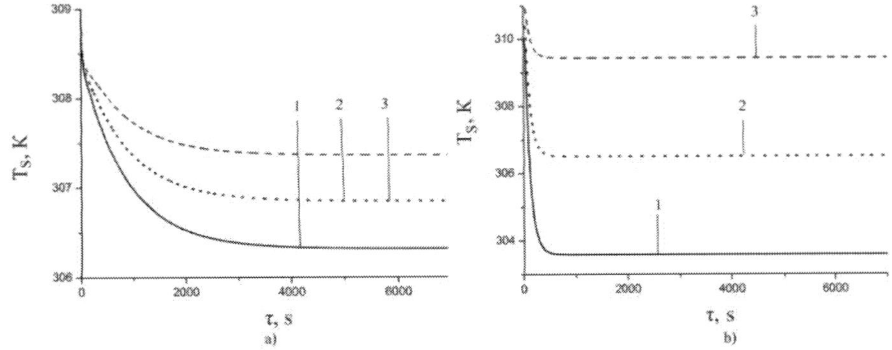

Fig. 1. Dependence of T_S on time at: 1 – T_A=291 K; 2 - T_A=294 K; 3-T_A=302 K: a) a first-kind boundary condition for x=0; b) a boundary condition of the second kind for x=0.

Dependences of skin surface temperature on time at different ambient temperatures for the adopted formulations of the problem are shown in Fig. 1. It is established that the T_S decreases significantly faster and depends more on T_A in a system with a boundary condition of the second kind.

"Comfortable" air temperatures for easily dressed people are given in [1], obtained in experimental studies. It was found that T_A is in the range of *17-24 °C.* Dependences of the surface temperature of the heat supply object on T_A are shown in Fig. 2 It can be concluded that a mathematical model with a boundary condition of the first kind (Fig. 2 a, 1, 2) at $x=0$ corresponds more exactly to experimental data than with a boundary condition of the second kind.

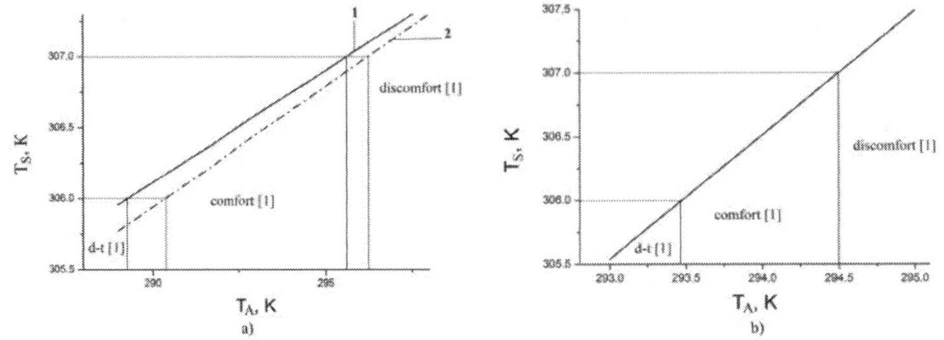

Fig. 2. Dependence of T_S on ambient temperature (single-layer wool clothing): 1 – T_A=291 K; 2 - T_A=294 K; 3-T_A=302 K: a) a first-kind boundary condition for x=0 (1 - results obtained in this paper, 2 - obtained in [15]); b) a boundary condition of the second kind for x=0.

The thickness of clothes plays a big role in choosing the air temperatures of production premises. The connection between the "comfort" T_A and the thickness of the layer of external thermal insulation (woollen clothing) for the two tasks discussed above is shown in Fig. 3. It is established that for a system with a boundary condition of the second kind at x=0, the influence of the clothing thickness on the "comfort" T_A is greater than for conditions of the first kind at this boundary.

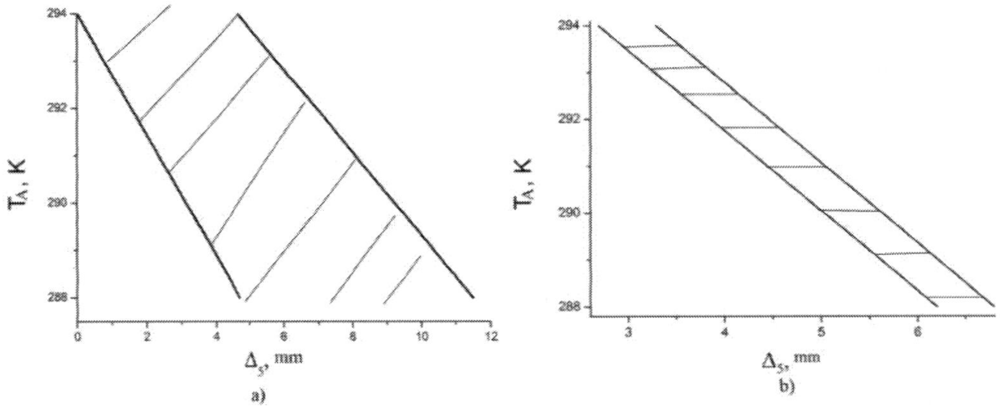

Fig. 3. Dependence of "comfortable" air temperature on the thickness of wool clothing: a) a first-kind boundary condition for x=0; b) a boundary condition of the second kind for x=0.

4 Main results and conclusions

1. The problems of thermal conductivity for the system "heat supply object - the layer of thermal insulation - the external environment" with allowance for natural convection and radiation at the external boundary with boundary conditions of the first and second kind at x = 0 are solved.
2. Dependences of the surface temperature of the skin on the ambient temperature for the two formulation of the problem are obtained. It is established that conditions of the first kind provide a better fit to the experimental data [1].
3. The connection of "comfortable" air temperature and the thickness of woollen clothing is revealed.

The work was supported by the Russian Federation President's grant (NSH project -7538.2016.8).

References

1. V. N. Platova, Biologia, **8** (2008) [in Russian]
2. P. Domingues, P. Carreira, R. Vieira, W. Kastner, Comp. St. Int. **45** (2016)
3. M. Wallance, R. McBride., P. Mhaskar, , J. House, T. Salsbury, Ch. Eng. Sc. Ener. **69** (2012)
4. M. Taleghani, M. Tenpierik, A. Dobbelsteen, Ren. Sust. En. R. **26** (2013)
5. A. Machkashi, L. Bahdidi, *Radiant heating* (Stroyizdat, Moscow, 1985)
6. F. Xu, T. J Lu, K. A., Sef. Acta Mech. S. **24** (2008)
7. Robert E. Simons, Electroning cooling, **6** (2001)
8. F. P. Incropera, D. P. DeWitt, *Fundamentals of Heat and Mass Transfer* (John Wiley and Sons, NY, 1985)
9. O.V. Vysokomornaya, G.V. Kuznetsov, P.A. Strizhak, Rus. J. Ph. Ch. B. **5** (2011)
10. V.L. Strakhov, A.N. Garashchenko, G.V. Kuznetsov, V.P. Rudzinskii, Comb., Exp. Sh. W. **37** (2001)
11. I.K. Kikoin, *Tables of physical quantities* (Atomizdat, Moscow, 1976) [in Russian]
12. A.P. Babichev, A.P. Babushkin, A.M. Bratkovsky *Physical values*, (Eneroatomizdat, Moscow, 1991) [in Russian]
13. M.V. Nazarova, S. Yu. Boyko, Modern problems of s. and ed. **5** (2009)
14. U. Kukelev, A. Ustinov, Resources and Technology, **3** (2005)
15. M.A. Ulchiekov, E.E. Bulba, MATEC Web of Conf. **8** (2017)

Heat and mass transfer in a liquid film flowing down the bottom side of inclined surface

Maria V. Bartashevich[1,*], *Maxim G.* Vlasenko[1], *Andrey A.* Pil'nik[1], and *Andrey A.* Chernov[1]

[1]Kutateladze Institute of Thermophysics, 630090 Novosibirsk, Russia

Abstract. The work presents derivation of the equation describing the shape of the free surface of a liquid film flowing along the bottom wall of an inclined surface, taking into account conjugate heat and mass transfer under the condition that the temperature and concentration on the free surface are unknown in advance and interrelated.

1 Problem statement

This work deals with investigation of combined heat and mass transfer on a thin film of liquid (water solution) moving along the bottom side of an inclined surface. The similar problems were considered in [1-6]. So, the analytical solutions to the simplest problems on combined heat and mass transfer at absorption on free surfaces of various geometries are presented in [1]. Heat and mass transfer in the entrance region of the falling axisymmetric semi-infinite film surrounded by the pure vapour was studied in [2]. A stationary axisymmetric flow of a thin liquid film along a confuser is considered in [3], the rivulet flow on the bottom wall of an inclined cylinder is analyzed in [4]. The exact solution to the problem of isothermal diffusion to a wave film was obtained in [5]. The mathematical model of combined heat and mass transfer on a vaporizing laminar liquid film of a constant thickness, flowing along a vertical cylinder, is presented in [6]. To solve the heat and mass transfer equations, Laplace transformations were used, and comparisons with experimental data were made. In this paper, we present an equation describing the shape of the free surface of a liquid film flowing down the bottom side of an inclined surface with consideration of conjugate heat and mass transfer.

We consider the two-dimensional stationary flow of a laminar liquid film of thickness h over a plate inclined at angle α to the horizon, as schematically shown in Fig. 1. The surface of liquid is in contact with immobile vapor. The temperature and concentration on the free surface are not known in advance and interrelated. The thermal-physical properties of solution are assumed constant.

* Corresponding author: bartashevichmv@gmail.com

DOI: 10.1051/matecconf/201711001087

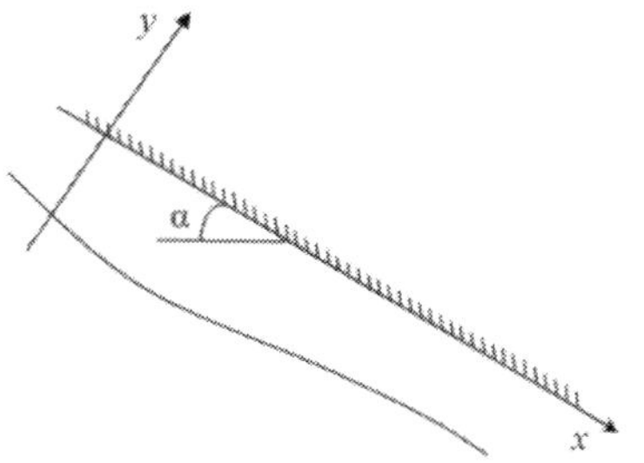

Fig. 1. Scheme of the problem of heat and mass transfer in a liquid film flowing down the bottom side of inclined surface.

2 Film thickness equation

Let us introduce the Cartesian coordinate system with Ox axis in the direction of the flow and Oy axis directed along the normal to the plate. It is assumed that in the region of liquid flow the following equations are satisfied:

$$u\frac{\partial u}{\partial x}+v\frac{\partial u}{\partial y}=-\frac{1}{\rho}\frac{\partial p}{\partial x}+\nu\frac{\partial^2 u}{\partial y^2}+g\sin\alpha\ , \tag{1}$$

$$-\frac{1}{\rho}\frac{\partial p}{\partial x}-g\cos\alpha=0\ , \tag{2}$$

$$\frac{\partial u}{\partial x}+\frac{\partial v}{\partial y}=0\ . \tag{3}$$

The boundary conditions for systems (1) – (3) take the following form.
At $y=0$

$$u=v=0\ . \tag{4}$$

At $y=-h$

$$\frac{\partial u}{\partial y}=0\ . \tag{5}$$

$$p=p_a-\sigma\frac{\partial^2 h}{\partial x^2}\ . \tag{6}$$

$$u\frac{\partial h}{\partial x}-\frac{\lambda}{\rho r}\frac{\partial T}{\partial y}-\frac{D}{\rho}\frac{\partial C}{\partial y}=v\ . \tag{7}$$

It follows from (2) and (6) that

$$p=p_a+\rho g(h(x)-y)\cos\alpha-\sigma\frac{\partial^2 h}{\partial x^2}\ . \tag{8}$$

From (1) and (8) we obtain

$$u\frac{\partial u}{\partial x}+v\frac{\partial u}{\partial y}=-g\frac{\partial h}{\partial x}\cos\alpha+\sigma\frac{\partial^3 h}{\partial x^3}+\nu\frac{\partial^2 u}{\partial y^2}+g\sin\alpha\,. \qquad (9)$$

Integrating (9) with consideration of (3) and (7), we obtain equation

$$\frac{\partial J}{\partial x}+gh\cos\alpha\frac{\partial h}{\partial x}-\nu\int_0^h\frac{\partial^2 u}{\partial y^2}dy-gh\sin\alpha+\frac{\sigma h}{\rho}\frac{\partial^3 h}{\partial x^3}-\frac{u_i\lambda}{\rho r}\frac{\partial T}{\partial y}-\frac{u_i D}{\rho}\frac{\partial C}{\partial y}=0\,, \qquad (10)$$

where $J=\int_0^h u^2 dy$. Let the velocity profile is

$$u=-\frac{gh^2\sin\alpha}{2\nu}\left(\frac{2y}{h}+\frac{y^2}{h^2}\right). \qquad (11)$$

We should note that conditions (4) and (5) are satisfied. Then, the surface equation will take form:

$$\frac{57}{40}\frac{\partial}{\partial x}\left(\frac{q^2}{h}\right)+gh\frac{\partial h}{\partial x}\cos\alpha+\frac{3q\nu}{h^2}+\frac{\sigma h}{\rho}\frac{\partial^3 h}{\partial x^3}-u_i\frac{\lambda}{\rho r}\frac{\partial T}{\partial y}-u_i\frac{D}{\rho}\frac{\partial C}{\partial y}=0\,, \qquad (12)$$

where $q=\int_0^h u dy$, $u_i=\frac{3}{4}\frac{q}{h}$.

3 Conclusions

The work presents derivation of the equation describing the shape of the free surface of a liquid film flowing along the bottom wall of an inclined surface, taking into account conjugate heat and mass transfer under the condition that the temperature and concentration on the free surface are unknown in advance and interrelated. The numerical solution of the resulting equation is a separate complex and non-trivial problem and it will be the subject of further research of the authors.

The work was supported by the grant MK-6384.2016.8.

References

1. V.E. Nakoryakov, N.I. Grigorieva, *Non-Isothermal Absorption in Thermal Transformers* (Nauka, Novosibirsk, 2010)
2. V.E. Nakoryakov, N.I. Grigoryeva, M.V. Bartashevich, Int. J. Heat Mass Transfer. **54**, 21 (2011)
3. L. I. Sen', A. M. Te, O. Yu. Tsvelodub, J. Appl. Mech.Tech. Phys. **24**, 3 (1983)
4. S.V. Alekseenko, P.I. Geshev, P.A. Kuibin, Doklady Physics, **42** (1997)
5. P.I. Geshev, A.M. Lapin, J. Appl. Mech.Tech. Phys. **24**, 6 (1983)
6. Y. Wu, Int. J. Heat Mass Transfer. **93** (2016)

Computational study of the effect of fuel element geometry on pellets' maximum temperature

Alexander Vorobiev[1,*], *Ksenia* Pavlova[1]

[1]National Research Tomsk Polytechnic University, 634050 Tomsk, Russia

Abstract. One of the factors that determine reliable operation of fuel elements in a nuclear reactor is maximum temperature of fuel pellets. This paper presents the computational study results of the effect of the pellet's central hole on its maximum temperature when using different types of fuel.

1 Introduction

Fuel elements are the most important and strained nodes in the core of a modern nuclear power plant reactor. The design of a fuel element must ensure its long-term operational reliability under extremely severe operating conditions.

At present most nuclear reactors are fueled with a compound known as uranium dioxide (UO_2), since this fuel does not interact with water and steam even at high temperatures, is compatible with fuel rod's cladding material, and is characterized by a a high melting point. Among its downsides are low thermal conductivity and brittleness.

Uranium carbides (UC and UC_2) are seen as potential nuclear fuels. They are characterized by higher density as compared to UO_2 and have a high thermal conductivity. The disadvantage of uranium carbides is a relatively high activity with many reactor materials. In the future the use of uranium nitrides (UN) and silicides is also possible. They are similarly characterized by high density and thermal conductivity but at the same time they are more compatible with reactor materials [1].

One of the most significant factors that determine operational reliability of a fuel element is maximum temperature of the fuel pellets. This temperature can be decreased by using fuel with higher thermal conductivity as well as by enhancing heat-transfer properties of a pallet-clad gas gap. Another method for lowering the peak temperature of the pellet involves application of fuel pellets with a central hole.

The objective of this work is to study the effect of a central hole and its dimensions on the pellets' maximum temperature when using different types of fuels.

2 Calculations and methodology of the study

* Corresponding author: worob@tpu.ru

The model of the object of the computational study is given in Fig.1. The assumed geometry and mode characteristics of a conventional fuel element are presented in Table 1.

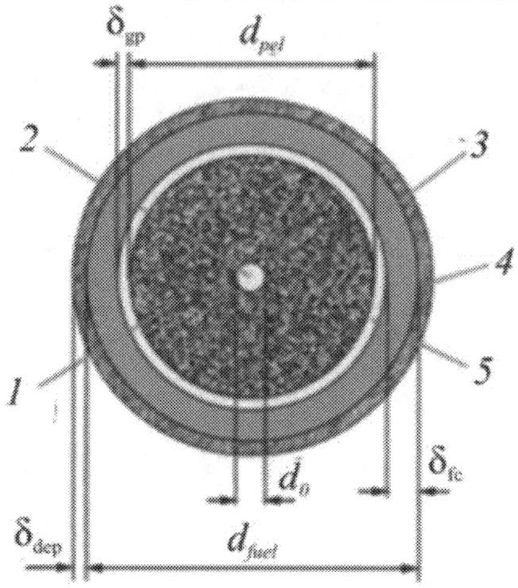

Fig 1. Cross-section of a conventional fuel rod: 1 – pellets; 2 – hole; 3 – cladding; 4 - deposits; 5 – gap.

Table 1. Main characteristics of a conventional fuel element.

Parameter	Description
Coolant	non-boiling water
Gas in pellet-clad gap	helium
Cladding material	Zr alloy
Coolant pressure p_c	16 MPa
Coolant temperature t_c	305 °C
Coolant velocity w_c	6 m/s
Spacer arrangement	triangular
Fuel rod pitch S_{pitch}	1.4
Pellet diameter d_{pel}, mm	7.53 mm
Cladding thickness δ_{cl}	0.685 mm
Gap thickness δ_{gap}	0.1 mm

The algorithm for determining pellet's maximum temperature includes well-known expressions [2]. This algorithm suggests the use of different nuclear fuels; their thermal conductivity is calculated by means of approximating expressions [1, 4].

Temperature distribution in fuel pellets including peak temperature can be calculated in the process of solving one-dimensional heat conduction problem by Seidel iteration method [3].

To implement the algorithm mentioned above and perform calculations of the temperature characteristics of the fuel composition the *FUEL* computer program has been developed in the Department of Nuclear and Thermal Power Plants of the Tomsk Polytechnic University. The program is made in the Borland Turbo Delphi 2006 intergated development environment and is actually a Windows Service application. The program's executable module allows graphic interpretation of the calculation results and their import to MS Excel.

Screenshot of the *FUEL* program's main window is presented in Fig. 2.

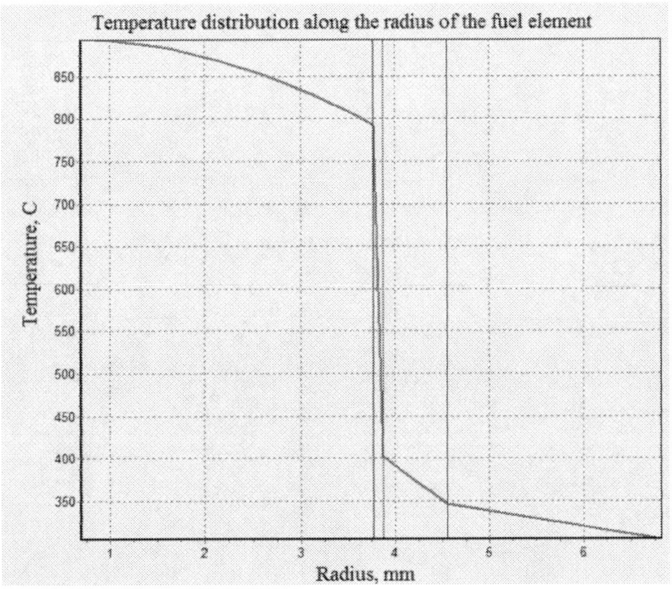

Fig. 2. Screenshot of the *FUEL* main window.

Two cycles of the variational calculations of the fuel rod's temperature characteristics were performed using the *FUEL* program (Fig.1). In each calculation cycle we varied the values for central hole diameter (d_0=0…2 mm) and fuel type (UO_2, UC, UN, and U_3Si_2).

The conditions for the first calculation cycle were: q_s = const and q_l = var (i.e. output of a fuel rod N_{fuel}=var); for the second cycle: q_s = var and q_l = const (N_{fuel}=const).

The main calculation results for uranium dioxide are given in Fig. 3.

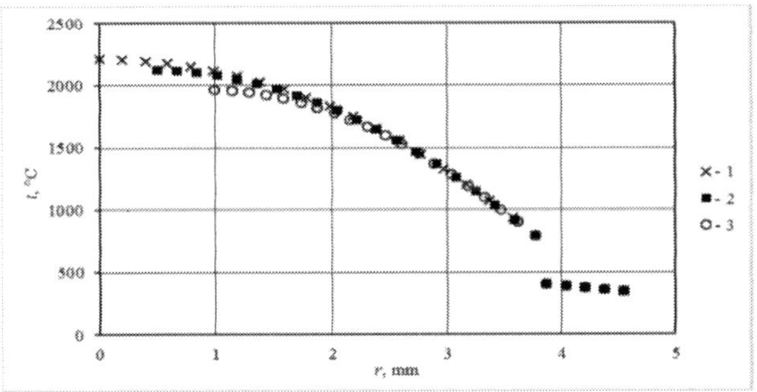

Fig. 3. Temperature distribution in a fuel rod (pellet material is UO_2) at different diameters of the central hole. q_l = 440 Wt/cm: 1 – d_0 = 0; 2 – d_0 = 1 mm; 3 - d_0 = 2 mm.

The main calculation results for other fuels are presented in Fig. 4.

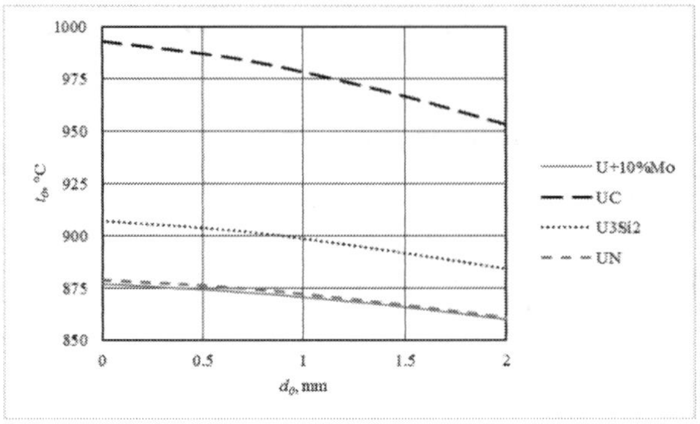

Fig. 4. Dependence of the pellets' maximum temperature on the diameter of a central hole for different fuels. $q_l = 440$ Wt/cm.

3 Conclusion

1. The presence of a central hole in the fuel pellets decreases their maximum temperature t_0 for any of the common fuel types. Yet, the amount of the decrease in temperature is largely dependent on the type of fuel.
2. When changing the central hole diameter from 0 to 2 mm the most considerable decrease in t_0 is observed for the pellets made of uranium dioxide (~250 °C). For other fuels changes in temperature t_0 are in the range of 17...40 °C, i.e. the decrease in temperature is much smaller.
3. Meanwhile, drop in the maximum temperature of the pellets due to the presence of a central hole is observed in both cases: when $q_s =$ const and when $q_l =$ const.

References

1. B.A. Kalin, *Physical materials science: textbook* (Izd. MEPhI, Moscow, 2008)
2. B.A. Dementiev, *Nuclear power reactors: textbook. 2nd edition* (Energoatomizdat, Moscow, 1990)
3. N.M. Belyaev, A.A. Ryadno, *Thermal conductivity theory methods: coursebook* (Vysshaya shkola, Moscow, 1982)
4. P.L. Kirillov, *Thermal physical properties of materials for nuclear equipment: reference book* (Izdat, Moscow, 2007)

MATEC Web of Conferences **110**, 01089 (2017) DOI: 10.1051/matecconf/201711001089
HMTTSC-2017

Research of composite fuels thermophysical properties based on low-grade coals with addition of fine sawdust and flour industry wastes

Stanislav Yankovsky[1,*], *Anton* Tolokolnikov[1], *Alexander* Matveev[1], and *Yana* Marysheva[1]

[1]National Research Tomsk polytechnic university, 634050 Tomsk, Russia

Abstract. Experimental studies have been carried out to determine the energy and technical characteristics of composite fuels from coal grade 3B (Balakhtinskoye deposit), fine waste and flour from the milling industry. An effective concentration of 50% / 50% composite fuel has been established without the addition of a flour component, in which the combustion heat is reduced by less than 8%, with the flour component added up to 15%, the heat of combustion further decreases less than 2% of the heat of combustion of homogeneous coal. At the same time, the ash content is reduced to 11%, the yield of volatile components remains at the concentration level of the composite fuel of coal and wood. It is established that this component composition of three types of fuel is suitable for the formation of fuel briquettes suitable for burning in furnaces of small power boilers.

1 Introduction

It is known [1-4], that coal is a basic fuel for power generation in small-scale power engineering [4-5]. Significant expenditures of funds for its supply to remote regions of the Russian Federation lead to an increase in final cost of energy resources in two or more times, depending on territorial location [5-8]. In this case, the use of local resources in the form of solid composite fuel made of low-grade coals with addition of waste from timber industry and impurities of flour waste as binder is promising for the near future. Studies in these areas are actively conducted; their results are published in papers [1-8]. But results of [1-8] are not the basis for analyzing dependencies of composite fuels thermophysical properties based on low-grade coals with addition of fine sawdust and flour industry wastes. Analysis of composite fuels formation with operational and energy characteristics close to homogeneous coal, but with high environmental parameters is relevant. The goal is experimental determination of binder proportion influence on composite fuel characteristics.

2 Experimental results and discussion

* Corresponding author: jankovsky@tpu.ru

Technical and energy characteristics of three composite fuels have been determined experimentally [9-10]. Samples of 3B grade coal (Balakhtinskoe deposit, particle size less than 80 microns), fine sawdust and flour industry waste (particle size less than 200 microns) were mixed in different volume concentrations in a galvanized drum of Pulverisette 6 planetary mill with spherical grinding bodies with 5 mm diameter at required volume ratios. Mixing process was being conducted for 7 minutes at rotation speed of 500 rpm [10].

Concentration compositions are presented in Table 1.

Table 1. Concentration compositions of composite fuels.

Fuel	Concentration, %		
3B brown coal	47.5	45	42.5
Fine sawdust	47.5	45	42.5
Flour waste	5	10	15

Figure 1 presents energy characteristics of fuels' main components. Experimentally obtained dependences of composite fuel heat of combustion on its composition are presented in Figures 2-3.

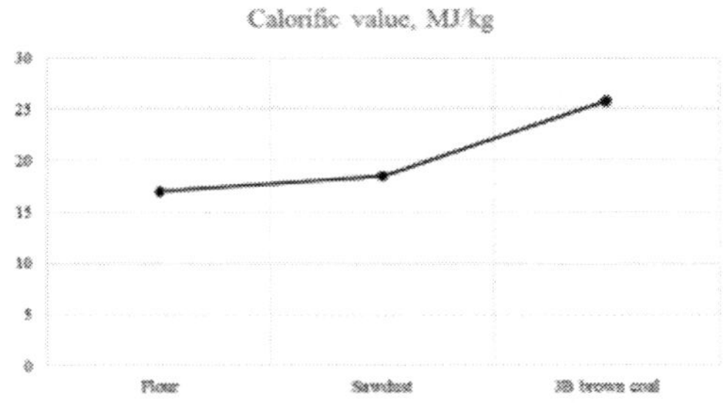

Fig. 1. Calorific content of each component of composite fuel.

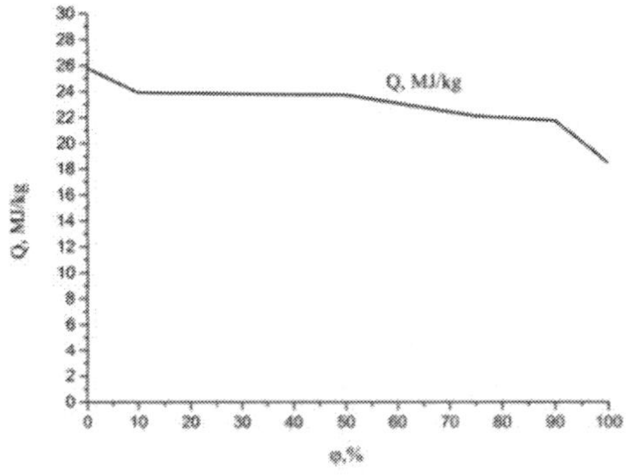

Fig. 2. Sawdust concentration effect on calorific value of composite fuel in relation to homogeneous coal (φ – sawdust content).

DOI: 10.1051/matecconf/201711001089

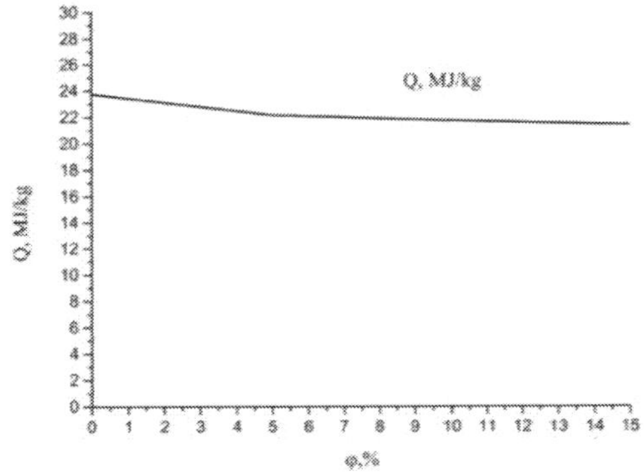

Fig. 3. Flour concentration effect on calorific value of composite fuel in relation to mixture composition of coal and sawdust 50 %/50 % (φ – flour content).

Analysis of the results shown in Figure 2 allows determining optimal concentration composition of composite fuel made of coal and sawdust in a ratio of 50 % / 50 %, in which the reduction in calorific content is less than 8 %.

Figure 3 illustrates that increase of flour concentration to 5 % in composite fuel reduces its calorific value by less than 6.7 %, but addition of 15 % of flour reduces calorific content by 10%.

Figure 4 provides dependence of change in fuel ash content with an increase of flour industry waste concentration.

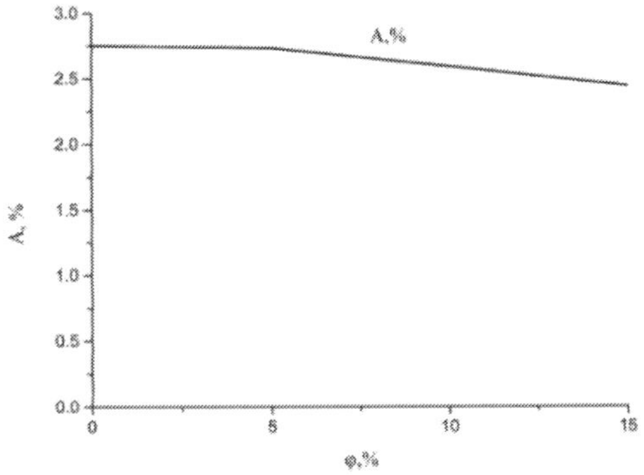

Fig. 4. Flour concentration effect on composite fuel ash content in relation to mixture composition of coal and sawdust 50 % / 50 % (φ – flour content)

Figure 4 shows that with an increase in flour concentration to 5 % ash content decreases by less than 6 %; at concentration of 15 % of flour products in composite fuel ash content is reduced to 11 %.

Dependence of composite fuel volatiles yield from its composition is shown in Figure 5.

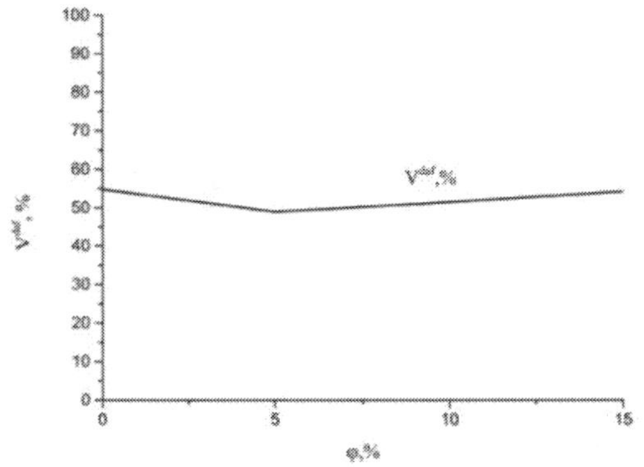

Fig. 5. Flour concentration effect on composite fuel volatiles yield in relation to mixture composition of coal and sawdust 50 % / 50 % (φ – flour content)

Analysis of volatiles yield dependence suggests that an increase of flour concentration in composite fuels does not lead to significant changes.

3 Conclusion

As a result of conducted experiments, it was determined that influence of flour industry wastes concentration is insignificant. Energy and technical characteristics of composite fuels are reduced by less than 10%. As a result of the foregoing, it can be assumed that this component compound is suitable for fuel briquettes formation by cold pressing, suitable for combustion in furnaces of small-scale energy boilers.

The research has been accomplished with financial support of the Russian Federation Ministry of Education and Science within implementation of the project No 13.7644.2017/BCH.

References

1. K. Y. Vershinina, Scientific Journal, **90**, 1(2017)
2. M Jumoke, Applied Energy, **188**, 215-225 (2017)
3. A.V. Fedyukhin, I.L. Maikov, V.A. Sinelshchikov, Book of Abstracts of International Conference on Interaction of Intense Energy Fluxes with Matter, 114 (2011)
4. W. Yan, S. Wang, D.G. Zhou, C. Xing, Y. Zhang, G.M. Pharr, Wood and fiber science, **44**, 1 (2009)
5. A.G. Korotkikh, K.V. Slyusarskiy, K.B. Larionov, V.I. Osipov, J. Phys. **754**, 5 (2016)
6. A.A. Abramovskikh, U.A. Ulyanova, S.V. Lavrinenko, A.A. Matveeva, D. Malishev, MATEC Web Conf. 91, 01002 (2016)
7. N. Galashov, S. Tsibulskiy, A. Kiselev, MATEC Web Conf. 72, (2016)
8. M.P. Baranova, V.A. Kulagin, *Fiziko-khimicheskiye osnovy polucheniya toplivnykh vodougol'nykh suspenziy (*Monograph, Siberian Federal University, 2011) [In Russian]
9. K. Larionov, I. Korjashov, S. Jankovskij, V. Gubin, A. Zenkov, MATEC Web Conf. **37** (2015)
10. A. Zenkov, S. Yankovsky, A. Matveeva, S. Lavrinenko, A. Gromov, MATEC Web Conf. 01131 (2016)

MATEC Web of Conferences **110**, 01090 (2017) DOI: 10.1051/matecconf/201711001090
HMTTSC-2017

The dynamics of ice melting in the conditions of crybot movement

Ekaterina Zakharova [1,*], *Arkadii* Zakharevich[1], and *Nadegda* Gutareva[1]

[1]National Research Tomsk Polytechnic University, 634050 Tomsk, Russia

Abstract The mathematical modeling results of the simultaneous processes of heat and mass transfer under the conditions of intense phase changes (melting of ice) during the movement of cryobot have been given. The spatial unevenness of the melting rate of ice has been taken into account. It has been established that the rate of passage of the cryobot depends essentially on its temperature. According to the results of the numerical simulation, considerable cooling of the cryobot sheath has been established. The latter is due to the high endothermic effect of melting ice.

1 Introduction

The study of the modern continental glaciation and all types of glaciers is of great importance for microbiology, geophysics, geology and other natural sciences [1]. At the moment, there are several directions in the study of glaciation. The most important for today is the study of the current state of the glaciers in Antarctica. Due to the fact that the problem of warming is global every year the situation worsens. About $30{\cdot}10^6$ km^3 ice is concentrated in Antarctica, and with such a rapid increase in temperature in the near future, a large-scale melting of these glaciers is possible. The latter will affect significantly the ecosystem, both the mainland and the entire world ocean. The level of the world ocean will also change significantly. In this regard, a careful study of the structure and the state of the glaciers is necessary [2].

One of the more effective ways of studying the structure, the material composition and the dynamics of ice deposits is drilling wells and analyzing ice cores. This process can be carried out in three ways: the first variant involves the thermal destruction of ice structures [3]; the second variant involves the use of mechanical action on the layers of ice ("classical" drilling) [2]; the third variant is based on the combined chemical and thermal effects [1]. To obtain an ecologically pure sample of the ice core, it is expedient to use exclusively the thermal method of drilling a well [4].

When studying the large ice arrays and objects located below them, the apparatus must meet the criteria: the minimum possible size and weight, the presence of a powerful energy source. These requirements are fully consistent with the cryobot — a specialized probe-robot of deep glacial penetration. The main advantages of a cryobot with respect to other devices are the smaller vulnerability of the device to damage, relatively small energy costs. The movement of the drilling machine is carried out by heating the ice to the formation of

* Corresponding author: eaz15@tpu.ru

melt water in front of the bow. Water flows around the cryobat and freezes behind it, while the cryobot itself continues to move under the influence of gravity. In this connection, it is of interest to study the heat transfer in the conditions of melting of the large ice masses by local heat sources.

In the conditions of Antarctica (the significant distance, the heterogeneous structure of ice, the extremely severe climate, the total impassability and scale of the objects of study) carrying out the field experiments will be either very expensive, or in some cases even an impossible process. In these conditions, the most acceptable is the application of the mathematical modeling. However, by the present time, a mathematical apparatus has not been developed practically that provides the possibility of solving the problem of modeling the motion of cryobot in the thickness of the extended ice structures. The known mathematical models, as a rule, are based on a number of substantially simplifying assumptions.

For example, in [2] the results of the computer simulation of the processes of heat and mass transfer during the motion of the cryobot through the water ice massifs are presented. In this case, the mathematical formulation [2] does not imply the presence of an ice-water system (the fusion front) in the vicinity of the apparatus. The absorption of heat as a result of the recrystallization of the ice structures is described in the framework of the enthalpy heat transfer model. In [3, 5], the mathematical model of heat transfer processes in ice structures is shown in the melting of the latter. However, the model [3, 5] is based on the hypothesis of the uniformity of the distribution of the fusion front over the entire ice sheet. The most complete is the model [6]. However, the latter describes the melting process in the framework of the approach based on the dynamically adaptive difference grids. This approach is very complicated and requires the use of a special mathematical apparatus. In addition, the use of the method of [6] requires the special control functions, the significance of which can only be found from the experiment. Such an insignificant number of the works on modeling of the motion of cryobot is associated with the high complexity of modeling the processes of phase transformations (melting of ice) in the multidimensional formulation of the problem (the so-called multi-front problems of Stefan) [5, 7]. It can be said that the development of a mathematical apparatus that provides the predictive modeling of the processes of heat transfer in the conditions of melting ice during the movement of cryobot in ice formations is an actual and unresolved task to date.

The aim of this work is the mathematical modeling of heat transfer processes that occur together under ice melting conditions during the movement of the cryobot.

2 Physical statement

At the initial moment of time, a massive heat source acts on the ice surface on a limited area. As a result, the ice layer starts to heat up. When the ice structure reaches the temperature equal to the phase transition temperature, the process of melting of the ice cover is initiated. The melting front moves deeper into the ice. As a consequence, in the solution area of the problem, a heterogeneous structure is formed, consisting of a solid and a liquid phase of water, with the significantly different thermal properties [7, 8]. In solving the problem, ice has been assumed to be homogeneous. At the initial time, the temperature of the entire volume of ice has been assumed to be the same.

3 Mathematical statement

The mathematical formulation of the problem, corresponding to the above physical model, consists of the following system of the equations. The energy equation for the "ice-water" system, taking into account the endothermic effect of the melting process (by analogy with [5,9,12]):

$$c(T)\rho(T)\frac{\partial T}{\partial t}=\frac{\partial}{\partial x}\lambda(T)\frac{\partial T}{\partial x}+\frac{\partial}{\partial y}\lambda(T)\frac{\partial T}{\partial y}-\frac{QW}{h_d}\delta(x_{melt},y_{melt}), \quad (1)$$

where c –heat capacity $kJ/(kg\cdot К)$; ρ – density, kg/m^3; λ – thermal conductivity, $W/(m\cdot K)$; h_d – Smearing front smearing parameter; W – Melting rate $kg/(m^2s)$; Q – Melting heat of ice J/kg; T – temperature, K; δ – Dirac delta function. [11] The value of thermophysical properties is determined from the conditions for localization of the fusion front.

The equation energy for cryobot:

$$c_c\rho_c\frac{\partial T}{\partial t}=\lambda_c\left(\frac{\partial^2 T}{\partial x^2}+\frac{\partial^2 T}{\partial y^2}\right)+Q_v, \quad (2)$$

where $c_c\rho_c,\lambda_c$ – heat capacity, density and thermal conductivity of cryobot; Q_v – power heat source, W/m^3. [10] The mathematical statement also includes the corresponding boundary conditions (Fig. 1):

$$t=0;\ 0<x<l;\ 0<y<H;\ T(x,y)=T_f$$
$$t=0;\ 0<x<l;\ 0<y<H;\ T(x,y)=T_{cr}$$

$$\left.\frac{\partial T}{\partial x}\right|_{x=0}^{0<y<H}=0;\ \left.\frac{\partial T}{\partial y}\right|_{y=0}^{0<x<l}=0;\ \lambda(T)\left.\frac{\partial T}{\partial y}\right|_{0<x<x_1}^{y=H}=\alpha(T_e-T);$$
$$\lambda(T)\left.\frac{\partial T}{\partial y}\right|_{x_2<x<l}^{y=H}=\alpha(T_e-T);\ \lambda(T)\left.\frac{\partial T}{\partial y}\right|_{y=H}^{x_1<x<x_2}=\lambda_c\left.\frac{\partial T}{\partial y}\right|_{y=H}^{x_1<x<x_2};$$
$$\lambda(T)\left.\frac{\partial T}{\partial x}\right|_{H<y<H_1}^{x=x_1}=\alpha(T_e-T);\ \ \lambda(T)\left.\frac{\partial T}{\partial x}\right|_{H<y<H_1}^{x=x_2}=\alpha(T_e-T);$$

$$\lambda(T)\frac{\partial T}{\partial x}=\lambda_c\frac{\partial T}{\partial x};\ \lambda_c\frac{\partial T}{\partial x}=\lambda(T)\frac{\partial T}{\partial x}$$

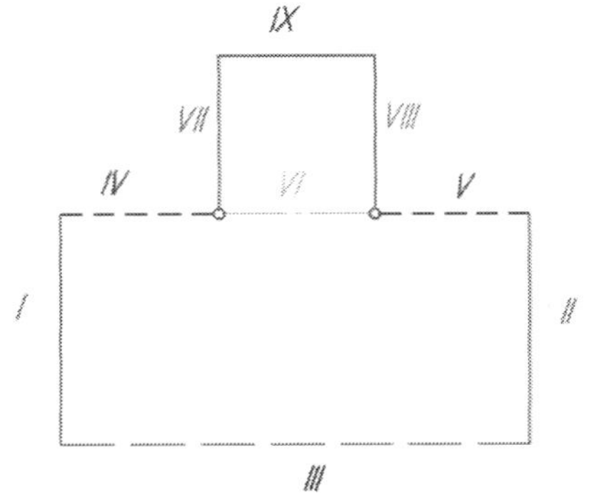

Fig.1 Areas for solving the problem.

4 The results

According to the results of the mathematical modeling of cryobot movement, the formation of a vertical shaft with even edges has been established. Figures 2 a. 2b and 3a, 3b show the temperature fields in the "cryobot-ice" system with varying widths and heights of a cryobot. The initial temperature of the apparatus is T = 500K. The analysis of figure 2 shows that with increasing width of the thermal drilling machine, the temperature distribution in the latter becomes uniform. In this case, the values of T at each point of space are substantially reduced (up to T = 350K). This is due to the influence of the phase transformations (melting of the ice structures) on the characteristics of heat transfer in the cover of the apparatus.

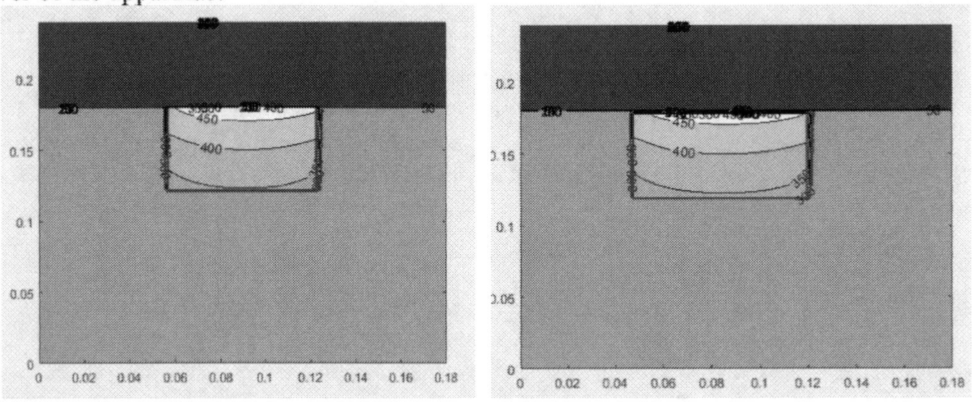

(a) (b)

Fig. 2. (a) The temperature field in the "ice-cryobot" system with a width of the last 0.06 m. (b)The temperature field in the "cryobot-ice" system at a width of 0.084 m.

Figures 3a and 3b show the temperature distributions in the ice-cryobot system under the conditions of its movement through the ice thickness with varying altitude apparatus. The analysis of figures 3a and 3b shows that an increase in the height of the cryobot leads to less cooling of the cover of the apparatus. Accordingly, it can reasonably be assumed that for a stable movement, the cryobot must have the geometric characteristics that ensure a slow cooling of the cover.

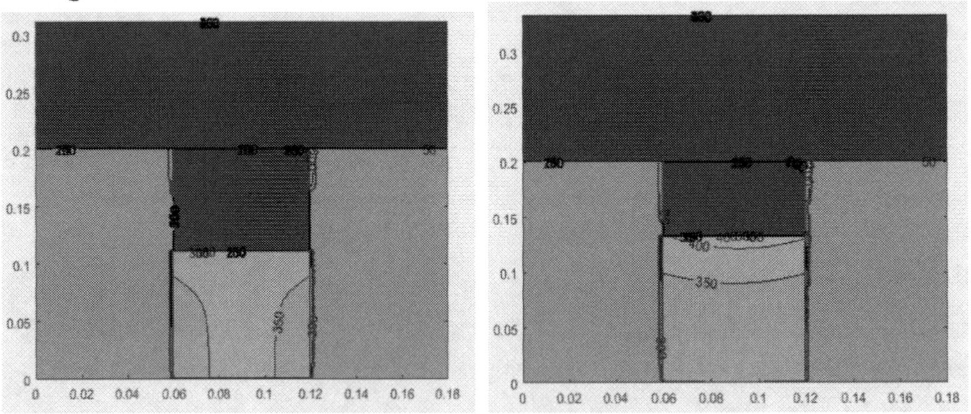

(a) (b)

Fig. 3. (a) The temperature field in the "cryobot-ice" system at a cryobot height of 0.1m; (b)The temperature field in the "cryobot-ice" system at a cryobat height of 0.12 m.

5 Conclusion

According to the results of the mathematical modeling, it has been established that the processes of the phase transformations (melting of ice) during the movement of a cryobot through ice structures affect significantly the temperature fields of the ice-cryobot system. It can be noted that even at a high initial temperature (T=500 K) of the cover of the apparatus, it cools down. The results of the numerical simulation have also showed that the geometric dimensions of a cryobot have a significant effect on the characteristics and conditions of motion of the thermal drilling machine. So an increase in height leads to a significant increase in the cooling period. Accordingly, it can reasonably be assumed that, for stable drilling, the cryobot must possess the geometric characteristics providing the optimal thermal regime.

The work was supported by the scientific schools grant NSH-7538-2016.8.

References

1. N.I. Vasiliev, Problems of the Arctic and Antarctic, **76** (2007)
2. E.N. Chumachenko, D. Dunham, R.R. Nazirov, V.P. Kulagin, Vestnik of the Nizhny Novgorod University named after. H, I. Lobakogo, **6** (2011)
3. P. Weiss, K.L. Yung, Ng T.C., Planetary and Space Science, **56** (2008)
4. P.G. Talalay, Nature. **9** (2006)
5. G.V. Parfenova, E.N. Chumachenko, New information technologies in automated systems, **13** (2010)
6. A.S. Ovcharova, Computational technologies, **4** (1999)
7. O.N. Korolyova *Dynamic adaptation in the multi-front problems of Stefan.* (Abstract, 2006)
8. J.Jonsson, J. Sundqvist, H.Nguyen, Acta Astronautica, **79** (2012)
9. L. Zelenyi, O. Korablev, M. Martynov, Advances in Space Research, **48** (2011)
10. S.V. Syrodoy, N.Y. Gutareva R.I. Taburchinov, MATEC Web Conf. **72** (2016)
11. S.V. Syrodoy, N.Y. Gutareva, R.I. Taburchinov, MATEC Web Conf. **72** (2016)
12. S.V Syrodoy, N. Y. Gutareva., K.A. Bugaeva, MATEC Web Conf. **72** (2016)

Review of coal-water fuel pulverization technology and atomization quality registration methods

Andrey Zenkov[1,*], *Kirill* Larionov[1], *Vladimir* Gubin[1]

[1]National Research Tomsk Polytechnic University, 634050 Tomsk, Russia

Abstract. Possibilities of coal-water fuel application in industrial power engineering are considered and described. Two main problems and disadvantages of this fuel type are suggested. The paper presents information about liquid fuel atomization technologies and provides data on nozzle type for coal-water fuel pulverization. This article also mentions some of the existing technologies for coal-water slurry spraying quality determination.

1 Coal-water fuel

Technologically advanced countries are conducting efforts to improve coal fuel utilization technologies to provide an increase of their ecological and economical parameters, as well as reliable and safe power equipment operation [1]. One of the advanced solutions in energy industry is development and implementation of coal-water fuel technology (CWF) [2].

CWF represents composite liquid fuel: mixture of pulverized coal and water with or without surfactants. It is capable of replacing gas and oil at boiler stations and TPPs [3].

CWF technology is not only a way of coal transportation through pipelines, but also a way of coal sludge recycling. These sludges are now recycled as waste, but they can be used for suspension preparation and used as fuel at power generating facilities. Moreover, moistened sludges are actually a semi-finished product for CWF preparation, which reduces cost of slurry production. It would be more economically advantageous to use coal-water fuel prepared from sludges directly on coal-mining fields to provide their auxiliary power. However, due to convenience of CWF transporting (in tanks or pipelines), it is possible to use this technology at energy facilities.

It is also important to note ecological aspect of sludge accumulation. There is already huge amount of waste in places where coal is extracted and enriched, and this figure continues to grow every day. For example, the Kemerovo Region (95.5 thousand square kilometers) occupies the first place in Russia in terms of industrial waste generation. More than 700 thousand tons of solid wastes are produced only in Kemerovo annually, most of which are sludges from coal industry. Area of disturbed land near the city is 500 hectares. Thus, processing of coal slurries will allow using them as raw materials, reducing the

*Corresponding author: andreyzenkov@mail.ru

amount of waste to a minimum. Clearance of the areas occupied today by this type of waste will allow reducing man-caused impact on the environment and amount of dust in the air [4].

The main advantages of coal-water fuel (comparing to coal) are:

- high burnout degree – up to 99%;
- possibility to use both high-quality and low-grade coal;
- large emission improvement of nitrogen oxides, carbon monoxide and fly ash;
- utilization of coal-preparation plants wastes (coal slack);
- fire and explosion safety;
- convenience of transportation (by pipe line, in tankers) and storage (no dusting).

Besides CWF application for direct combustion in boilers, there is a possibility of its gasification and further implementation of the obtained synthesis gas for energy production [5]. For example, it can be used at Combined Cycle Gas Turbine Plant with triple cycle in extreme north regions [6].

However, considering advantages listed above, CWF technology has some unsolved problems which influence its application in energy industry. First of all it concerns nozzles operation [7], which is affected by initial fuel metamorphic properties, granulometric and concentration compositions because an increase of coal part in CWF leads to significant increase of viscosity. It, in turn, causes combustion efficiency reduction [8]. It is proved [9] that viscosity hinges on coal concentration in CWS. Consequently, it is necessary to decrease coal concentration to obtain suspension with an acceptable viscosity, which leads to CWF calorific value decrease. Therefore, the main obstacle of coal-water fuel common usage is its rheological properties (viscosity and sedimentation stability).

2 Existing technologies of coal-water fuel atomization

There are many different technologies for pulverization of liquid fuel including coal-water slurry. They are hydraulic, mechanical, pneumatic, acoustic atomization and pulverization with radial-spray nozzles. Each of these methods is divided into several subtypes [10].

Atomization efficiency is strongly influenced by correct choice of pulverization device. It is believed, that it is viable to implement mechanical and pneumatic nozzles for coal-water fuel atomization because this type of fuel is highly viscous. Now, combination of these two nozzle types is applied for CWF injection into a combustion chamber – pneumatic-mechanical (fig. 1).

These injectors operate with high fuel pressure in nominal mode, but at low flow rate – with supply of spray agent (air). The initial spraying is conducted in mechanical nozzle, and then additionally with compressed air, which is fed through the holes of swirler. In most designs, tangential speed is imparted to the fuel. Atomizer is constructed by one of the simplest schemes of centrifugal nozzles: with circular or rectangular input channels, with distributing washer or without it, etc. Such injectors are divided into two types: with external and internal interaction of flows [11].

However, due to CWF unstable rheological characteristics and presence of coal particles in its composition these atomizers are exposed to erosive wear, which in turn leads to necessity of this technology improvement.

3 Methods of atomization quality analysis

Stability and completeness of fuel combustion in combustion chamber are characterized by degree of its atomization and perfection of gas flow aerodynamics inside the chamber. These qualities provide reliable conditions for ignition and combustion of spray cone, reliable mixing of fuel and oxidizer and temperature distribution, as well as optimal conditions for burning fuel throughout the chamber. Quality of coal-water suspension atomization in combustion chamber depends on fuel properties and design of atomizing device [12].

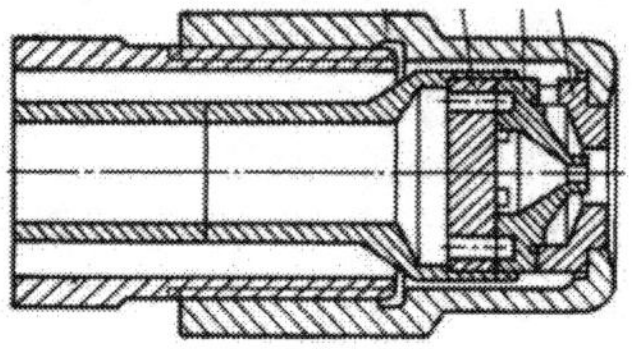

Fig. 1. Pneumatic-mechanical atomizer.

It is pointed out in work [13] that coal-water fuel combustion differs significantly from pulverized coal burning process.

In [7], it was suggested that when coal-water suspension is sprayed, a polydisperse droplet stream is formed which contains only carbon particles “freed” from liquid phase due to high speed and drops of coal-water fuel consisting of fine coal particles and liquid phase. At the same time, the amount of coal particles in CWS can reach 25-30%, depending on particle size distribution.

According to [12], spraying of coal-water fuel with air or steam is conducted in two stages. In the first stage, when CWS and atomizing agent mixing occurs in the nozzle, jet of suspension is crushed due to kinetic energy of the latter. At the second stage, crushing of CWS drops happens due to resistance forces of surrounding gaseous medium, velocity of which is many times less than speed of droplet. As a result of dynamic effects, droplets are flattened and torn into smaller ones.

Paper [7] represents experimental results on CWF pulverization with pneumatic-mechanical atomizer. Special test-bench was created for it (fig. 2). CWF pulverization was conducted onto a screen with tray. Then configuration of spray cone was determined: it was intercepted by aluminum platen; atomization quality was defined by the print on the platen.

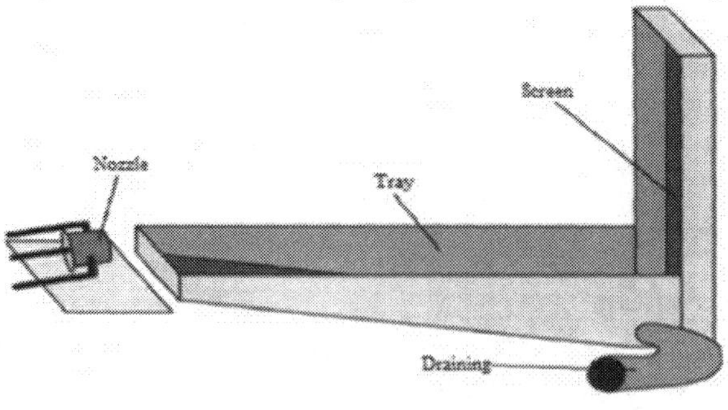

Fig 2. Test bench for coal-water fuel pulverization.

However, this method of atomization quality registration is obsolete and inexact. Nowadays, there are new methods for shaded flows atomization analysis.

One of them is PIV-method (Particle Image Velocimetry). Small particles, called tracers, which are placed in the flow, are used in this method. Their speed and direction are defined by a laser and a high speed camera [14].

Daviault et al. [15] used laser-Doppler anemometry (LDA) to determine quantitative characteristics of pulverization. LDA is noncontacting measurement method of direction and velocity of suspended particulate matters in a liquid or gas flow. It is based on two convergent laser beams directed at the flow so that they intersect in the flow and form control measurement area. Interference fringes are formed in this area (fig. 3). The distance between them can be calculated. Particles intersect these fringes and produce light pulses, the time of which is registered. Then, knowing distance and time, velocity of particles in the flow is calculated [16, 17].

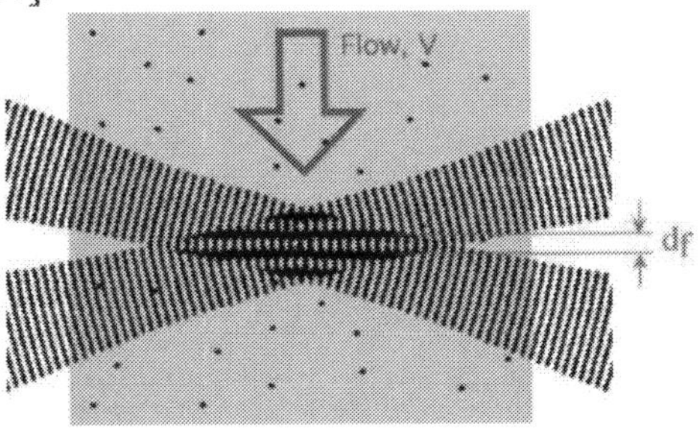

Fig 3. Control measurement area with interference fringes.

4 Conclusion

Conducted review showed that application of coal-water fuel in power engineering is a promising direction. However, there are several unsolved issues associated with CWF production, storage and pulverization. Therefore, one of the most urgent goals in this area is development of more effective atomizers, as well as evaluation and selection of CWF production method in each specific case due to nonuniformity of coal characteristics even within one grade.

The research has been accomplished with financial support of the Russian Federation Ministry of Education and Science within implementation of the project N 13.7644.2017/BCH.

References:

1. B.M. Vorobiev, Bulletin of Russian academy of sciences, **11,** 65 (2011) [in Russian]
2. G.S. Hodakov, Therm. Eng. **54**, 36 (2007)
3. Y. Wei, J. Wang, Fuel Proc. Tech. **143**, 18 (March 2016)
4. E.S. Zlobina, A.V. Papin, A.Yu. Ignatova, IV International Scientific and Practical Conference, 213 (2015) [in Russian]
5. K.V. Slyusarskiy, K.B. Larionov, V.I. Osipov, V.E. Gubin, A.A. Gromov, Fuel **191**, 383 (2017)
6. N. Galashov, S. Tsibulskii, A. Matveev, V. Masjuk, EPJ Web Conf. **110**, 01019 (2016)
7. U.A. Senchurova, V.I. Murko, V.I. Fedyayev, D.A. Dzyuba, E.M. Puzyrev, Bulletin Tomsk polytech. univ. **312**, 37 (2008) [in Russian]
8. M.P. Baranova, V.A. Kulagin (Siberian Federal University, 2011) [in Russian]
9. K.B. Larionov, A.V. Zenkov, S.A. Yankovsky, A.A. Ditc, IFOST 2016. 7884323, 568 (Novosibirsk, 2017)
10. D.G. Paji, V.S. Galustov, *Atomizers of liquid* (Chemistry, Moscow, 1979) [in Russian]
11. L.V. Kulagin, M.Ya. Moroshkin, *Nozzles for spraying heavy fuels* (Mashinostroenie, Moscow 1973) [in Russian]
12. U.A. Senchurova, E.V. Zarechneva, Inf. tech. **122** (2012) [in Russian]
13. B.N. Smetannikov, Investigation of ignition and combustion of coal-water suspension drop, **556** (1969) [in Russian]
14. R. Ben Haj Slama, B. Gilles, M. Ben Chiekh, J-C. Bera, Ultrasonics, **76**, 217 (2017)
15. S.G. Daviault, O.B. Ramadan, E.A. Matida, P.M. Hughes, R. Hughes, Fuel, **98**, 183 (2012)
16. J.J. Chinn, D Cooper., A.J. Yule, G.G. Nasr, Heat Mass Trans. **52**, 2037 (2016)
17. S.V. Poplavski, V.M. Boiko, A.U. Nesterov, ICMAR 2016. **1770**, 030016 (2016)

MATEC Web of Conferences **110**, 01092 (2017) DOI: 10.1051/matecconf/201711001092
HMTTSC-2017

The ignition and burning of mixed compositions with heavy fillers

Alexander Zhukov, *Valeriy* Kuznetsov, and *Sergey* Timoshensky*

National Research Tomsk State University, Tomsk, Russia

Abstract. The problem of increasing the efficiency of metal components in the composition of high-energy fuel compositions by complete or partial replacement of aluminum by metal borides is considered. In this paper, we present and analyze the experimental results of the investigation of the ignition and combustion of mixtures based on a hydrocarbon and active binder with inorganic oxidants and energy-intensive fillers.

Introduction

A promising trend of solution to the problem of increasing the efficiency of the metal componentsas part of high energy fuel compositions is seen as full or partial substitution of aluminum for high-energy borides of metals. In this case, the high energy characteristics of metallic fuel compositions is achieved due to the increased heat of combustion components and by increasing the chemical conversion rate of metals into oxides [1, 2].

It is known [2] that the heat of combustion of boron is almost twice the calorific value of aluminum and the oxidation onset temperature of boron is much lower than that of aluminum. However, due to the specific properties of boron oxide B_2O_3 (high value of boiling temperature and viscosity of liquid oxide) the oxidation degree of elemental boron at its introduction into the fuel is small. Even in oxygen-containing environment, the combustion efficiency of boron is only 26,5 %(3). One of the solutions to implement high heat of combustion of boron in the fuel composition is its use in the form of borides of aluminum [3, 4].

Borides of aluminum has low binding energy between the atoms Al and B (67 kJ/mol). Therefore, they can be considered as alloys, and in the preflame area, they will decompose into Al and b, and separately enter into a chemical reaction with oxygen. Aluminum that forms high temperature combustion products, will contribute to the efficient combustion of boron [5, 6, 7].

* Corresponding author: Timosh13@mail.ru

Results of calculations

In the present work the results of experimental studies of ignition and burning of mixed compositions on the basis of hydrocarbon (HB) and active (AB) binder with inorganic oxidizing agents and heavy fillers (Al, B, AlB_2, AlB_{12}, TiB_2) are presented.

As a hydrocarbon binder (HB) the divinyl rubber solution in the petroleum oil in a ratio of 20/80 (SKDM-80) was used. Active binder (AB) consists of a solution of the methylprednisolone copolymer in eutectic mixture of nitrate ester with nitramines (NWT-ASPECTS) [8]. Oxidizers were ammonium perchlorate bidispersed fraction (60% size 165-315 microns and 40% less than 50 microns) and ammonium nitrate with a particle size of 165-315 microns.The coefficient of excess for oxidant of model compounds on HB is 0.5, and on Al – 0.64. Samples of cylindrical shape were produced by the method of through passage pressing. The diameter of samples is 10 mm. Thecontent of components in the model formulations is given in table 1

Table 1. The compositions of energy-intensive compositions.

Composition	The content of components, wt.%				
	PHA	**ON**	**SKDM-80**	**NWT-ASPECTS**	**Metal, mark**
Al	$69.0_{(b/d)}$		16.0		15.0 Al
A2	$74.0_{(b/d)}$		11.0		15.0 B
A3	$71.4_{(b/d)}$		13.6		15.0 AlB_2
A4	$73.2_{(b/d)}$		11.8		15.0 AlB_{12}
A5	$70.2_{(b/d)}$		14.8		15.0 TiB_2
B1	21.3	49.8	13.9		15.0 Al
B2	22.5	52.5	10.9		15.0 B
B3	21.9	51.3	11.8		15.0 AlB_2
B4	22.5	52.3	10.0		15.0 AlB_{12}
B5	21.6	50.4	13.0		15.0 TiB_2
C1	15	35		20	30Al
C2	18	42		25	15 B
C3	15.6	36.4		23	25 AlB_2
C4	17.4	40.6		25	17 AlB_{12}
C5	15.9	37.1		23	24 TiB_2

Studies of the ignition of model mixed compositions were performed in air at atmospheric pressure through heat flux radiation exposure from a xenon lamp in an optical furnace "Uranium -1" [9].

The radiation from the xenon lamp is focused by an elliptical mirror into a spot with a 12 mm diameter. The installation provides at least 95% light flux stability in the radiation intensity range $Q = (5 \div 400)$ W/cm^2. The intensity of thermal radiation was measured with a copper calorimeter with an error not exceeding 10%. The ignition delay time, t_{ign}, was determined by the appearance of a flame recorded by means of photodiodes on an H117/1 oscilloscope. The error in measuring the delay time of ignition of the mixture compositions did not exceed 7%.

For all compositions (Table 1), the values of the ignition delay time from the intensity of thermal radiation in the range $(15 \div 100)$ W/cm^{2}are determined.

Since in the investigated range of heat flux experimental data, presented in logarithmic coordinates, reasonably well fit to a slope line, which correspond to the exponent in the chart of $t_{ign}=aq^{-n}$, the experimental results can be conveniently represented in the form of parameters ($-n$ and t_{ign} for a given q) in table 2.

Table 2. Characteristics of ignition and combustion of the mixed compositions.

Composition	t_{ign}, s with 54W/cm^2	$\|-n\|$ $t_з=Aq^{-n}$	Q, J/cm^2 with 54 W/cm^2	$K_I=Q_{al}/Q_i$	U, mm/s	$K_I=U_I/U_{Ac}$
A1	0.33	1.61	17.8	1.00	1.06±0.01	1.00
A2	0.21	1.45	11.3	1.57	1.71±0.02	1.61
A3	0.25	1.55	13.5	1.32	1.15±0.01	1.08
A4	0.34	1.25	18.4	0.97	1.08±0.02	1.02
A5	0.55	1.36	29.7	0.60	1.21±0.04	1.14
B1	0.50	1.58	27.0	1.00	0.20±0.02	1.00
B2	0.24	1.50	13.0	2.08	1.00±0.02	5.00
B3	0.38	1.50	20.5	1.32	0.60±0.02	3.00
B4	0.39	1.45	21.1	1.28	0.33±0.02	1.65
B5	0.62	1.43	33.5	0.81	0.18±0.02	0.9
C1	0.75	1.48	40.5	1.00	1.47±0.07	1.00
C2	0.45	1.20	24.3	1.67	1.93±0.07	1.31
C3	0.44	1.27	23.8	1.70	2.40±0.20	1.63
C4	0.46	1.20	24.8	1.63	1.42±0.08	0.96
C5	0.69	1.25	31.3	1.29	0.31±0.03	0.21

In addition, the table shows the values of energy required for ignition (Q), the values of the stationary burning rate of compositions in the air at atmospheric pressure (U), the performance parameters of the boron and borides chemical activity in comparison with aluminum under the conditions of ignition (K1) and combustion (K2).

Conclusion

Analysis of experimental results showed that the direct correlation of the ranks of the compounds reactivity in terms of ignition and combustion is not observed. However, it is

DOI: 10.1051/matecconf/201711001092

impressive to see that in the investigated conditions, the boron and the borides of aluminum accelerate the process of ignition and combustion of model compositions.

This work was supported financially by the Ministry of Education and Science of the Russian Federation in terms of state order, project code 9.9036.2017.

References

1. V. N. Aleksandrov, V. M. Byatskevich, V. K. Verkholomov *Integral ramjet air-jet engines on solid fuels* (ICC Akademkniga, Moscow, 2006) [in Russian]
2. M. K. King, Journal of spacecraft and Rockets, **19**, 4 (1982)
3. J.Lui, J.Xi, W. Yang, Acta Astronaut **96**, 89 (2014)
4. S. Xu, Yu. Chen, X. Chen, D. Wu, and D. Lui, Combust. Explos. Shock Waves. **52**, 3 (2016)
5. G. Young, K. Sullivan, M.R. Zachariah, K. Yu, Combust. Flame **156**, 322 (2009)
6. D. A. Yagodnikov, A. V. Voronetskii, and V. I. Sarab'ev, Combust. Explos. Shock Waves. **52**, 3 (2016)
7. A.B. Vorozhtsov, *Novel micro-and nanofuels: Production characterization, and applications for High-Energy materials* (Springer International Publishing, Switzerland, 2017)
8. V. F. Komarov, N. I. Popok, G. V. Sakovich, *Fundamental and applied problems of technical chemistry* (Science, Novosibirsk, 2011)
9. V. A. Arkhipov, S. E. Bondarchuk, A. G. Korotkih, V. T. Kuznetsov, Combustion explosion, and shock waves, **48**, 5 (2012)

DOI: 10.1051/matecconf/201711001093

Optimization of the stand for test of hybrid rocket engines of solid fuel

Nikolay Zolotorev*, *Valeriy* Kuznetsov , and *Aleksey* Konovalenko

National Research Tomsk State University, Tomsk, Russia

Abstract. In the paper the laboratory experimental stand of the hybrid rocket engine of solid fuel to study ballistic parameters of the engine at burning of high-energy materials in flow of hot gas is presented. Mixture of air with nitrogen with a specified content of active oxygen is used as a gaseous oxidizer. The experimental stand has modular design and consists of system of gas supply, system of heating of gas, system for monitoring gas parameters, to which a load cell with a model engine was connected. The modular design of the stand allows to change its configuration under specific objective. This experimental stand allows to conduct a wide range of the pilot studies at interaction of a hot stream of gas with samples high-energy materials.

Introduction

Experimental study of interaction of hot stream of gas with samples of high-energy materials (HEM) is used for projection of perspective rocket engines, and for experimental verification of the existing theoretical models and calculation data executed on their basis. In the course of experiment it is important to control not only thermodynamic and gas-dynamic parameters of process (temperature, pressure, gas consumption) but also ballistic parameters (rocket engine thrust) [1].

1 Experimental stand

To carry out experimental studies based on the analysis of literature sources, the design of a laboratory stand for a hybrid rocket engine solid fuel (HRESF) was developed. Sample of HEM as is used the studied material with the central channel on the centre of which the gaseous oxidizer arriving from the block of preparation of gas mix is blown is used [2, 3]. High rates of motion of the combustion products of the charge assume that not all fuel has time to react in the combustion chamber, in this regard HRESF is complemented by an afterburner [4–8].

The scheme of the experimental stand for research of burning of samples of high-energy materials in the hot gas stream is shown in figure 1.

* Corresponding author: nikzolotorev@mail.ru

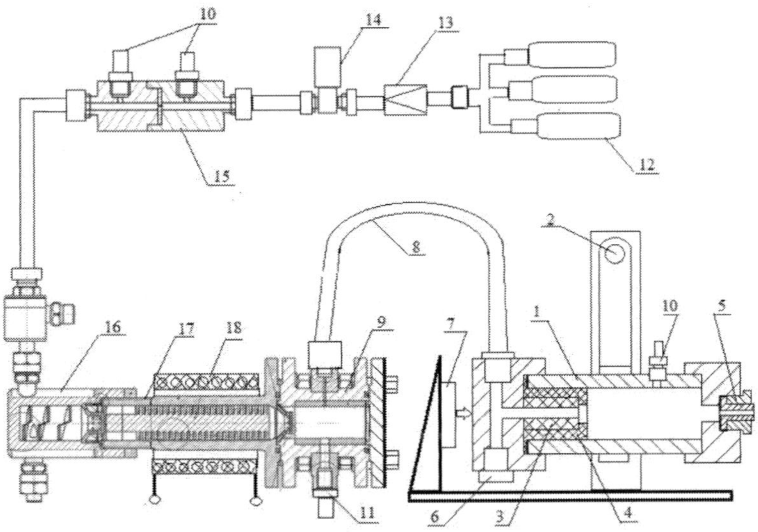

Fig. 1. The scheme of the experimental stand with heating of air. 1 – model engine; 2 – pendular suspender of the stand; 3 – sample of HEM; 4 – container for sample HEM; 5 – nozzle; 6, 10 – pressure sensors; 7 – sensor thrust; 8 – flexible metal sleeve; 9 – antechamber; 11 – temperature sensor; 12 – cylinder with gas; 13 – reducer; 14 – electropneumatic valve; 15 – differential flowmeter; 16 – mixer; 17 – oven of heating of gas; 18 – inductor.

The experimental stand consists of two blocks: the block of preparation of gas mix and the measuring block executed in the form of the model rocket engine. The model rocket engine is installed at the stand for measuring the thrust. Air was used as the working gas in the experimental stand. This experimental stand allows using different mixes of gases by means of the mixer. Heating of gas is carried out in the labyrinth furnace by means of the 40 kVA induction heater (furnace heating temperature to 900 K). Gas parameters at the exit of the block of preparation of gas mix (in the antechamber) made: pressure is up to 0.3 MPa, temperature is up to 600 K, gas consumption 5÷30 of g/s.

The modular design of the stand allows changing its configuration under specific objective. For receiving hot gas stream with temperature over 550 K in installation use of solid-fuel gas generator which is established instead of the induction heater is provided. The scheme of the block of preparation of gas on the basis of solid-fuel gas generator is presented on figure 2.

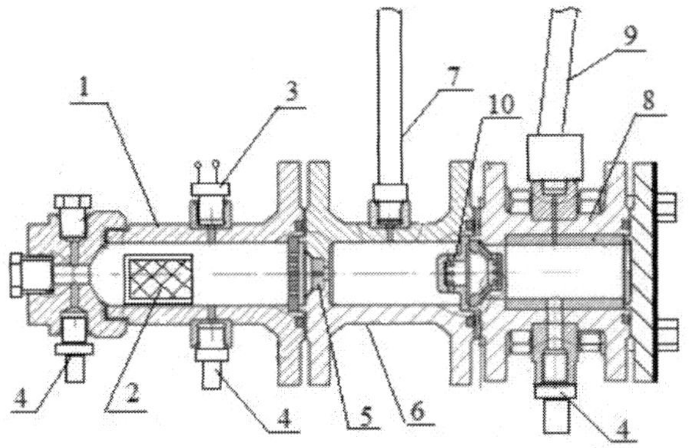

Fig. 2. Gas generator of solid-fuel with the camera of mixing and the antechamber. 1 – gas generator body; 2 – charge of solid-fuel; 3 – igniter; 4 – pressure sensor; 5 – nozzle; 6 – mixing camera; 7 – air supply; 8 – antechamber; 9 – supply gas mixture to a model engine; 10 – diaphragm.

DOI: 10.1051/matecconf/201711001093

In this case between gas generator and the antechamber the camera of mixing of gases (figure 2) is installed [9, 10]. Air for obtaining necessary temperature of gas mix and the set content of oxygen, moves in the mixing camera across gas stream from gas generator and mixes up, being filtered through system of diaphragms. As fuel for gas generator the charge of gunpowder of brand "H" was used. Temperature of products of burning of gunpowder "H" is 2368 K. The gas rate made of gas generator 20÷80 of g/s with depending on combustion chamber pressure.

The studied samples HEM in the form of cylindrical checker height 50mm diameter 40 mm and the internal channel 10 mm were placed in the textolite container in front part of the model engine coaxially with the channel with a diameter of 10 mm of supply of gas mix. Burning of sample of HEM is carried out on the internal channel. The engine rigidly fastened in the pendulum suspender of the stand for measurement of thrust and front cover leaned on the thrust measurement sensor. Both blocks of the stand connected by means of the flexible metal pipeline of the bellows type providing free swing of pendulum.

Measurement of pressure at the stand was taken by means of tensometric LX-412 and MBS-3000 sensors. Intrinsic error of measurement of pressure for LX-412 sensors no more than 0.8 %, and for MBS-3000 sensors no more than 0.5 %. Temperature was taken by thermocouples of BP 5/20 with diameter of electrodes of 0.35 mm. The thrust was measured using force sensors DST1909 and MLB20. The gas rate was measured on differential pressure on the calibrated diaphragm. As data acquisition system digital unit were used: tensometric AT1-8 amplifier and programmable SIMATIC S7-1200 controller. The tensometiric AT1-8 amplifier has sampling rate to 5 kHz, measurement error of rated input signal no more than 0.2 % at nonlinearity of transformation no more than 0.1 %. Accuracy of transformation of analog signal for the programmable SIMATIC S7-1200 controller makes 3 %. Signals from measuring apparatuses were transmitted to the personal computer and registered in the form of spreadsheets.

2 Results of experimental study

The temperatures and pressure received during experiment are shown in figure 3.

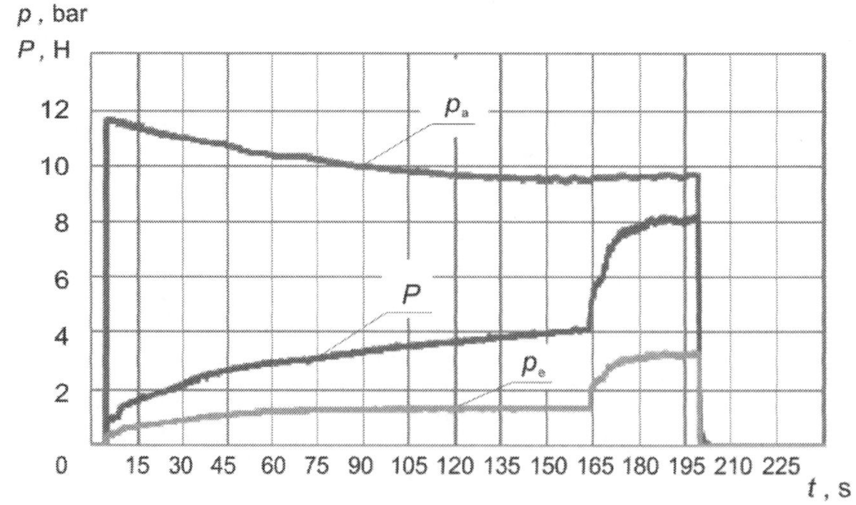

 DOI: 10.1051/matecconf/201711001093

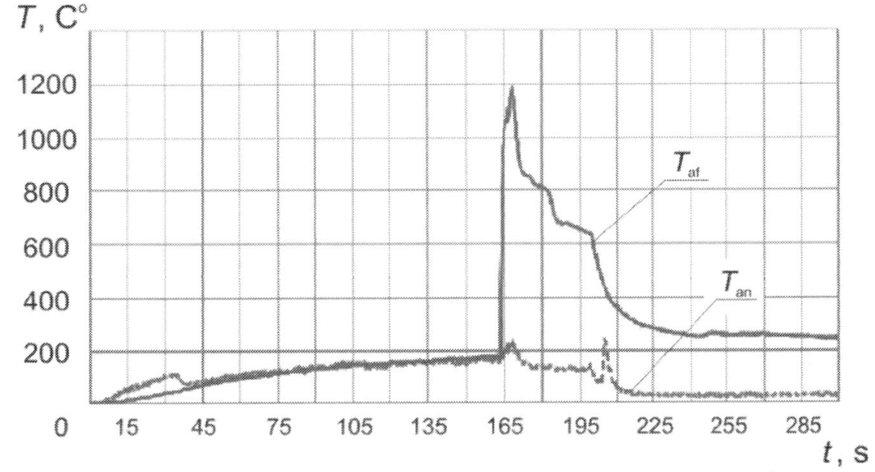

Fig. 3. Typical oscillograms of the process experimental study.

The data obtained from pressure sensors where p_a – pressure of air, P – thrust of the model engine, p_e – the chamber pressure of afterburner of the model engine are presented on figure 3. The data obtained from thermocouples T_{af} – gas temperature in afterburner of the model engine, T_{an} – gas temperature in the antechamber are presented on figure 3.

At air supply, temperature and gas pressure in afterburner of the model engine, and also its thrust gradually increase. After ignition of sample of HEM gas parameters and thrust of the engine sharply increase in the combustion camera. In process of burning out of heated-up layer parameters of gas and thrust of the model engine aim at quasistationary state. At air supply shutdown burning of sample of HEM stops both parameters of gas and thrust of the model engine sharply decrease. The appearance of the sample of HEM before and after blowing with hot gas stream, shown in figure 4.

Fig. 4. Sample of HEM before tests at the stand (at the left) and after tests.

Conclusions

On the basis of the developed outline design documentation the HRESF laboratory stand is made. Test experimental studies of samples of HEM at the stand are conducted. Taking into account results of experimental studies testing of the software for calculation intra chamber and the HRESF power traction parameters is executed. The HRESF this laboratory stand allows providing carrying out experimental studies under different conditions of interaction of more hot gas stream with HEM samples.

This work was supported financially by the Ministry of Education and Science of the Russian Federation in terms of state order, project code 9.9036.2017.

References

1. V.T. Volkov, D.A. Yagodnikov, *Studies and bench testing of solid-propellant rocket engines* (BMSTU Publ., Moscow, 2007)
2. D.A. Yagodnikov, A.V. Voronetskii, V.I. Sarab'ev, J. Combust. Expl. Shock Waves, **52**, 300 (2016)
3. V.A. Arkhipov, S.S. Bondarchuk, A.B. Vorozhtsov, A.S. Zhukov, B.V. Pevchenko, L.A. Savelyeva, Patent Rus. No 2014149746 (2015)
4. J. Liu, D. Liang, J. Xiao, B. Chen, Ya. Zhang, J. Zhou, K. Cen, J. Combust. Expl. Shock Waves, **53**, 55 (2017)
5. M.A. Karabeyoglu, D. Altman, B.J. Cantwell, J. Propul. Power, **18**, 610 (2002)
6. M.A. Karabeyoglu, D. Altman, B.J. Cantwell, J. Propul. Power, **18**, 621 (2002)
7. G.A. Marxman, M. Gilbert, Internat. Symp. on Combust., **9**, 371 (1963)
8. V.A. Arkhipov, S.S. Bondarchuk, A.B. Vorozhtsov, L. De Luca, A.G. Korotkikh, L. Galfetti, G. Colombo, A. Bandera, Izv. Vyssh. Uchebn. Zaved., Fiz., **9**/2, 3 (2007)
9. G.P. Sutton, O. Biblarz, *Hybrid propellant rockets* (Rocket Propul. Elements. John Wiley & Sons, New York, 2001)
10. L.T. De Luca, *Hybrid rocket engines* (In: Energetic problems in aerospace Propulsion, Milano, 2009)

Increase of energy efficiency in proportional adjusting of flow rate in the boiler circuit

Pavel A. Artamonov[1,*], *Nikolai I.* Kurilenko[1], and *Gennady Ya.* Mamontov[2]

[1] Tyumen Industrial University, 625000 Tyumen, Russia
[2] National Research Tomsk Polytechnic University, 634050 Tomsk, Russia

Abstract. The article presents the results of theoretical studies in the field of the boiler circuit operating modes for the boiler rooms operating by the independent heat supply scheme. The 3D model of a boiler circuit for a boiler room with 3 MW rated output was developed, based on which there was made an estimation of the boiler pump performance indicators. There is proposed a method for reducing energy costs for the operation of the pumping equipment of the boiler circuit.

1 Introduction

The strong policy in the field of increasing energy efficiency in the world as well as in Russia has led to the strict requirements for energy facilities. The major priority is not the volume increase of the energy complex production, but the modernization of the country's energy sector [1]. At the same time over the past few years there has been an increase in the number of individual boiler rooms [2]. The reason for it is the studies devoted to assessing the positive effect of the decentralization of the heat supply systems (such as [3], [4], etc.). While the number of boiler rooms is increasing and the cost of energy resources is going up the development of ways to increase the heat production efficiency is an urgent task of theoretical and practical research.

2 Problem statement and solution methods

This article presents the result of an analytical evaluation of the effectiveness of the control method which is based on a change of the flow rate in the boiler circuit in proportion to the heat load of consumers connected through an independent heat supply scheme.

Generally the release of heat energy in the boiler circuit can be changed by adjusting the combustion parameters (flame modulation), by changing the temperature and flow rate of the heat transfer fluid. Consider the basic scheme of the boiler house, presented in Figure 1.

* Corresponding author: P.Artamonovv@yandex.ru

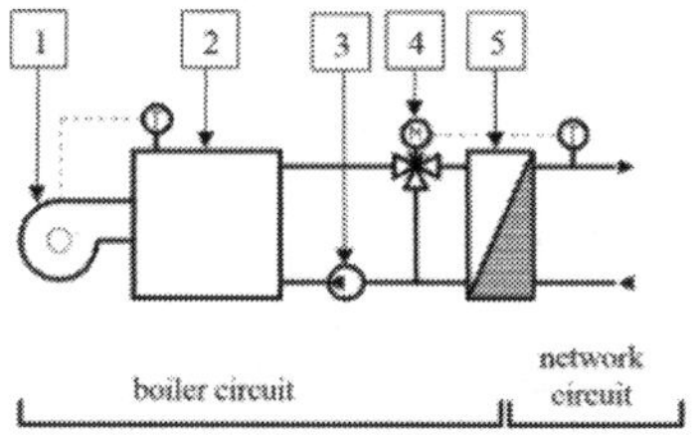

Fig.1. The schematic diagram of the boiler circuit operation.

The circulation of the heat transfer fluid through the boiler (pos. 2) and heat exchanger (pos. 5) is carried out by the boiler pump (pos. 3), and the amount of the heat supplied to the consumer and the temperature control of the network circuit according to the temperature chart is carried out by the three-way valve (pos. 4). This scheme of the boiler room operation is generally used during design and construction. The activation of the boiler pump (pos. 3) occurs when the burner starts (pos. 1). At the same time regardless of the capacity of the burner, the total flow of the heat transfer fluid through the boiler remains conditionally constant because the pump operates at a constant speed of the impeller as determined by the design of the electric motor. The flow deviation can be caused only by the displacement of the operating point according to the pump characteristic as a result of the change in the hydraulic resistance in the boiler circuit (changing the position of the three-way valve, deposits in the pipelines, shutting off the stop valves, etc.) [5]. The considered operation principle of the boiler pump is simple and reliable for the automation of the technological cycle. However, this solution is not optimal for the energy efficiency.
In order to evaluate the efficiency indicators consider the boiler circuit assemble model designed according to the scheme in Figure 1 (Figure 2).

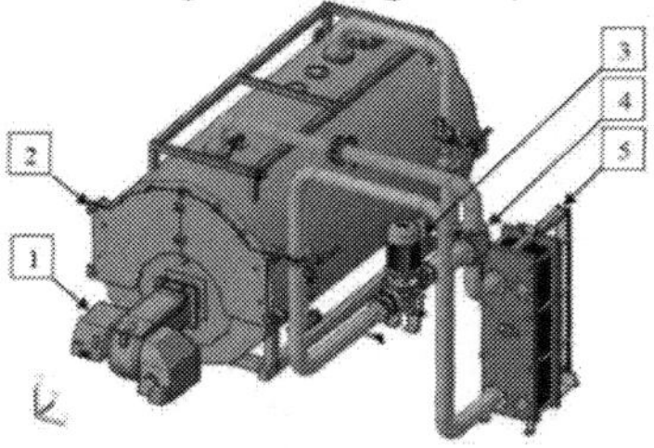

Fig.2. The model of the boiler circuit assemble.

The designation of the equipment corresponds to the numbering shown in Figure 1. The power (Q) of the considered circuit is 3 MW (boiler model KSV-3.0 manufactured by MAXTERM (pos.2)). The calculated temperature difference at the boiler inlet and outlet is 25oC. The temperature at the boiler outlet is 105 ^{0}C. The heat carrier is water with a heat capacity of 4.208 kJ / kg·^{0}C and a dynamic viscosity of $0.3078 \cdot 10^{-3}$ Pa.s at an average temperature. The calculated flow rate of the heat carrier under the given conditions is 102.7 m^3 / h. There were used a heat exchanger NN-62-16 / 2-78-TKTM40 (pos. 5) manufactured by Ridan and a three-way valve DR150GFLA (pos. 4) manufactured by Honeywell. Rotary shutters were used as shut-off valves. In order to select the pump (pos. 3) the calculation of the pressure loss was made considering all local resistances and equipment characteristics. When modeling the hydraulic mode the pressure loss due to friction by the length was determined by the Darcy-Weisbach equation, the losses in local resistances were determined by the Weisbach equation. The losses on the equipment (dampers, return and three-way valves, air collectors, etc.) were determined by the capacity factor specified in the manufacturer's technical documentation. The estimated hydraulic losses during the operation on the rated power mode amounted to 7.4 mm H_2O. The unit was equipped with the pump IL100/170-3/4 manufactured by Wilo with the following characteristics at the operating point: consumption – 105.5 m^3/h, pressure – 7.76 m^3/h, energy consumption – 3.15 kW.

3 Results and discussion

As noted earlier, while regulating the amount of heat given out to the consumer using the three-way valve (pos. 4), the total hydraulic resistance of the system changes and the operating point of the pump shifts. In order to assess the effect of heat consumption on the electric power of the pump we will calculate the operating parameters of the boiler circuit model when the heat load of consumers varies from 300 kW to 3000 kW in 300 kW steps. The value of the electric power consumption of the boiler pump (P1) depending on the heat load of the consumers, expressed as a percentage of the installed heat output (Q), is shown on Figure 3 (dotted line).

The results of the calculation show that the consumed electric power of the pump varies from 3.35 kW to 3.15 kW when the heating power of the circuit varies from 10% to 100%. The reduce of energy consumption occurs due to the displacement of the operating point because of the change in the hydraulic resistance when the three-way valve is adjusted (pos. 4). Let us consider the relation between the generated power output (Q), the electric power consumption of the pump (P1) and the heat load expressed in percentage (Figure 4, dotted line).

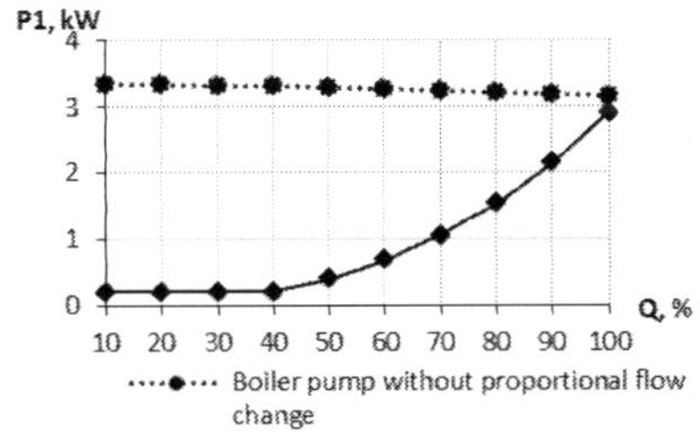

Fig.3. Consumed electrical power of the boiler pump, depending on the heat load of consumers.

It is obvious from the presented dependence that when the heat load decreases this ratio decreases as well. Therefore more electricity is consumed to produce each one megawatt of heat which means that the operating costs of the boiler room are increasing. The system reaches the most efficiency when operating at nominal heat output (at 100%). However, the graph of consumption of heat energy has a curvilinear characteristic during the year and the concept of peak loads is normatively established. Analyzing this graph under various climatic conditions it can be concluded that most of the time the boiler room does not work at the nominal power. Thus, when the heat load decreases the technical solution with the installation of a boiler pump operating at a constant engine speed becomes inefficient.

To improve the energy efficiency of the boiler circuit it seems reasonable to change the performance of the pump, particularly the adjusting the flow rate of the pump in proportion to the heat load of consumers. Changing the characteristics of the pump is possible by varying the motor speed.

Let us calculate the efficiency of the boiler circuit with a proportional change in flow rate. At the same time it is necessary to consider that the minimum flow rate of the boiler must be at the 40% level of the flow rate determined at the nominal power and the difference between outlet and inlet temperature of the heat carrier to be 25°C according to the boiler's specification. For example consider the minimum flow limited at 41.1 m^3/h. Thus, if the heat output varies from 1200kW to 3000 kW the flow through the boiler will vary from 41.1 m3/h to 102.7 m3/h respectively, and if a heat output is less than 1200 kW, the flow through the boiler will be 41.1 m^3/h. Changing of the pump's flow characteristic will be carried out by a frequency converter according to a control algorithm based on the dependence of the flow rate from the motor speed: the boiler room controller calculates the required value of the boiler's heat output based on the parameters of the network circuit heat carrier and the graph of the temperature dependence of the supply pipeline at the heat generating unit output from the outside air temperature.

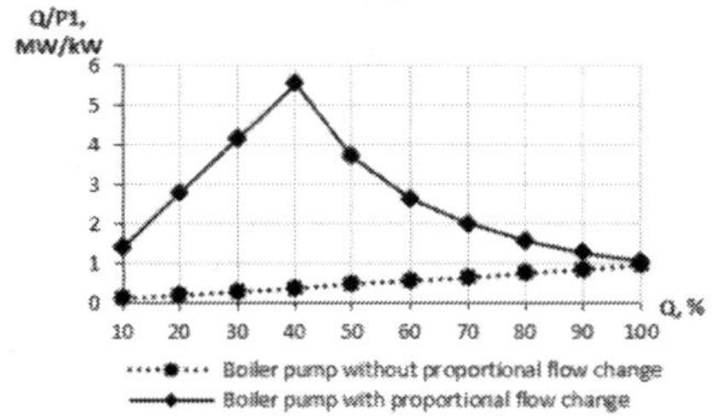

Fig. 4. Consumed electrical power of the boiler pump, depending on the heat load of consumers.

According to the flowmeter installed on the supply line of the boiler circuit and the temperature sensors installed on the supply and return lines of the boiler circuit, the controller calculates the actual value of the boiler's heat output. The difference between the required and actual values should be close to or equal zero. If the difference is less than zero, i.e. the value of the generated heat energy is more than required, the controller sends a signal to reduce the rotational speed of the pump motor. In the case when the difference is greater than zero the frequency converter should receive a signal to increase the frequency. In order to avoid an abrupt change in flow, the control logic must include a proportional-integral-differentiating (PID) regulator [6, 7]. In parallel with this, there should be a smooth regulation of boiler heat output and their cascade connection. The required temperature difference on the sensors installed on the supply and return lines of the boiler circuit is set constant. Otherwise, the pumps will compensate for the low temperature drop by increasing the flow rate, which will lead to over-consumption of electricity. This algorithm for controlling the boiler circuit for a given temperature difference on the supply and return pipelines makes it possible to change the flow rate of the heat carrier in proportion to the change in the heat load of the consumers. All other parameters, the circuit diagram (Fig. 1), the calculation model (Fig. 2) and the equipment are similar to those taken for the first calculated example. As can be seen from the data obtained (Figures 3 and 4, the solid line) the electric power consumption of the pump decreases along with the reduce of the heat output of the circuit throughout the proportionally adjusting range from 40% to 100% heat load, which corresponds to a range from 1200kW to 3000 kW (Figure 3). In this case, in comparison to the linear nature of change in the heat load the energy consumption graph in the proportionally adjusting range is an exponential. This is caused by the fact that when the flow rate decreases the pressure loss decreases as well. This decrease is not linear, because the loss of pressure in particular case is a function of the velocity squared and other nonlinear dependences. The simultaneous decrease of both the pump flow rate and the head of pump due to the rotational speed reduction leads to a decrease in the consumed electric energy [8, 9]. In the range with a constant flow rate (range from 10% to 40% (Figure 3)) there is no change in the electric power consumption. The proportional adjusting of the boiler pump flow rate has made it possible to increase the value of the ratio of the generated heat energy to the consumed electric power of the pump during the heat load decreasing. Thus, the boiler room circuit operation with proportional flow adjusting becomes highly efficient over the entire range of the heat load. Thereby, the energy consumption of the simulated boiler circuit in the boiler room operating at 40% heat output (1200 kW) was reduced by 15 times.

Let us make a calculation of the electricity consumption by the boiler circuit during the heating season for the pump control scheme with a proportional flow rate change and without it. For this purpose there was built a chart of the average daily heat load of consumers in the climatic conditions during the heating season in Tyumen city in 2015. Depending on the average daily outside temperature the average daily heat load was determined by the equation (1):

$$Q_{Oi}^{\text{day}} = Q_O \frac{t_{\text{in}} - t_i^{\text{day}}}{t_{\text{in}} - t_{\text{o}}} \tag{1}$$

where Q_{Oi}^{day} – average daily heat load of consumers during the particular day of the heating season, kW; Q_o – calculated load of the heating system for a simulated boiler circuit, assumed to be 3000 kW; t_{in} – calculated internal air temperature, assumed to be 20°C; t_{o} – the coldest five-day period temperature, assumed equal to minus 35°C; t_i^{day} – average daily temperature during the i-th day of the heating season, taken as statistical data of average daily temperature during the particular day according to the data of [10].

The electricity consumed per i-th day of the heating season was determined by the equation (2), kWh:

$$P_i = 24 \cdot P_{1i}, \tag{2}$$

where P_{1i} – electric power consumption of boiler pump under average daily heat load of consumers, kW.

Based on the calculation results of the boiler circuit operation modes shown in Figures 3 and 4, approximation equations were found to determine the electric power consumption of the boiler pump with the particular values of the determination coefficient R2.

For the calculating case at a constant electric motor rotation speed of the boiler pump the electrical power can be found by the system of equations (3):

$$\begin{cases} P_{1i} = 0.003G_y + 3.456, \text{at } 92.4 \leq G_y \leq 102.7 \text{ and } R^2 = 0.99, \\ P_{1i} = 8 \cdot 10^{-8} G_y^3 - 2.6 \cdot 10^{-5} G_y^2 - 6 \cdot 10^{-5} G_y + 3.349, \text{at } 0 < G_y < 92.4 \text{ and } R^2 = 0.99 \end{cases}, \tag{3}$$

where G_y = conditional flow rate of the boiler pump, m^3/h, is found by the equation (4) with the temperature difference Δt = 25°C:

$$G_{\text{y}} = \frac{3.6 \cdot Q_{\text{O}i}^{\text{day}}}{c \cdot \Delta t} \tag{4}$$

For the calculating case with proportional boiler pump flow adjusting the electrical power can be found by the equation (5):

$$\begin{cases} P_{1i} = 0.215, \text{at } 0 < G_{\text{y}} < 41 \text{ and } R^2 = 1, \\ P_{1i} = 2 \cdot 10^{-6} G_{\text{y}}^3 + 8.7 \cdot 10^{-5} G_{\text{y}}^2 - 3 \cdot 10^{-3} G_{\text{y}} + 0.036, \text{at } 41 < G_{\text{y}} 102.7 \\ \qquad \text{and } R^2 = 0.99. \end{cases} \tag{5}$$

The electricity consumed during the heating season will be found by the equation (6):

$$P = \sum_{i=1}^{220} P_i \quad (6)$$

where, i – ordinal number of the day during heating season.

Using the equations (1) – (6) there was determined the value of the boiler circuit power consumption for the heating season. For the calculating case of using proportional pump flow adjusting this value was 2332.9 kWh, in the case with a constant engine speed of the boiler circuit this value was 17407.4 kWh. The obtained results show that using the proportional flow adjusting allowed to reduce the energy consumption of the considered boiler circuit by almost 7.5 times which indicates a high potential for increasing energy efficiency by the proposed method for controlling boiler pumps.

4 Conclusion

The analytical evaluation of the proposed method of proportional adjusting of the flow rate has shown a high potential for reducing the energy consumption of heat generating units and becomes the most attractive for investment opportunities.

The example analyzed in this article gives an understanding of the general prospects of the proportional adjusting of the boiler circuit flow rate in boiler rooms. Despite of the high potential for energy saving the use of the proposed method should be based on a feasibility study, because the decrease of the designated boiler room heat output leads to a decrease of the absolute value of the saved resources and a change of the payback period.

The work was supported by the Russian President's grant (Scientific School project - 7538.2016.8).

References

1. *Energy strategy of Russia for up to 2035* Available at: http://minenergo.gov.ru/
2. P.A. Khavanov, *Heat sources of autonomous heat supply systems: monograph* (MGSU, Moscow, 2014)
3. P.A. Khavanov, Bulletin of MGSU, **11**, 219 (2012)
4. N.S. Oleinik, Energy saving and water treatment, **6**, 92 (2014)
5. S.V. Kolesnikov, Bulletin of the Samara State Technical University, **1**, 127 (2014)
6. J. Quevedo, T. Escobet, *Digital control: past, present and future of PID control Proceedings of the IFAC Workshop* (Terrassa, Spain, 2000)
7. K.J. Astrom, T. Hagglund, The Instrumentation, System, and Automation Society, **460** (2006)
8. V.A. Shabanov, Electrotechnical and Information Complex and Systems, **2**, 34 (2013)
9. B. Janusz, Energy, **34**, 9 (2009)
10. *Daily average air temperature in Tyumen* Available at: http://weatherarchive.ru/Temperature/Tyumen/

Development of the mathematical model of heat and mass transfer for researching the processes of evaporation of inhomogeneous water droplets

Dmitry Antonov[1], *Ivan* Voytkov[1], and *Maxim* Piskunov[1,*]

[1]National Research Tomsk Polytechnic University, 634050 Tomsk, Russia

Abstract. On the basis of experimental date the heat and mass transfer models are developed in ANSYS Fluent software package for researching the processes of evaporation of inhomogeneous water droplets. The influence of the temperature of the gases (550-850 K) on the evaporation of inhomogeneous water droplets is estimated. Times of complete liquid evaporation from free surface of inhomogeneous water droplets and explosive vaporization of water droplets at different gas temperatures are established.

1 Introduction

In the recent years, in the development of chemical and energy technologies based on inhomogeneous and two-phase gas and vapor-drop streams, much attention is paid to the study of the intensification of heat transfer processes, the establishment of appropriate mechanisms for phase transformations of liquid droplets with solid inclusions, the conditions for realizing different regimes of phase transformations at the "liquid–solid opaque particle" interfaces [1-2]. This approach contributes the development of a large range of practical applications (for example, flame cleaning of liquids [3] and polydisperse firefighting [4]).

The results of investigations [3, 4] are shown that the introduction of carbon particles into water droplets intensifies several times the phase transformations of the latter in a high-temperature (more than 700 K). It is established [3] that the evaporation of a liquid droplet containing a commensurate opaque inclusion in high-temperature gases is realized according to one of two schemes: 1) evaporation only from the free (external) surface of the drop; 2) evaporation from a free surface and intensive vaporization at internal interfaces. In the experiment, the range of various parameters (temperatures gases, volumes and shapes of droplets) is limited. Therefore, it is expedient to create a mathematical model of heat and mass transfer for studying the evaporation processes of inhomogeneous water droplets.

The aim of this work is development of the mathematical model of heat and mass transfer for researching the processes of evaporation of inhomogeneous water droplets.

* Corresponding author: piskunovmv@tpu.ru

DOI: 10.1051/matecconf/201711001095

2 Mathematical model

In the statement of the heat and mass transfer problem we assumed the initial water temperature T_{w0} = 300 K, gas temperature T_g = 550 – 850 K. Inhomogeneous droplets of water are heated due to thermal conductivity when they are flowed by a stream of high-temperature gases.

In the numerical modeling we assumed that the droplets have a time-constant spherical form (Fig. 1), are immobile, and are flown around by a high-temperature gas flow with the known velocity $V_g \approx 0.3$ m/s.

Figure 1 represents schemes of the domains of solution of the heat and mass transfer problems for researching the processes of evaporation of inhomogeneous water droplets.

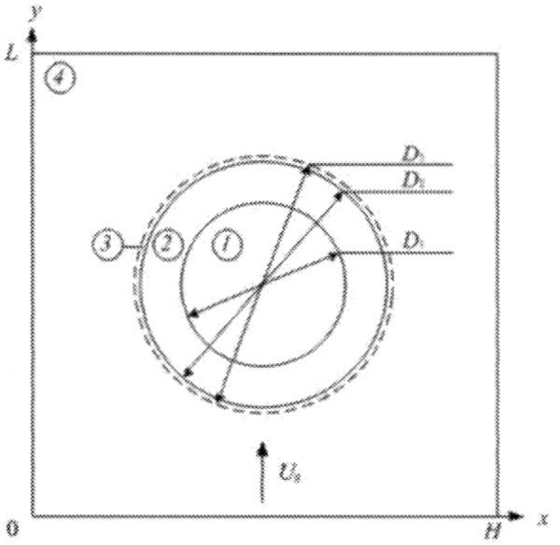

Fig. 1. Scheme of solution area: water + graphite inclusion (*1* – graphite inclusion; *2* – water liquid; *3* – water vapor; *4* – high-temperature gases).

The system of nonstationary differential equations in partial derivatives for the considered systems within the scope of the assumptions has the following form.
For the vapor–gas mixture ($0<x<H$, $0<y<L$):
the continuity equation

$$\frac{\partial u}{\partial x}+\frac{\partial w}{\partial y}=0. \tag{1}$$

the motion equation

$$\rho_3\left(\frac{\partial u}{\partial t}+u\frac{\partial u}{\partial x}+w\frac{\partial u}{\partial y}\right)=-\frac{\partial P_3}{\partial x}+\nu_3\left(\frac{\partial^2 u}{\partial x^2}+\frac{\partial^2 u}{\partial y^2}\right); \quad 2)$$

$$\rho_3\left(\frac{\partial w}{\partial t}+u\frac{\partial w}{\partial x}+w\frac{\partial w}{\partial y}\right)=-\frac{\partial P_3}{\partial y}+\nu_3\left(\frac{\partial^2 w}{\partial x^2}+\frac{\partial^2 w}{\partial y^2}\right)+\rho_3\beta g(T_3-T_0). \tag{3}$$

the energy equation

$$\rho_3 C_3\left(\frac{\partial T_3}{\partial t}+u\frac{\partial T_3}{\partial x}+w\frac{\partial T_3}{\partial y}\right)=\lambda_3\left(\frac{\partial^2 T_3}{\partial x^2}+\frac{\partial^2 T_3}{\partial y^2}\right). \tag{4}$$

the water vapor diffusion equation

$$\rho_2(\frac{\partial C_w}{\partial t}+u\frac{\partial C_w}{\partial x}+w\frac{\partial C_w}{\partial y})=\rho_2 D_2(\frac{\partial^2 C_w}{\partial x^2}+\frac{\partial^2 C_w}{\partial y^2}). \quad (5)$$

the balance equation

$$C_g + C_w = 1. \quad (6)$$

For water droplets ($0<r<R_d$, $0<\varphi<2\pi$):
the equation of thermal conductivity in the spherical system of coordinates

$$\frac{\partial T_1}{\partial t}=a_2\left[\frac{1}{r^2}\frac{\partial}{\partial r}(r^2\frac{\partial T_1}{\partial r})+\frac{1}{r^2\sin(\varphi)}\frac{\partial}{\partial\varphi}(\sin(\varphi)\frac{\partial T_1}{\partial\varphi})\right]. \quad (7)$$

Initial conditions ($t=0$): $T=T_0$, $C_w=0$, $C_g=1$, $u=w=0$, $P=P_0$ при $0<x<H$, $0<y<L$; $T=T_0$ at $0<r<R_d$, $0<\varphi<2\pi$.

Boundary conditions at $t>0$:

$T=T_g$, $C_w=0$, $C_g=1$, $V=V_g$ at $y=0$, $0<x<H$;

$\frac{\partial^2 T}{\partial y^2}=0$, $\frac{\partial^2 C_w}{\partial y^2}=0$, $\frac{\partial^2 C_g}{\partial y^2}=0$ at $y=L$, $0<x<H$;

$\frac{\partial T}{\partial t}=0$, $C_w=0$, $C_g=0$ at $x=0$, $0<y<L$;

$\frac{\partial T}{\partial t}=0$, $C_w=0$, $C_g=0$ at $x=H$, $0<y<L$.

Boundary conditions for "liquid – gas" were set taking into account vaporization:

$$R=R_1,\ 0<\varphi<2\pi,\ \lambda_1\frac{\partial T_1}{\partial R}=\lambda_2\frac{\partial T_2}{\partial R}+W_{e2}Q_{e2},\ \rho_2 D_2\frac{\partial C_w}{\partial r}=W_{e2}. \quad (8)$$

The components of the velocity vector (u and w) are determined using the following equations:

$$u=\left(\frac{2}{\rho_3}\frac{\partial P}{\partial x}\right)^{1/2},\quad w=\left(\frac{2}{\rho_3}\frac{\partial P}{\partial y}\right)^{1/2}. \quad (9)$$

We used steps in time $\Delta t=0.1$ s and in coordinate grid $\Delta x=\Delta y=0.05$ mm. Near the phase transition boundaries the coordinate grid became thicker (up to $\Delta x=\Delta y=\Delta r=0.001$ mm).

2.1 Nomenclature

a – temperature conductivity, m^2/s; C – specific heat capacity, J/(kg·K); D – diffusion coefficient, m^2/s; C_g – concentration of combustion products ($0<C_g<1$); C_w – concentration of water vapor; g – acceleration of gravity, m/s^2; H, L – dimensions of solution area, mm; P – pressure, N/m^2; R_d – initial droplet radius, mm; t – time, s; T – temperature, K; T_0 – initial temperature, K; T_g – temperature of gases, K; Q_e – water vaporization heat, J/kg;

W_e – evaporation rate, kg/(m^2·s); V_g – velocity of gases (combustion products), m/s; x, y – coordinates of the Cartesian coordinate system, m; u, w – x and y components of convection velocity of the gas-vapor mixture, m/s; Δt – time step size, s; Δx, Δy, Δr – coordinate steps, mm; ρ – density, kg/m^3; λ – thermal conductivity, W/(m·K); ν – kinematic viscosity, m^2/s; β – thermal expansion coefficient, K^{-1}; r, φ – coordinates of the Spherical coordinate system, m; indices: *1* – water droplets, *2* – water vapor, *3* – gas-vapor mixture.

3 Results

Figures 2 represents typical temperature fields of the vapor–gas mixture for the above statements. Simulations are estimated for droplets with volume 5, 10 and 15 mkl.

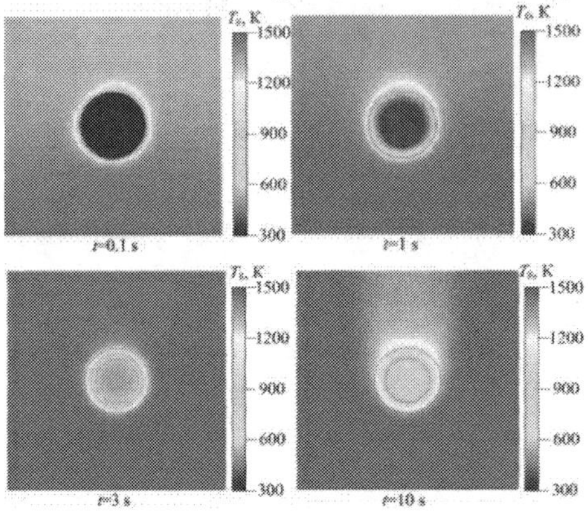

Fig. 2. Temperature fields for inhomogeneous water droplet (water + graphite inclusion) for different times at T_g=1500 K (volume of droplet V_d=10 mkl).

The results of numerical studies show the dependence of the times of complete evaporation (Fig.3) and the explosive vaporization (Fig.4) of inhomogeneous water droplets at different gas temperatures (T_g = 550-850 K). A satisfactory correlation of the results of numerical studies with experiment is shown.

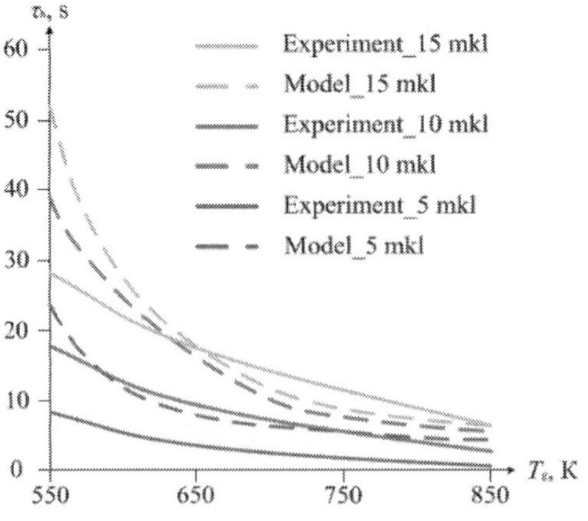

Fig. 3. Times of complete liquid evaporation τ_h from free surface of inhomogeneous water droplets at different gas temperatures.

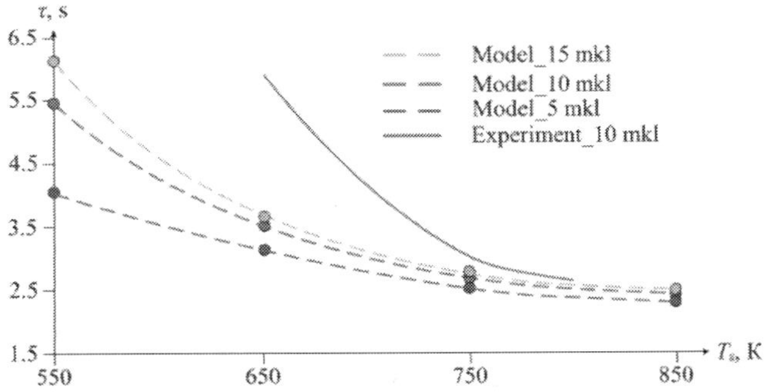

Fig. 4. Times of explosive vaporization of water droplets τ at different gas temperatures.

The obtained dependences (Fig.3) illustrate, as it was possible to assume, with the increasing gas temperature the evaporation of the liquid is more intensively.

It should be noted that when the inhomogeneous droplets of water are heated in a gas with temperature above 650-850 K, the effect of their "explosive" destruction with "boiling" of the liquid and the decay of the film into smaller droplets is recorded. Such an effect arises from the considerable warming up of the solid particle on the side of contact with the gases, as well as due to the transmission of the energy radiation from the high-temperature gases to the internal interface of the "inclusion-liquid". The destruction of an inhomogeneous droplet was preceded by evaporation from the free surface of the liquid film, formation, a significant increase in volume, and the fusion of vapor bubbles at the internal interface. The presented times of explosive vaporization of inhomogeneous water droplets at the temperature gases (Fig. 4) illustrate that the times of explosive vaporization decrease with increasing temperature.

4 Conclusion

Based on the results of the experiments, a mathematical model of heat and mass transfer in the Ansys Fluent package has been developed, which can be used to predict the times of complete evaporation and the times of explosive vaporization of inhomogeneous water droplets. A satisfactory correlation between simulation results and experimental results is shown.

Research was supported by Foundation of President of Russian Federation (MD-1221.2017.8).

References

1. D.H. Doh, J.H. Yum, G.R. Cho, M.H. Kim, G.W. Ryu, M. Takei, J. Therm. Sci. **22**, 5 (2013)
2. I. Gherasim, G. Roy, C.T. Nguyen, D. Vo-Ngoc, Int. J. Therm. Sci. **50**, 3 (2011)
3. D.V. Antonov, R.S. Volkov, M.V. Piskunov, P.A. Strizhak, Technical Physics Letters, **42**, 3 (2016)
4. R.S. Volkov, G.V. Kuznetsov, P.A. Strizhak, Int. J. Therm. Sci. **88**, 1 (2015)

Mathematical and physical modeling of heat transfer through window with heat-reflecting screens to determine the potential of reducing thermal costs for microclimate parameters maintaining

Nikolay Smirnov[1,*], *Vladimir* Tyutikov[1] , and *Vadim* Zakharov[1]

[1]Ivanovo State Power Engineering University, Heat Power Engineering Department, 153003 Ivanovo, Russia

Abstract. A mathematical model describing the process of heat transfer through windows with heat-reflecting screens, taking into account physical and geometrical parameters of the building construction, was developed. The computer program for the calculation of heat transfer through the window has been developed and results of numerical heat transfer simulation for different parameters of indoor and outdoor air are given. Expressions to determine the thermal resistance for the specific types of glazing with a heat-reflecting screens installed outside were derived on the basis of the developed method. The adequacy of the suggested mathematical model was confirmed by experimental data.

Introduction

Pursuant to Russian legislation, annual specific consumption of energy in buildings as of 1 January 2020 shall be reduced by 40% of the basic level.

Among the costs of fuel and energy resources at an industrial enterprise, the cost of maintaining the necessary parameters of the microclimate is from 10% (the oil refinery, chemical industry) to 40-50% (machine-building, electrotechnical enterprises) [1]. According to the data of the Central Research Institute of Housing [2] a significant proportion of heat loss in residential buildings falls on windows. So, for 5-storey residential buildings (series 1-511 and 1-447) the thermal transmission losses through the windows make up from 46 to 54%. European Union legislatures stipulate a coefficient of heat transfer resistance for windows by 2020 of 1.67-2.0 $m^2 °C/W$, while Russian official bodies stipulate 1.0-1.05 $m^2 °C/W$ by 2016. Therefore, the development of new windows with increased heat-shielding properties is an actual task.

The scientists from Ivanovo State Power Engineering University (ISPU) and National Institute of Applied Sciences in Strasbourg (INSA de Strasbourg) developed [3] and patented window designs with panel, roll and blinds type heat-reflecting screens, which are

* Corresponding author: nsmirnov@bk.ru

DOI: 10.1051/matecconf/201711001096

made of metal and significantly reduce heat losses. Screens can be positioned indoors, outdoors or in a space between glass panes. The use of screens is desirable during nighttime or in the absence of people. Screens may be placed inside or outside buildings, or between window panes.

An example of outside panel heat-reflecting screen installation in the window (position 3) in room, where the maintenance of the necessary microclimate parameters is carried out with the help of an air conditioning system, is shown in Fig. 1.

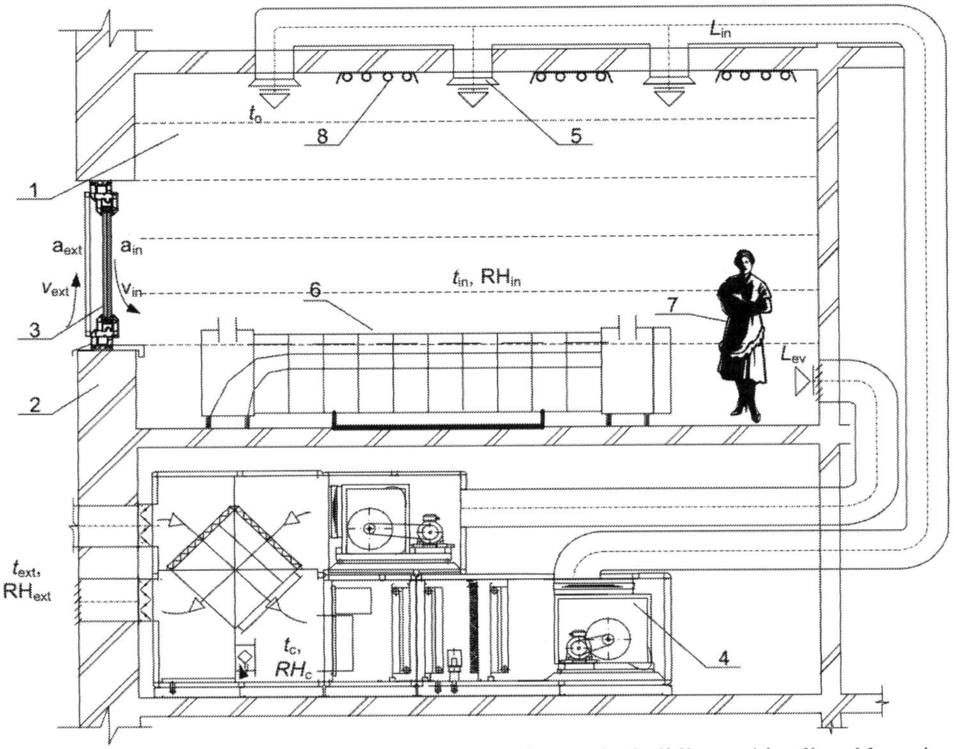

Fig. 1. System for maintaining of technological microclimate for buildings with adjustable resistance windows: 1 - production room; 2 - wall; 3 - a window with a panel-type heat-reflecting screen installed outside; 4 - the conditioner; 5 - air distributor; 6 - production equipment; 7 - staff; 8 - lighting fixtures.

The use of screens not only reduces losses related to heat transfer but also permits ambient temperature reduction in setback heating mode [3].

It is necessary to know the actual resistance to heat transfer of windows with heat-reflecting screens for calculating the real savings of thermal energy for the compensation of transmission losses.

1 Development and software implementation of mathematical model of heat transfer through a window with screens

A computational scheme was developed for the model describing the process of heat transfer through the translucent part of the window using heat-reflecting screens. The mathematical model of heat exchange through the window system with screens is considered as a multi-layer inertialless system consisting of glazing layers, metal screens and air gaps. Each layer has its own physico-geometric parameters determined by the glazing and screens material, the type of filling of the air gap, the thickness of the layers (h, δ_{gl}), etc. On the interfaces between layers, the conditions for equality of temperatures and heat fluxes are specified. In each air gap, as well as on inner and outer surfaces of the structure, there was a coefficient of heat transfer $\alpha_{i(j)}$, which is the sum of convective $\alpha_{i(j)}^{konv}$ and radiant $\alpha_{i(j)}^{rad}$ heat transfer.

$$\alpha_{i(j)} = \alpha_{i(j)}^{rad} + \alpha_{i(j)}^{konv} \tag{1}$$

Coefficients of convective and radiant heat exchange were determined by well-known empirical formulas [3].The system of equations was solved numerically in the program Matlab. The developed mathematical model was implemented in the program for calculating of heat transfer process for a glazed unit with metal heat-reflecting screens "Glass + Screens". The results of numerical simulation in the "Glass + Screens" program can be seen in Fig. 2.

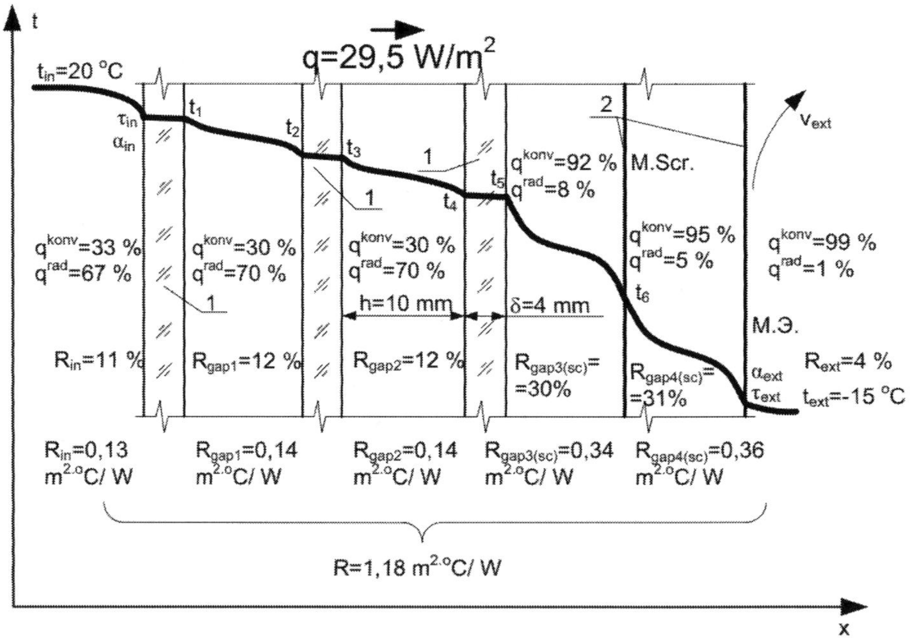

Fig. 2. The results of simulating the process of heat transfer through a triple-glazed window with screens in the developed program: 1 - glass, 2 - aluminum screens.

We used a glass unit (hereinafter G.U.) as 4M1x10x4M1x10x4M1. At the same distance from each other, h = 10 mm, we simulated the installation of heat-reflecting metal screens (hereinafter M.Scr.), made of polished aluminum. The discrepancy in the calculation of the surface temperatures varied from 0.04 to 0.18 ° C (or 0.1 to 0.5%). It should be noted that a significant part of the resistance of this design was given by the thermal resistance of the air gap between the glass and the screen (R_{gap} = 0.34 (m2 ° C) / W or 30%) and between the screens (R_{gap} = 0.36 (m2 ° C) / W or 31%), and in these gaps, because of the high reflectivity of polished aluminum, the radiant heat flux q^{rad} was only 8 and 5%, respectively.

With the help of regression analysis, after processing the data of the computer program "Glass + Screens", empirical formulas were obtained [3] for calculating the resistance to heat transfer in the middle (central) window zone with heat-reflecting screens. The change of the temperature difference between the internal and external air Δt was set in the range from 10 to 60 °C (at t_{in}= 20 °C), the speed v_{ext} – from 1 to 5 m / s.

Thus, for a triple-glazed window with the formula 4M1x10x4M1x10x4M1 and two heat-reflecting screens installed from the outside at a distance of 10 mm from each other, in case of natural convection at the inner surface, the resistance to heat transfer can be determined by formula

$$R(\Delta t, v_{ext}) = 0.889 + \frac{14.1}{\Delta t + 17.6} + \frac{1}{7.34 \cdot v_{ext}^{0.656} + 3.78 \cdot e^{-1.91 \cdot v_{ext}}}, \quad (2)$$

where Δt – the difference between temperatures of indoor and outdoor air, °C;

v_{ext} – speed of outside air, m / s.

If the recirculation ventilation unit is operating in a room or the inflow to a room is effected by mechanical ventilation, that is, forced convection is present at an inner surface of window. The resistance to heat transfer of a 2-screen window can be expressed

$$R(\Delta t, v_{in}, v_{ext}) = 0.749 + \frac{1}{9.6 + 3.8 \cdot v_{in}} + \frac{14{,}1}{\Delta t + 17.6} + \frac{1}{7.34 \cdot v_{ext}^{0.656} + 3.78 \cdot e^{-1.91 \cdot v_{ext}}}, \quad (3)$$

where v_{in} is the air velocity at the inner surface of the window, m / s.

On the basis of the derived empirical relationships for G.U. with heat-reflecting screens, in the case of natural convection inside, graphs of the dependences of the resistance to heat transfer R from Δt, v_{ext} were obtained (see Fig. 3).

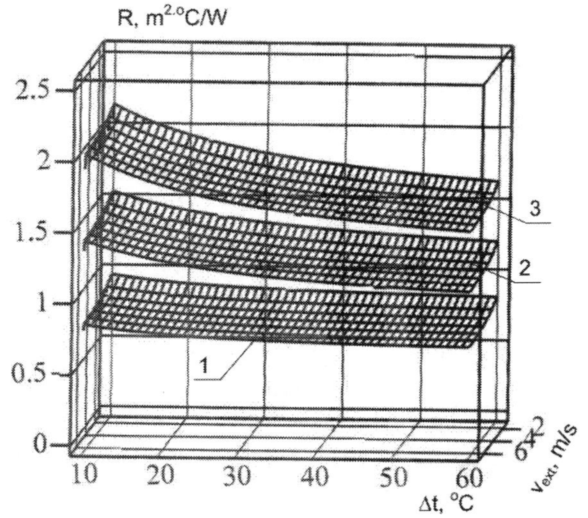

Fig. 3 Dependence of the resistance to heat transfer R of the central zone of window when using the G.U. by the formula 4M1x10x4M1x10x4M1 and installed outside 1 - one metal screen; 2 - two screens; 3 - three screens.

As can be seen from the simulation data (Fig. 3), the resistance to heat transfer depends more on the temperature difference between the internal and external air. The maximum resistance is obtained with the minimum temperature difference and the minimum speed of the outside air. Thus, the resistance to heat transfer for windows with heat-reflecting screens is a nonlinear dependence.

2 Verification of a mathematical model using physical experiment

For the good verification of the heat transfer model through a window with a heat-reflecting screen, experimental data from a field experiment conducted in a certified climate chamber of Ivanovo Construction Research Center were used. The study was carried out with different variants of the arrangement of heat-reflecting screens relative to the triple-glazed unit. As a control sample we used a triple-glazed window with the formula 4M1x10x4M1x10x4M1 and aluminum screens. The system carried out 20 measurements of the values for each steady-state regime. Experimental data were processed using mathematical statistics.

The temperature inside the warm compartment of the climatic chamber was maintained at 20-21 °C. Thermograms and temperature graphs along the cutting line of the central part of glass can be seen in Figure 4. Aluminum screens are installed on the left outside at a distance of 10 mm from the outer glass of the triple-glazed window, to the right - only the glass unit.

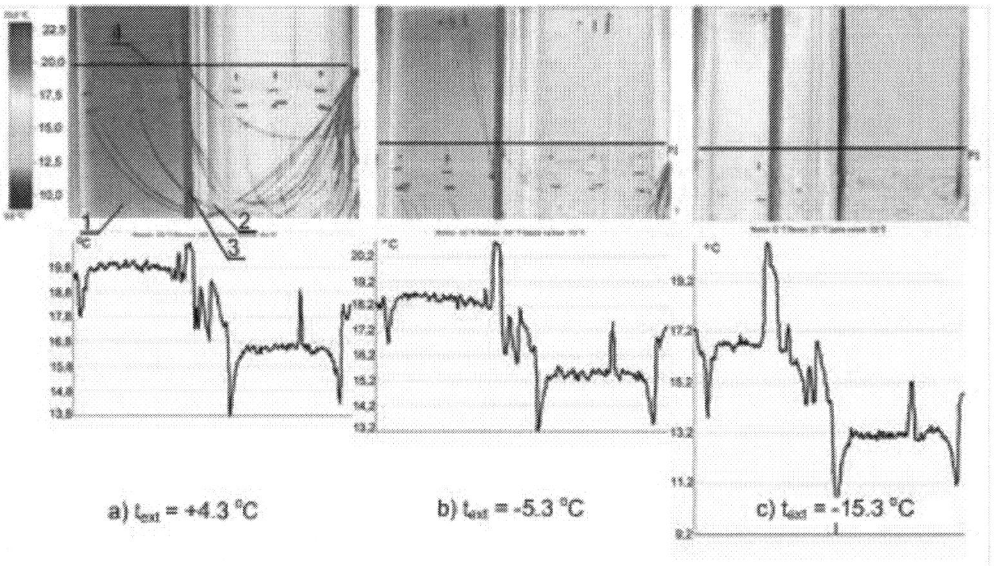

Fig. 4. Thermograms and temperature graphs along the cutting line on the inner surface of the central part of glass unit while keeping the air temperature in the cold compartment of the air chamber equal to +4.3 °C (a), -5.3 °C (b) and -15.3 °C (c): 1 - the internal surface of the window; 2 - binding from plastic pro-fillet; 3 - a tire for fixing thermocouples measuring; 4 - cutting line.

As expected, the temperatures on the surface of the glass pane are more reduced when the outside air temperature changes in the control version (only the triple-glazed unit). In the zone where the metal screen is installed from the outside, the temperature drop on the inner glass is less when the change is made. Thermograms clearly show the edge zones of the glass unit, where, due to the formation of "cold bridges" due to internal aluminum frames, the surface temperature drops by 2-4 °C in the control, and by 2-3 °C in the screen mounting option.

In Fig. 5 a comparison of computer calculation data and values of a physical experiment for the resistance of heat transfer for window with metal screens installed outside 10 mm apart from each other in the case of a change in the temperature difference between internal and external air is made. The air velocity in the cold compartment of the climatic chamber was maintained within 3 m/s.

As can be seen from Fig. 5, the experimental values of the resistance to heat transfer is occupied in the confidence interval of mathematical modeling data obtained from empirical formulas. The data of the physical experiment also support the hypothesis that the resistance to heat transfer of a window with heat-reflecting screens is not a constant (without taking into account the correction for speed), as in the case of using only a triple-glazed unit, but depends on the difference between the temperatures of internal and external air.

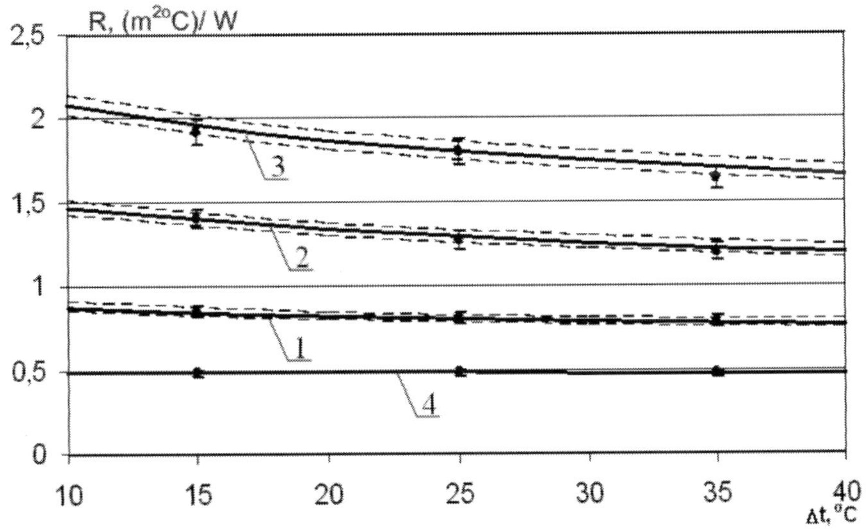

Fig. 5. Comparison of the experimental and calculated values of the resistance to heat transfer of the central zone of window, depending on the difference between the temperatures of internal and external air for: 1 – G.U. and 1 M.Scr.; 2 - G.U. and 2 M.Scr.; 3 - G.U. and 3 M.Scr.; 4 - G.U. (control). Solid line - the results of calculation by empirical formulas; dashed lines - the boundaries of the confidence interval due to the relative error of the mathematical model ± 3%; points are experimental values with a relative error of measurement ± 4%.

Conclusion

The results of these studies can be used to determine the potential for using heat-reflecting screens in windows for systems of intermittent heating of buildings, including the case of applying the technology of preliminary air drying for wet or normal conditions of premises of industrial enterprises.

References

1. A.V. Sinilova, Russian foreign economic bulletin, **10** (2012)
2. V.S. Belyaev, S.K. Esengabulov, Building physics in the XXI-st century (NIISF RAASN, Moscow, 2006)
3. N.N. Smirnov, B. Flaman, M. Barba, V.M. Zakharov, V.V. Tyutikov, D.A. Lapateev,Vestnik of Ivanovo State Power Engineering University, **2** (2015)

AUTHOR INDEX

AUTHOR INDEX

AUTHOR INDEX